LIBERATION ECOLOGIES,
SECOND EDITION

"Linking urban and first world concerns to rural and third world struggles, this volume covers a whole new ground. Bound to remain a central text in the field."
Paul Robbins, *Department of Geography, Ohio State University*

"*Liberation Ecologies* is an outstanding volume rich in empirical material and theoretical sophistication that makes a landmark contribution to political ecology. A definite classic!"
Raymond Bryant, *Department of Geography, King's College, London*

At the beginning of the twenty-first century, the environment and the future of development continue to be issues of crucial importance. Most explanations of environmental crisis emphasize the role of population growth, thus focusing their attention on the poor. By comparison, *Liberation Ecologies* elaborates a political–economic explanation drawing from the most recent advances in social theory.

The new edition has been extensively revised to reflect recent changes in debates over the real definitions of "development" and "environment," and contains nine completely new chapters. Bringing together some of the leading researchers in the development field, the book discusses the theory, growth and impact of political ecology. In-depth case material drawn from across the developing world is used to explore the realities and impact of sustainability, whilst emphasizing both the environment and social theory as areas of contention and struggle.

Liberation Ecologies highlights new theoretical engagements between political ecology and poststructuralism, and challenges many conventional notions of development, politics, democracy and sustainability.

Richard Peet is Professor of Geography, Clark University, Worcester, Massachusetts.

Michael Watts is Director of the Institute of International Studies, the University of California at Berkeley.

LIBERATION ECOLOGIES, SECOND EDITION

Environment, development, social movements

Edited by Richard Peet and Michael Watts

Routledge
Taylor & Francis Group

LONDON AND NEW YORK

First published in 1996
by Routledge
This second edition published in 2004
by Routledge
11 New Fetter Lane, London EC4P 4EE

Simultaneously published in the USA and Canada
by Routledge
29 West 35th Street, New York, NY 10001

Routledge is an imprint of the Taylor & Francis Group

Typeset in Garamond by LaserScript Ltd, Mitcham, Surrey

Printed and bound in Great Britain by
TJ International, Padstow, Cornwall

British Library Cataloguing in Publication Data
A catalogue record of this book is available from the British Library

Library of Congress Cataloging in Publication Data
Liberation ecologies : environment, development, social movements /
edited by Richard Peet and Michael Watts.–2nd ed.
p. cm.
Includes bibliographical references and index.
1. Human ecology–Developing countries–Case studies. 2. Economic
development–Environmental aspects–Case studies. I. Peet, Richard.
II. Watts, Michael.
GF900.L52 2004
304.2–dc22
2003017832

ISBN 0–415–31235–3 (hbk)
ISBN 0–415–31236–1 (pbk)

This book is dedicated to Elaine, Eric
and Anna and Mary Beth, Ethan and Nan
with love

CONTENTS

ILLUSTRATIONS

FIGURES

TABLES

CONTRIBUTORS

Anthony Bebbington, Institute for Development Policy and Management, University of Manchester.

Judith Carney, Department of Geography, University of California, Los Angeles.

Mike Davis, Department of History, University of California, Irvine.

Navroz K. Dubash, World Resources Institute, Washington DC.

Tim Forsyth, Department of Geography, London School of Economics.

Michael Goldman, Department of Sociology, University of Illinois, Urbana.

Susanna B. Hecht, School of Public Policy, University of California, Los Angeles.

Jake Kosek, Department of Anthropology, Stanford University.

Tania Murray Li, Department of Anthropology, Dalhousie University.

Roderick P. Neumann, Department of Geography, Florida International University.

Dara O'Rourke, Department of Environmental Sciences and Policy Management, University of California, Berkeley.

Richard Peet, Graduate School of Geography, Clark University, Massachusetts.

Haripriya Rangan, School of Geography and Environmental Science, Monash University-Clayton, Melbourne.

Richard A. Schroeder, Department of Geography, Rutgers University, New Jersey.

Krisnawati Suryanata, Department of Geography, University of Hawaii, Honolulu.

Michael Watts, Department of Geography and Institute of International Studies, University of California, Berkeley.

Karl S. Zimmerer, Department of Geography, University of Wisconsin, Madison.

PREFACE TO THE SECOND EDITION

The concept of progress is to be grounded in the Idea of catastrophe. That things "just go on" is the catastrophe.

(Walter Benjamin)

For twenty years or more, many fine minds were preoccupied by poststructural philosophy applied to understanding of what was widely assumed to be a new post-modern era. Now that the original brilliance of this obsession has tarnished, we might ask: what was it all about? The elusive reply is a great deal, spelled out in suitably obscure language of course. Centrally we might suggest that it was an obsession with reflexivity as the main constituent of late-modern hegemony. That is, the building of responses to criticisms anticipated in advance into the actions of states, corporations and global governance institutions. Proposals for the Iraqi War of 2003 came replete with multiple level's of legitimation in cascading orders of priority: if weapons of mass destruction does not work, move to the psychopathologies of the Hussein regime, the beautifully devious notion of in-bedding reporters with front line troops, and subject positions for military heroes waiting to be filled by . . . whoever. Advertisements for corporate products are fantasized around the presumed feeble critical reactions of couch-potato consumer-viewers. And governance institutions have learned not to say what they are doing, directly, but to emphasize instead what they are not doing, indeed what they are under-mining, as though they had a conscience. Poststructural philosophy, in many of its social–theoretic guises, tried to understand this version of perverse reflexivity by emphasizing words more than deeds, discourse rather than development, hype over reality. And perhaps, as we tried to show in the first edition of this book, poststructuralism tried also to reflect on alternative reflexivities, to endeavor to open new spaces in which alternatives could be thought and maybe practiced – hence "liberation ecologies."

Now that poststructuralism's moment has passed or, rather since some of its essential insights are now taken for granted, we can plausibly ask: what next? Usually this is taken, in academic circles, to mean what next *theoretically* in the context in which social struggles occur in a reflexive world where it seems that everything has happened before. So a critique of the present has to include a phenomenology of reflexivity. But we think that projecting the next fascination from the dynamic of thought alone, in contemplation lonely for experience, might summon up a theoretical detour replete with proposal and counterproposal that ultimately leads down the dead end street of impractical politics. And the theories we have to think with, analyze and criticize with, are good enough at least for the moment . . . else why is the military-industrial complex so anxious to deploy poststructural notions in launching post-modern wars? So we might gesture towards the next moment by asking not how to think, but what to think about. This particularly seems germane for the case of political–ecological research, grounded in struggles over resources and livelihood, especially its liberatory version, that tries to listen to what social movements are saying without naively believing in the inherent wisdom of the "traditional."

Struggles over resources lie at the center of struggles over power. Power is unequally held by institutions, clustered at centers, or dispersed across peripheries. Discourses with hegemonic depth originate in political and economic command centers and achieve their hegemony by extending persuasion, coercion and power over spatial fields of influence. Counter-hegemonic alternatives have their own bases in power complexes situated, often, in social movements, trade unions or the forces of civil society, and are distinctive in their use of more informal media of thought, discussion and dissemination. Contention takes multiple political, social and discursive forms. We suggest that critical theorists might look at these forms, and that liberation ecologies are moments of wider, global struggle over power and widely contested resources. Let us briefly present a couple of examples of what we mean through global governance institutions, their neoliberal policies and the incorporation of critique through terms like "sustainable development."

Two central, yet contradictory, positions recur in contemporary, neoliberal capitalism: trade-led, economic growth; and regulation, by states and governance institutions, of the environmental impacts of economic growth. Economic growth produces the profits and jobs that in turn provide incomes and consumption for all classes of economic actors. Hence, the forces behind growth are generalized and integral to the neoliberal economic process. Growth is also integral to the political process, particularly in the following way: multinational and large national corporations increasingly pay the mounting expenses of media-led political campaigns that elect the leadership of the democratic states of all the advanced capitalist countries. Thus contemporary neoliberalism is structured by an alliance between the multinational corporations and business, trade and growth-oriented political parties, whether these be of the "left" or "right" and the popular interest that must be concerned about growth because it provides ever-growing consumption and employment. Yet the environmental contradictions of unregulated growth have also produced powerful environmental movements and concerned public opinion. Therefore, the political and economic tensions between growth and environmental degradation have to be relieved through a number of "strategies," a term we use to imply both conscious institutional intention and a less conscious process of groping for a workable answer.

The first strategy involves the cultural production of conceptual devices employing appealing terms that appear to bridge the impassable divide between growth and environment, as with the notion of "sustainable development." In this term, we suggest, "development" essentially means the economic growth that fuels ever-increasing consumption, the main marker of mass happiness and contentment in neoliberal societies, that also provides jobs and incomes. "Sustainable" we think, has now to be understood ideologically, as the effects of growth that the majority of people can be persuaded to find tolerable (so, "sustainable in the sense that growth can go on), as the necessary social, cultural or environmental consequences of an even more necessary growth process. Sustainability in this sense has a number of meanings that range from keeping growth going using state intervention, through swapping pollution rights in the market, to minimizing pollution effects so that public concern does not result in too much organized political action (unorganized action being dismissed as deranged).

The second strategy involves a displacement of regulatory power "upwards" to unelected and only partially responsible global governance institutions. These relieve pressure on nation states and provide the thin regulatory context for the smooth operation of global capitalism. This "regulatory context" has to pretend reconciliation between global economic growth and global environmental destruction. Yet all global institutions are now dominated by neoliberal, trade-oriented, growth-based policy regimes. All have learned to turn a sympathetic eye on environmentalism. As Bill Clinton once said "we should be leveling environmental protections up, not down." Or as the Rt. Hon. Sir Leon Brittan QC, Vice-President of the European Commission, put it: we should be "mainstreaming sustainability." In this "mainstreaming" global governance institutions like the WTO play increasingly significant roles. These involve the production, but more importantly the enforcement, of a global ideological framework, for membership in the WTO has now become essential for nation states wishing to participate in the "global community." An examination of the GATT/WTO ideological apparatus reveals a fundamental commitment to trade-induced growth that most basically denies its environmental consequences. This denial is achieved by reversing the real position taken by the WTO, so that trade-based growth solves, rather than causes, pollution, and by appealing to environmental policy regimes established at the national or increasingly the multilateral level (MEAs). We suggest that the "trade solves pollution" position may be undercut by the more compelling argument that trade-induced growth is the main cause of pollution. And we suggest that the appeal to multilateral environmentalism is wishful thinking in that MEAs lack agreement and enforcement governance institutions with anything like the power exerted by the WTO. The WTO's most audacious strategic gambit involves a claim that its dispute settlement system, mainly used to adjudicate inter-nation quarreling over trade matters, provides exactly the required enforcement mechanism for deciding inter-national differences over issues of global environmental regulation. Yet the

WTO dispute settlement system is a secretive practice that finds consistently against environmental restrictions on the freedom of trade. Yet again, however, massive opposition to the WTO, particularly at the Seattle ministerial meeting in 1999, but also against every global economic summit since, has shaken governance institutions, frightening them into partial concession, mainly the production of ever-more sophisticated masking devices.

On this note we make the following political conclusion. The contradictions embedded in terms like "sustainable development" can be employed in the interests of environmental protection through a combination of radical political interpretation and mass political action. In terms of interpretation, the meaning of "sustainable" can be directed through critique of its use by organizations like the WTO towards its radical extreme – zero environmental damage. Likewise, the meaning of "development" can be turned from consumptive growth to the satisfaction of basic human needs. And political action, as we have seen especially since 1999, has to be directed at the new global centers of institutional power, the World Bank, the IMF, the WTO, the World Economic Forum, the European Union. In such a radical political action we need to combine two types of counter-force: the thousands of protestors willing to face systematic violence by the police and military protecting the existing global order; and the hundreds of research institutions and non-governmental organizations dedicated to uncovering the sophisticated lies that global governance organizations persist in telling us. In such a radicalized environment does liberation ecology flourish?

What underlines the struggles operating under the sign of globalization, global governance and what Perry Anderson has called the "neol-liberal grand slam" of the last two decades, is the reality of twenty-first-century primitive accumulation and enclosure as a starting point for understanding the relations between environment and development. In Marx's analysis, the process of primitive accumulation was the violence of creating a "free labor force," with an eye on the role that British enclosures, and imperialism, played in the process of uprooting the peasantry. But primitive accumulation is not a singular or one-time event, it is an on-going process of dispossession as the market, and private property relations, enters new arenas, often driven by both the "opening" of new market-friendly states (the socialist bloc since 1989) and by techno-scientific innovation. In early March 1998, for example, the US Department of Agriculture and the Delta and Pine Land Company were granted US patent number 5,723,765. It involves a method of producing sterile seeds in second-generation plants by the transfection of a so-called lethal gene that produces seeds incapable of germination. As the inventor pointed out, the patent prevents the escape of a plant by natural seed dispersal; it is a way of self-policing the unauthorized use of American technology. The terminator gene condenses the novel essence of the so-called molecular or rDNA revolution: namely the capacity to vastly expand the taxonomic range available for gene transfer, and the ability to target specific processes in plants and animals. The fact that the genome of humans or fruitflies or tomatoes or chickens can now be mapped presents the possibility that the new frontier of accumulation and profitability has opened up: namely, the building blocks of human, plant and animal life. As Karl Polanyi might have put it, the very stuff of life has now been embedded in the marketplace – and disembedded from Nature. Many of the case studies in *Liberation Ecologies* address the multiform ways in which this enclosure is occuring under global neoliberalism and the manifold ways in which it is resisted.

The centrality of the gene as the distinctively modern map of Nature reaffirms the power of the enclosure as a way of exploring the contemporary landscape of what Clarence Glacken in his great book *Traces on the Rhodeian Shore* called interpretations of nature and culture. To return to the terminator gene, the rDNA revolution has provided the technocratic capacity to enhance precision, control and goal specificity within the agricultural and indeed the entire biological realm. Molecular biology now can make use of a complex legal and juridical apparatus – the patenting industry – by which the genes or germplasm may itself be subject to the dull compulsion of the marketplace. The debate over the trade in germplasm (a pre-requisite for the life science companies like Norvartis and Monsanto) is being subject to a massive genetic enclosure. All manner of germplasm is now owned privately and traded through the marketplace; bioprospecting for using genetic material is now at the forefront of life science accumulation. There is a profound sense in which communities in Ethiopia or New Guinea have lost their common rights to their own genetic and plant material.

In the same ways that the eighteenth-century English enclosures were an object of popular scorn and resistance, so it is to be expected that the appearance of genetically modified crops and animals has produced a generation of "luddites." During 1997 a number of actions against the 300 or so field trial sites for GE crops in the British Isles were undertaken by the likes of the Lincolnshire Loppers, Captain Chromosome, the Kenilworth Croppers and Superheroes Against Genetics. In the fall of 1997 Gaelic Earth Liberation trampled Monsanto's modified maize field in County Carlow; in West Sussex a genetically engineered rapeseed was uprooted, and there has been significant sabotage – eliciting much popular sympathy – in France and Germany, most memorably against the Norvartis drying plant in Nerca, France. At least thirty-six field sites were "visited" by these luddites in the UK during the 1998 growing season. As Iain Boal has shown, the surfacing of someone like Captain Chromosome in Norfolk cannot be understood outside of the fact that the region has a long history of both resistance to enclosures and an association with agricultural innovation. The story could be repeated many times over, not least in the South as GMOs are widely adopted. Captain Chromosome gestures to an earlier history of enclosure but cannot claim to return to an uncontaminated or natural landscape of the past either.

Captain Chromosome and the terminator gene confront us with the twenty-first century enclosure as the forces of the market and of capital are now massing on the frontiers of the building blocks of life itself. New forms of confinement, of dispossession and loss of rights are already at work backed by the forces of the state but now armed with what Iain Boal calls "the promethan mythos of science and technics." Recombinant DNA and genetically modified organisms also returns us to Glacken's triad of historical ideas of culture/nature: design, influence and modification. Of course the capacity to transform is now incomparably vast in relation to Glacken's account of the Hellenistic world or early modern Europe. Indeed the very idea of design has now scaled heights that even some of the early nineteenth-century scientific boosters like Saint Simon could not possibly have anticipated. But curiously this has made Glacken's reference to environmental influence even more compelling. There is no simple Promethean victory for science or capitalist industry for the very good reason that the likes of Mad Cow disease, or the global resurgence of malaria, or the possibility of a genetically modified apocalypse cast a long shadow over the enhanced powers of human modification of nature. Enclosure was the final blow to peasants in the commoning economy of England. But the result was a landscape that legitimized and sharpened class politics. Perhaps there is a moral in the story for the new millennium, and for the readers of *Liberation Ecologies* around the world.

Michael Watts and Richard Peet
July 2003

ACKNOWLEDGMENTS

Richard Peet would like to thank Clark University for providing a supportive environment for my sometimes alarmingly critical thinking, even to the point of awarding me a Senior Faculty Fellowship in 2002, knowing that it would underwrite such nefarious activities. At Clark too I am surrounded by progressive graduate and undergraduate students in numbers far too large to mention by name. (You know who you are!) However, all this pales before the real source of my contentment, at last: my family, expanded since the first edition of this book with the birth of Eric, now almost 3 years old, and rambunctious to the core, and Anna, 1-year old on the day this was written, and an utter delight to all who meet her, especially me. Most of all, I thank Elaine for her fierce political commitment and her loving care. Elaine provided every moment of relative quiet in which to work on this second edition.

Michael Watts would like to acknowledge the assistance of Abby Lambert, Dena Shupe Woolwine, Terry Yee and Andrew Lauck who provided yeoman's work in the assemblage and editing of the book. The Berkeley Workshop on Environmental Politics at Berkeley has provided the testing ground for virtually all of the ideas that appear in *Liberation Ecologies*, and I am especially grateful for the solidarities expressed in that forum by the likes of Don Moore, Nancy Peluso, Iain Boal, Candace Slater, Louise Fortmann, Kate O'Neill, Ben Gardener, Joe Bryan, Elana Ripps and the regular attendees. My own research has been funded by the Guggenheim Foundation, the National Geographic and the MacArthur Foundations whose generosity I wish to acknowledge. Mary Beth, Ethan and Nan are the most compelling embodiments of red and green that I know. They are also the people I love most.

Part I

RENEWING POLITICAL ECOLOGY

1

LIBERATING POLITICAL ECOLOGY[*]

Michael Watts and Richard Peet

Even society as a whole, a nation, or all existing societies put together, are not owners of the Earth. They are merely its occupants, its users; and like good caretakers, they must hand it down improved to subsequent generations.

(Marx, *Capital*, vol. 1)

In this world which is so respectful of economic necessities, no one really knows the real cost of anything which is produced. In fact the major part of the real cost is never calculated; and the rest is kept secret.

(Debord, *The Society of the Spectacle*)

We begin by recalling some recent stories, distinguished more than anything by their ubiquity and familiarity. First, the occupation by a militant youth wing of the Istekiri people of a number of Chevron oil flow-stations in the Nigerian Niger Delta. Over the last five years, increasingly militant ethnic minorities in the oil-producing Delta have aggressively occupied oil installations operated by transnational petroleum companies. This comes in the wake of a growing clamor over the control of local petro-revenues by impoverished oil producing communities, and claims for compensation for the ecological destruction associated with forty years of commercial drilling and pumping. A second story speaks to the question of economic globalization. The United Nations Development Program says that globalization is dangerously polarizing the "haves" and "have-nots" with little in the way of regulatory structures to counter its risks and threats. Central to the UN agenda is the need for a new multilateral environmental agency to regulate the global commons (the seas, ozone and so on). Third, there is a report about escalating conflicts between, on the one hand, the Brazilian federal ministry of agriculture and coalitions of regional states (led by the Marxist-oriented Rio Grande do Sul), and on the other, local agro-cooperatives. Contention centers on the potential environmental and social consequences of the widespread introduction into Brazil of genetically modified soy by the Monsanto corporation. And not least, there is a story of massive water-poisoning in Bangladesh.

Environmental issues of this sort speak to the goals of *Liberation Ecologies* in two important ways. First, they are very much the object of study for a field of political ecology that seeks to understand the complex relations between Nature and Society through careful analysis of social forms of access and control over resources – with all their implications for environmental health and sustainable livelihoods. And second, they display vividly what geographers refer to as the politics of scale. These news stories encompass a number of political arenas, from *the body* to the locally imagined *community* to *state and intra-state struggles* to new forms of *global governance*.

Struggles over biotechnology or corporate responsibility may appear commonplace and pedestrian. But it is precisely their quotidian character that marks the extent to which "Nature" is now so deeply embedded in our twenty-first century political identities. This "green reading" of the popular press comes at a moment when we recently celebrated the thirtieth birthday of a foundational moment in environmental activism, namely the first Earth Day in 1970, and subsequently two years later the United Nations Stockholm Conference on the Environment. But has the politics of the environment changed since these defining moments? One obvious difference is the enhanced knowledge of, and sensitivity to, transborder and global forms of environmental harm (ozone depletion, global climate change, toxic dumping), and the extent to which green issues are legislated through inter-state agreements (the Rio Agenda 21 and the Biodiversity Convention of 1992 for example) and multilateral (inter-governmental) organizations. Indeed one of the striking trends in the last decade has been the "greening" – with limited success it needs to be said – of multilateral institutions like the World Bank (e.g. the Global Environmental Facility), the World Trade Organization, and regional associations such as the European Union and the North American Free Trade Association (NAFTA).

Another difference turns on the restructuring of global capitalism itself, and quite specifically the profound environmental changes associated with the rapid growth and maturity of the newly industrializing countries and the collapse of the Socialist bloc since 1989. The chickens of rapid industrialization in Brazil and Taiwan, and fifty years of Stalinist hyper-industrialization in the former sovietsphere, came home to roost in the 1990s. And not least, the deepening reach of transnational capital, marked incidentally by the rise of a massive corporate and transnational environmental technology industry (Pratt and Montgomery 1998), has its counterpoint in a proliferation of social movements which typically link economic and ecological justice (the *politics of distribution*) with human rights and cultural identity (the *politics of recognition*). New social movements can be understood as an effort by national and global civil society – social networks and transnational coalitions – to impose some sort of control over transnational corporations and irresponsible or rogue states, most especially the environmental externalities (e.g. toxic dumping) and distributional conflicts generated by the export of industry to the Third World via an increasingly deregulated world economy. The road from Stockholm to Rio is littered, then, with new ecological problems and different ecological politics.

4

Driven by momentous political and economic changes and by apocalyptic visions of impending global ecological doom, the environmental question now very much occupies center stage. The World Bank Report for 2003 addresses the question of sustainable development in the context of enormous pressures on "local and global common property resources" (water, soil, fisheries) as well as "local and global sinks" (the ability of the biosphere to absorb waste and regulate climate). Nature and the economy evolve in different ways and at different speeds, says the World Bank, and in an era of globalization the ability of natural life support systems to adapt is in question. The challenge, they say, "is daunting" (World Bank 2003: 1). The meanings of terms like sustainability are hotly contested (M. O'Connor 1994). But the new lexicon is so endemic that it appears with as much frequency in the frothy promotional literature of the World Bank as in the rhetoric of the Sierra Club, the US military, or the myriads of Third World grassroots environmental and community movements. Whatever its semantic ambiguity, sustainability has the effect of linking three hitherto relatively disconnected discourses. It is now taken for granted that the *global environmental crisis*, and renewed concern with *global demography* (the return of the Malthusian specter), are inseparable from the terrifying map of *global economic inequality*. In sharp contrast to the 1960s, even conventional views confirm that eradicating poverty through enhancing and protecting livelihood strategies is as much an environmental sustainability issue and a fertility question (in which women's employment and education figures feature centrally), as it is a "simple" asset or resource endowment question (World Bank 1992).

Located on this expansive canvas of intellectual and political–economic ferment, *Liberation Ecologies* explores, through a series of thematic chapters and rich case-studies drawn from Latin America, Africa, and East, Southeast and South Asia, the current debates over development and the environment. In choosing this title we seek to emphasize a number of concerns. Obviously we wish to mark the potential liberatory or emancipatory potential of current political activity around environment and resources. However we also wish to signal the fact that the proliferation of environmental concerns linked to questions of development has other profound theoretical and practical consequences. One is that the politics of the environment seem to embrace a wide terrain including not just new social movements, but transnational environmental alliances and networks as with the World Social Forum, multilateral governance through for example the Global Environmental Facility of the World Bank, and a sensitivity to a panoply of local conflicts and resistances that may not warrant the term "movement." Another is that theories about environment and development – political ecology in its various guises – have been pushed and extended both by the realities of the new social movements themselves, and by intellectual developments associated with green Marxism, cultural and social theory, discourse theory and poststructuralism. These exciting new developments – many of which appear in the chapters which follow – represent, for us, the possibility of a more robust political ecology which integrates politics more centrally, draws upon aspects of discourse theory which demand that the politics of meaning and the construction of knowledge be taken seriously, and engages with the wide ranging critique of development and modernity particularly associated with Third World intellectuals and activists such as Vandana Shiva, Arturo Escobar and Victor Toledo. *Liberation Ecology* highlights, in other words, new theoretical engagements between political ecology, Marxism and social theory on the one hand, and a practical political engagement with new movements, organizations and institutions of civil society challenging conventional notions of development, politics, democracy and sustainability on the other.

In this Introduction, we address ways in which environmental problems have been addressed in the last thirty years, with particular attention to the field of political ecology. We provide a history of a field that now contains a large body of work, possesses its own electronic journal, and (as one might expect of a "mature" science) contains substantial debates within its ranks, together with providing an overview of its conceptual toolkit and its theoretical claims. We show how, since its formation in the 1970s, political ecology has been challenged and deepened both by "internal" theoretical debates and by the "external" environmental and political economic realities it seeks to explain. Political ecology, over the last decade, has grappled with environmental politics by way of a broader and more sophisticated sense of the *forms* of political contention and a deeper conception of *what* is contended, what we refer to in shorthand as "liberation ecology." Central to the new political ecology is a sensitivity to environmental politics as a process of cultural mobilization, and the ways in which such cultural practices – whether science, or "traditional" knowledge, or discourses, or risk, or property rights – are contested, fought over and negotiated.

INTELLECTUAL ORIGINS OF POLITICAL ECOLOGY

What, then, is political ecology? The origins of the couplet – politics and ecology – is instructive. It dates back to the 1970s (Watts 1983a) when a variety of commentators – journalist Alexander Cockburn, anthropologist Eric Wolf, and environmental scientist Grahame Beakhurst – coined the term as a way of thinking about questions of access and control over resources (that is to say the toolkit of political economy), and how this was indispensable for understanding both the forms and geography of environmental disturbance and degradation, and the prospects for green and sustainable alternatives. The fact that such writers were concerned to highlight politics and political economy – that is to say a sensitivity to the dynamics of differing forms of, and conflicts over, accumulation, property rights, and disposition of surplus – reflects a concern for distancing themselves from other accounts of the environmental crisis which sought to locate the driving forces in technology, or population growth, or culture or poor land use practice.[1]

6

Political ecology's originality resided in its efforts to integrate human and physical approaches to land degradation, through an explicitly theoretical approach to the ecological crisis capable of addressing diverse circumstances (soil erosion in Nepal, water pollution in Delhi) and capable of accommodating both detailed local study and general principles. As a defining text puts it: "[T]he phrase 'political ecology' combines the concerns of ecology and a broadly defined political economy. Together this encompasses the constantly shifting dialectic between society and land-based resources, and also within classes and groups within society itself" (Blaikie and Brookfield 1987: 17). Less a problem of poor management, inappropriate technology, or overpopulation, environmental problems were *social* in origin and definition. Analytically, the fulcrum of any Nature–Society study had to be the "land manager" whose relationship to nature must be considered in a "historical, political and economic context" (1987: 239). Hence, rapid deforestation in eastern Amazonia, to take one example, needed to be understood in terms of why those who were clearing tropical rainforests did so in the pursuit of economically inefficient and environmental destructive cattle ranching, and how these social forces – ranchers, peasants, workers, transnational companies – were shaped by larger political–economic forces, not the least of which was the Brazilian government acting through subsidies, class alliances, and the military. In the first generation of political ecology, however, land managers were almost wholly male, rural, Third World subjects, and curiously un-political in their practices and intentions.

What set of ideas and events "produced" this welding together of ecology and political economy in the first place? To simplify one can say that efforts to link culture and environment in anthropology and geography arose in part through a combination of Darwinian or evolutionary thinking, the new sciences of ecosystems and cybernetics, the growing political visibility of Third World peasantries (in China and in Vietnam), and the consequences of the Cold War and the atomic bomb. We emphasize a post-1945 confluence between three sets of ideas. First, the important connection between on the one hand cybernetics and systems theory – which derived from the theory of machines and from artificial intelligence developed particularly during the Second World War – and on the other community ecology. The central figures were Gregory Bateson and Howard Odum who, while different in intellectual orientation, provided languages and concepts for thinking about humans in eco- and living systems, the flows of matter, information, and energy that coursed through human practice with respect to the environment, but also the mechanisms – homeostasis, equilibrium, flexibility – by which "adaptive structure" could be maintained in ecosystems (see Watts 2002).

Second, within anthropology and geography the twin themes of cultural evolution and cultural materialism provided a powerful Darwinian framework for thinking about not only historical change but also patterns of resource use and human adaptation in different environments. In geography this approach was referred to as cultural ecology, but it was the Columbia school of ecological anthropology which provided the most sophisticated ideas. Peter Vayda and Roy Rappaport (1967) in the 1960s showed how tribal subsistence people in isolated regions could maintain an "adaptive structure" with respect to their environment. In Rappaport's (1968) terms the native's "cognized model" of the environment – embodied in various ritual, symbolic and religion practices – could elicit adaptive behavior understood in terms of the "operational" model of Western ecology. The pig killing rituals of the Tsembaga Maring of highland Papua New Guinea could function as a thermostat preventing overpopulation by pigs and maintaining some sort of environmental balance with their fragile ecology. Much of this ecological anthropology of the 1960s sought the "hidden" adaptive functions of culture with respect to the ecosystem, in order to build an abstract model of adaptive structure which existed in all living systems (see Bateson 1972, Wilden 1972).

The third lineage is rooted in the social science of the nuclear age and the post-war development of human responses to hazards and disasters. The immediate threat was of course atomic and the deepening of the Cold War. These produced a number of government funded studies on the perception of, and responses to, environmental threats. Geographers – Gilbert White, Ian Burton, Robert Kates – were very much part of this work in the 1950s and 1960s, focusing on differing sorts of "natural" perturbations – tornadoes, earthquakes, floods – in the United States, and on the perceptions and behaviors of threatened communities and households. Disaster studies centers appeared around the country and sociologists and geographers schooled in survey research, cognitive studies, and behavioralism sought to understand why individuals misperceived or ignored environmental threats, and how communities responded to threats from tornadoes versus those from floods or droughts. By the 1970s Clark University, the University of Colorado and Ohio State University were centers of hazard or disaster research. Much of this work also drew upon organic analogies – adaptation and response – but was also sensitive to cultural perceptions and questions of organizational capacity and information access and availability. Systems theory was again central to the intellectual architecture of this body of scholarship (Watts 1983a).

These three approaches – ecosystems/cybernetics, ecological anthropology/cultural ecology, and natural hazards/disaster research – differed in terms of theoretical approach, points of emphasis and method, and geographical sites, but collectively defined a ground from which political ecology emerged. What triggered debate within these fields in the 1970s was contention over the limits of organic analogies, adaptation, and systems/organization theory. In geography and anthropology the challenge came from two related sources: first, a proliferation of what one might call "peasant studies" (Shanin 1970) in which questions of exploitation, social differentiation, and the role of the market among the Third World rural poor were central; second, and relatedly, the growth of Marxism within social sciences, especially in development studies in a variety of guises (world systems theory, dependency, structural Marxism) during the late 1960s and 1970s (Bryant 1998). These two tendencies confronted cultural ecology and ecological anthropology by examining not isolated or subsistence communities in harmony with their physical environment, but rather peasant societies marked by the presence of the markets, social inequalities, conflict and forms of social and cultural disintegration associated with their integration into a modern world system. Here maladaptation, rather than adaptation, was the order of the day. Disequilibrium and positive feedback ruled, rather than balance or community maturity. While some geographers tried to understand the development of capitalism within peasant communities in ecological terms (Nietschmann 1972, Grossman 1984), political economy provided a different set of questions and answers with more purchase than the evolutionary and Darwinian toolkit. Marginalization, surplus appropriation, relations of production, and exploitation displaced the old lexicon of self-regulation, adaptation, homeostasis, and system response.[2]

FOUNDATIONS: THE POLITICAL ECOLOGICAL TOOLKIT AND ITS LIMITS

> Environmental degradation is created ... by the rational response of the poor households to changes in the physical, economic and social circumstances in which they define their survival strategies
>
> (de Janvry and Garcia 1988: 3)

From its inception, political ecology was never a coherent theoretical position for the very good reason that the meanings of ecology and political economy, and indeed politics, were often in question. For Watts (1983b) political economy drew upon a Marxian vision of social relations of production as an arena of possibility and constraint; for Blaikie and Brookfield a "broadly defined political economy" (1987: 17) meant a concern with effects "on people, as well as on their productive activities, of on-going changes within society at local and global levels" (1987: 21). For Martinez Alier, political economy was synonymous with "economic distributional conflicts" (see Alier and Guha 1998: 31). Forged in the crucible of Marxian or neo-Marxian development theory, this new "political ecology" was not inspired by the isolated rural communities studied by Rappaport, but by peasant and agrarian societies in the throes of complex forms of capitalist transition. Market integration, commercialization and the dislocation of customary forms of resource management – rather than adaptation and homeostasis – became the lodestars of a critical alternative to the older cultural or human ecology.

Notwithstanding this diversity of opinion, the work of Blaikie and Brookfield (1987), and their notion of political ecology, can plausibly be taken as exemplary. In their view, political ecology makes three essential assumptions. The first is interactive, contradictory and dialectical: society and land-based resources are mutually causal in such a way that poverty can induce, via poor management, environmental degradation which itself deepens poverty (1987: 48). Second, Blaikie and Brookfield argue for regional or spatial accounts of degradation which

9

link through "chains of explanation" (1987: 46), local decision-makers to spatial variations in environmental structure (stability and resilience as traits of particular ecosystems specifically). Locality studies are, thus, subsumed within multi-layered analyses pitched at a variety of regional scales. And third, land management is framed by "external structures" which, in the lexicon of their political ecology, means the role of the state (1987: 17), the core-periphery model (1987: 18), and "almost every element in the world economy" (1987: 68).

What then was the political ecology conceptual toolkit? The first was a refined concept of *marginality* in which political, ecological and economic aspects might be mutually reinforcing: "land degradation is both a result and a cause of social marginalization" (1987: 23). Second, *pressure of production on resources* was transmitted through social relations that impose excessive demands on the environment (see Watts 1983b on the "simple reproduction squeeze"). And third, the inadequacy of environmental data of historical depth linked to a chain of explanation analysis compelled a *plural approach*. One must, in short, accept "plural perceptions, plural definitions ... and plural rationalities"(1987: 16). Implicit was a sense that one person's profit was another's toxic dump. While explored in depth by the authors, political ecology opened the possibility of a serious discussion of how Nature and environmental problems were represented and how discursive formations shaped policy and practice (Peet and Watts 1996).

Blaikie and Brookfield's important intervention represents, within geography and social theory more generally, a sophisticated extension of previous efforts to integrate questions of access and control over resources – relations of productions as realms of possibility and constraint – with human ecology. But there were also complementarities between *Land Degradation and Society* and other social theorists working on questions – of ecological crisis and rehabilitation. Like Little and Horowitz (1987), regional political ecology focused on the producer and ecological pressure points; it shared with Redclift (1987) an emphasis on the contradictions of development; and with Jane Collins (1987) a sensitivity to the social causes of degradation and the need for a rethinking of development itself. Like Bunker's (1985) Amazonian study they employed a regional analysis sensitive to spatial variation and environmental heterogeneity; like Perrings (1987) they raised the suggestion that the market-price system as a means of regulating the environment was limited by the time perspectives of economic agents under capitalism, and by the presence of uncertainty. And not least they shared with Martinez Alier and Schlupmann (1987) a belief that value in land (what Blaikie and Brookfield called "landesque capital") was inconsistent with both neo-classical economic theory and Marx's labor theory of value.

Collectively this body of work punched a huge hole in the "pressure of population on resources" view, and the market distortion or mismanagement explanation of degradation. Instead it affirmed the centrality of *poverty* as a major cause of ecological deterioration (de Janvry and Garcia 1988, Mellor 1988, Martinez Alier 1990; see Blaikie and Brookfield 1987: 48). This represented an important advance in our thinking about nature–society relations. But it nonetheless required a much greater refinement, and an explicit theorization which is typically lacking because of political ecology's frequent appeal to plurality. What then are some of the limits and weaknesses of the political ecology that emerged in the late 1970s and early 1980s?

First, an undue emphasis on poverty and poor peasants must recognize that impoverishment is no more a cause of environmental deterioration than its obverse, namely affluence/capital. Hecht and Cockburn (1989) make this point with respect to the rates of deforestation in the Amazon basin. The danger lies in neglecting the obvious power of capital as a material force in degradation and, as a consequence, this comes close to blaming the victim, albeit in terms of the situational rationality of the land manager compelled to mine the soil or fell the forest. Second, the focus on poverty is perhaps not unrelated to the bias toward rural, agrarian and Third World matters in *Degradation and Society*, and indeed in political ecology more generally. How, for example, might poverty or political ecology help explain worker injuries in the maquila plants in northern Mexico, toxic dumping in Nigeria, or urban water pollution in Penang? (Chapters in this book, for example by Forsyth and O'Rourke, address such questions; see also Evans 2002; Gandy 2002.) And third, Blaikie and Brookfield privilege land – with good reason in view of its special significance in largely agrarian Third World states – as opposed to other "resources." The point we emphasize is that a poverty-centered analysis is, as the authors concede, only part of the story: there are other stories to tell of worker health and safety, air pollution, the decay of Third World cities, and of the restructuring of capitalism and so on. The extent to which this partiality is of analytical consequence rests, of course, on the theory which grants to poverty its causal powers.

Poverty is, at best, only a proximate cause of environmental deterioration. In other words, one has to have a theory capable of explaining how the poverty of specific land managers is reproduced through determinate structures and by specific relations of production. Blaikie and Brookfield move some way toward this by isolating production, but in an extremely diffuse and inconsistent way. Specifically, they invoke marginalization (an awkward label for several complex and contradictory processes) and an absence of control over resources. In short they make the land manager, and occasionally the production unit, the fulcrum, trapped within complex webs of relations, all held together by a political economy which, in a rather unhelpful way, is lumped together as "exogenous" (1987: 70). These exogenous, and largely untheorized, inputs into the political ecology decision-making model (i.e. how and why the land-manager acts) purportedly explain declines in land quality, a process which, according to Blaikie and Brookfield, elicits three responses: perception-correction, "change the social data," and migration. A broad and untheorized exogenous cause seems to deterministically produce quite specific outcomes. In other words this is hardly the sort of dialectical analysis which they themselves suggest should be on offer.

In spite of the fact that Blaikie and Brookfield talk of the selection of strategic factors that have causal power, we are not given a theory that guides the act of selection. Rather we are provided with "a chain of explanation" (beginning with Nepalese farmers and ending with Nepal's relationship to India) in which there is no sense of how, or why, some factors become causes. Coupled with their emphasis on plurality, the authors actually produce not "a theory which allows for . . . and identifies complexity" (1987: 239), but an extremely diluted and diffuse series of explanations. Degradation can arise under falling, rising or stable population pressures, under an upswing or downswing in the rural economy, under labor surplus and labor shortage; in sum, under virtually any set of conditions. The best that Blaikie and Brookfield provide is what they call a "conjunctural" explanation which it seems operates under all empirical circumstances.

Despite their claims for theory construction and the importance of social structural antecedents, Blaikie and Brookfield actually present a voluntarist view of degradation. Political ecology is radically pluralist, largely without politics or an explicit sensitivity to class interest and social struggle. Yet any analysis of land-based resources must surely confront and incorporate politics inscribed in various social arenas: familial-patriarchal, production-labor process, and the state (Burawoy 1985). The politics of ownership and control must be central to political ecology in order to give the bare bones of poverty some sort of flesh if it is to be employed analytically. Political ecology comes closest to theory when it invokes surplus extraction. And yet the authors on occasion seem more inclined to abandon theory altogether. Rather than outlining an explicit theory of production or political economy, with an arsenal of middle-level concepts, Blaikie and Brookfield provide only a plurality of disconnected linkages and levels. Hence their discussion of degradation in socialist economies can only conclude that it exists, and cannot offer any insights into the question they pose, namely "is there a distinctive socialist environmental management" (1987: 208), which presupposes a theorization of socialist political economy (see Gille 1997; O'Rourke 2002).

In short, for Blaikie and Brookfield, political ecology's conception of political economy appears fuzzy – "almost every element in the world economy" (1987: 68) – and diffuse. Their emphasis on plurality comes perilously close to voluntarism, while their chains of explanation seem incapable of explaining how some factors become causes. Particularly striking is the fact that political ecology has an undeveloped sense of politics (an issue taken on by the chapters by Schroeder and Suryanata, Rangan and Bebbington in this book). There is no serious attempt at treating the means by which control and access of resources or property rights are defined, negotiated and contested within the political arenas of the household, the workplace and the state.

ITERATIONS: NEW DIRECTIONS AND NEW QUESTIONS IN THE 1990s

A number of loosely configured areas of scholarship have extended the frontiers of political ecology and have elaborated and developed the important work of

Blaikie, Brookfield and others. The first attempts to refine political economy within the circumference of political ecology: in other words to make rigorous and explicit the causal connections between the logics and dynamics of capitalist growth and specific environmental outcomes. Some of the most exciting new work centers on efforts at explicitly re-theorizing political economy and environment at several different levels. At the philosophical level there are debates over Marxism and ecology (Benton 1989, Grundemann 1991; see also Leff 1995) and whether the labor process is compatible with eco-regulation and the notion of biological limits. The work of O'Connor (1988; 1999) and the journal *Capitalism, Nature, Socialism* starts from the "second contradiction of capitalism." In this view, Marx identifies production conditions (nature, labor power and communal conditions of production) which capital cannot produce for itself as commodities. The state mediates, and hence politicizes, conflicts around these conditions (environmental movements, feminism, and social movements) in an effort at maintaining capitalist accumulation. Many contributions explored these ideas in various parts of the Third World. Also there are attempts at harnessing specific concepts drawn from political economy as a way of linking the two structures of nature and society. For example how the simple reproduction squeeze compels self-exploitation among peasants who mine the soil; or how functional dualism can facilitate labor migration which undermines local conservation or constrains sustainable herding practices (Faber 1992; Garcia Barrios and Garcia Barrios 1990; Little and Horowitz 1987; Stonich 1989; Toulmin 1992; Watts 1987). Much remains to be done, however, in theorizing the specific dynamics of actually existing socialisms and the environment (Herskovitz 1993). Here of course the devastating ecological consequences of socialist political economy must be located not with respect to markets and profit but in relation to what Janos Kornai (1992) calls "the economics of shortage," that is to say the complementary and contradictory rationalities of centralized state planning (and its attachments to industrial gigantism and heavy goods) on the one hand and the reciprocities and networks at the enterprise level on the other (see Shapiro 1999).

A second broad thrust questions the absence of a serious treatment of politics in political ecology. Efforts at integrating political action – whether everyday resistance, civic movements or organized party politics – into questions of resource access and control have proven especially fruitful (Broad 1993, Kirby 1990). It is against this backdrop that "feminist political ecology" emerged (see Rocheleau *et al.* 1996), exploring the ways in which environmental concerns are traced through gender roles, knowledges and practices. Perhaps the most compelling work is drawn from Africa. MacKenzie's book (1998) traces both the erosion – what she calls the "silencing" – of women's environmental knowledge in central Kenya after 1890, and the ways in which women organized and struggled to resist the impact of colonial conservation on their economic liberty, not least through male appropriation of property rights. Richard Schroeder's book, *Shady Practices* (1999), focuses on the ways in which efforts to create sustainable development projects in drought-prone Gambia – local forest and fruit tree projects – precipitated struggles within the household and often over the obligations and reciprocities of conjugality. Local "traditional" women's work groups become the vehicle for local protest as resistance to male claims over property and access rights spill into a larger public domain. At the household management level, several studies focus on gender and domestic politics and struggles around the environment (Aggarwal 1992; Guha 1987; MacKenzie 1998; see also Carney and Schroeder in this book) specifically focused on property rights labor and the micropolitics of access and control within the domestic sphere.

At other levels of analysis – the state, inter-state and multilateral institutions, and local, i.e. community level resource control – important new work has forged analytical links between power relations, institutions and environmental regulation and ecological outcomes. Rich's (1994) book on the World Bank, and more generally the ecological establishment, Peter Haas's studies of transnational scientific communities (1993) – "epistemic communities" in his lexicon – and international environmental agreements as well as Peter Sand's (1995) analysis of post-UNCED legal frameworks, all illustrate David Harvey's suggestion that "control over resources of others in the name of planetary health [and] sustainability . . . is never too far from the surface of many western proposals for global environmental management" (1993: 25). Peluso's brilliant study (1993) links the historiography of criminality with everyday resistance to show how state power and forest management institutions are contested by Indonesian peasants and raises the larger issue of the colonial legacy – and of coercive patterns of conservation. In subsequent work on Kenya, Peluso shows how the militarization of environmental and resource conservation can be legitimated by international conservation groups (Peluso 1993). What is at stake here is the more general question of participation, community rights and local needs in environmental protection and conservation strategies (Utting 1994). Finally, the emancipatory potential which unites nature with social justice is a key theme in the emerging body of work on the ecology of the poor (Martinez Alier 1990; Broad 1993; Cockburn and Hecht 1990; Gadgil and Guha 1995) and in the large body of work on Indian environmental movements (see IICQ 1992). Contained within this work is a sensitivity to the panoply of political forms – movements, domestic struggles over property and rights, contestations within state bureaucracies – and the ways in which claims are made, negotiated and contested.

A third focus has been the complex analytical and practical association of political ecology and the institutions of civil society. The growth of environmental movements largely unregulated by, and distinct from, the state, poses sharply the question of the relations between civil society and the environment. There are two obvious facets of these relations, both having received some attention. The first is the origins, development and trajectories of the environmental associations and organizations (see Escobar 1995; Ghai 1992; *Socialist Review* 1993). What are the spaces within which these movements develop and how, if at all, do they articulate with other organizations, and resist the predations of the state (see Bebbington in this volume)? The second facet draws on the substantial literature on local knowledges and ecological populisms (Richards 1985; Warren 1991). The concern is not simply mounting a salvage operation – recovering disappearing knowledges and management practices – but rather obtaining a better understanding of both the regulatory systems in which they inhere (see the literature on common property, Ostrom 1990) and the conditions under which knowledges and practices become part of alternative development strategies. In this latter sense we return to the politics of political ecology but more directly to the institutional and regulatory spaces in which knowledges and practices are encoded, negotiated and contested and ultimately of the relation between democracy and environmentally sound livelihoods.

A fourth theme employs discursive approaches to tackle head-on Blaikie and Brookfield's point about the plurality of perceptions and definitions of environmental and resource problems. Several new lines of thinking are important. One draws upon the critical studies of science as a way of exploring the politics of what one might call "regulatory knowledge"; why particular knowledges are privileged, how knowledge is institutionalized, and how the facts are contested. Beck's (1994) work on risk and reflexive modernization, and Shrader-Frechette's (1991) work on risk and rationality are important illustrations of this sort of research. Another traces the history of particular institutions – say forestry – and how particular knowledges and practices are produced and reproduced over time (Sivaramakrishnan 1995). The genesis and transmission of conservation ideas and the institutions of national parks and their management have been explored productively in this way (Beinart and Coates 1995; Grove 1993; Neumann 1992). Another line of work examines the globalization of environmental discourse and the new languages and institutional relations of global environmental governance and management. Taylor and Buttel (1992) for example trace the moral and technocratic ways in which the new global discourse on the environment is privileged, and how in the formulation of environmental science some courses of action are facilitated over others.

The question of doing environmental history represents a fifth aspect of an invigorated political economy (see the journal *Environmental History* edited by Richard Grove). In providing much needed historical depth to political ecology, environmental historians raise important theoretical and methodological questions for the study of long-term environmental change. The obvious theoretical contrasts between Worcester (1977), Merchant (1993) and Cronon (1992) point to an extraordinary heterogeneity in the field. Contained within each is the idea of writing alternative histories from the perspective of long-term ecosystemic changes that cannot be captured with the clumsy unilinear models of agricultural and environmental change. The relations between agrarian intensification and the environment are rarely so simple. Tiffen and Mortimore's (1994) study of Machakos District in Kenya shows how population increased five fold between 1930 and 1990 but the environmental status actually improved over the same period. Soil structure improved and even wood-fuel was sustained. Similarly Fairhead and Leach (1994) working in Sierra Leone locate forest quality and biodiversity in the influence of past land use practices. Like new work in Amazonia and South Africa, they show how habitation and cultivation can improve soil and support denser woodlands. As they put it, "vegetation patterns are the unique outcomes of particular histories not predictable divergences from characteristic climaxes" (1994: 483). In a sense the new environmental historians meet on the same ground as a quite different intellectual tradition, derived from the so-called agrarian question (cf. Kautsky 1906), which attempts to chart the ways in which the biological character of agriculture shapes the trajectories of capitalist development (Kloppenburg 1989). Opportunities for exploring the long-term capitalization of nature through "appropriation" and "substitution" (Goodman *et al.* 1990; M. O'Connor 1994), and their environmental ramifications, can, and should be, readily seized by political ecologists.

Finally, there was the much-needed interrogation of "ecology" in political ecology and the extent to which political ecology is harnessed to a rather outdated view of ecology rooted in stability, resilience and systems theory (Zimmerer 1994). Botkin (1990) and Worster (1977) among others describe the relatively new ecological concepts which pose problems for the theory and practice of political ecology. The shift from 1960s systems models to the ecology of chaos, that is to say chaotic fluctuations, disequilibria and instability, suggests that many previous studies of range management or soil degradation resting on simple notions of stability, harmony and resilience may have to be rethought (Zimmerer 1994). The new ecology is especially sensitive to rethinking space–time relations to understand the complex dynamics of local environmental relations in the same way that the so-called dialectical biologists (Levins and Lewontin 1985) rethink the evolutionary dynamics of biological systems. Notwithstanding Worster's (1977) warning that disequilibria can easily function as a cover for legitimating environmental destruction, some of the work on agro-ecology (Altieri and Hecht 1990; Gleissman 1990; see also Zimmerer in this volume) suggests that the rethinking of ecological science can be effectively deployed in understanding the complexities of local management (for example in intercropping and pest management).

All of these new directions were not necessarily of a theoretical piece. What is striking, however, is the extent to which these new directions attempt to engage political ecology with certain ideas and concepts derived from poststructuralism, and cultural and social theory (see Bryant 2001; Goldblatt 1996; Keil *et al.* 1998; Raffles 2002; Redclift and Benton 1994). There is, in other words, an extraordinary vitality within the field reflecting the engagements within and between political economy, the power–knowledge field and critical approaches to ecological science itself. As shorthand we refer to these confluences and engagements as "liberation ecology." The implication in this notation is to recognize the emancipatory potential of the environmental imaginary and to begin to chart the ways in which natural as much as social agency can be harnessed to a sophisticated treatment of science, society and environmental justice. Of course, a major site of such engagement is in the analysis of social and environmental movements, a field which draws together the explosive growth of organizations and civic movements around sustainability with an implicit critique of (and an alternative vision) of "development." It is to the philosophical and social theoretical underpinnings of development and the environmental imaginary that we now turn.

CRITIQUE: POLITICAL ECOLOGY JETTISONED?

The measure of maturity and solidity of any field of knowledge is that it can generate debate and dissent over its approach, its internal consistency, its conceptual apparatuses and its ability to wrestle with the problems thrown up by history, to "reconstruct" its theory as Burawoy (1991) puts it. Political ecology bears all the hallmarks of an established inter-disciplinary sub-field; it has foundational texts (the work of Piers Blaikie, Susanna Hecht, Karl Zimmerer, Priya Rangan, Nancy Peluso, Mike Davis, Richard Schroeder, Bill Durham, Rod Neumann); it has its "in-house" journals (*Journal of Political Ecology, Land Degradation and Rehabilitation*); it has a history of theoretical reconstruction and extension; it possesses something like a conceptual toolkit, and has a breadth of basic research encompassing a multiplicity of regions and disciplines (the international citation index reveals a forty fold increase between 1980 and 2000). There are at least three undergraduate/graduate texts, with a handbook in preparation. Political ecology might even have reached the hallowed status of a normal science (see Bryant 2001).

No surprise then that political ecology has been attacked, vociferously, by Vayda and Walters (1999) among others. In their account, "self-styled political ecologists" make "*a priori* judgments" and produce work characterized by "bias," intellectual closure, and intrinsic "question begging." Vayda and Walters accuse "many" political ecologists of a limitless capacity to neither verify their object of scrutiny nor to have understood, or indeed have any interest in, understanding, the "complex and contingent interactions of factors whereby environmental changes are produced" (1999: 67). Political ecology is, in their account, a sort of anti-science, convinced of its veracity and truth claims long before the research begins.

What critical claims do they make? First, political ecology emerged, they say, as a reaction to "the neglect of the political dimensions of human/environment interaction." Second, they claim, "self-styled" political ecologists deal "only with politics."[3] For Vayda and Walters, however, privileging politics is only part of the problem. In doing so, political ecologists have first undermined the tenets of science by deciding on causal significance *a priori*, have second used their political biases to promote normative outcomes, and specifically, a "populist political agenda" and third, are so driven by their own political preoccupations that have either abandoned *any* interest in the environment (they only study "political controls ... over natural resources") or are utterly disinterested in even determining the veracity of the environmental process or outcome that politics purportedly explains.

What sort of critique of political ecology does this represent? First, if political ecology has an epistemological position in common it is critical realist (see Sayer 1980); which is to say it examines relations between events, structures and mechanism through a stratified sense of reality in which theory-building and theory-reconstruction precisely deploys a toolkit to explain the world. In this sense, all theory makes prior assumptions and judgments; this is not a flaw but a necessity. It is in this way that a theory disposes the research to generate particular hypotheses or to identify particular anomalies to be explained. Vayda and Walters, of course, reject such a model of science – adopting, instead a rather crude empiricism – but this is hardly consistent with their own claims of "openness" and, in any case, vastly mischaracterizes the nature of the scientific project of theory-building that political ecology represents. In place of critical realism they see a sort of theoretical Stalinism.

Second, the claim for populism and green romanticism is rather extraordinary. Virtually all the work on access and control that Vayda and Walters so clearly dislike has been directed to exposing the limits of a naïve invocation of the local community as a theatre of governance. Priya Rangan's work (2000 and this volume) – to take one recent example, but there are many others – is a compelling critique and exposure of the populism of Chipko. Far from the mythic community of tree hugging, unified, undifferentiated women articulating alternative subaltern knowledges for an alternative development – forest protection and conservation by women in defense of customary rights against timber extraction – we have three or four Chipko's each standing in quite different relationship to development, modernity, sustainability, the state and local management. It was a movement with a long history of market involvement, of links to other political organizing in Garawhal, and with aspirations for regional autonomy. Ribot's (1998) work on forest management is precisely about the pitfall of an assumed invocation of community powers or regulation. One of the discoveries of this political ecological work is that the community expresses quite different sorts of social relations and forms: from a nomadic band to a sedentary village to a confederation of Indians to an entire ethnic group. It is usually assumed to be the natural embodiment of "the local" – configurations of households, lineages, longhouses – which has some territorial control over resources which are historically and culturally constructed in distinctively local ways. A community, then, typically involves a territorialization of history ("this is our land and resources which can be traced in relation to these founding events"), and a naturalized history ("history becomes the history of my people and not of our relations to others"). Communities fabricate, and re-fabricate through their unique histories, the claims which they take to be naturally and self evidently their own. This is why communities have to be understood in terms of hegemonies: not everyone participates or benefits equally in the construction and reproduction of communities, or from the claims made in the name of community interest. Tradition or custom hardly captured what is at stake in the definition of the community. The same might be said of so called feminist political ecology, which shows how custom and tradition can mask undemocratic forms of environmental governance. MacKenzie's book (1998) for example traces both the erosion – what she calls the "silencing" – of women's environmental knowledge in central Kenya after 1890, and the ways in which women organized and struggled to resist the impact of colonial conservation on their economic liberty not least through male appropriation of property rights. Work in anthropology has also exposed the pitfall of eco-populism in terms of the ways in which environmental movements *create* a sense of community and tradition for political purposes. Brosius (1997) shows how activists can be guided by self-centered interests of program building which rest on misleading stereotypes of the community, just as Tsing (1999) documents the ways in which Meratu community leaders play to a "fantasy" of tribal green wisdom to mobilize international attention.

Finally there is the very important question of the environment in political ecology. Vayda and Walters' claim that the environment has disappeared is curious on its face since so much of political ecology in the last decade has turned increasingly to nature itself. The questions are, of course, *what passes for the environment? What form nature takes as an object of scrutiny?* And here Vayda and Walters really show their colors, always draped of course in the hues of "openness." For them the *only* expression of environment can be the biophysical events of environmental change. Their event ecology rests precisely on identifying environmental events. But political ecology rests on the dialectics of Nature and Society in which environment can be approached in a number of ways. One way is certainly what Vayda and Walters call "events." Karl Zimmerer's work (in this book and elsewhere) documenting environmental change indeed derives from biology and ecology. But what political ecology has done obviously is to open up the category of the environment itself and explore its multiform representations. Knowledge of the environment itself is examined – why particular forms of knowledge predominate, circulate and how. What really is at stake is the fact that Vayda and Walters are not prepared to acknowledge – in the name of their own openness – the variety of ways in which environment can be explored; what is so striking is their rigidity and limited intellectual horizons in terms of (to deploy their language) "what is to be explained."

What then do Vayda and Walters have to offer? Events, open questions and causal sequences rather than "factors privileged in advance by the investigators." All of this starts from concrete events – as if political ecologists had ignored such concrete events such as game theft, or poaching, or forest felling – and discovering what antecedent events produced particular outcomes. But what distinguishes this approach from empiricism or induction? Vayda and Walters say openness should not be taken to mean "considering all causes." But why are some causes selected or identified over others? And how does this selection avoid the stain of *a priori* judgments? Vayda and Walters, speaking in the name of science provide no transparency for the process by which their causal chains are constituted; and show no concern whatsoever for the hermeneutic complexities of exploring such causal analyses through (presumably) talking to people or interrogating texts. What sort of alternative, one may reasonably ask, is this?[4]

We have spent some time on this critique both because it reveals something of the vitality of the field, but also because Vayda and Walter's argument exposes and clarifies precisely what in our view are the strengths and insights of what passes as political ecology in its contemporary variants.

LIBERATING POLITICAL ECOLOGIES[5]

As a response to this internal critique, political ecology has moved forward substantially in the last fifteen years along a number of key fronts which for convenience we shall discuss under two broad headings: *knowledge, power and practice*, and *justice, governance and ecological democracy.*

Knowledge, power and practice

Underlying this new work on knowledge is the recognition that any sophisticated political ecology must contain a phenomenology of nature. That is, it must take seriously Blaikie's (1985) point that the environmental problem can be "perceived" in a variety of ways. The newer political ecology, however, draws from poststructuralism's concern with knowledge, power and discourse (see Peet and Watts 1996). Much of this newer scholarship turns especially on what individuals and groups (and *de facto* communities) know and practice with respect to their local environments (so-called indigenous technical knowledge – ITK – which harkens back to earlier studies of ethno-botany). Perhaps the best political ecological study addressing the question of peasant experimentation and practice, and the threats which this world confronts, is Zimmerer's *Changing Fortunes* (1996) on biodiversity and peasant livelihoods in the Peruvian Andes. ITK has been widely explored (and there are a number of international organizations devoted to its generation, propagation and use) and now is widely understood within academic and activist circles (Brush 1996; Richards 1985). In problematizing environmental knowledges, political ecology has identified a number of core issues. First, a recognition that environmental knowledge is unevenly distributed *within* local societies; second, that it is not necessarily right or best just because it exists (i.e. it can be often wrong or inappropriate); and third, that traditional or indigenous knowledge may often be of relatively recent invention (which is to say these knowledges are not static or stable but, as Paul Richards (1985) suggests, may be predicated on forms of experimentation). Indeed, it may not be indigenous as such but really is *hybrid* (see Aggrawal 1999; Gupta 1998). Indigenous knowledge is a tricky idea because most knowledges are not simply local but complex hybrids drawing upon all manner of knowledges –

farmers in India may simultaneously employ concepts from Hindu religion and modern Green Revolution technologies. ITK can also take on mystical and ideological forms as in Vandana Shiva's account (1989) of Indian women as "natural" peasant scientists. Insofar as local actors know a great deal about local ecology and that this knowing is typically culturally "institutionalized" and "embedded" in a variety of persons, offices, rituals, and customary practices, the questions become: (i) why has this knowledge been so difficult to legitimate, (ii) under what circumstances can such knowledge/practice be institutionalized without co-optation or subversion and (iii) how might it be systematized in some way? Some of the most compelling work addressing these questions has emerged with the assistance of the tools of the science and technology field and the debates surrounding the constitution of nature (Braun and Castree 1998; Castree and Braun 2001; Demeritt 2001; Whatmore 2002).

Candace Slater's (2001) excellent work on Amazonia reveals another aspect of the knowledge question. Her work focuses on how there is a popular imagery of the region (perhaps transnational in appeal), how this imagery has a history, and how literature, media and other cultural machinery contributes to we have elsewhere called a "discursive ecological formation" (Peet and Watts 1996; see also Guthman 1997). In Slater's account, the Edenic – or naturalized narrative – always silences (the Indians have no voice or no voice of their own), and these tropes exclude or distort. Slater ends with the provocation that there is an absence of competing images of Amazonia. True (perhaps?), but under what conditions can competing images *really* compete? In a quite different context Kuletz's (1997) account of the nuclear damage to the US West also turns on how the landscape is constructed, in part by science, as "worthy" of being subject to nuclear attack (its desirability for the state was its "undesirability" and of course the invisibility of Native American communities). But do these images and constructions of landscape really have the power and effect implicit in these accounts of narratives? Are they "just" images and irrelevant to the hard edges of political economy and environmental destruction (see Redfield 2000; Stepan 2000)?

The idea of *reflexive modernization*, associated with the work of Ulrick Beck and Anthony Giddens (Beck *et al.* 1994), and *ecological modernization* associated with Maarten Hajer (1997) (see Macnaghten and Urry 1999), also draw on a concern with modernity and green discourses, both of which are distinguished by their efforts to link social theory and the environment. The focus is on the self-reflexive qualities of modernization and on the ways in which the ecological costs and consequences of capitalist modernity are built reflexively into modernity itself; that is to say, it is the environmental consequences of modernity and its scientific understanding of them, which constitutes a defining quality of modernity itself. Discourses of "risk" or "uncertainty," what Rom Harre (1999) calls "greenspeak," often constitute the powerful languages in which this self-reflexivity is constituted. These approaches often employ linguistic and discursive analysis rooted in social studies of science, and institutional analyses of regulation. Ecological modernization has been overwhelmingly urban and First World in orientation and has the great merit, like political ecology, of focusing on politics. It draws, however, from a heady mix of social studies of science and discursive institutional analysis (Lash *et al.* 1996), and has the advantage of examining the corporate sector and firm which has been largely neglected by political ecology (see Mol's research on "refinement of contemporary production technologies" (1994)). If early political ecology had as its cornerstones in vulnerability, marginalization and access, then ecological modernization has its risk, uncertainty and discourse (see Dupuis 2002).

Another approach to environmental knowledge production targets environment science and policy-making through work on epistemic communities, or communities of expertise. Here the knowledge is western science, and more properly the cosmopolitan scientist-expert-policy maker. Peter Haas (1990) has argued in the context of understanding regional (European Union) and global (multilateral) conventions that the process of consensus building and collective action more generally is *knowledge-based* and *interpretive*. That is to say, international regulatory co-operation is fueled by fundamental scientific uncertainty about the environment which ensures that governments seek out authoritative advisors (experts) who, to the extent they are part of epistemic communities, are more important to the political solution than the content of the ideas *per se*. Cross-national differences in state behavior are determined by the variation in the penetration and institutionalization of experts (epistemic communities). Biodiversity and stratospheric ozone co-operation are seen in this way as instances of the cognitive and bureaucratic power of scientific experts. An argument has also been made for NAFTA by Redclift and Benton (1997) who argue that the trade and environmental constituencies brought together around tariff reduction actually created a dialogue – a transnational community of experts – which had not hitherto existed.

The epistemic community idea is not unrelated to new political ecological work which examines particular scientific policy discourses, "conventional models" as Leach calls them, rooted in particular institutions and practices, which become hegemonic and are then subsequently contested. Some of the most interesting research has examined the politics of colonial and post-colonial conservation. For example, work in Africa has traced debates over soil erosion and land conservation in the 1930s to the complex political struggles among and between the colonial state, white settlers and the Native Reserves (MacKenzie 1997). Neumann's excellent book (1999; also this volume), *Imposing Wilderness*, on the creation of the Arusha national park in Tanzania and the ideas of landscape and nature which lay behind state appropriations of land from local peasant communities is an especially compelling illustration of how cultural and historical representations of nature intersect with colonial and post-colonial rule. Fairchild and Leach's (1996) reinterpretation of the forest-savanna mosaic is a careful deconstruction of a conventional model in which historical studies coupled with detailed local analysis of agro-ecology confirms what the new "non-equilibrium" ecology posits, namely that climax models of ecological stasis are unhelpful. These static models however do enter into administrative practice (colonial and post-colonial) which reinforces the idea of Guinee's forest cover as "relic" (which Mearns and Fairchild see as the basic for driving "repressive policies designed to reform local land use practice") rather than as the outcome of intentional local management practice. Similarly, Swift (1997) has shown how assumptions about desertification not only rest on remarkably spare evidence, but on questionable models of the dynamics of semi-arid rangelands – their resiliency and stability in other words – which are (i) expression of linear, cybernetic models of ecological structure and temporalities, and (ii) attached to neo-Malthusian models of social change. The key here is that conventional wisdom is challenged as an embodied form of knowledge, and the challenge itself reflects a peculiar unity of local knowledge and practice with non-linear models of new ecology. Out of this emerges a concern with pluralism (at the level of truth claims), with democracy (to open up the practices of policy-making to other voices), and complexity/flexibility (of local conditions and historical dynamics).

It remains an open question however whether these epistemic communities or conventional model have real power, in contrast to power politics approaches in which inter-state rivalries dominate or indeed, as Raustialla (1997) has shown in examining the differences between Britain and the US at the Biodiversity Convention, that what matters is domestic regulatory and political structures and the differing influence of business, not scientists and experts (see O'Neill 2000). The epistemic community is nonetheless especially relevant to the "greening" of multilateral organizations. Kingsbury (1997) shows in his account of the incorporation of environmental issues into the WTO and trade debates, that the process of greening theses institutions has only just begun. Robert Wade's (1997) work on the greening of the World Bank and McAfee's work (1999) on what she calls "green developmentalism" shows precisely how discourses (like gender and development) are institutionalized in quite specific ways with quite specific institutional powers. Of course much of this discourse turns on how the idea that nature has to be sold to be saved is constructed and legitimated (see Chapters 5 and 6 in this book).

Politics, justice and governance: toward an ecological democracy

Political ecology's concern with knowledge, representation, imaginary and imagery addresses politics – the politics of knowledge. But politics of another sort, what one might call "ecological democracy," has been addressed by political ecologists explicitly in a number of ways in the 1980s and 1990s (Goldman 1999; Zerner 1999). Perhaps the most influential studies linked questions of cultural studies and everyday resistance with gender. Nancy Peluso's path-breaking political ecological study of timber and forestry in Indonesia, *Rich Forests, Poor People* (1992), showed how local communities resisted the incursions of the state, and how the state in turn attempted to "criminalize" local customary rights over access to, and control over, local forest products. Politics, community and state were also central to Hecht and Cockburn's (1989) account of Amazonian deforestation in which state subsidies and powerful ranchers and timber companies were key to understanding the dynamics of the frontier violence, and, in turn, relevant to understanding the panoply of social movements (Chico Mendes most visibly) – often with links to left political parties and to transnational green NGO's – which resisted loss of local autonomy. The state figures centrally in these accounts: as an instrument through which conservation takes on a coercive or military cast (Peluso 1992), and as the means by which land becomes a geo-strategic matter (for example the Brazilian military government accelerating deforestation to "secure" the country's borders).

The community looms large in the new political ecology (see Chapters 5, 13 and 14 in this volume). But the community turns out to be – along with its lexical affines, namely tradition, custom, and indigenous – a sort of keyword whose meanings (always unstable and contested) are wrapped up in complex ways with the problems that it is used to discuss. The community is important because it is typically seen as: a locus of *knowledge*, a site of *regulation* and management, a source of *identity* and a repository of "tradition," the embodiment of various *institutions* (say property rights) which necessarily turn on questions of representation, power, authority, governance and accountability, an object of *state control*, and a theater of *resistance* and struggle (of social movement, and potentially of *alternate visions of development*). It is often invoked as a unity, as an undifferentiated entity with intrinsic powers, which speaks with a single voice to the state, to transnational NGOs or the World Court. Communities, of course, are nothing of the sort.

One of the problems is that the community expresses quite different sorts of social relations and forms: from a nomadic band to a sedentary village to a confederation of Indians to an entire ethnic group. It is usually assumed to be the natural embodiment of "the local" – configurations of households, lineages, longhouses – which has some territorial control over resources which are historically and culturally constructed in distinctively local ways. A community, then, typically involves a territorialization of history ("this is our land and resources which can be traced in relation to these founding events"), and a naturalized history ("history becomes the history of my people and not of our relations to others"). Communities fabricate, and refabricate, through their unique histories, claims which they take to be naturally and self evidently their own. This is why communities have to be understood in terms of hegemonies: not everyone participates or benefits equally in the construction and reproduction of communities, or from the claims made in the name of community interest. And this is exactly what is at stake in the current political ecological work on the infamous tree-hugging or Chipko movement in north India (see Rangan 2000; Sinha *et al.* 1997). Far from the mythic community of tree hugging, unified, undifferentiated women articulating alternative subaltern knowledges for an alternative development – forest protection and conservation by women in defense of customary rights against timber extraction – we have three or four Chipko's each standing in quite different relationship to development,

24

modernity, sustainability, the state and local management. It was a movement with a long history of market involvement, of links to other political organizing in Garawhal, and with aspirations for regional autonomy. Tradition or custom hardly captured what is at stake in the definition of the community.

The community-politics focus has also been central to the work – largely based in the advanced capitalist states – on economic justice, particularly in regard to toxic dumping and hazardous exposure in minority and working class communities. Pulido's *Environmentalism and Economic Justice* (1996) is an excellent example of how a sensitivity to community struggles over environment and health meets up with larger claims over economic justice and class politics. These sorts of movements were in no simple sense "environmental" since they typically combined human rights, ethnicity/identity, and questions of social justice (Escobar 1995; Pellow and Park 2003; Pellow 2002).

A number of implications stem from the community and justice approaches addressed by political ecologists. First, and most obviously, the forms of community regulations and access to resources are invariably wrapped up with questions of identity. Second, these forms of identity (articulated in the name of custom and tradition) are not stable (their histories are often shallow), and may be put to use (they are interpreted and contested) by particular constituencies with particular interests.[6] Third, images of the community, whether articulated locally or nationally, can be put into service as a way of talking about, debating and contesting various forms of property (and therefore of claims over control and access). Fourth, to the extent that communities can be understood as differing fields of power – communities are internally differentiated in complex political, social and economic ways – we need to be sensitive to the internal political forms of resource use or conservation (there may be three or four different Chipkos or Love Canal movements within this purportedly community struggle). Fifth, communities are rarely corporate or isolated, which means that the fields of power are typically non-local in some way (ecotourism working through local chiefs, local elites in the pay of the state or local logging companies and so on). And not least, the community – as an object of social scientific analysis or of practical politics – has to be rendered politically; it needs to be understood in ethnographic terms as consisting of multiple and contradictory constituencies and alliances (Li 1999, 1996 and this volume). This can be referred to as identifying "stakeholders" – a curiously anemic term – but often what is at stake is something that comes close to class analysis or at the least identification of wildly different forms of political power and authority.[7]

Inevitably, in its concern with the community and environmental politics, political ecology has turned to institutions as a necessary starting point to linking socially differentiated communities with biologically differentiated environments. Institutions – understood not simply as the "rules of the game" but as the habituated and regularized "rules-in-use" maintained by human practice and investment performed over time – are typically distinguished from organizations understood as actors or players brought together for a particular purpose. One way to approach institutions and their character is through Amartya Sen's (1980, 1990) theory of entitlement. Leach *et al.* (1997) suggest that "environmental entitlements" provide a way of linking what Sen calls "capabilities" with institutional design and performance. Entitlements refer to effective command over alternative commodity bundles, which derive from a person's endowments (i.e. through direct access to land I can command commodities produced on my land). Environmental entitlements can be seen as the "sets of benefits derived from environmental goods and services over which people have *legitimate effective command* and which are instrumental in achieving well-being" (Leach *et al.* 1997: 9, emphasis added). Environmental entitlements are thus a subset of a larger group of entitlements collectively providing the means by which basic human needs are met and people experience well-being. All of this sounds very abstract but it highlights the means by which differentiated social actors gain access to and control over resources through institutionalized practices. And not least there is the field of environment and rights. The attraction here is that it compels political ecology to both dig further into philosophies of nature (Soper 1995), and to link this political philosophy to questions of rights. The emerging geography of animal rights for example (Philo and Wilbert 2000; Wolch and Emel 1995) and the history of the zoo (Berger 1980; Hanson 2002; Rothfels 2002) in relation to capitalism and the law, is especially suggestive.

What might all this mean for environmental governance? This turns in part on rather different meanings of governance. Ribot's work (1998), which draws upon the work of political scientists and a neo-Polanyian perspective (see also Chapter 8 in this book), opens a number of important avenues for analysis. He examines state institutional arrangements shaping access and control to fuel-wood in Senegal. In his view the state deploys law as a form of rural control. Local appointed authorities backed by the state create fictions in which there is no local representation. Community participation is in fact disabled by forms of state intervention – and in his view by the continuance of the colonial model of rule through "decentralized despotism." Ribot argues that participation without locally accountable representation is no participation at all. As he puts it "when local structures have an iota of representativity no powers are devolved to them, and when local structures have powers they are not representative but rather centrally controlled" (1998: 4). What passes in Mali or Niger or Senegal as community participation is circumscribed by the continuing power of chiefs backed by state powers, by the lack of open and free elections, and by the decentralized despotism of post-colonial regimes. In the case of institutions which involve state–community linkages, it is influence and prestige, coupled with authority and money which fundamentally frame the forms of governance and hence who participates and who benefits (see also Gibson *et al.* 2000; Pellow 2002).

A second approach addresses the role of NGO and civic action around green concerns (Princen and Finger 1994). The work of Peter Evans (1996) on social capital is especially relevant here because his concern with what he calls *public–private synergies* speaks to the ways in which multiple institutions of control and access associated with the state and with civil society, operating at different scales and levels, operate synergistically. It poses the question sharply of how public institutions can be coherent, credible and have organizational integrity and how the institutions of civil society can engage in accountable ways with the public sphere. In the case of environmentalism, however, these public–private synergies crosscut international boundaries and pose difficult questions for both multilateral regulation and for transaction activism. The work of Keck (1995) on NGO activism and community watershed management in Brazil (and her work on the Acre rubber tappers which shows how the movement gained power precisely by presenting their interest as "worker" interests rather than as ethnic or tribal) and Pezzoli's (1998) book on community activism in Mexico City pay testimony to both the powers and the limits of local green activism, and to the difficulty of building new forms of public–private contract. Baviskar's (1995) account of the Narmada Dam movement reveals the tensions between sustainable development activists and the community attracted to a collaboration with the state as "tribal" peoples. Along the way some constituencies – the migrant laborers – are left out completely.

Transnational advocacy groups (TNGOs) – and transnational environmental organizations in particular such as World Wildlife Fund or Nature Conservancy – also highlight questions of governance and institutional politics (Bryant and Bailey 1997; Keck and Sikkink 1998). Brosius (1997) shows how activists can be guided by self-centered interests of program building which rest on misleading stereotypes of the community, just as Tsing (1999) documents the ways in which Meratu community leaders play to a "fantasy" of tribal green wisdom to mobilize international attention. A number of large TNGOs have themselves been shaped by the changing political and market-driven winds in the West producing a sort of in-house corporate environmentalism ("green corporatism") within the larger TNGO community. This itself raises the question of large TNGOs as major donors (i) changing the domestic politics and structure of the local NGOs communities in the South, (ii) how foreign and local NGOs actually build political strategy and alliances; and (iii) how social capital is constructed in North–South inter-NGOs collaborations. Bailey's (1998) work on the activities of the WWF in Ecuador highlights the tensions between transnational and local NGO green activism – that there is a necessary unity of interests between North and South environmentalism.

A third stream sees governance as "green governmentality" (see Chapters 5, 10 and 13 in this book) deploying the ideas of Michel Foucault. Government for Foucault (2000) referred famously to the "conduct of conduct," a more or less calculated and rational set of ways of shaping conduct and of securing rule through a multiplicity of authorities and agencies in and outside of the state and at a variety of spatial levels. In contrast to the forms of pastoral power of the Middle Ages, from which evolved a sense of sovereignty, Foucault charted how an important historical shift, beginning in the sixteenth century, moved toward government as a right manner of disposing things "so as to not lead to the common good ... but to an end that is convenient for each of the things governed" (2000: 211). The new practices of the state, as Mitchell Dean (1999: 16) says, shape human conduct by "working through our desire, aspirations, interests and beliefs for definite but shifting ends." Foucault revealed the genealogy of government, and the origins and modern power, the fabrication of a modern identity. The conduct of conduct – governmentality – could be expressed as pastoral, disciplinary or as bio-power. Modern governmentality was rendered distinctive by the specific forms in which the population and the economy was administered, and specifically by a deepening of the "governmentalization of the state" (that is to say how sovereignty comes to be articulated through the populations and the processes that constitute them). Key for Foucault was not the displacement of one form of power by another, nor the historical substitution of feudal by modern governmentality, but the complex triangulation involved in sustaining many forms of power put to the purpose of security and regulation.[8]

The work on green governmentality has taken up two interrelated aspects of governmentality.[9] One is what Foucault explicitly refers to as relations between men and resources/environment as an expression of his complex notion of the governance of things. As he put it:

> On the contrary, in [the modern exercise of power], you will notice that the definition of government in no way refers to territory: one governs *things*. But what does this mean? I think this is not a matter of opposing things to men, but rather of showing that what government has to do with is not territory but, rather, a sort of complex composed on men and things. . . . What counts is essentially this complex of men and things; property and territory are merely one of its variables.
>
> (2000: 208–9)

The other is taken from Rose's notion of "governable spaces" as they emerge from the four analytics of government detailed above. For Rose governable spaces, and the spatialization of government, are "modalities in which a real and material governable world is composed, terraformed, and populated" (1999: 32). The scales at which government is "territorialized" – territory is derived from *terra*, land, but also *terrere*, to frighten – are myriad: the factory, the neighborhood, the commune, the region, the nation. Each of these governable spaces has its own topology and is modeled – as Rose puts it – through systems of cognition and remodeled through government practice – in a way that frames how such topoi have emerged: the social thought and practice that has territorialized itself upon the nation, the city, the village or the factory. The map has been central to this process as a mode of objectification, marking and inscribing but also as "a little machine for producing conviction in others" (1999: 37). In this way it is possible to explore how the environment enters into the constitution of regulable object and political subjects, and indeed into the creation of particular sorts of governable spaces. Braun's book (2003) on forests and the Canadian imaginary, Richard Drayton's book (2001) on nature and institutions of empire, and Koerner's monograph (1999) on nature, Linneaus and nationalism are powerful examples of this genre of thinking about green governance.

Finally, there is a body of work that focuses on the relations between environment, geopolitics and violence – namely the field of "environmental security" (see Dalby 1996, 2002). The central ideas – that the environment is the post-Cold War security issue and that environmental change can cause war and violence – has a long pedigree dating at least to Malthus and Hobbes. In the 1960s, the return of apocalyptic views of food shortage, and of oil scarcity a decade later, brought environmental concerns onto a larger Cold War geo-strategic landscape. There is now a substantial industry around environmental security which arose around a nexus of geo-political conditions: namely, the end of the Cold War, the need of over-funded militaries to legitimize their existence in the face of the clamor for the "Peace Dividend," and the emergence of "new" forms of violence often articulated as identity politics (or the "clash of civilizations") within putatively weak or rogue states which represent "threats" (Islamic terrorism, ethnic cleansing) to peace and security. The most rigorous and systematic effort to theorize environmental security, however, has been provided in Homer-Dixon's book *Environment, Scarcity and Violence* (1999). Here, the debt to Malthus is clear and explicit, and the entire argument rests on a more differentiated notion of "scarcity." The essence of Homer-Dixon's argument is that environmental scarcity (which means, to him, scarcity of renewable resources) has three forms, namely degradation, increased demand, or unequal distribution. The presence of any of these "can contribute to civil violence" especially through "resource capture" (generally by "elites"), and subsequent "ecological marginalization" of vulnerable or disenfranchised people as a result of resource capture (1999: 177). The language of this analysis is replete with ecological systems theory of old – interactive effects, adaptability, thresholds and so on – and contains a simple model of social friction and conflict. But environmental security does raise important questions – violence and mass conflict – on which political ecology has been remarkably silent but it is now becoming an important topic (see Peluso and Watts 2001). Several contributions in this volume (see Watts, Hecht) explore environmental violence using the toolkit of political ecology.

These confluences and inter-mixings suggest, as Peter Taylor (1997) says, that "appearances notwithstanding, we are all doing something like political ecology." Mike Davis's account in *Ecology of Fear* (1997) of the environmental foibles of Los Angeles, or Daniel Weiner's (1999) story of the survival of independent scientist-led citizen's movement for nature protection in the Soviet Union from Stalin to Gorbachev, all confirm, for example, a "family resemblance" to political ecology. Like any family there are complex interdependencies, interactions, conflicts and negotiations among members. And yet this dialectics of ideas reflects precisely the dialectics of nature and society itself.

STRUCTURE OF THE BOOK

The book is divided into five parts. This introduction has dealt mainly with the literature of political ecology and the position of liberation ecology within it – for a broader consideration of the development and social movements literature see the introduction to the first edition (Peet and Watts, 1996) and for a guide to our vision of the contemporary structure of political ecology see Figure 1.1. In the rest of Part I, on Renewing Political Ecology, Mike Davis and Susanna Hecht continue modifying and deepening the traditional political ecology toolkit in important ways. Davis takes on the question of globalization and empire, and links these with the equation of the history of science via El Niño and global climate change. Hecht builds on her foundational Brazilian work, with a sensitivity to globalization and the "new ecology." More specifically, in Chapter 2, Davis argues that great Third World famines that accompanied El Niño events were the missing pages in virtually every contemporary account of the Victorian era. At issue, he argues, is not simply that tens of millions of poor rural people died in these famines, but that they died for reasons that contradict much of the conventional understanding of the economic history of the nineteenth century. Hence, we are not dealing with "lands of famine" becalmed in stagnant backwaters, but with the fate of tropical humanity as its labor and products were incorporated into the "modern world system." From the perspective of political ecology, the vulnerability of tropical agriculturalists to extreme climatic events after 1870 was magnified by simultaneous restructuring of household and village linkages to regional production systems, world commodity markets, and the colonial (or dependent) state. Davis finds three

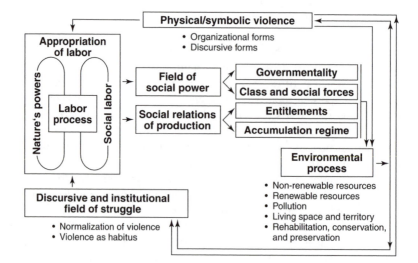

Figure 1.1 Political ecology.

points of articulation with larger socio-economic structures to be especially decisive for rural subsistence:

1   The forcible incorporation of smallholder production into commodity and financial circuits controlled from overseas tended to undermine traditional food security.
2   The integration of millions of tropical cultivators into the world markets during the late nineteenth century was accompanied by a dramatic deterioration in their terms of trade.
3   Formal and informal Victorian imperialism, backed up by the super-national automatism of the Gold Standard, confiscated local fiscal autonomy and impeded state-level developmental responses, especially investments in water conservancy and irrigation, that might have reduced vulnerability to climatic shocks.

Davis sees this as a "political ecology of famine" because it takes the viewpoint both of environmental history and Marxist political economy.

Hecht argues that extensive deforestation occurred throughout the New World Tropics during much of the last century, roughly parallel with the period of imperialism that Davis documents. But, today, she says, the "doomed forest" narrative needs to be rethought: many rural areas show robust resurgence of anthropogenic and successional forests. Farmers, she argues, balance a broad array of economic and social elements in the construction livelihoods that shape the use of natural resources. Global processes (including international migration, remittances, commodity prices, international environmental ideologies), regional dynamics (such as the Central American Common Market, the Meso American Biological Corridor), national structural adjustment and development policies (like decentralization, credit, agrarian reform policies, restructured labor markets) and local socio-environmental circumstances (household strategies, tenurial structures, access rights, gender of farmers, ethnicity, traditional beliefs, natural resources endowment) all affect land use decisions. So, while globalization is among the most powerful of the forces affecting farmers, in some literatures it is assumed to unfold through a uniform set of social and institutional structures, like de-territorialized economic waves washing over regions. Hecht's position is that globalization is mediated by local institutional arrangements, local assets, and a wide array of ethnographic and household factors that modify larger processes. Observed natural resource use reflects not just globalization "from above" but also its modification and manipulation "from below." So, regenerating forests are a profound outcome of socio-economic as well as bio-geographic features: they are truly a socially "constructed" nature. These are foundational political ecology arguments building on, and showing the strengths of, the political economy–environment connection in critical analyses.

Part II, Discourse and Practice, offers some compelling analysis derived from the poststructural turn in political ecology, especially a series of engagements with ideas and terms developed by Michel Foucault. Karl Zimmerer looks at discourses in relation to knowledge production; Jake Kosek takes on issues of race, an area that political ecology has ignored until now and links this with Foucault on governmentality; Michael Goldmann looks at the greening of development institutions. More specifically, Zimmerer says that a widely held concern about soil erosion in Bolivia has not resulted in public consensus. Individuals and institutions hold a variety of divergent views on the causes, as well as the preferred solutions, for the problem of soil degradation. During his fieldwork, Zimmerer came to realize that people and institutions did not form their environmental discourses as islands of self-contained dialogue. Instead, they expressed their viewpoints through interaction within and among groups. Processes of resistance and contestation as well as accommodation and agreement guided their elaboration of environmental ideas. An historical approach and sensitivity to discursive alterations over time were crucial to gaining insights on their views of changing nature and efforts to conserve it. Also their environmental discourses are regularly articulated in diverse ways, sometimes subtle though often not, with global environmental policy and environmental science. These articulations highlight a perspective on political ecology that takes global environmental policies or prevailing scientific interpretations as a point of departure. This political ecological approach to environmental discourse is distinctively cross-scalar, rather than singularly local-scale. Typically it follows linkages of ideas, policies, and resources that are North–South, or rural–urban within advanced industrial countries. In general, Zimmerer seeks to renew consideration of local knowledges and peasants' personal or "everyday" perspectives on the soil erosion dilemma as part of an effort to invigorate political ecology through the analysis of discourse: i.e. as part of a more open "liberation ecology."

Kosek examines how notions of wilderness have been infused with racialized notions of purity and pollution. Using links between contemporary New Mexico and the rise of particular racialized notions of nature around the turn of the twentieth century, he investigates how the movement to protect forests from degradation and pollution draws on national metaphors regarding the contamination of pure white bodies and unsoiled bloodlines. He traces the entanglement of eugenicist conceptions of bodily purity with wilderness protection, and demonstrates how past formations of whiteness articulate with current struggles over wilderness in New Mexico. Finally, he argues that local Hispano activists' animosity towards environmental groups espousing strict preservation of forests is not as mysterious as it may seem to some environmentalists: it has to do with the ways in which forest preservation activities are haunted by exclusionary rhetoric of purity and entrenched fears of racial pollution. Kosek is concerned with the First World – or more properly the South within the North – but his concerns with the relations between race, nature and identity has a much wider resonance in the global south as the paper by Li and Rangan later in *Liberation Ecologies* reveals.

Goldman presents the case that there has been a change in the character and increase in intensity of activity around constructing global truth and rights regimes on the environment and natural resource use. He argues that the World Bank has strengthened itself during re-invention as a catalyst for green institutional change: the Bank has been able to enlist scores of social actors and institutions to help generate a new development regime that is coherently green as well as neoliberal. That is, the institution has been at the center of constituting a hegemonic, albeit fragile, project of "green neoliberalism." For illustration, Goldman turns to the Mekong River in Laos, where the World Bank represents its recent interventions as reflecting a new *modus operandi:* "environmentally sustainable development." For analytical depth he uses Foucault's concept of governmentality including, on the one hand, the making of the modern rational subject and the efficient state which s/he would help build, and, on the other, the intensified regulation of the relation of these subjects to their natural territory. He calls these productive relations of government – with their emphasis on "knowing" and "clarifying" relationships with nature and the environment as mediated through new institutions – eco-governmentality. We can learn a great deal about relations of power through an inquiry into the co-production of regimes of territorial rights and discourses of environmental truth. Through what he calls the "green neoliberal project" in which neocolonial conservationist ideas of enclosure and preservation and neoliberal notions of market value and optimal resource allocation find common cause that institutions such as the World Bank have made particular natures and natural resource-dependent communities legible and accountable.

Part III, Institutions and Governance, approaches similar topics from a rather different set of perspectives, engaging the current debate over the role of communities and participation in environmental governance, the understanding of community dynamics and the role of the state. Roderick Neumann makes an innovative comparison between national park formation in the United States and in Tanzania. As the millennium turned, scholarship on US national parks turned in a critical direction, that led, among other explorations, to a reexamination of the process of park establishment in the western US. New histories, he says, challenge the notion that iconic natural landscapes were "virgin wilderness," and instead suggest that "uninhabited wilderness had to be created before it could be preserved." Scholars linking US national parks to the dispossession, and near genocide, of hundreds of thousands of Native American peoples offer a different morality tale, in which local commons are enclosed by the state, sovereign nations are disposed of their territory and written out of history, and the ecological consequences of park establishment are not altogether positive. These alternative narratives of dislocation and enclosure offer new possibilities for geographical and historical comparative analysis. Wilderness preservation and national park establishment are now standardized practices in virtually every Third World state, based largely on the US model and experience. His analysis concentrates on "traditional" forms of state conservation territories – national parks, game reserves, and officially designated wildernesses, the "fortress conservation" models that require the absence of human occupation and impact and reflect the priorities of national conservation agencies and international organizations. Using case studies of two areas in southeastern Tanzania, Neumann presents a preliminary theorization of the role of conservation territories in the construction of the modern state. Wild areas of national parks and reserves, as products of the creation of the modern nation state, are as much an expression of modernism as skyscrapers he claims. They are integral parts of the practice of modern statecraft and result from plans to divide and contain the central antinomies of modernity: nature and culture, consumption and production, wilderness and civilization. The establishment of protected areas represents an historic transition under modernity wherein human civilization becomes the caretaker of a wild nature that poses no threat beyond the threat of disappearing. Nature, in a phrase, becomes a ward of the state.

Navroz Dubash looks at the process by which the state withdrew from water control and provision in the 1990s, ceding this space to the market, as part of the widespread adoption of "Washington Consensus" policies. Implementation of market-approaches to water, he says, spurred an outbreak of protest, particularly at the municipal level and in reaction to large water projects. Based on a detailed examination of rural groundwater use and systems of exchange for groundwater in western India, he illustrates the ways in which water and social practice are intertwined. The so-called "water markets" in the state of Gujarat in Western India are far from the abstract price controlled institutions of neoclassical economic theory. Instead, they are complex systems of exchange shaped by historical and natural context, and regulated at the local level using social practices. By exploring the micro-politics through which local regulatory institutions are forged, he shows how groundwater exchange flourishes because of, local customary practice, rather than in spite of it. Communities and their supporters struggle to articulate an alternative framing of water management, around community rights, social risks, and local control rather than around supply, demand, and corporate control. At the root of global water politics today, he finds a Polanyian tension between the commoditization of water and the counter-pressures arising from its embeddedness in social practice and institutions. Political ecology brings a tradition of analysis to the discussion of this tension that focuses on the local and the particular, provides a basis for questioning the abstract market as a uniformly valid mechanism of coordination. Political ecology serves to re-embed water politics in daily practice through its focus on management questions and the articulation of these with the changing context of state and market roles. Political ecology, in short, helps counter the abstraction of the apolitical and self-regulating market. Instead, it argues for detailed empirical attention to the local micro-politics of institutional change. Political ecology in Dubash's hands is especially attentive to the environmental context in which history is made: attention to the natural resource characteristics of water as a commodity requires an appreciation for how nature and society interact in shaping and constraining markets. Political ecology provides a powerful counter-narrative to the dis-embedded market.

Dara O'Rourke examines whether the socialist "transition to the market" is supportive or destructive of so-called "sustainable development" using the cases of China and Vietnam. Market triumphalists, he says, argue that rising wealth, technological advances, and efficiency gains and will support gradual improvements in environmental quality and reductions in pollution. By comparison, market critics argue conversely that the rapidly transforming (and increasingly polluted) landscapes of China and Vietnam represent the birth of a hybrid model of development that combines the worst of central planning and state control, with the worst of capitalist exploitation. From this perspective, the post-socialist transition, while good for GDP growth may be worse for the environment, communities, and workers. The chapter draws on political ecology to analyze transitions in post-socialist development and provides a powerful optic for analyzing struggles around environmental resources and the complex dynamics between states, firms, and civil society actors influencing environmental degradation and response. However, this literature has rarely directed its light on industrial environmental issues, or on the most dynamic region of industrial development (and environmental degradation) in the world, namely the socialist transition economies of East Asia. O'Rourke's central argument is that on-going state and market transitions in countries such as Vietnam and China drive industrial expansion and result in adverse environmental impacts. Yet concurrent transitions in the public sphere and in the politics of development are creating space for community and worker resistance to the worst impacts of this industrialization. Local actors challenge the environmental and social impacts of what is now seen by many as the most successful economic development model in the world. Communities and workers, he concludes, make surprising demands on the state, often unsuccessfully, but occasionally effectively, to constrain industrial activities and protect workplace and community environments.

Following from this, Part IV, Conflict and Struggle, examines some social forms of contention in particular resource settings. Michael Watts tells of the long, ugly history of petrolic violence. He says that protection by American armed forces keeps the oil flowing, working hand-in-hand with a phalanx of African dictators and political psychopaths on the one side, and supermajors like ExxonMobil and ChevronTexaco on the other. In Nigeria the discovery of oil, and annual oil revenues of $40 billion, has ushered in a miserable, undisciplined, decrepit, and corrupt form of "petro-capitalism" that leaves per capita income at $290 per year, with the oil-producing states of the Niger Delta benefiting least from oil-wealth. Devastated by the ecological costs of oil spillage and the highest gas flaring rates in the world, the Niger Delta has become a political tinderbox, marked by a generation of militant restive youth, deep political frustrations among oil-producing communities, and electoral thuggery. In virtue of the geo-strategic significance of oil to contemporary capitalism, relations among natural resources (oil in particular) and economic growth, democracy, and civil war have emerged as an object of substantial scholarly attention (operating under the sign of "resource politics"). One conclusion reached by economists and political scientists is that oil is a "resource curse." Watts finds this kind of "resource politics" rendering invisible both transnational oil companies and the specific forms of rule associated with petro-capitalism. His analysis charts the relations between oil and violence by examining how forms of governable (or non-governable spaces) are created through the analytics of an "authoritarian governmentality" growing out of petro-capitalism. Rather than see oil-dependency as a source of predation, or as a source of state military power, he explores how oil capitalism produces particular sorts of enclave economies and particular sorts of governable spaces characterized by violence and instability. He traces the varieties of violence engendered by oil, elaborating ways in which resources, territoriality and identity can constitute forms of rule (or unrule), so as to understand the genesis of economies of violence emerging from different governable (or ungovernable) spaces.

Richard Schroeder and Krisnawati Suryanata turn to a different kind of supposedly benign resource setting – agroforestry systems, widely acclaimed for their beneficial environmental and economic effects. However, their chapter challenges the assumption that environmentalist policies and development practices related to agroforestry are universally beneficial to local interests. Following the lead of scholars who have combined fine-grained political analysis with a detailed understanding of socialized ecological processes, they redirect attention to agroforestry as a site of contentious political struggle. Agroforestry sometimes opens up critical options for otherwise disenfranchised groups: women mobilize agroforestry strategies to make the best use of the minimal landholdings allotted to them; indigenous peoples use agroforestry systems to perpetuate livelihood practices and safeguard key components of cultural identity; the complexity of so-called "home garden" agroforestry systems provide peasant groups with the means to effectively resist the extractive propensities of the state. Focusing on gender, cultural, and class relations therefore requires alternate readings of agroforestry. Their chapter applies these insights to two contemporary agroforestry initiatives in The Gambia and upland Java, which illustrate the importance of understanding both the political and the ecological dimensions of environmental interventions. In both cases, an environmental discourse has served to mask the exclusionary objectives of fruit tree holders – male mango growers in The Gambia, and a new class of "apple lords" in Java – ultimately directed at entrepreneurial gain and control over key production resources. While agroforestries often contribute in some measure to ecological goals, they nonetheless can also be seen as deliberate strategies of dispossession and private accumulation. In the two case studies, the commoditization of tree cropping has driven a wedge between holders of tree and land/crop rights, which in turn produced a range of agro-ecological and social contradictions. Such dynamics erode moral economies and replace them with a morally indifferent stance, that elevates profit taking above all other objectives, including ecological stability.

Judith Carney finds a growing association of environmental change with female-based social movements and gender conflict within rural households. This suggests the need for an improved understanding of gender relations and the domestic sphere, since class as well as non-class struggles over resources are frequently mediated in the idiom of gender. Additionally, a central insight of poststructuralist research in Third World agrarian studies over the past fifteen years is the need to extend the definition of politics from the electoral politics of state and class to one that includes the political arenas of the household and workplace. This emphasis brings attention to the crucial role of family authority relations and property relations in structuring the gender division of labor and access to rural resources. As development interventions, environmental transformations, and markets place new labor demands and value on rural resources, these socially constructed relations of household labor and property rights often explode with gender conflict. Struggles over labor and resources reveal deeper conflicts over meanings in the ways that property rights are defined, negotiated, and contested within the political arenas of household, workplace, and state. Gender informs the non-capitalist class relations that frequently attend projects of modernization in Third World societies. By linking property rights and gender conflict to environmental change, her chapter extends the poststructuralist concern with power relations and discourse to political ecology. She uses the case of the environmental transformation of the wetlands of Gambia, specifically an examination of the gender-based resource struggles accompanying the development of wetlands to irrigation schemes. Using multiple case studies of two forms of wetland conversion, irrigated rice schemes and horticultural projects, she traces the disputes surfacing over the past fifty years in Mandinka households over women's land rights. The analysis reveals repeated gender conflicts over rural resources as male household heads concentrate landholdings in order to capture female labor for surplus production. Mandinka gender conflicts involve disputes over women's traditional land rights within the common property system, thereby illustrating the significance of struggles over meaning for contemporary struggles over labor and resources. Carney's analysis is an exemplary case of so-called "feminist political ecology" at work.

Part V, Movement, examines the complex relations between environment, identity (indigeneity), social movements and the market (and relatedly movements and regionalism, movements and state power, urban versus rural movements and movements and class). Tania Murray Li asks how a transnational category, such as "indigenous people," comes to mobilize people in a particular place? How does it enter their struggles and reconfigure alliances? How do the "environmental imaginaries" associated with indigeneity translate on the ground? Who is privileged by the indigenous frame, and who is left out? She focuses on Indonesia, where answers to the question "who is indigenous" are far from obvious, yet a social movement supporting indigenous people is taking hold. Her argument is that a group's self-identification as tribal or indigenous is neither natural nor inevitable, nor is it simply invented, adopted, or imposed. It is, rather, a positioning that draws upon historically sedimented practices, landscapes, and repertoires of meaning, one that emerges through particular patterns of engagement and struggle. The conjunctures at which some people come to identify themselves as indigenous, realigning the ways they connect to the nation, the government and their own, unique tribal place, are the contingent products of agency and the cultural and political work of articulation – a key word in political ecology because it recognizes the structured character of distinct entities (means of production, social groups), at particular conjunctures. She draws on two case studies from the hilly interior of Central Sulawesi, where in one, Lake Lindu, a collective, indigenous identity has been persuasively articulated. The tense politics around indigeneity and its serious consequences for rights, resources, and the formation of alliances highlight, she argues, new ways to conceptualize identity that are theoretically more adequate to the diversity of conditions and struggles in the Indonesian countryside.

Haripriya Rangan records a decade of popular protest that demanded statehood for the Himalayan regions of Garhwal and Kumaon. For the men and women involved in the struggle, the new state of Uttaranchal within the Indian Union was the necessary condition for extricating their region from its backwardness. They claimed that the hill regions were afflicted by poverty and high unemployment because the previously dominant politicians and bureaucrats had been mainly concerned with the interests of the plains and cared little for the distinctive culture and needs of the mountains. What she finds curious about this claim for statehood is its apparent disjuncture from the rhetoric commonly associated with Chipko, a movement that had also emerged in the region during the mid-1970s, one that was hailed by academics and environmental activists as a powerful critique of development, and described as a grassroots struggle against a modernizing Indian state whose pursuit of development destroyed local ecologies and traditional ways of peasant life. Chipko, it was claimed, symbolized a new ecological consciousness that challenged the conventional, reductionist approaches to development and shone an alternative path towards a green earth and a true civilization. Rangan points out that in the two decades following, the region that gave rise to the Chipko movement was consumed by another struggle that, rather than calling for environmental protection or "alternatives to development," positively embraced conventional notions (i.e., modernization-as-development) in demanding the creation of a separate state for promoting economic growth. She goes on to show how subsequent protests and the movement for statehood, rather than being different from Chipko continued to struggle for the same demands that were initially articulated by the Chipko movement: the need for policies and conditions for generating local employment opportunities and fostering regional economic development. Rangan also takes a critical look at the anti-, post-, and alternative development literatures, specifically claims that new social movements in poor countries grassroots struggles seek alternatives *to* development. She argues, on the contrary, that social movements in most poor regions of the world do not dismiss or subvert development, but make it an integral part of an expanded "moral economy." Contemporary social movements in poor countries of the world, she suggests, are not against the idea of development, they are part of it.

In a similar way, Tony Bebbington thinks that the idea behind writing development stories is not simply to understand the world but to contribute to changing it. Our ability to contribute to such change is greater if we build bridges across the gulf between the languages of social movements and popular organizations, and those of activists, intellectuals, and "policy makers." He relates this to conceptual discussions of "alternative" development that draw on concepts of "indigenous technical knowledge," "farmer-first" agricultural development, new social movements, and civil society. The assumption of much alternative development writing is that "alternative actors," such as indigenous peoples' organizations, will carry forward "alternative" agendas. This is not always the case. His case study tries to make sense of the work of popular and non-governmental organizations in the Central Andes of Ecuador. Specifically, it considers how federations of indigenous, highland communities have emerged since land reform in the region, the agricultural and rural development strategies they have pursued, and the ways they have engaged in formal political processes. Its rhetorical and analytical strategy is to sustain a conversation between the ways in which these federations seem to have pursued agricultural and political change with the ways in which indigenous agriculture and social movements have been conceptualized by certain authors. Exploring the dissonances and convergences between these conceptualizations and the strategies of these federations sheds light on the "liberatory" possibilities embodied by these federations as well as on the broader project of liberation ecology. In particular, a commitment to native, traditional, and agro-ecological techniques found in intellectual currents in social science and development activism is often missing among some of these organizations, who instead seem to focus on reforming dominant models of development while also keeping hold of principles of local control, democratization, and community-based sustainable development. Thus while intellectual concepts and popular practice may differ at the level of strategy, they converge at the level of wider political objectives. To organize, be innovative, and create "decentered autonomous spaces" is simply not enough – alternatives must also make a material difference in livelihoods. If we are interested in "liberation ecologies," the alternative proposals of social movements and intellectuals alike fall well short of the practical challenge.

In the concluding chapter, Tim Forsyth productively extends the discussion of liberation ecologies to Thailand's urban and industrial social movements, an area previously neglected by this predominantly rurally focused area of study. Yet the chapter also challenges liberation ecology. While acknowledging the crucial role played by social movements in livelihood struggles, Forsyth argues that liberation ecology might be too optimistic in assuming that social movements may successfully revise environmental discourses in favor of marginalized people. Instead of focusing only on the political agency of social movements, he says, political ecologists must also assess how movements interact with, and even may become, exclusive discursive structures. Movements may not be "liberatory" if, rather than helping poor people, they replicate or impose new hegemonic discourses. The chapter considers these arguments in relation to social movements concerning suspected lead and lignite poisoning in Thailand. In addition, the chapter explores other epistemological challenges to liberation ecology. Many industrial and urban environmental risks are new to localities in developing countries, and consequently local experience and scientific certainty about their causes are low. Under such circumstances, local social movements may easily be dominated by outside expertise, or framed in ways that do not reflect local experience of risks. His point is not to dismiss liberation ecologies, or the importance of brown environmental social movements, but instead, the aim is to understand each better. Much political ecology to date has focused on the marginalizing impacts of capitalist development, and on the struggles of poor people against alliances of industry and state. The case studies in Forsyth's chapter show that such activism is also linked with the production of discourses that do not always assist the most vulnerable people. Forsyth concludes that a science-policy approach of understanding the co-production of environmental activism and discourses needs to be integrated with the political economy approach of understanding social marginalization under rapid industrialization.

All of these contributions are clearly not of a theoretical piece but they reveal the creative ways in which political and cultural economy is being brought to bear upon environment construed in a broad (and discursive) sense. All of the chapters reveal important confluences drawing upon different threads and conceptual tools. These confluences and inter-mixings suggest, as Peter Taylor (1997: 111) says, that "appearances notwithstanding, we are all doing something like political ecology." Mike Davis's account in *Ecology of Fear* (1997) of the environmental foibles of Los Angeles, or Daniel Weiner's (1999) story of the survival of independent scientist-led citizen's movement for nature protection in the Soviet Union from Stalin to Gorbachev, all confirm, for example, a "family resemblance" to political ecology. Like any family there are complex interdependencies, interactions, conflicts and negotiations among members. And yet this dialectics of ideas reflects precisely the dialectics of nature and society itself.

NOTES

* Parts of this chapter previously appeared in M. Watts (2002) and are reprinted by permission from the copyright holder, Blackwell Publishing Company. Figure 1.1 is reprinted from *Violent Environments*, edited by Nancy Lee Peluso and Michael Watts. Copyright © 2001 by Cornell University. Reprinted by permission of the publisher, Cornell University Press.
1 The concern to link Nature and Society in some fashion was of course not new – Richard Grove (1995) has traced the origins of western conservationist thinking back to the seventeenth century, and Clarence Glacken's (1967) magisterial work *Traces on the Rhodeian Shore* pushes back the frontier of thinking about the culture and nature much further.

2 In this transition from cultural ecology (ecological anthropology) to political economy the Marxist inspired critical work of anthropologist Jonathon Friedmann (1976) was key, and the work of geographer Bernard Nietschmann (1972) was an important bridge revealing the power of cultural ecology in understanding how Miskito fisherman and farmers adapted to differing niches in their aquatic and land based tropical environments, but also its limited ability to grasp how these coastal communities were changing and confronting serious ecological problems.

3 This is a total nonsense. The theoretical turn provided by the foundational work on political ecology was to replace human with Marxian inflected political economy, not a narrow construal of the political. Indeed it has often been said that the first generation of political ecological work was curiously bereft of a serious treatment of politics. Now it is certainly the case that what passed as "broadly defined political economy" (the language is taken from Blaikie) was not of a piece. For some it meant a concern with the labor processes of the land manager, for others the forms of surplus appropriation and disposition, and the social relations of production in which property figured centrally. But to reduce all of this to politics (also deployed in a blanket fashion by Vayda and Walters) is obviously inaccurate and unhelpful. Neither incidentally, was political ecology simply about the deployment of wealth and poverty as causal mechanisms for the explanation of environmental outcomes as Vayda and Walters suggest. If political ecology stood for anything from the beginning it was not about blaming the poor or the rich for environmental change, but examining the social relations of production and the broader political economy of which actors, particularly but not exclusively land-based managers, were part. Our own work published almost twenty years ago was precisely about using concepts derived from marxian agrarian studies – processes of differentiation, the process of commodification, forms of surplus extraction – not about "politics." One can certainly contest the nature of these arguments, the evidence adduced, and the logic of the argument but to say that it overdetermined explanation by "insisting" on the centrality of political influences and "political influences from the outside" is a travesty.

4 The case study provided by Vayda and Walters is itself quite astonishing. The case does indeed confirm the salience of political economy on a number of levels as they concede. They argue however that multiple causality would have "quite likely been overlooked" by political ecologists, especially the causality of biophysical events and species specific selection. We pass over the rather obvious fact that the likes of Susanna Hecht, Karl Zimmerer and Nancy Peluso documented ten years ago and more all of the sorts of findings identified by Vayda and Walters (planting and land claims, species selection, the role of reproductive biology of tree species) and simply note that there is *nothing* in their case that is in the least bit surprising from the vantage point of political ecology; indeed it is all rather well trodden territory. Of course they conclude that politics is indeed important in their story – they refer to elites in a way that would be embarrassing in its lack of specificity for a political ecologist – and that rich and poor can both be implicated in mangrove destruction and despoliation. Well that's alright then.

5 For overviews of the "new" political ecology the reader is recommended to read Peet and Watts (1996), Bryant and Bailey (1997), Keil *et al.*, (1997), Alier and Guha (1998).

6 Kingsbury (1997) has shown beautifully how the contested nature of the community has its counterpoint in international environmental law over the cover term "indigenous" (and one might as well add tribe or ethnicity). The UN, the ILO and the World Bank have, as he shows, differing approaches to the definition of indigenous peoples. The complexity of legal debate raised around the category is reflected in the vast panoply of national, international and inter-state institutional mechanisms deployed, and the on-going debates over the three key criteria of non-dominance, special connections with land/territory, and continuity based on historical priority. These criteria obviously strike to the heart of the community debate which we have just outlined, and carry the additional problems of the

normative claims which stem from them (rights of indigenous peoples, rights of individual members of such groups, and the duties and obligations of states). Whatever the current institutional problems of dealing with the claims of non-state groups at the international level (and there are knotty legal problems as Kingsbury (1997) demonstrates) the very fact of the complexity of issues surrounding "the indigenous community" makes for at the very least what Kingsbury calls "a flexible approach to definition," and at worst a litigious nightmare.

7 Here we might refer to the excellent work of Brosius (1997) and Li (1996, 1999) in Indonesia who conducted comparative community work in two seemingly similar local communities to show how (i) the type and fact of resistance varied dramatically between the two communities which were in many respects identical "cultural" communities, and (ii) how these differences turn on a combination of contingent but nonetheless important historical events. Broisus found that the radical differences in resistance to logging companies between two communities turned on their histories with respect to colonial forces, their internal social structure, their autonomy and closed, corporate structure, and the role of transnational forces (environmentalists in particular). The point is that some communities do not resist (which disappoints the foreign or local academic) and may not have, or have any interest in, local knowledge. By the same token, as Zerner shows (1999), local "traditions" can be discovered (not necessarily by the community and often driven by academic work of local traditions drawn from elsewhere) which can be put to the service of the new political circumstances in which villages and states find themselves. Indeed, we know that some groups within communities are happy to take on board essentialism and wrong headed "local traditions" pedaled by foreign activists of investors, in order to further local struggles.

8 "Accordingly, we need to see things not in terms of the replacement of a society of sovereignty by a disciplinary society and the subsequent replacement of a disciplinary society by a society of government; in reality one has a triangle, sovereignty–discipline–government, which has as its primary target the population, and as its essential mechanism the apparatuses of security.... *I want to demonstrate the deep historical link between movement that overturns the constants of sovereignty on consequence of the problem of choices of government; the movement that brings about the emergence of population as a datum, a field of intervention ... the process that isolates the economy as a specific sector of reality; and political economy as the science and the technique of intervention of the government in that reality.* Three movements – government, population, political economy – that constitute from the eighteenth century onward a solid series, one that even today has assuredly not been dissolved." Foucault 2000: 219, emphasis added.

9 Some of these Foucaldian ideas have already been productively deployed in the understanding of nature and resource management – what one might call green governmentality – and the relations between nature and nationalism, see B. Braun, "Producing vertical territory," *Ecumene* (July 2000) pp. 7–46.

REFERENCES

Aggrawal, A. (1999) *Greener Pastures*, Durham: Duke University Press.

Alier, J. and Guha, R. (1998) *Varieties of Environmentalism*, London: Earthscan.

Altieri, M. and Hecht, S. (1990) *Agroecology and Small Farm Development*, Boca Raton, Florida: CRC Press.

Bailey, J. (1998) "Green Corporatism," unpublished manuscript, University of California, Berkeley.

Bateson, G. (1972) *Steps toward an Ecology of Mind*, New York: Ballantine.

Baviskar, A. (1995) *In the Belly of the River*, Delhi: Oxford University Press.

Beck, U. (1994) *Risk and Modernization*, Boulder: Westview.

Beck, U., Giddens, A. and Lash S. (1994) *Reflexive Modernization*, London: Polity.

Beinart, W. and Coates, P. (1995) *Environment and History*, London: Routledge.

Benton, Ted (1989) "Marxism and Natural Limits," *New Left Review*, 178: 51–86.

Berger, J. (1980) *About Looking*, New York: Pantheon.

Blaikie, P. (1985) *The Political Economy of Soil Erosion*, London: Methuen.

Blaikie, P. and Brookfield, H. (1987) *Land Degradation and Society*, London: Methuen.

Botkin, D. (1990) *Discordant Harmonies*, New York: Oxford University Press.

Braun, B. (2003) *The Intemperate Rainforest*, Minneapolis: University of Minnesota Press.

Braun B. and Castree, N. (eds) (1998) *Remaking Reality*, London: Routledge.

Broad, R. (1993) *Plundering Paradise*, Berkeley: University of California Press.

Brosius, P. (1997) "Prior Transcripts, Divergent Paths," *Comparative Studies in Society and History*, 39/3: 468–510.

Brush, S. (1996) *Valuing Local Knowledge*, San Francisco: Island Press.

Bryant, R. (2001) "Political Ecology," in N. Castree and B. Braun (eds) *Social Nature*, Oxford: Blackwell, 151–70.

Bryant, R. (1998) "Power, Knowledge and Political Ecology in the Third World," *Progress in Physical Geography*, 22/1: 79–94.

Bryant, R. and Bailey, S. (1997) *Third World Political Ecology*, London: Routledge.

Bunker, S. (1985) *Underdeveloping the Amazon*, Chicago: University of Illinois Press.

Burawoy, M. (1985) *The Politics of Production*, London: Verso.

Blum, J., George, S., Gille, Z., Gowan, T., Haney, L., Klawiter, M., Lopez, S., O'Riain, S. and Thayer, M. (1991) *Global Ethnography*, Berkeley: University of California Press.

Castree, N. and Braun, B. (2001) *Social Nature*, Oxford: Blackwell.

Cockburn, A. and Hecht, S. (1990) *The Fate of the Forest*, London: Verso.

Collins, J. (1987) *Unseasonal Migrations*, Princeton: Princeton University Press.

Cronon, W. (1992) *Nature's Metropolis*, New Haven: Yale University Press.

Dalby, S. (1996) "The Environment as Geopolitical Threat," *Ecumene*, 3: 472–96.

—— (2002) *Environmental Security*, London: Routledge.

Davis, M. (1997) *Ecology of Fear*, New York: Vintage.

Dean M. (1999) *Governmentality*, London: Sage.

de Janvry, A. and Garcia, R. (1988) "Rural Poverty and Environmental Degradation in Latin America," paper presented to the IFAD Conference on Smallholders and Sustainable Development, Rome.

Demerritt, D. (2001) "Being Constructive about Nature," in N. Castree and B. Braun (eds) *Social Nature*, Oxford: Blackwell, 22–40.

Drayton R. (2001) *Nature's Government*, New Haven: Yale University Press.

Dupuis, M. (2002) *Nature's Perfect Food*, New York: New York University Press.

Escobar, A. (1995) *Encountering Development*, Princeton: Princeton University Press.

—— (1999) "After Nature," *Current Anthropology*, 40/1: 30.

Esty, D. (1994) *Greening the GATT*, Washington, DC: Institute for International Economics.

Evans, P. (1996) "Government Action, Social Capital and Development," *World Development*, 24/6: 1119–32.

—— (2002) *Livable Cities*, Berkeley: University of California Press.

Faber, D. (1992) *Environment Under Fire*, New York: Monthly Review.

Fairchild, J. and Leach, M. (1996) *Misreading the African Landscape*, Cambridge: Cambridge University Press.

Foucault, M. (2000) *Power*, New York: Pantheon.

Gadgil, M. and Guha, R. (1995) *Ecology and Equity*, New Delhi: Oxford University Press.

Gandy, M. (2002) *Concrete and Clay*, Cambridge, MA: MIT Press.

Garcia Barrios, R. and Garcia Barrios L. (1990) "Environmental and Technological Degradation in Peasant Agriculture," *World Development*, 18: 1569–85.

Ghai, D. (1992) "Conservation, Livelihood and Democracy" Discussion Paper no. 33 Geneva: UNRISD.

Gibson, C., McKean, M. and Ostrom, E. (2000) *People and Forests*, Cambridge, MA: MIT Press.

Gille, Z. (1997) "Social and Environmental Inequalities in Hungarian Environmental Politics," in P. Evans (ed.), *Livable Cities*, Berkeley: University of California Press, 132–61.

Glacken, C. (1967) *Traces on the Rhodean Shore*, Berkeley: University of California Press.

Gleissman, S. (ed.) (1990) *Agroecology*, Berlin: Springer Verlag.

Goldblatt, S. (1996) *Social theory and the Environment*, London: Routledge.

Goldman, M. (1999) *Privatizing nature*, London: Routledge.

Goodman, D., Sorj, B. and Wilkinson, J. (1990) *From Farming to Biotechnology*, Oxford: Blackwell.

Grossman, L. (1984) *Peasants, Subsistence Ecology and Development in the Highlands of Papua New Guinea*, Princeton: Princeton University Press.

Grove, R. (1993) *Green Imperialism*, Cambridge: Cambridge University Press.

Grundemann, R. (1991) *Marxism and Ecology*, Oxford: Oxford University Press.

Guha, R. (1987) *Unquiet Woods*, Berkeley: University of California Press.

Gupta, A. (1998) *Postmodern Development*, Durham: Duke University Press.

Haas, P. (1990) *Saving the Mediterranean*, New York: Columbia University Press.

—— (1993) "Epistemic Communities and the Dynamics of International Environmental Co-operation," in V. Rittberger (ed.) *Regime Theory and International Relations*, London: Oxford University Press.

Hajer, M.A. (1997) *The Politics of Environmental Discourse*, Oxford: Oxford University Press.

Hanson, E. (2002) *Animal Attractions*, Princeton: Princeton University Press.

Harre, R., Mulhausler, P. and Brockmeier, J. (1999) *Greenspeak*, London: Sage.

Harvey, D. (1993) "The Nature of Environment," *Socialist Register*, London: Merlin, 1–51.

—— (1996) *Justice, Nature and the Geography of Difference*, Oxford: Blackwell.

Hecht, S. and Cockburn, A. (1989) *The Fate of the Forest*, London: Verso.

Herskovitz, L. (1993) "Political Ecology and Environmental Management in Loess Plateau, China," *Human Ecology* 21: 327–53.

Homer-Dixon, T. (1999) *Environment, Scarcity and Violence*, Princeton: Princeton University Press.

IICQ (1992) "Indigenous Vision," *India International Center Quarterly* 1–2.

Kautsky, K. (1906) *La Question Agraire*, Paris: Maspero.

Keck, M. (1995) "Social Equity and Environmental Politics in Brazil," *Comparative Politics*, 27/4: 409–24.

Keck, M. and Sikkink, K. (1998) *Activists Without Borders*, Ithaca: Cornell University Press.

Keil, R., Fawcett, L., Penz, P. and Bell, D. (eds) (1998) *Political Ecology*, London: Routledge.

Kingsbury, B. (1994) "Environment and Trade," in A. Boyle (ed.) *Environmental Regulation and Economic Growth*, Oxford: Clarendon Press.

—— (1997) "The International Concept of Indigenous Peoples in Asia," in D. Bell and J. Bauer (eds) *Human Rights and Economic Development in East Asia* (forthcoming).

Kirby, A. (ed.) (1990) *Nothing to Fear*, Tucson: University of Arizona Press.

Kloppenburg, J. (1989) *First the Seed*, New York: Cambridge University Press.

Koerner, L. (1999) *Linneaus: Nature and Nation*, Cambridge, MA: Harvard University Press.

Kornai, J. (1992) *The Socialist System*, Princeton: Princeton University Press.

Kuletz, V. (1997) *The Tainted Desert*, London: Routledge.

Lash, S., Szersynski, B. and Wynne, B. (eds) (1996) *Risk, Environment and Modernity*, London: Sage.

Leach, M. and Mearns, R. (1997) "Challenging the Received Wisdom in Africa," in *The Lie of the Land*, London: Currey.

Leach, M., Mearns, R. and Scoones, I. (1997) *Environmental Entitlements*, IDS Working Paper, University of Sussex.

Leff, E. (1995) *Green Production*, New York: Guilford.

Lewis, M. (1993) *Green Delusions*, Durham: Duke University Press.

Levins, R. and Lewontin, R. (1985) *The Dialectical Biologist*, Cambridge, MA: Harvard University Press.

Li, T. (1996) "Images of Community," *Development and Change*, 27: 501–27.

——— (1999) *Transforming the Indonesian Uplands*, London: Harwood.

Little, P. and Horowitz, M. (eds) (1987) *Lands at Risk in the Third World: Local Level Perspectives*, Boulder: Westview.

McAfee, K. (1999) "Selling Nature to Save it?" *Society ad Spaces*, 17: 133–54.

MacArthur (1995) MacArthur Foundation Conference on "Collectivization, Decollectivization and the Environmental Record in the Socialist World," Havana, Cuba, 20–26 June.

MacKenzie, F. (1998) *Land, Ecology and Resistance in Kenya*, London: IAI.

Macnaghten, P. and Urry, J. (1999) *Contested Natures*, London: Sage.

Martinez Alier, J. (1990) "Poverty as a Cause of Environmental Degradation," unpublished manuscript prepared for the World Bank (Latin American Division), Washington DC.

Martinez Alier, J. and Schulpmann, K. (1987) *Ecological Economics*, Oxford: Blackwell.

Mellor, J. (1988) "Environmental Problems and Poverty" *Environment 30*, 9: 8–13

Merchant, C. (1993) *Major Problems in American Environmental History*, Boston: Heath

Mol, A. (1994) *The Refinement of Production*, Utrecht: Van Arkel.

Moore, D., Kosek, J. and Pandian, A. (eds) (2003) *Race, Nature and the Politics of Difference*, Durham: Duke University Press.

Neumann, R. (1992) "Resource Use and Conflict in Arusha National Park, Tanzania," Ph.D. Dissertation, University of California, Berkeley.

——— (1999) *Imposing Wilderness*, Berkeley: University of California Press.

Nietschmann, B. (1972) *Between Land and Water: The Subsistence Ecology of the Miskito Indians*, New York: Academic.

O'Connor, J. (1988) "Capitalism, Nature, Socialism: A Theoretical Introduction," *Capitalism, Nature, Socialism*, 1: 11–38.

——— (1999) *Natural Causes*, New York: Guilford.

O'Connor, M. (1994) *Ecology and Capitalism*, New York: Guilford.

O'Neill, K. (2000) *Waste Trading Among Rich Nations*, Cambridge, MA: MIT Press.

O'Rourke, D. (2002) "Community Driven Regulation," in P. Evans (ed.) *Livable Cities?*, Berkeley: University of California Press, 95–131.

Ostrom, E. (1990) *Governing the Commons*, Cambridge: Cambridge University Press.

Peet, R. and Watts, M. (1996) "Liberation Ecology: Development, Sustainability and Environment in an Age of Market Triumphalism," in R. Peet and M. Watts (eds), *Liberation Ecologies: Environment, Development, Social Movements*, London: Routledge, 1–45.

Pellow, D. (2002) *Garbage Wars*, Cambridge, MA: MIT Press.

Pellow, D. and Park, L. (2003) *The Silicon Valley of Dreams*, New York: New York University Press.

Pelsuo, N. (1992) *Rich Forests, Poor People*, Berkeley: University of California Press.

—— (1993) "Coercing Conservation?," *Global Environmental Change*, 199–217.

—— (1996) "Fruit Trees and Family Trees in an Anthropogenic Forest," *Comparative Studies in Society and History*, 38/3: 510–48.

—— and Watts, M. (2001) *Violent Environments*, Ithaca: Cornell University Press.

Perrings, C. (1987) *Economy and Environment*, Cambridge: Cambridge University Press.

Pezzoli, K. (1998) *Human Settlements and Planning for Ecological Sustainability*, Cambridge, MA: MIT Press.

Philo, C. and Wilbert, C. (eds) (2000) *Animal Spaces and Beastly Places*, London: Routledge.

Pratt, L. and Montgomery, W. (1998) "Green Imperialism," *Socialist Register 1997*, London: Monthly Review, 75–95.

Princen, T. and Finger, M. (1994) *Environmental NGOs in World Politics*, London: Routledge.

Pulido, L. (1996) *Environmentalism and Economic Justice*, Tucson: University of Arizona Press.

Raffles, H. (2002) *In Amazonia*, Princeton: Princeton University Press.

Rangan, P (2000) *Of Myths and Movements*, London: Verso.

Rappaport, R. (1968) *Pigs for the Ancestors*, New Haven: Yale University Press.

Raustiala, K. (1997) "Domestic Institutions and International Regulatory Cooperation," *World Politics*, 49: 482–509.

Redclift, M. (1987) *Sustainable Development*, London: Methuen.

—— and Benton, T. (eds) (1997) *Social Theory and the Global Environment*, London: Routledge.

Redfield, P. (2000) *Space in the Tropics*, Berkeley: University of California Press.

Reichel Dolmatoff, G. (1972) *Amazonian Cosmos*, Chicago: Aldine.

Ribot, J. (1998) "Theorizing Access," *Development and Change*, 29: 307–41.

Rich, R. 1994 *Mortgaging the Earth*, Boston: Beacon Press.

Richards, P. (1985) *Indigenous Agricultural Revolution*, London: Hutchinson.

Rocheleau, D., Thomas-Slayter, B., Asamba, A. and Jama, M. (1996) *Feminist Political Ecology*, London: Routledge.

Rose, N. (1999) *Powers of Freedom*, Cambridge: Cambridge University Press.

Rothfels, N. (2002) *Savages and Beasts*, Baltimore: Johns Hopkins University Press.

Sand, P. (1995) "Trusts for the Earth," in W. Lang (ed.) *Sustainable Development and International Law*, London: Graham and Trotman, 167–84.

Sayer A. (1980) *Method in Social Science*. London: Methuen.

Schroeder, R. (1999) *Shady Practice*, Berkeley: University of California Press.

Sen, A. (1980) *Poverty and Famines*, Oxford: Clarendon.

—— (1990) "Food, Economics and Entitlements," in J. Dreze and A. Sen (eds) *The Political Economy of Hunger*, vol. 1, London: Clarendon, 34–50.

Shanin, T. (ed.) (1970) *Peasants and Peasant Societies*, London: Penguin.

Shapiro, J. (1999) *Mao's War Against Nature*, Berkeley: University of California Press.

Shiva, V. (1989) *Staying Alive*, London: Zed Press.

—— (1991) *Ecology and The Politics of Survival*, New Delhi: Sage.

Shrader-Frechette, K.S. (1991) *Risk and Rationality*, Berkeley: University of California Press.

Sinha, S., Guruani, S. and Greenberg, B. (1997) "The New Traditionalist Discourse of Indian Environmentalism," *Journal of Peasant Studies*, 24/3: 65–99.

Sivaramakrishnan, K. (1995) "Colonialism and Forestry in India," *Comparative Studies in Society and History*, 37/1: 3–40.

Slater, C. (1994) *The Dance of the Dolphin*, Chicago: University of Chicago Press.

—— (1995) *The Nature of Amazonia: The Amazon as Metaphor for the Natural World*, Berkeley: University of California Press (forthcoming).

—— (2001) *Entangled Edens*, Berkeley: University of California Press.

Socialist Review (1993) *Environment as Politics, 2*, special issue.

Soper, K. (1995) *The Problem of Nature*, Oxford: Blackwell.

Stepan, N. (2000) *Picturing Tropical Nature*, Ithaca: Cornell University Press.

Stonich, S. (1989) "The Dynamics of Social Processes and Environmental Destruction," *Population and Development Review*, 15, 269–96.

Swift, J. (1997) "Narratives, Winners and Losers," in M. Leach and R. Mearns (eds) *The Lie of the Land*, London: Currey, 73–90.

Taylor, P. (1997) "Notwithstanding Appearances, we are all Doing Something Like Political Ecology," *Social Epistemology*, 11/1: 111–27.

Taylor, P. and Buttel, F. (1992) "How do we know we have Environmental Problems?," *Geoforum*, 23, 405–16.

Tiffen, M. and Mortimore, M. (1994) *More People, Less Erosion*, New York: J. Wiley.

Toulmin, C. (1992) *Water, Women and Wells*, Oxford: Oxford University Press.

Tsing, A. (1999) "Becoming a Tribal Elder and Other Green Development Fantasies," in T. Li (ed.) *Transformation of the Indonesian Uplands*, London: Harwood, 159–202.

Utting, P. (1994) "Social and Political Dimensions of Environmental Protection in Central America," *Development and Change*, 25: 231–59.

Vayda, P. and Rappaport, R. (1967) "Ecology, cultural and non-cultural," in J. Clifton (ed.) *Introduction to Cultural Anthropology*, Boston: Houghton & Mifflin, 477–97.

Vayda, P. and Walters (1999) Against Political Ecology, *Human Ecology*, 27/1, 1–18.

Wade, R. (1997) "Greening the Bank," in R, Kanbur, J. Lewis and R. Webb (eds) *The World Bank*, Washington DC: Brookings Institution, 611–734.

Warren, M. (ed.) (1991) "Indigenous Knowledge Systems and Development," *Agriculture and Human Values*, VII, special issue.

Watts, M. (1983a) "The Poverty of Theory," in K. Hewitt (ed.) *Interpretations of Calamity*, London: Allen and Unwin, 231–62.

—— (1983b) *Silent Violence: Food, Famine and Peasantry in Northern Nigeria*, Berkeley: University of California Press.

—— (1987) "Drought, Environment and Food Security," in M. Glantz (ed.) *Drought and Hunger in Africa*, Cambridge: Cambridge University Press, 171–212.

—— (2002) "Political Ecology," in T. Barnes and E. Sheppard (eds) *Reader in Economic Geography*, Oxford: Blackwell, 257–75.

Weiner, D. (1999) *A Little Corner of Freedom*, Berkeley: University of California Press.

Whatmore, S. (2002) *Hybrid Geographies*, London: Routledge.

Wilden, A. (1972) *System and Structure*, London: Tavistock.

Wolch, J. and Emel, J. (1995) *Animal Geographies*, London: Verso.

Wolf, Eric (1972) "Ownership and Political Ecology," *Anthropological Quarterly*, 45: 201–5.

World Bank (1992) *The Development Report*, Washington DC: World Bank.

Worster, D. (1977) *Nature's Economy*, Cambridge: University Press.

—— (2003) *The Development Report*, Washington DC: World Bank.

Zerner, C. (ed.) (1999) *People, Plants and Justice*, New York: Columbia University Press.

Zimmerer, K. (1994) "Human Geography and the New Ecology," *Annals of the Association of American Geographers*, 108–25.

—— (1996) *Changing Fortunes*, Berkeley: University of California Press.

2

THE POLITICAL ECOLOGY OF FAMINE

The origins of the Third World[*]

Mike Davis

> The failure of the monsoons through the years from 1876 to 1879 resulted in an unusually severe drought over much of Asia. The impact of the drought on the agricultural society of the time was immense. So far as is known, the famine that ravished the region is the worst ever to afflict the human species.
>
> (John Hidore, *Global Environmental Change*)[**]

It was the most famous and perhaps longest family vacation in American history. "Under a crescendo of criticism for the corruption of his administration," the newly retired president *of* the United States, Ulysses S. Grant, his wife Julia, and son Jesse left Philadelphia in spring 1877 for Europe. The ostensible purpose of the trip was to spend some time with daughter Nellie in England, who was married (after the fashion that Henry James would celebrate) to a "dissolute English gentleman." Poor Nellie, in fact, saw little of her publicity-hungry parents, who preferred red carpets, cheering throngs and state banquets. As one of Grant's biographers has put it, "much has been said about how Grant, the simple fellow, manfully endured adulation because it was his duty to do so. This is nonsense." Folks back home were thrilled by *New York Herald* journalist John Russell Young's accounts *of* the "stupendous dinners, with food and wine in enormous quantity and richness, followed by brandy which the general countered with countless cigars." Even more than her husband, Mrs Grant – but for Fort Sumter, a drunken tanner's wife in Galena, Illinois – "could not get too many princely attentions." As a result, "the trip went on and on and on" – as did Young's columns in the *Herald*.[1]

Wherever they supped, the Grants left a legendary trail of gaucheries. In Venice, the General told the descendants of the Doges that "it would be a fine city if they drained it," while at a banquet in Buckingham Palace, when the visibly uncomfortable Queen Victoria (horrified at a "tantrum" by son Jesse) invoked her "fatiguing duties" as an excuse to escape the Grants, Julia responded: "Yes, I can imagine them: I too have been the wife of a great ruler."[2]

With James Gordon Bennett Jr. of the *New York Herald* paying the bar tab and the US Navy providing much of the transportation, the ex-First Family plotted an itinerary that would have humbled Alexander the Great: up the Nile to Thebes in Upper Egypt, back to Palestine, then on to Italy and Spain, back to the Suez Canal, outward to Aden, India, Burma, Vietnam, China and Japan, and, finally, across the Pacific to California.

Vacationing in famine land

Americans were particularly enthralled by the idea of their Ulysses in the land of the pharaohs. Steaming up the Nile, with a well-thumbed copy of Mark Twain's *Innocents Abroad* on his lap, Grant was bemused to be welcomed in village after village as the "King of America." He spent quiet afternoons on the river reminiscing to Young (and thousands of his readers) about the bloody road from Vicksburg to Appomattox. Once he chastized the younger officers in his party for taking unsporting potshots at stray cranes and pelicans. (He sarcastically suggested they might as well go ashore and shoot some "poor, patient drudging camel, who pulls his heavy-laden hump along the bank.") On another occasion, when their little steamer had to pull up for the night while the crew fixed the engine, Grant's son Jesse struck up a conversation with some of the bedouin standing guard around the campfire. They complained that "times are hard," forcing them far from their homes. "The Nile has been bad, and when the Nile is bad, calamity comes and the people go away to other villages."[3]

Indeed the Grants' idyll was soon broken by the increasingly grim conditions along the river banks. "Our journey," reported Young, "was through a country that in a better time must have been a garden; but the Nile not having risen this year all is parched and barren." Although so far the Grants had only basked in the warmth of peasant hospitality, there had been widespread rioting in the area south of Siout (capital of Upper Egypt) and some of the *fellahin* had reportedly armed themselves and headed into the sand hills. At the insistence of the governor, the Americans were assigned an armed guard for the remainder of their journey to Thebes and the First Cataract. Here the crop failure had been nearly total and thousands were dying from famine. Young tried to paint a picture of the "biblical disaster" for *Herald* readers: "Today the fields are parched and brown, and cracked. The irrigating ditches are dry. You see stumps of the last season's crop. But with the exception of a few clusters of the castor bean and some weary, drooping date palms, the earth gives forth no fruit. A gust of sand blows over the plain and adds to the somberness of the scene."[4]

Young, who had become as enchanted with Egypt's common people as with its ancient monuments, was appalled by the new British suzerains' contemptuous attitude toward both. "The Englishman," he observed, "looks upon these people as his hewers of wood and drawers of water, whose duty is to work and to thank the Lord when they are not flogged. They only regard these monuments [meanwhile] as reservoirs from which they can supply their own museums and for that purpose they have plundered Egypt, just as Lord Elgin plundered Greece." Young noted the crushing burden that the country's enormous foreign debt, now policed by the British, placed upon its poorest and now famished people. The ex-President, for his part, was annoyed by the insouciant attitude of the local bureaucrats confronted with a disaster of such magnitude.[5]

A year later in Bombay, Young found more evidence for his thesis that "English influence in the East is only another name for English tyranny." While the Grants were marveling over the seeming infinity of servants at the disposal of the sahibs, Young was weighing the costs of empire borne by the Indians. "There is no despotism," he concluded, "more absolute than the government of India. Mighty, irresponsible, cruel ..." Conscious that more than 5 million Indians by official count had died of famine in the preceding three years, Young emphasized that the "money which England takes out of India every year is a serious drain upon the country, and is among the causes of its poverty."[6]

Leaving Bombay, the Grant party passed through a Deccan countryside – "hard, baked and brown" – that still bore the scars of the worst drought in human memory. "The ride was a dusty one, for rain had not fallen since September, and the few occasional showers which usually attend the blossoming of the mango, which had not appeared, were now the dread of the people, who feared their coming to ruin the ripening crops."[7] After obligatory sightseeing trips to the Taj Mahal and Benares, the Grants had a brief rendez-vous with the viceroy. Lord Lytton, in Calcutta and then left, far ahead of schedule, for Burma. Lytton would later accuse a drunken Grant of groping English ladies at dinner, while on the American side there was resentment of Lytton's seeming diffidence towards the ex-president.[8] Grant's confidant, the diplomat Adam Badeau, thought that Lytton had received "instructions from home not to pay too much deference to the ex-President. He believed that the British Government was unwilling to admit to the half-civilized populations of the East that any Western Power was important, or that any authority deserved recognition except their own." (Grant, accordingly, refused Badeau's request to ask the US ambassador in London to thank the British.)[9]

A magnificent reception in China compensated for Lytton's arrogance. Li Hongzhang, China's senior statesman and victor over the Nian rebellion (which Young confused with the Taiping), was eager to obtain American help in difficult negotiations with Japan over the Ryukus. Accordingly, 100,000 people were turned out in Shanghai to cheer the Grants while a local band gamely attempted "John Brown's Body."

En route from Tianjin to Bejing, the Americans were wearied by the "fierce, unrelenting heat" compounded by depressing scenery of hunger and desolation.[10] Three years of drought and famine in northern China – officially the "most terrible disaster in twenty-one dynasties of Chinese history" – had recently killed somewhere between 8 million and 20 million people.[11] Indeed nervous American consular officials noted in their dispatches that "were it not for the possession of improved weapons mobs of starving people might have caused a severe political disturbance."[12]

THE SECRET HISTORY OF THE
NINETEENTH CENTURY

After Beijing, Grant continued to Yokohama and Edo, then home across the Pacific to a rapturous reception in San Francisco that demonstrated the dramatic revival of his popularity in light of so much romantic and highly publicized globetrotting. Throat cancer eventually precluded another assault on the White House and forced the ex-President into a desperate race to finish his famous *Personal Memoirs*. But none of that is pertinent to this preface. What is germane is a coincidence in his travels that Grant himself never acknowledged, but which almost certainly must have puzzled readers of Young's narrative: the successive encounters with epic drought and famine in Egypt, India and China. It was almost as if the Americans were inadvertently following in the footprints of a monster whose colossal trail of destruction extended from the Nile to the Yellow Sea.

As contemporary readers of *Nature* and other scientific journals were aware, it was a disaster of truly planetary magnitude, with drought and famine reported as well in Java, the Philippines, New Caledonia, Korea, Brazil, southern Africa and the Mahgreb. No one had hitherto suspected that synchronous extreme weather was possible on the scale of the entire tropical monsoon belt plus northern China and North Africa. Nor was there any historical record of famine afflicting so many far-flung lands simultaneously. Although only the roughest estimates of mortality could be made, it was horrifyingly clear that the million Irish dead of 1845–7 had been multiplied by tens. The total toll of conventional warfare from Austerlitz to Antietam and Sedan, according to calculations by one British journalist, was probably less than the mortality in southern India alone.[13] Only China's Taiping Revolution (1851–64), the bloodiest civil war in world history with an estimated 20 million to 30 million dead, could boast as many victims.

But the great drought of 1876–9 was only the first of three global subsistence crises in the second half of Victoria's reign. In 1889–91 dry years again brought famine to India, Korea, Brazil and Russia, although the worst suffering was in Ethiopia and the Sudan, where perhaps one-third of the population died. Then in 1896–1902 the monsoons again repeatedly failed across the tropics and in northern China. Hugely destructive epidemics of malaria, bubonic plague, dysentery, smallpox and cholera culled millions of victims from the ranks of the famine weakened. The European empires, together with Japan and the United States, rapaciously exploited the opportunity to wrest new colonies, expropriate communal lands, and tap novel sources of plantation and mine labor. What seemed from a metropolitan perspective the nineteenth century's final blaze of imperial glory was, from an Asian or African viewpoint, only the hideous light of a giant funeral pyre.

The total human toll of these three waves of drought, famine and disease could not have been less than 30 million victims. Fifty million dead might not be unrealistic. (Table 2.1 displays an array of estimates for famine mortality for

Table 2.1 Estimated famine mortality (millions).

India	1876–79	10.3	(Digby)
		8.2	(Maharatna)
		6.1	(Seavoy)
	1896–1902	19.0	(*The Lancet*)
		8.4	(Maharatna/Seavoy)
		6.1	(Cambridge)
India total		12.2–29.3	
China	1876–79	20	(Broomhall)
		9.5–13	(Bohr)
	1896–1900	10	(Cohen)
China total		19.5–30	
Brazil	1876–79	0.5–1.0	(Cunniff)
	1896–1900	n.d.	
Brazil total		2	(Smith)
Total		31.7–61.3 million	

Source: William Digby, *"Prosperous" British India*, London 1901; Arap Maharatna, *The Demography of Famine*, Delhi 1996; Roland Seavoy, *Famine in Peasant Societies*, New York 1986; *The Lancet*, 16 May 1901; *Cambridge Economic History of India*, Cambridge 1983; A.J. Broomhall, *Hudson Taylor and China's Open Century, Book Six, Assault on the Nine*, London 1988; Paul Bohr, *Famine in China*, Cambridge, Mass. 1972; Paul Cohen, *History in Three Keys*, New York 1997; Roger Cunniff, "The Great Drought: Northeast Brazil, 1877–1880," Ph.D. diss., University of Texas, Austin 1970; and T. Lynn Smith, *Brazil: People and Institutions*, Baton Rouge, La. 1954.

1876–79 and 1896–1902 in India, China and Brazil only.) Although the famished nations themselves were the chief mourners, there were also contemporary Europeans who understood the moral magnitude of such carnage and how fundamentally it annulled the apologies of empire. Thus the Radical journalist William Digby, principal chronicler of the 1876 Madras famine, prophesized on the eve of Queen Victoria's death that when "the part played by the British Empire in the nineteenth century is regarded by the historian fifty years hence, the unnecessary deaths of millions of Indians would be its principal and most notorious monument."[14] A most eminent Victorian, the famed naturalist Alfred Russel Wallace, the codiscoverer with Darwin of the theory of natural selection, passionately agreed. Like Digby, he viewed mass starvation as avoidable political tragedy, not "natural" disaster. In a famous balance-sheet of the Victorian era, published in 1898, he characterized the famines in India and China, together with the slum poverty of the industrial cities, as "the most terrible failures of the century."[15]

But while the Dickensian slum remains in the world history curriculum, the famine children of 1876 and 1899 have disappeared. Almost without exception, modern historians writing about nineteenth-century world history from a metropolitan vantage-point have ignored the late Victorian mega-droughts and famines that engulfed what we now call the "Third World." Eric Hobsbawm, for example, makes no allusion in his famous trilogy on nineteenth-century history to the worst famines in perhaps 500 years in India and China, although he does mention the Great Hunger in Ireland as well as the Russian famine of 1891–92. Likewise, the sole reference to famine in David Landes's *The Wealth and Poverty of Nations* – a magnum opus meant to solve the mystery of inequality between nations – is the erroneous claim that British railroads eased hunger in India.[16] Numerous other examples could be cited of contemporary historians' curious neglect of such portentous events. It is like writing the history of the late twentieth century without mentioning the Great Leap Forward famine or Cambodia's killing fields. The great famines are the missing pages – the absent defining moments, if you prefer – in virtually every overview of the Victorian era. Yet there are compelling, even urgent, reasons for revisiting this secret history.

At issue is not simply that tens of millions of poor rural people died appallingly, but that they died in a manner, and for reasons, that contradict much of the conventional understanding of the economic history of the nineteenth century. For example, how do we explain the fact that in the very half-century when peacetime famine permanently disappeared from Western Europe, it increased so devastatingly throughout much of the colonial world? Equally how do we weigh smug claims about the life-saving benefits of steam transportation and modern grain markets when so many millions, especially in British India, died alongside railroad tracks or on the steps of grain depots? And how do we account in the case of China for the drastic decline in state capacity and popular welfare, especially famine relief, that seemed to follow in lockstep with the empire's forced "opening" to modernity by Britain and the other Powers?

We are not dealing, in other words, with "lands of famine" becalmed in stagnant backwaters of world history, but with the fate of tropical humanity at the precise moment (1870–1914) when its labor and products were being dynamically conscripted into a London-centered world economy.[17] Millions died, not outside the "modern world system," but in the very process of being forcibly incorporated into its economic and political structures. They died in the golden age of Liberal Capitalism; indeed, many were murdered, as we shall see, by the theological application of the sacred principles of Smith, Bentham and Mill. Yet the only twentieth-century economic historian who seems to have clearly understood that the great Victorian famines (at least, in the Indian case) were integral chapters in the history of capitalist modernity was Karl Polanyi in his 1944 book *The Great Transformation*. "The actual source of famines in the last fifty years," he wrote, "was the free marketing of grain combined with local failure of incomes":

> Failure of crops, of course, was part of the picture, but despatch of grain by rail made it possible to send relief to the threatened areas; the trouble was that the people were unable to buy the corn at rocketing prices, which on a free but incompletely organized market were bound to be a

reaction to a shortage. In former times small local stores had been held against harvest failure, but these had been now discontinued or swept away into the big market.... Under the monopolists the situation had been fairly kept in hand with the help of the archaic organization of the countryside, including free distribution of corn, while under free and equal exchange Indians perished by the millions.[18]

Polanyi, however, believed that the emphasis that Marxists put on the exploitative aspects of late-nineteenth-century imperialism tended "to hide from our view the even greater issue of cultural degeneration":

The catastrophe of the native community is a direct result of the rapid and violent disruption of the basic institutions of the victim (whether force is used in the process or not does not seem altogether relevant). These institutions are disrupted by the very fact that a market economy is foisted upon an entirely differently organized community; labor and land are made into commodities, which, again, is only a short formula for the liquidation of every and any cultural institution in an organic society.... Indian masses in the second half of the nineteenth century did not die of hunger because they were exploited by Lancashire; they perished in large numbers because the Indian village community had been demolished.[19]

Polanyi's famous essay has the estimable virtue of knocking down one Smithian fetish after another to show that the route to a Victorian "new world order" was paved with bodies of the poor. But he simultaneously reified the "Market" as automata in a way that has made it easier for some epigones to visualize famine as an inadvertent "birth pang" or no-fault "friction of transition" in the evolution towards market-based world subsistence. Commodification of agriculture eliminates village-level reciprocities that traditionally provided welfare to the poor during crises. (Almost as if to say: "Oops, systems error: fifty million corpses. Sorry. We'll invent a famine code next time.")

But markets, to play with words, are always "made." Despite the pervasive ideology that markets function spontaneously (and, as a result, "in capitalism, there is nobody on whom one can pin guilt or responsibility, things just happened that way, through anonymous mechanisms"),[20] they in fact have inextricable political histories. And force – *contra* Polanyi – is "altogether relevant." As Rosa Luxemburg argued in her classic (1913) analysis of the incorporation of Asian and African peasantries into the late-nineteenth-century world market:

Each new colonial expansion is accompanied, as a matter of course, by a relentless battle of capital against the social and economic ties of the natives, who are also forcibly robbed of their means of production and labour power. Any hope to restrict the accumulation of capital exclusively to "peaceful competition," i.e. to regular commodity exchange such as takes place between capitalist producer-countries, rests on the pious belief that capital ... can rely upon the slow internal process of a disintegrating natural economy. Accumulation, with its spasmodic expansion, can no more wait for, and be content with, a natural internal disintegration of non-capitalist formations and their transition to commodity economy, than it can wait for, and be content with, the natural increase of the working population. Force is the only solution open to capital, seen as an historical process, employs force as a permanent weapon...[21]

The famines that Polanyi abstractly describes as rooted in commodity cycles and trade circuits were part of this permanent violence. "Millions die" was ultimately a policy choice: to accomplish such decimations required (in Brecht's sardonic phrase) "a brilliant way of organizing famine."[22] The victims had to be comprehensively defeated well in advance of their slow withering into dust. Although equations may be more fashionable, it is necessary to pin names and faces to the human agents of such catastrophes, as well as to understand the configuration of social and natural conditions that constrained their decisions. Equally, it is imperative to consider the resistances, large and small, by which starving laborers and poor peasants attempt to foil the death sentences passed by grain speculators and colonial proconsuls.

"PRISONERS OF STARVATION"

Synchronous and devastating drought provided an environmental stage for complex social conflicts that ranged from the intra-village level to Whitehall and the Congress of Berlin. Although crop failures and water shortages were of epic proportion – often the worst in centuries – there were almost always grain surpluses elsewhere in the nation or empire that could have potentially rescued drought victims. Absolute scarcity, except perhaps in Ethiopia in 1889, was never the issue. Standing between life and death instead were new-fangled commodity markets and price speculation, on one side, and the will of the state (as inflected by popular protest), on the other. As we shall see, the capacities of states to relieve crop failure, and the way in which famine policy was discounted against available resources, differed dramatically. At one extreme, there was British India under viceroys like Lytton, the second Elgin and Curzon, where Smithian dogma and cold imperial self-interest allowed huge grain exports to England in the midst of horrendous starvation. At the other extreme was the tragic example of Ethiopia's Menelik II, who struggled heroically but with too few resources to rescue his people from a truly biblical conjugation of natural and manmade plagues.

Seen from a slightly different perspective, the subjects of this book were ground to bits between the teeth of three massive and implacable cogwheels of modern history. In the first instance, there was the fatal meshing of extreme events between the world climate system and the late Victorian world economy. This was one of the major novelties of the age. Until the 1870s and the creation of a rudimentary international weather reporting network there was little scientific apprehension that drought on a planetary scale was even possible; likewise, until the same decade, rural Asia was not yet sufficiently integrated into the global economy to send or receive economic shock waves from the other side of the world. The 1870s, however, provided numerous examples of a new vicious circle (which Stanley Jevons was the first economist to recognize) linking weather and price perturbations through the medium of an international grain market.[23] Suddenly the price of wheat in Liverpool and the rainfall in Madras were variables in the same vast equations of human survival.

The last quarter of the nineteenth century saw the malign interaction between climatic and economic processes. Most of the Indian, Brazilian and Moroccan cultivators, for example, who starved in 1877 and 1878 had already been immiserated and made vulnerable to hunger by the economic crisis (the nineteenth century's "Great Depression") that began in 1873. The soaring trade deficits of Qing China – artificially engineered in the first place by British *nar-cotraficantes* – likewise accelerated the decline of the "ever-normal" granaries that were the empire's first-line defense against drought and flood. Conversely, drought in Brazil's Nordeste in 1889 and 1891 prostrated the population of the backlands in advance of the economic and political crises of the new Republic and accordingly magnified their impact.

But Kondratieff the theorist of economic "long waves" and Bjerknes (the theorist of El Niño oscillations) need to be supplemented by Hobson, Luxemborg and Lenin. The New Imperialism was the third gear of this catastrophic history. As Jill Dias has so brilliantly shown in the case of the Portuguese in nineteenth-century Angola, colonial expansion uncannily syncopated the rhythms of natural disaster and epidemic disease.[24] Each global drought was the green light for an imperialist landrush. If the southern African drought of 1877, for example, was Carnarvon's opportunity to strike against Zulu independence, then the Ethiopian famine of 1889–91 was Crispi's mandate to build a new Roman Empire in the Horn of Africa. Likewise Wilhelmine Germany exploited the floods and drought that devastated Shandong in the late 1890s to aggressively expand its sphere of influence in North China, while the United States was simultaneously using drought-famine and disease as weapons to crush Aguinaldo's Philippine Republic.

But the agricultural populations of Asia, Africa and South America did not go gently into the New Imperial order. Famines are wars over the right to existence. If resistance to famine in the 1870s (apart from southern Africa) was overwhelmingly local and riotous, with few instances of more ambitious insurrectionary organization, it undoubtedly had much to do with the recent memories of state terror from the suppression of the Indian Mutiny and the Taiping Revolution. The 1890s were an entirely different story, and modern historians have clearly established the contributory role played by drought-famine in the Boxer Rebellion, the Korean Tonghak movement, the rise of Indian Extremism and the Brazilian War of Canudos, as well as innumerable revolts in eastern and southern Africa. The millenarian movements that swept the future "Third World" at the end of the nineteenth century derived much of their eschatological ferocity from the acuity of these subsistence and environmental crises.

But what of nature's role in this bloody history? What turns the great wheel of drought and does it have an intrinsic periodicity? Synchronous drought – resulting from massive shifts in the seasonal location of the principal tropical weather systems – was one of the great scientific mysteries of the nineteenth century. The key theoretical breakthrough did not come until the late 1960s, when Jacob Bjerknes at UCLA showed for the first time how the equatorial Pacific Ocean, acting as a planetary heat engine coupled to the trade winds, was able to affect rainfall patterns throughout the tropics and even in the temperate latitudes. Rapid warmings of the eastern tropical Pacific (called El Niño events), for example, are associated with weak monsoons and synchronous drought throughout vast parts of Asia, Africa and northeastern South America. When the eastern Pacific is unusually cool, on the other hand, the pattern reverses (called a La Niña event), and abnormal precipitation and flooding occur in the same "teleconnected" regions. The entire vast see-saw of air mass and ocean temperature, which extends into the Indian Ocean as well, is formally known as "El Niño–Southern Oscillation" (or ENSO, for short).

The first reliable chronologies of El Niño events, painstakingly reconstructed from meteorological data and a variety of anecdotal records (including even the diaries of the conquistadors), were assembled in the 1970s.[25] The extremely powerful 1982 El Niño stimulated new interest in the history of the impacts of earlier events. In the 1986 two researchers working out of a national weather research laboratory in Colorado published a detailed comparison of meteorological data from the 1876 and 1982 anomalies that identified the first as a paradigmatic ENSO event: perhaps the most powerful in 500 years.[26] Similarly, the extraordinary succession of tropical droughts and monsoon failures in 1896–97 1899–1900, and 1902 were firmly correlated to El Niño warmings of the eastern Pacific. (The 1898 Yellow River flood, in addition, was probably a La Niña event.) Indeed, the last third of the nineteenth century, like the last third of the twentieth, represents an exceptional intensification of El Niño activity relative to the centuries-long mean.[27]

If, in the eyes of science, ENSO's messy fingerprints are all over the climate disasters of the Victorian period, historians have yet to make much of this discovery. In the last generation, however, they have generated a wealth of case-studies and monographs that immeasurably deepen our understanding of the impact of world market forces on non-European agriculturalists in the late nineteenth century. We now have a far better understanding of how sharecroppers in Ceará, cotton producers in Berar and poor peasants in western Shandong were linked to the world economy and why that made them more vulnerable to drought and flood. We also have magnificent analyses of larger pieces of the puzzle: the decline of the Qing granary and flood-control systems, the internal structure of India's cotton and wheat export sectors, the role of racism in regional development in nineteenth-century Brazil, and so on.

STRUCTURAL VULNERABILITY

Over the last generation, scholars have produced a bumper-crop of revealing social and economic histories of the regions teleconnected to ENSO's episodic disturbances. The thrust of this research has been to further demolish orientalist stereotypes of immutable poverty and overpopulation as the natural preconditions of the major nineteenth-century famines. There is persuasive evidence that peasants and farm laborers became dramatically more pregnable to natural disaster after 1850 as their local economies were violently incorporated into the world market. What colonial administrators and missionaries – even sometimes creole elites, as in Brazil – perceived as the persistence of ancient cycles of backwardness were typically modern structures of formal or informal imperialism.

From the perspective of political ecology, the vulnerability of tropical agriculturalists to extreme climate events after 1870 was magnified by simultaneous restructurings of household and village linkages to regional production systems, world commodity markets and the colonial (or dependent) state. "It is, of course, the constellation of these social relations," writes Watts, "which binds house-holds together and project them into the marketplace, that determines the precise form of the household vulnerability. It is also these same social relations that have failed to stimulate or have actually prevented the development of the productive forces that might have lessened this vulnerability." Indeed, new social relations of production, in tandem with the New Imperialism, "not only altered the extent of hunger in a statistical sense but changed its very etiology."[28] Three points of articulation with larger socio-economic structures were especially decisive for rural subsistence in the late Victorian "proto-third world."

First, the forcible incorporation of smallholder production into commodity and financial circuits controlled from overseas tended to undermine traditional food security. Recent scholarship confirms that it was *subsistence adversity* (high taxes, chronic, indebtedness, inadequate acreage, loss of subsidiary employment opportunities, enclosure of common resources, dissolution of patrimonial obligations, and so on), not entrepreneurial opportunity, that typically promoted the turn to cash-crop cultivation. Rural capital, in turn, tended to be parasitic rather than productivist as rich landowners redeployed fortunes that they built during export booms into usury, rack-renting and crop brokerage. "Marginal subsistence producers," Hans Medick points out, "did not benefit from the market under these circumstances; they were devoured by it."[29] Medick, writing about the analogous predicament of marginal smallholders in "proto-industrial" Europe, provides an exemplary description of the dilemma of millions of Indian and Chinese poor peasants in the late nineteenth century:

> For them [even] rising agrarian prices did not necessarily mean increasing incomes. Since their marginal productivity was low and production fluctuated, rising agrarian prices tended to be a source of indebtedness rather than affording them the opportunity to accumulate surpluses. The "anomaly of the agrarian markets" forced the marginal subsistence producers into an unequal exchange relationship through the market.... Instead of profiting from exchange, they were forced by the market into the progressive deterioration of their conditions of production, i.e. the loss of their property titles. Especially in years of bad harvests, and high prices, the petty producers were compelled to buy additional grain, and, worse, to go into debt. Then, in good harvest years when cereal prices were low, they found it hard to extricate themselves from the previously accumulated debts; owing to the low productivity of their holdings they could not produce sufficient quantities for sale.[30]

As a result, the position of small rural producers in the international economic hierarchy equated with downward mobility, or, at best, stagnation. There is consistent evidence from north China as well as India and northeast Brazil of falling household wealth and increased fragmentation or alienation of land. Whether farmers were directly engaged by foreign capital, like the Berari *khatedars* and Cearan *parceiros* who fed the mills of Lancashire during the Cotton Famine, or were simply producing for domestic markets subject to international competition like the cotton-spinning peasants of the Boxer hsiens in western Shandong, commercialization went hand in hand with pauperization without any silver lining of technical change or agrarian capitalism.

Second, the integration of millions of tropical cultivators into the world markets during the late nineteenth century was accompanied by a dramatic deterioration in their terms of trade. Peasants' lack of market power vis-à-vis crop merchants and creditors was redoubled by their commodities' falling international purchasing power. The famous Kondratief downswing of 1873–97 made dramatic geographical discriminations. As W. Arthur Lewis suggests, comparative productivity or transport costs alone cannot explain an emergent structure of global unequal exchange that valued the products of tropical agriculture so differently from those of temperate farming. "With the exception of sugar, all the commodities whose price was lower in 1913 than in 1883 were commodities produced almost wholly in the tropics. All the commodities whose prices rose over this thirty-year period were commodities in which the temperate countries produced a substantial part of total supplies. The fall in ocean freight rates affected tropical more than temperate prices, but this should not make a difference of more than five percentage points."[31]

Third, formal and informal Victorian imperialism backed up by the supernational automatism of the Gold Standard, confiscated local fiscal autonomy and impeded state-level developmental responses – especially investments in water conservancy and irrigation – that might have reduced vulnerability to climate shocks. As Curzon once famously complained to the House of Lords, tariffs "were decided in London, not in India; in England's interests, not in India's."[32] Moreover, as we shall see in the next chapter, any grassroots benefit from British railroad and canal construction was largely canceled by official neglect of local irrigation and the brutal enclosures of forest and pasture resources. Export earnings, in other words, not only failed to return to smallholders as increments in household income, but also as usable social capital or state investment.

In China, the "normalization" of grain prices and the ecological stabilization of agriculture in the Yellow River plain were undermined by an interaction of endogenous crises and the loss of sovereignty over foreign trade in the aftermath of the two Opium Wars. As disconnected from world market perturbations as the starving loess provinces might have seemed in 1877, the catastrophic fate of their populations was indirectly determined by Western intervention and the consequent decline in state capacity to ensure traditional welfare. Similarly the depletion of "ever-normal" granaries may have resulted from a vicious circle of multiple interacting causes over a fifty-year span, but the *coup de grace* was certainly the structural recession and permanent fiscal crisis engineered by Palmerston's aggressions against China in the 1850s. As foreign pressure intensified in later decades, the embattled Qing, as Kenneth Pomeranz has shown, were forced to abandon both their traditional mandates: abandoning both hydraulic control and grain stockpiling in the Yellow River provinces in order to concentrate on defending their endangered commercial littoral.[33]

British control over Brazil's foreign debt and thus its fiscal capacity likewise helps explain the failure of either the empire or its successor republic to launch any anti-drought development effort in the sertão. The zero-sum economic conflicts between Brazil's rising and declining regions took place in a structural context where London banks, above all the Rothschilds, ultimately owned the money-supply. In common with India and China, the inability to politically regulate interaction with the world market at the very time when mass subsistence increasingly depended upon food entitlements acquired in international trade became a sinister syllogism for famine. Moreover in the three cases of the Deccan, the Yellow River basin and the Nordeste, former "core" regions of eighteenth-century subcontinental power systems were successively transformed into famished peripheries of a London-centered world economy.

The elaboration of these theses, as always in geo-historical explanation, invites closer analysis at different magnifications. Before considering case-studies of rural immiseration in key regions devastated by the 1870s and 1890s El Niño events or looking at the relationships among imperialism, state capacity and ecological crisis at the village level, it is necessary to briefly discuss how the structural positions of Indians and Chinese (the big battalions of the future Third World) in the world economy changed over the course of the nineteenth century. Understanding how tropical humanity lost so much economic ground to western Europeans after 1850 goes a long way toward explaining why famine was able to reap such hecatombs in El Niño years. As a baseline for understanding the origins of modern global inequality (and that is the key question), the herculean statistical labors of Paul Bairoch and Angus Maddison over the last thirty years have been complemented by recent comparative case-studies of European and Asian standards of living.

This is a "political ecology of famine" because it takes the viewpoint both of environmental history and Marxist political economy: an approach to the history of subsistence crisis pioneered by Michael Watts in his 1983 book, *Silent Violence: Food, Famine and Peasantry in Northern Nigeria*.[34] Although other umbrella terms and affiliations are possible, the fact that Watts and his co-thinkers label their ongoing work as "political ecology" persuades me to do the same, if only to express my indebtedness and solidarity. (Those familiar with Watts's book will easily recognize its influence in this work.)

Finally, there is David Arnold's indispensable emphasis on famines as "engines of historical transformation."[35] The great Victorian famines were forcing houses and accelerators of the very socio-economic forces that ensured their occurrence in the first place. A key thesis is that what we today call the "Third World" (a Cold War term)[36] is the outgrowth of income and wealth inequalities – the famous "development gap" – that were shaped most decisively in the last quarter of the nineteenth century, when the great non-European peasantries were initially integrated into the world economy. As other historians have recently pointed out, when the Bastille was being stormed, the vertical class divisions inside the world's major societies were *not* recapitulated as dramatic income differences *between* societies. The differences in living standards, say, between a French *sans-culotte* and Deccan farmer were relatively insignificant compared to the gulf that separated both from their ruling classes.[37] By the end of Victoria's reign, however, the inequality of nations was as profound as the inequality of classes. Humanity had been irrevocably divided. The origins of the Third World had been mercilessly put in place. And the famed "prisoners of starvation," whom the *Internationale* urges to arise, were as much modern inventions of the late Victorian world as electric lights, Maxim guns and "scientific" racism.

NOTES

* This chapter represents an amalgam of the Preface, (pp. 1–15) and sections (pp. 288–92) from Chapters 10, 11 and 12 of *Late Victorian Holocausts: El Niño Famines and the Making of the Third World*, London: Verso, 2000. It is reprinted by permission from the copyright holder, Mike Davis and the publisher, Verso.

** The epigraph is from John Hidore, *Global Environmental Change: Its Nature and Impact*, Upper Saddle River, N.J. 1996, p. 96.

1 W. McFeely, *Grant: A Biography*, New York: Norton, 1981, pp. 453, 457–60 and 471.

2 McFeely, op. cit., pp. 458–71.

3 J. Russell Young, *Around the World with General Grant*, subscription edn. (American News Company in 20 parts), New York, 1878–79, pp. 242 and 246.

4 Young, op. cit., pp. 266–7 and 274.

5 Young, op. cit., pp. 278 and 284–5.

6 Young, op. cit., p. 622.

7 Young, op. cit., p. 624.

8 "On this occasion 'our distinguished guest,' the double Ex-President of the 'Great Western Republic,' who got as drunk as a fiddle, showed he could also be as profligate as a lord. He fumbled Mrs. A., kissed the shrieking Miss B. – pinched the plump Mrs. C. black and blue – and ran at Miss D. intent to ravish her. Finally, after throwing all the … female guests into hysterics by generally behaving like a must elephant, the noble beast was captured by main force and carried (quatre pattes dans l'air) by six sailors … which relieved India of his distinguished presence. The marine office … reports that, when deposited in the public saloon cabin, where Mrs. G. was awaiting him … this remarkable man satiated there and then his baffled lust on the unresisting body of his legitimate spouse, and copiously committed during the operation. If you have seen Mrs. Grant you will not think this incredible" (Lytton quoted in McFeely 1981: 473).

9 A. Badeau, *Grant in Peace. From Appomattox to Mount McGregor. A personal memoir*, Hartford: S.S. Scranton & Co., 1887, pp. 310–11.

10 Young, op. cit., p. 414.

11 H. He, *Drought in Northern China in the Early Guang Xu* (1876–1879), Hong Kong, pp. 36–7 (in Chinese).

12 McFeely, op. cit., p. 557 fn43.

13 W. Digby, *"Prosperous"British India: A Revelation from Official Records*, London: Unwin, 1901, p. 118.

14 Digby, op. cit., p. 122.

15 A. Russel Wallace, *The Wonderful Century: Its Successes and Its Failures*, London: Swan Sonnenschein & Co., 1898, p. 341.

16 D. Landes, *The Wealth and Poverty of Nations*, New York: W. W. Norton, 1998, p. 437.

17 W. Arthur Lewis, *Growth and Fluctuations, 1870–1913*, London: G. Allen & Unwin 1978, pp. 29, 187 and 215.

18 K. Polanyi, *The Great Transformation*, Boston: Beacon Press, 1944, p. 160.

19 Polanyi, op. cit., pp. 159–60.

20 Slavoj Zizek, *The Spectre Is Still Roaming Around! An Introduction to the 150th Anniversary Edition of the Communist Manifesto*, Zagreb: Bastard Books, 1998, p. 17.

21 R. Luxemburg, *The Accumulation of Capital*, trans. Agnes Schwarzchild, London: Routledge, 1951 [1913], pp. 370–71.

22 Bertolt Brecht, *Poems 1913–1956*, London: Methuen, 1976, p. 204.

23 See Chapter 7.

24 J. Dias, "Famine and Disease in the History of Angola, c. 1830–1930," *Journal of African History* 22, 1981.

25 P. Wright, *An Index of Southern Oscillation*, University of East Anglia, Climate Research Unit Publication, Norwich, 1975; and William Quinn *et al.*, "Historical Trends and Statistics of the Southern Oscillation, El Niño, and Indonesian Droughts," *Fish. Bull.* 76, 1978.

26 G. Kiladis and H. Diaz, "An Analysis of the 1877–78 ENSO Episode and Comparison with 1982–83," *Monthly Weather Review* 114, June 1986. Although they "resist the temptation to compare the 'intensity'" of the two events, they point out that the 1876–78 event lasted longer and was associated with sea-level pressure anomalies across a larger area of the tropics (p. 1046).

27 P. Whetton and I. Rutherfurd, "Historical ENSO Teleconnections in the Eastern Hemisphere," *Climatic Change* 28, 1994, p. 243.

28 M. Watts, *Silent Violence: Food, Famine and Peasantry in Northern Nigeria*, Berkeley: University of California Press 1983, pp. 267 and 464.

29 H. Medick, "The Proto-Industrial Family Economy and the Structures and Functions of Population Development under the Proto-Industrial System," in P. Kriedte *et al.* (eds), *Industrialization Before Industrialization*, Cambridge, 1981, p. 45.

30 Medick, op. cit., pp. 44–5.

31 Lewis op. cit., p. 189.

32 Cited in Clive Dewey, "The End of the Imperialism of Free Trade," p. 35.

33 K. Pomeranz, *The Making of a Hinterland: State, Society, and Economy in Inland North China, 1853–1937*, Berkeley: University of California Press, 1993.

34 Ibid.

35 D. Arnold, *Famine: Social Crisis and Historical Change*, London: B. Blackwell, 1988.

36 A. Sauvy, "Trois Mondes, une planete," *L'Observateur* 118, 14 Aug. 1952, p. 5.

37 See the discussion in Chapter 9; also the much-awaited study by Kenneth Pomeranz, *The Great Divergence: China, Europe, and the Making of the Modern World Economy*, Princeton, N.J. 2000, which appeared while this book was in proof.

3

INVISIBLE FORESTS

The political ecology of forest resurgence in El Salvador

Susanna B. Hecht

In El Salvador, Nature has been extinguished.
(Terborgh, *Requiem for Nature*)

INTRODUCTION

Extensive deforestation was dramatic throughout the New World Tropics
during much of the last century. Whether one emphasized Malthusian processes,
social inequality, confused property regimes or misguided land use policy, Latin
American forests relentlessly fell. Today, however, the "doomed forest" narrative,
at least about Mesoamerica needs to be rethought. While deforestation continues
in some areas, especially near expanding cities and in sites apt for mechanized
agriculture, many rural areas are showing robust resurgence of anthropogenic
and successional forests. Recent field research using an array of techniques
including remote sensing and detailed land use surveys reveals a rising
proportion of diverse, forested or semi forested anthropogenic ecosystems in the
region (Hecht *et al.* 2002b; Hecht and Saatchi 2003), including, surprisingly,
the "environmental basket case" of El Salvador. This process of regeneration has
been reported elsewhere in Latin America (Aide *et al.* 2001; Klooster 2003;
Moran *et al.* 2000; Rudel *et al.* 2002; Zimmerman 2001) and is also described for
Asia (cf. Chokkalingam *et al.* 2001; de Jong 2001). This woodland recovery and
the types of forests that comprise it, often ignored because of the overarching
interest by many conservation scientists in "primary" ecosystems, has enormous
implications for the longer socio-ecological future of the region, and for resource
policy. This paper explores the explanations of forest trend, the political ecology
of woodlands in El Salvador, and examines how well these models account for
these changes. The final section focuses on the policy implications of these
dynamics.

Today, farmers balance a broad array of economic and social elements in the construction livelihoods that shape the use of natural resources. Global processes (including international migration, remittances, commodity prices, international environmental ideologies) regional dynamics (such as the Central American Common Market, the Meso American Biological Corridor) national structural adjustment and development policies (like decentralization, credit, agrarian reform policies, restructured labor markets) and local socio-environmental circumstances (household strategies, tenurial structures, access rights, gender of farmers, ethnicity, traditional beliefs, natural resources endowment) all affect land use decisions in very concrete ways.

Globalization – by which we mean the effects of trade liberalization, international labor markets, migration, capital flows and the "internationalization" of certain ideas – is among the most powerful of the forces affecting farmers. In some literatures it is assumed to unfold through a uniform set of social and institutional structures, rather like kinds of deterritorialized economic waves that wash over regions. We take the position, in line with a large literature in economic geography, political science and anthropology (cf. Bebbington 1999; Storper 1997, among many others), that globalization is mediated by local institutional arrangements, local assets, and a wide array of ethnographic and household factors that modify larger processes. Observed natural resource use reflects not just globalization "from above" but also its modification and manipulation "from below." Manifestation of these social processes include the physical forms of the landscape, especially in the resurgent forests of densely occupied rural areas and new institutional arrangements. Regenerating forests are a profound outcome of socio-economic as well as bio-geographic features. They are truly a socially "constructed" nature.

EXPLAINING FOREST TREND

The deforestation question

The dynamics of deforestation continue to be an active area of research, although most explanations of clearing processes have not integrated the impacts of current macroeconomics, and paid little attention to the impact of consolidation of both globalization and structural adjustment policies on resource management (but see Angelsen and Kaimowitz 2001; Barbier 1997; Hecht 2003). There are decades of summaries and discussions of the causes of deforestation but these will be rapidly summarized in the next few paragraphs.

Malthus and markets

These two approaches – Malthus and markets – population and prices – remain the dominant frameworks for explaining forest loss. While empirical and modeling data show ample reason to reject Malthusian premises (Fairhead and Leach 1998; Gibson *et al.* 2000; Kaimowitz and Angelsen 1999; Steininger *et al.* 2001; Templeton and Scheer 1999, among many others) population pressure is still seen as the key process in most of the biological literature (e.g. Myers *et al.* 2000; Terborgh 1999, most articles in the journal *Conservation Biology*). The simplicity of the correlation and the habit of two centuries have helped reinforce this approach.

Markets are seen to play a significant role in deforestation, either as enhancing demand for commodities that destroy forests or replace them (like timber, pasture or soybeans), or as stimulating markets for green products, correctly pricing tropical resources, and providing economic alternatives that promote more sustainable forest use, such as in non-timber forest products, certification of timber, and biodiversity friendly coffee (Hecht and Cockburn 1989; Perfecto *et al.* 1996; Shanley *et al.* 2003). Globalization of markets is conventionally viewed as an important driver in forest transformation as international demand for tropical commodities expands and produces forest depletion, and stimulates local farmers and other economic actors to respond to far flung market pressures (cf. Barbier 2000; Hecht 2003; Sunderlin *et al.* 2001). Another dynamic often mentioned of globalization of capital markets is the impact of debt on natural resources. In this approach tropical resources are plundered to pay off government obligations to international borrowers. The impact on natural resources of international migration and remittances remains largely unstudied (Hecht *et al.* 2002a) though it is central to the arguments advanced later in this paper.

Institutions: property and political structures

Institutional features associated with tenurial structures, property regimes and access rights are also viewed as key elements in deforestation (cf. Gibson *et al.* 2000). Insecurity over tenure, and the importance of clearing as a means of claiming lands has been widely analyzed as a stimulus for deforestation (cf. Downing *et al.* 1992; Hecht 1984, 1994; Schmink and Woods 1992). Weakness in the political institutions of environmental governance contributes to forest clearing (Didia 1997; Hecht *et al.* 2002b; Repetto and Gillis 1988).

Policy deficiencies, corruption and state cronyism – institutional rent seeking – are also recognized as stimulants to deforestation. Institutional rents associated with regional development policies and state initiatives that subsidized land clearance, particular commodities, timber cartels, and/or colonist programs that triggered clearing or speculative dynamics have also been blamed for fueling clearing (Hecht 1985, 1993; Lopez and Mitra 2000; Repetto and Gillis 1988). Corruption, especially in the timber and construction industries is implicated in the development of the policy and political climate that supported deforestation. Deeper macro-political processes involving geopolitical concerns, internal ethnic control, structural alternatives to agrarian reform, and larger effects of the political economy provided the overt ideological apparatus and the deeper justifications for massive clearing.

Intensification and technology

Technology arguments have also been widely advanced to explain patterns of clearing. This view suggests that increasing intensive production on its own will reduce clearing by keeping each hectare in cultivation and by increasing returns per area. This view, championed by advocates for alternative agricultures as well as the larger scale research organizations like the CGIAR[1] and agribusiness essentially views deforestation as a technical problem of extensive land uses. Allied with technology arguments have been those associated with infrastructure development, suggesting that better roads would assist in consolidating frontiers and reducing deforestation, even though historically infrastructure expansion has been the most straightforward correlate with clearing (cf. Dale *et al.* 1993; Kaimowitz *et al.* 2002). Discussions of Technology /intensification is usually unmoored from social contexts, and has been viewed as having a braking effect on forest clearance. Recent research suggests that this dynamic is far more complex. Intensive technologies, such as those associated with soybeans, sugarcane, industrial maize and cotton have, however, been decisive in massive deforestation processes (See Angleson and Kaimowitz 2001; Hecht 2003; Steininger *et al.* 2001; Utting 1993) and have been problematic elsewhere (See Angleson and Kaimowitz 2001).

Both endogenous and exogenous forces have been invoked to explain patterns of clearing, and there are case studies that support each of these larger models. Most of these models of deforestation derive however from the macroeconomic milieu that preceded deepening globalization and structural adjustment reforms. These larger scale macroeconomic impacts have been poorly theorized in the deforestation literature. But regardless of the nature of the drivers of forest destruction, the overwhelming policy choice for maintaining forests has centered on conservation set asides – national parks and reserves.

The last policy stand: parks and forest persistence

The literature on forest clearing gives a depressing and unidirectional sense of forest trend, and provides a sense that most deforested landscapes are forever barren and degraded. This version of a post-clearing world of holocaust has justifiably fueled the eagerness for conservation set aside throughout the tropics, and given the National Park or national reserve model pride of place as the best possible response to the seemingly inevitable, irreversible ravages of tropical deforestation. The lion's share of conservation funding has been applied to this model, in part due to its conceptual simplicity as policy, as well as the large theoretical and applied intellectual apparatus has emerged with the study of Island Biogeography (cf. MacArthur and Wilson 1967; Terborgh 1999), and early institutional analysis (Hardin 1967). This framework supports the idea of reserves as the last feeble hope in a landscape of despair. Its related field, fragment ecology, has proved to be far more contradictory, since the quality and type of the matrix in which forest fragments occur are seen as having a decisive impact on the diversity parameters of both the forest fragments *and* the matrix. Nonetheless, informed by theoretical models of species/area curves and tested in various contexts (islands in reservoirs, cf. Terborgh 1999) the Central Amazonian minimal critical size project (Laurence *et al.* 2001) among others, seemed to argue that large set asides in areas of significant diversity (hot spots) would provide the best protection for maintaining ecological processes and biodiversity (cf. van Shaik *et al.* 1997; Myers *et al.* 2000).

While increasingly advocated and adored by conservationists, a large body of research is emerging that places reserves in a social context that explains why, to the bafflement of many ecologists, local populations often detest parks. The Reserve model has proved controversial everywhere for a number of reasons outlined by Neumann (1999, among others). These issues often revolve around the creation of mythology of a European image of "wild" uninhabited nature in areas already occupied by people, and the criminalization of traditional subsistence and livelihood activities like gathering firewood, grazing, extraction of building materials, hunting and fishing (Hecht and Cockburn 1989; Neumann 1999, and this volume Spense 1999). Local peoples often had their reserve in – holdings expropriated and were "resettled" in ways not always especially beneficial for them.[2] In Latin America these set asides also carried with them other reasons for resistance. More than 70 percent of Latin American parks were established in the 1970s, and 1980s, (van Shaik *et al.* 1997), a period associated with military and authoritarian regimes, intense violence against rural populations, and a period especially associated with enclosure and rural dispossession. Even when the transition to democratic governance was underway, rural policies, including many environmental ones, still reflected this authoritarian legacy. In places like Brazil, and Bolivia, large scale conservation set asides were seen as part of the ideology and practice of authoritarian planning models of regional integration, and enclosure in the tropics that marginalized vast numbers of traditional peoples (Hecht and Cockburn 1989; Schmink and Woods 1992). Today, conservation initiatives have become important drivers of centralized land use planning, and often underpinned large Amazonian planning exercises such as "Planofloro" in Rondonia Brazil, and "Plus" in Bolivia and the Meso American biological corridor. Indeed, many community-based conservation initiatives have been rooted in a critique of these large scale modernization and land rationalization (Scott 1997) programs (Hecht and Cockburn 1989; Lima 1999).

Predicated on the idea of empty landscapes, the architects of forest reserves, simply viewed non-conservation, social landscapes with disdain – a fragmented forest, or successional form, home to the living dead in their shrinking little habitat. Thus as ecologists whizzed from the capital to their favorite conservation areas, the matrix in which the conservation landscapes occurred was largely overlooked. In many areas of the world, this landscape was increasingly wooded, as anthropogenic forests and successions replaced and supplemented earlier pastures and grain crops. As Fairhead and Leach (1998) have pointed out, the dominance of the scientific discourse of degradation/deforestation made it difficult to imagine that humans might have positive impacts on the environment. Meanwhile, many rural social movements began to articulate position that saw a "convivencia" and closed adaptation between forests, and people – a position increasingly supported by the archeological and ethnographic record (cf. Denevan 2001). In the midst of these heated debates over people and parks and whether even the idea of people and reserves was legitimate (cf. Sanderson and Redford 1997; Schwartzman *et al.* 2001) another dimension of forest processes remained mostly unobserved. Forest resurgence was occurring throughout the region.

Model of forest persistence

While complex sets of theories and explanations have emerged on why forests fall, few explanations are on offer for why forests stand. Six frameworks: geographic isolation, livelihood, institutional contexts, the environmental Kusnets curve, forest transition, and equity models explain the dynamics of forest persistence outside of parks.

The first model, *isolation*, simply argues that the forests are so distant or access is so difficult that the forests exist by default away from the realm of human transformation. Here, forests exist in geographical areas "outside" of history.[3] The standard set aside approach is a socially created "isolation" from human impacts.

In the *livelihood* analysis, traditional populations are often understood as highly forest dependent and require forest products as subsidies to subsistence and as essentials for market. Well known examples such as Brazilian Rubber tappers, Xateiros in Guatemala (among others) inform this "livelihood" logic of why forests stand. But for most forest dwellers, successional forests or earlier successional elements are especially important even within "primary" forest models. The manipulation of palms, fruit trees of many types, and successional formations by forest people has been extensively documented (cf. de Jong 2001; Padoch *et al.* 1999; Posey and Balee 1989, among many others). The high utility of secondary forests for "non forest" people has also been widely described (see Chazden and Coe 1999; de Jong 2001; Hecht *et al.* 1988; Padoch *et al.* 1999, among many others). In short, this utility argument emphasizes forest products in livelihood as the over-arching explanation for forest protection, and by extension, construction or reconstruction by local people. The livelihood model has obvious economic and policy dimensions to it, and is very sensitive to economic and institutional dynamics. This livelihood approach is often linked to ideas of territoriality, identity, and indigenous knowledge systems as well.

Social movements

While social movements are often organized around economic/livelihood concerns, they can be mobilized, and regularly are in the developed world around non-economic goals. The national parks movements, and various wildlife preserve advocates usually make their cases on the basis of a moral or scientific discourse rather than one that emphasizes utility. But there are also spiritual arguments that may cause local movements to agitate for protection of sacred spaces or sacred forests, or places of special significant for local identities.

Property and political institutions

The *institutional* approaches *vis-à-vis* forest persistence takes two main forms: those pertaining to property regimes and those associated with political institutions. The dynamics of property regimes are usefully summarized by Gibson *et al.* (2000) who review the controversies and outcomes for forest management through several case studies that explore the issues in local regulation of access and ownership regimes. Institutional dynamics are highly contingent and must be carefully specified at the historical, ethnographic as well as policy levels. The research on property regimes (whether private, state, or communal ownership) does show that institutions of property, whether collective or private have a significant effect on forests, and can contribute to their protection or their demise, depending on the context (cf. Gibson *et al.* 2000; Klooster and Masera 2000; Orstrom 1991). The data support both sides of the argument for and against collective and private regimes, although mixed systems (private household with collective holdings) give the impression of somewhat better performance when institutions of social regulation are in place. While this is perhaps not a satisfactory conclusion, it reflects the deeply conjunctural and social nature of resource control.

The impact of property regimes on diverse forest "creation" is far less studied. The conventional arguments usually weigh in on the side of private property because this forest "construction" literature focuses so heavily on plantation development which is heavily dependent on clear ownership to assure investment over long time frames. But there are some studies that focus on communal rights that emphasize non private tenurial forms and strategies of forest recuperation and expansion, mainly as an outcome of sustainable development, forest co-management and watershed recuperation activities as parts of Aid programs (cf. Klooster and Masera 2000; Primack *et al.* 1998, among many others). Property institutional issues are contentious due to their intense ideological content, the volatility of power relations and policies that affect them especially in the face of national and international effort to "rationalize" communal holdings and to reduce land conflict through privatization as part of recent neoliberal policy interventions.

Another institutional approach specifically emphasizes the idea that more democratic political institutions and local accountability produce more environmentally sound outcomes in resource management (cf. Didia 1997; Torras and Boyce 1998). This was the crux of the argument advanced by leftist movements and environmentalists just prior to the massive implementation of decentralization programs associated with structural adjustment policies. Local accountability, democratic access and decision-making could, it was argued, provide flora that would permit local communities to articulate their environmental needs more effectively and thus slow degradation and promote recuperation. The literature on *ejidos* in Mexico, extractive reserves, and some of the Bolivian decentralization efforts have shown a generally positive relation of more accessible political institutions with the maintenance of forests. Overall, however, the decentralization literature shows a far more ambiguous set of outcomes. Local management can be hijacked by native or outside elites, local communities can be unorganized and thus unable to articulate a coherent program, localities may have groups or ethnicities with highly different goals in terms of resource uses (See Ribot 2001). In addition, programs that emphasize an "environmental" nature – like potable water projects, often eclipse the more "ecological" terrain of natural resource management.

Decentralization in general has not usually been associated with budgets necessary for management or regulation of natural resources and thus depends a great deal on local social capital for successful implementation (cf. Hecht *et al.* 2002b; Ribot 2002). In short, political institutional dimensions of resource management is clearly important, but like the literature on property regimes, must be highly specified. The questions of social capital, as well as institutional structure and access lie at the heart of this debate.

History-ing the making (of forests)

The next approaches – the environmental Kuznets curve and the forest transition models are historical in nature, and view forest degradation as a phase in the development process. The central question that the Environmental Kuznets Curve (EKC) addresses is whether environmental degradation increases monotonically with rising GDP, or whether there is some point where it decreases. Inspired by Kuznets (1956) work on inequality, the EKC suggests that environmental damage increases up to a certain income point after which the degradation parameters decline as economies become more efficient, less resource based, and tastes change. Graphically this is represented as an "inverted U." In its simplest formulations it is meant to correlate with GDP, but modeling exercises increasingly incorporate a broader range of variables. Approaches to the EKC include "brown" studies that emphasize the dynamic of pollutants (Torras and Boyce among others) and "green" studies that specifically focus on deforestation (Bhattarai and Hammig 2001; Koop and Tole 2001, among others). One issue that needs to be underscored is that there is an analytic difference between EKC for externalities like atmospheric emissions and those for deforestation. The conflation in the EKC models for control of "brown" pollutants with "green" EKCs is problematic for a number of reasons. "Brown" externalities like sulfur emissions, urban waste, etc. are classic cases of externalities where traditional approaches for reduction, like taxes, regulation, pollution markets and technical change can emerge, and are not as institutionally complex or as ecologically unpredictable. Deforestation or forest maintenance and/or recovery, as we have seen, expresses a much more complex set of biotic, ecological and social processes. The socio-ecological terrain, really that of political ecology, focuses on the complex intersections of biology, land use, history, social context, institutional, economic, cultural, symbolic and power relations. Thus, the data for deforestation EKCs is much "noisier" at all analytic levels when forest trend and the EKC are analyzed.

The deforestation EKC

The deforestation EKC studies are based on panels of FAO forest statistics (often highly questioned by local field researchers), and then the trends of forest change are evaluated using a range of variables including population, population density, GDP, debt, institutional factors (such as enhancement of democracy) and policy factors to test whether the curves exist, and at what income level the inflection point occurs (Bhattarai and Hammig 2001; Ezzati *et al*. 2001; Koop and Tole 2001; Stern *et al*. 1996; Usivuori *et al*. 2002). Forest EKC modeling efforts, which involve all tropical countries where the basic data sources are highly questionable, do not easily assess "informal," successional and anthropogenic forests. For example, the data sets used for the Mesoamerican biological corridor recognize some 133 native vegetation types, but conflate all anthropic landscapes from cotton fields to agroforestry into one category, making the array of regenerating woodlands largely "invisible."

The final results of the models are contradictory: Bhattarai and Hammig (2001) focusing on institutional dimensions of clearing and EKC find that a deforestation EKC exists for Latin America, but its inflection point occurs with a mean income of $6,600, well above the mean income for Latin America of $3,500. Bhattarai and Hammig emphasize that political institutions (based on criteria provided by Freedom House data)[4] were the key elements of the model that explained the emergence of an environmental Kuznets curve. The impact of population (density and rates of growth) in their model was not significant when the political institutions and macroeconomic structures (in this case presence or absence of a black market in natural resources and the percent of debt) were specified.

Usivuori *et al*. (2002) also assert the existence of an EKC but links this to population density and income per capita and places the inflection point at around $2,500, although with many caveats. Koop and Tole (2001) on the other hand, find no EKC for Latin America – arguing that policy contexts and other factors affecting land use differ so much as to defy useful comparisons.

Critiques of the EKC point to problems with the data base on which assertions are made, comparability between countries and between types of ecosystems, and the statistical methods used to address the questions (Ezzati *et al*. 2001; Koop and Tole 2001, among others). Others cite the sanguine epistemology that devalues current environmental degradation as part of "growing pains" in the transition to a more ecologically sound future (Stern *et al*. 1996), in spite of the fact that serious, and perhaps irrevocable degradation might occur, as with extinctions associated with deforestation. Another problem with the EKC is that structural change is implicit, but not specified, presumably reflected in the rise in income. That is, the processes that lead to inflection, whether growth, redistribution, institutional change etc., remain a "black box."

The next historical model, that of forest transition, is really the flip side to the more widely known phenomenon of the urban transition. Its analysts, (cf. Mather and Needle 1998) argue that as countries become more urbanized, small holders abandon their agriculture and move to cities. Largely based on the history of the US and Europe, the forest transition model is intimately tied to ideas of long term structural change in the economy with permanent out migration. It remains informed by the Harrod-Domar two sector economy equations that informed a great deal of rural development policy throughout the twentieth century, and depends on a definitive differential labor market and more or less permanent out migration. In his study of Puerto Rico, Rudel *et al.* (2000) argues strongly in favor of a forest transition as an outcome of the aggressive industrial development via Operation Bootstrap, and as a result of the special relation with the United States which provided both labor markets and US "safety net transfer payments" which drew off labor from farm activities and raised its costs. The lack of migration barriers to the United States also was a key dynamic, as was the lack of competitiveness of PR agriculture. The ensemble undermined most agricultural production by devaluing rural production and by creating a competing and relatively highly paid labor market. In this approach, the rural exodus reduces rural population and agricultural enterprises and results in land abandonment. Yet he argues against the forest transition model in his study of Ecuador, suggesting both the dynamics of cattle and intensive small holder agriculture do not imply deserting the land, which is implicit in most forest transition models.

The EKC and transition models are useful because of their historical sweep, and because they stimulate questions about specification of the actual processes that produce the "inflection" and forest resurgence. They suffer from problems of over aggregation and data quality.

The final model is the "Equity" model. Torras and Boyce (1998) empirically analyzed the relationship between air and water qualities indicators that improve with income. Their study argued that it was not so much the dynamics of economic growth, but rather changes in the distribution of income – especially transfers to the "average citizen" that was most significant, because these empowered citizenry to demand better environmental management. Greater income inequality was associated with more pollution and less safe water at lower income levels, and the patterns continued at higher income countries. Expanding on this approach, Koop and Tole (2001) modeled the relationship between income inequality, growth and environment (deforestation in this case) by essentially linking the Kuznets curve with the environmental Kuznets curve to test "to what degree social welfare policy mediates the environmental effects of economic growth" that is, are the effects of economic expansion on the environment influenced by a country's distributive policies. It is, after all quite possible to raise the average GDP/capita without improving living standards of most of the population. In their model they used indicators of inequality such as the Gini coefficients for income and the Gini coefficients for land. Their findings showed that in countries with a high degree of income inequality, increasing GDP and deforestation was positively related, while in those countries with greater equality in their land distribution, GDP had only a marginal effect on deforestation. More equitable resource distribution resulted in lower rates of forest clearing. These results complement some of the findings of the political institutional impacts, and help explain why the deforestation EKC results are variable. These results are important because there are several situations in Latin America that appear to support this finding – the *ejidos* of Mexico, inhabited reserves of various kinds, and some situations of colonization. Indeed, studies of deforestation in Bolivia show that the smaller holdings have slower rates of deforestation than large ones (Hecht 2003; Steininger et al. 2001). But while the Koop and Tole model help us understand forest persistence, can it illuminate forest resurgence? Overall, how well do these models of forest persistence and resurgence explain the new patterns of forests in El Salvador, Central America's poster child for environmental degradation?

EL SALVADOR

In the popular consciousness, El Salvador is notorious for its degree of deforestation, as summarized in Terborgh's cheerless sentence. Numerous articles assert that only 2 percent to 5 percent of its forests remain, giving the impression of a blasted landscape with barely a tree in sight (FAO 2000; MARN 1999; Terborgh 1999)[5]. This view is faulty, and informed by the extension of older deforestation trends whose logic and dynamics no longer hold given the enormous structural changes in El Salvador's and the global economy (Hecht et al. 2002a). This view also misses several processes that have encouraged forest recuperation and overlooks the widespread anthropogenic and regenerating woodlands that are significant in their total area, and in their ecological and social impacts. While some might revile these types of forests, it is worth pointing out that the major "wild" national park – El Impossible – with its impressive biodiversity was a former coffee plantation, and has at least eight major archeological sites within it.

This section outlines the dynamics of forest recovery in El Salvador and reviews the global processes, national policies, local politics and varying ideologies of environment and environmental practices that now shape rural landscapes. The last section of the paper discusses how well the models on offer help explain these changes, and what they imply for conservation policy in the new global context.

Biodiversity in El Salvador

While often derogated for a lack of biological richness (FAO 2000)[6] recent ecological research in several fields (cf. Berendsohn 1995; Dull 2001; Komer 1998; Ramirez 2001) contradicts this inaccurate, and unfortunately, oft repeated impression. The view of "forest Free" El Salvador misreads forest trend, and devalues the importance of anthropogenic and disturbed forests which are significant for the maintenance of biodiversity and other ecosystem services. While conservation science often fetishizes "virgin" forests, there is some question in a region as geologically and biologically dynamic as El Salvador, whether the idea of "primary" versus "disturbed" ecosystems even makes sense as "categories" especially, given the antiquity of intensive occupation and the ubiquity of human impact over at least eight millennia. (See for example Bush *et al.* 1992; Daugherty 1969; Dull 2001; Flannery 1982; Pohl *et al.* 1996; Sharer 1978; Sheets 1979, 1982, 1984). Recent research on the Holocene vegetation history of El Salvador suggests that secondary vegetation types are the more "characteristic" formations of the region, and that the region has never been completely forested (Dull 2001).

Biological collectors and tropical ecologists generally prefer areas with extensive wild areas (good laboratories) and so have largely avoided El Salvador, with its dense population and dearth of facilities. With little international attention, and not much emphasis in local natural resources training due to twelve years of civil war (and decades of rural unrest), there is a ubiquitous impression that ES is depauperate in biodiversity which is reinforced by comments like those of Terborgh, and a general sneery attitude to the biotic gifts of the country (Myers *et al.* 2000). But recent studies of trees (Berendsohn 1995; Ramirez 2001) and birds (Komar 1998) have discovered so many unrecorded species that the old perception must be rethought. Amazingly, 580 bird species have been reliably recorded here, and another 75 are expected to occur. Northern Central America is a center for avian endemism and this suggests that other taxa will also be highly endemic (Komer 1998). The patchiness of the habitats themselves, and the types and the resource rich structure of anthropogenic forests have undoubtedly contributed to the maintenance of overall biodiversity. Coffee farms suit many generalist species including a large array of international migrants because of the permanent nature of the crop and its shade trees (Perfecto and Vandemeer 2001). The widespread planting of hedgerows, fruit trees, and extensive domestic agro-forests provide resource islands throughout the landscape. In addition secondary vegetation and arboreal diversity in abandoned pastures plays an increasingly important role in the maintenance of El Salvador's biotic complexity. While it is true that the charismatic fauna of Central America – jaguars, tapirs and some primates – are rare, the region does embrace significant diversity.

Table 3.1 compares the diversity of El Salvador with other countries in Central America. It is worth pointing out that "diversity" is a feature of ecosystems as well as an artifact of biological collection. Panama with its long association with the Smithsonian Institution, and Costa Rica's venerable tropical ecology study

Table 3.1 Biotic diversity in Central America (number of species per 10,000 km^2).

Country	Forest Area	Mammals	Birds	Reptiles	Amphibians	Higher Plants
El Salvador	167,000	106	365	57	18	1,956
Guatemala	4,253,000	114	304	105	45	3,638
Honduras	4,608,000	78	308	68	25	2,252
Nicaragua	6,027,000	86	322	69	25	3,003
Panama	2,123,000	112	477	116	84	4,618
Costa Rica	1,569,000	120	496	125	95	6,421

Source: World Resources Institute (1996).

site, La Selva, have meant that both places have undergone extensive taxonomic analysis and had thousands of students collecting specimens for more that 40 years The diversity of El Salvador is especially impressive given the "small" (indeed, underestimated) forest area of the country, and the prevalent view that it has almost no worthwhile forests. How then, does one explain such diversity in a country that is seen as having no significant forests?

Secret forests of El Salvador

Table 3.2 outlines land uses in El Salvador and provides classification and area of woody vegetation from three different sources.

Secondary forests of various ages and forms (which is often classified as "pasture"), and with different degrees of density cover more than a third of El Salvador. Pastures are either largely diverse silvo pastoral systems[7] since during the war cattle were largely sold off or eaten, beef imports from Honduras have sharply reduced local market prices, and the current level of banditry results in such high rates of cattle rustling that grazing without close supervision produces spectacular losses.[8] As such, secondary growth and advanced pasture successions represent the largest forest types in the country, and given the millennial anthropogenic and disturbance dynamics in El Salvador, secondary formations are probably its characteristic vegetation.

In spite of its reputation for "extinguished" nature ES has substantial forest cover. All the sources indicate that the country has at least 600,000–700,000 ha of tree cover of all types – at least a third of the country is forested, and in its montane zones – Chalatenago, Morazon, La Union, Cabanas the proportion may increase to close to 60 percent.

Deforestation drivers in El Salvador are now substantially different from those of the past and mainly include urbanization especially in the urban fringes of San Salvador, vacation and tourism in the mangrove forests and sugar cane with its protected markets on the coastal plain and Lempa valley. The factors that drove

Table 3.2 Post-1995 estimates of land use according to various sources (in Ha).

	MARN	Hecht	Komer
Forest	320,761	320,000***	
Secondary forests	–	300,000	
Mangroves	24,382	25,000	38,000
Coffee	230,000	170,000*	195,000
Plantations	–	7,000	7,000
Coco	5,000	5,000	1,314
Commercial orchards	46,863	35,450	
Domestic	–	50,000–100,000**	
Protected areas	–	28,000 (+48,000 ha to still be added)	
Urban forests	–	8,000	
Hortalizas	9,014	–	
Basic grains	499,000	300,000*	
Pasture	1,160,738	–	
Urban area	44,261		
Total forest	627,761		

Source: Hecht 1999; Komer 1998; MARN 1999.

Notes:
* MAG (1998).
** Estimates of forests in household orchards and agroforests, land demarcations and hedgerows.
*** Includes Pasture.

clearing before have shifted, and numerous processes now support the return of woodlands. These are outlined in the next section.

FACTORS OF FOREST RECUPERATION

Forest recuperation in El Salvador reflects several processes that are the outcomes of political and economic globalization, structural adjustment politics, and processes of democratization and decentralization. These include:

- the impacts of El Salvador's civil conflicts as they reflected hemispheric cold war politics. These had effects on the agricultural frontier, migration and agrarian reform;
- the outcomes of regional and international economic integration and trade liberalization on grain prices and the volatility of international coffee prices;
- the effects of structural adjustment policies on rural credit and subsidies, and the implementation of decentralization programs;

- the emergence of local and regional environmental politics as an outcome of political opening after 50 years of authoritarian regimes and decentralization.

Conflict and clearing

El Salvador was a "hotspot" in the cold war. A long history of civil uprising marked the history of the twentieth century in this country, but among the most severe episodes was the civil war that lasted from 1980–1992. The main impacts of the war on natural resource management involved its chilling effects on the expansion of the commercial and peasant agricultural frontiers, its role in stimulating international and urban migration (and later, the economy of remittances), its ensuing agrarian reform and its impact on holding size and the rural economic structure. While war should not be advocated as a resource policy, from the environmental perspective, these outcomes all tended to reduce pressures on forests and permitted large areas of cultivated landscapes to revert to successional formations.

Deforestation in El Salvador

In the period from 1960 to 1980, El Salvador began to diversify its export base and transform the structure of agriculture. In an effort to modernize the agro-export sector, extensive land uses like cattle, and intensive industrial, high input agriculture like cotton and sugar cane exploded (Conroy *et al*. 1996; Faber 1993; Paige 1999). This dynamic was stimulated by an ensemble of fiscal incentives that were reproduced in many Central American countries, including subsidized credit for land, animal and machinery acquisition, duty free equipment importation, tax holidays, land grants and infrastructure development. Protected and preferential markets further buffered these activities from external competition, and until the economic downturns of the 1980s, agricultural modernization created;

1   a powerful, corrupt agro-commercial elite;
2   concentrated land ownership among this coterie; and
3   marginalized an increasingly landless population who sought to produce their livelihoods on ever more precarious holdings at the mountainous agricultural frontier.

The combination of the export agricultural and marginal subsistence production generated an extremely aggressive deforestation frontier (Faber 1993; Paige 1997; Williams 1986). The array of institutional subsidies also produced immense distortions in the agro-export economies as owners sought institutional rents of various kinds and engaged in land speculation (Hecht 1994). The effect of the war can be usefully disaggregated to clarify why, now, forests return.

The end of the agricultural frontier

Impact on large producers

While Malthusian drivers are often invoked to explain El Salvador's deforestation in the 1960s and 1970s, other analysts have pointed to the mechanization of the traditional crops of sugar cane and cotton on the coastal plain and the central Lempa Valley. This reduced the demand for labor, and also marginalized share croppers and other types of informal access, leading to a structural dispossession. As these populations moved into the mountains, they were increasingly displaced by an expanding livestock frontier, stimulated by cheap credits and a variety of subsidies to the sector that characterized much of Latin American at the time (cf. Kaimowitz 1995; Leonard 1987; Paige 1997; Utting 1993). Because of the geographic dynamics of displacement, mangrove coastal zones and the mountain areas became strongholds of the FMLN. As war theatres, commercial activities were curtailed, and subsistence producers also were forced to abandon cultivation (cf Pearce 1986).[9] These skirmish zones were often left alone for more than 20 years, and the continued existence of ordnance in these sites can make them still very hazardous to clear. In the coastal areas, also controlled by the FMLN, the production of cotton, the land use that was most responsible for chemical pollution of the landscape and waters was also decisively curtailed. With its extremely high use of biocides of all types, cotton was central in contaminating local water bodies and in the pollution of mangroves. (Murray 1994; Williams, 1987).

The civil war had important effects on coffee cultivation, the historic source of wealth in El Salvador. As Perfecto *et al.* (1996) show, during the 1980s, throughout Central America, coffee producers began to switch from the ecologically sound and "biodiversity friendly" shade coffee to the more sun tolerant varieties being promoted through various development agencies because they were somewhat more productive and as a control measure for coffee rust. The war precluded this changeover, and as a consequence 85 percent of El Salvadorian coffee is grown under traditional highly diverse shade canopies, and is organic, since chemical cultivation and harvesting were also inhibited by the conflict in this period (Procafé 1998).

Campesino production

The small farmer agricultural frontier, based on the cultivation of maize and beans that had been progressively pushed into montane areas by land scarcity during the 1970s (Faber 1993; Paige 1999; Utting 1993) was also reduced in the guerrilla zones by the periodic military skirmishes, the threats and realities of civilian massacres, forced resettlement, migration and widespread instability and marauding. While some areas limped along, the expansion of small-scale cultivation was increasingly difficult.

The War thus had the effect of curtailing the changeover to sun coffee, the expanding agro-industrial, livestock and peasant agricultural frontiers at exactly the moment that both these exploded in other parts of Latin America and were significant drivers of deforestation (cf. Downing *et al.* 1992; Hecht and Cockburn 1989; Perfecto *et al.* 1996; Schuuman and Partridge 1989).

Migration, remittances and environment

Another significant effect of the war was the out-migration of roughly one sixth of El Salvador's population as the war and its human rights abuses accelerated. Internal and international migration increased sharply during the 1980s when war ravaged the countryside, and has continued due to the economic declines in agriculture in the 1990s (Lungo and Kandel 1999).

The problem with Malthus

Although only 3 percent of the national territory, 32 percent of the population resides in the Metro Area San Salvador, which experienced a 13 percent gain in population since 1971. The northern third of the country, absolute numbers of the population remained constant. In Southeastern El Salvador, the proportion of national population dropped from 28 percent to 20 percent, but its absolute numbers increasing its absolute numbers by 200,000. In the southwest, once the Metro area of San Salvador is excluded, the percentage of the population has remained constant since 1971, although in absolute numbers the population more than doubled, largely due to peri-urban development around San Salvador, and the Maquila industrial economy near the airport (PRISMA 2002). Thus, rural population remains at roughly the same densities as the high deforestation phase in the 1971. (See Figure 3.1.)

Roughly 2.5 million Salvadorans live outside the country, some 94 percent in the US (Kandel 2002). The expatriates are the main source – 66 percent – of El Salvador's foreign exchange. Table 3.3 shows magnitude of the dollars sent by households to their relatives and also illustrates the profound structural changes that have occurred in the agrarian economy. What is most noticeable is the collapse in tradition agro-exports, and the preponderance of remittances.

The "economy of affection" thus eclipses all other hard currency sources and accounts for 13 percent of the National GDP. The average amount of these remittances sent to rural households is roughly US $121/mo, about the equivalent of one minimum salary. Roughly one fifth of all rural households receive them. The average income per capita in El Salvador is US $1,990, (although rural incomes are roughly half that) so the average remittance contribution of US $1,452/year is a significant, and direct welfare subsidy.

The spatial distribution of remittances is also variable as Figure 3.2 suggests, ranging from a low in the areas of Ahuachapan – coffee production areas largely buffered from the war, versus areas like Morazan where the war, massacres,

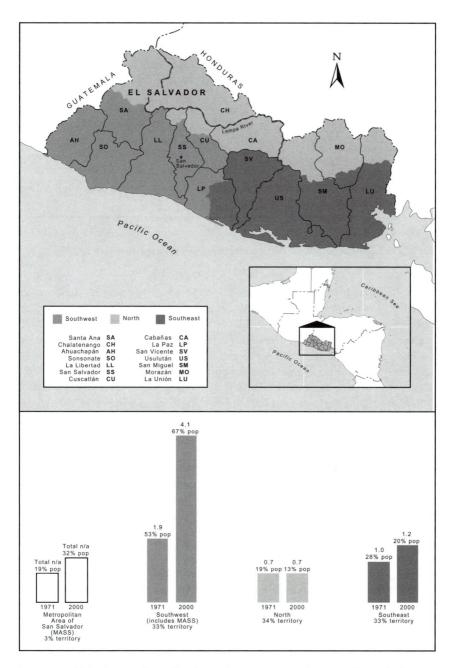

Figure 3.1 El Salvador: population distribution by zones, 1971 and 2000 (millions of habitants).

Source: PRISMA, based on population census.

Table 3.3 El Salvador: changes in the primary sources of foreign exchange, 1978 and 2000.

	Millions of dollars		% of traditional agro-exports		Structure (%)	
	1978	2000	1978	2000	1978	2000
Traditional agro-exports*	514	292	100	100	81	11
Non-trade exports outside Central America	54	145	11	50	8	5
Maquila (net income)	21	456	4	156	3	17
Remittances	51	1,750	10	599	8	66
Total	640	2,643			100	100
Total excluding remittances	589	893				

Source: PRISMA (2002) based on data from the Central Reserve Bank of El Salvador.

Note: *Coffee, cotton, sugar, shrimp. The table does not include exports to Central America.

resettlement and economic collapse fed large migration flows. Remittances correlate with the zones of forest recuperation: the mountain zones of Chalatenango (25 percent households get about US $125/mo), Cabanas (28 percent about US $100/mo), Morazan (30 percent about US $135/mo), La Union (41 percent US $145/mo), and northern San Miguel (28 percent, US $125/mo). The effect of remittances has been to buffer the incomes of the poor so that relentless environmental exploitation is less necessary, and to permit the purchases of food, (so that it does not need to be grown) medicine and improved shelter and education (Lungo and Kandel 2000).

The collapse of agricultural prices meant that the costs of production often exceed their return. In a series of group interviews on the impact of remittances in rural households carried out in Chalatenango and Ahuachapan, respondents were unanimous in their assessment that many such households had stopped cultivating, and would rather wait for remittances (Hecht *et al.* 2002a). The effect of declining grain prices worked against peasants as producers, but benefited them, through low prices, as consumers with cash from remittances wages. This has, however undermined food security.

Agrarian reform

Agrarian reform has been one of the central political questions in Central and Latin America throughout the last century. Indeed, the roots of the civil war were inflamed by demands for equitable land distribution. Redistributive agrarian reforms in El Salvador have undergone several distinct phases, each with a specific set of legal characteristics. Two main reform periods are especially relevant to our concerns: the Agrarian reform of 1980 that was put into place as an effort to stave off the war, and the PTT (Land Transfer Program) of 1992 inaugurated with the

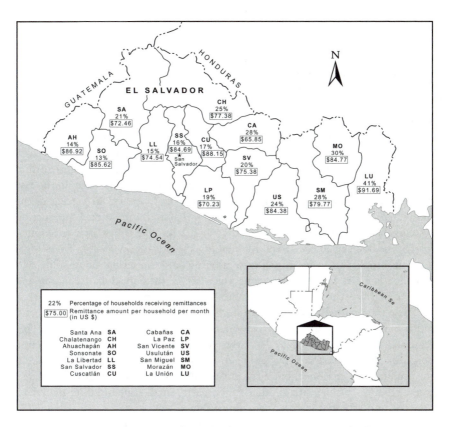

Figure 3.2 El Salvador: percentage of households that receive remittances by department.

Source: PRISMA.

Peace Accords. While the Agrarian reform enabled collective property through cooperatives, the PTT beneficiaries took advantage of the "proindiviso" phase to define and legalize common areas. In both these transfers, land could be held in single plots, but provision were also made for collective holdings and communal management.

The agrarian reforms of the 1980s distributed 295,000 ha among 84,000 beneficiaries. In the early nineties 106,232 ha were distributed amongst 36,597 beneficiaries – mostly ex-combatants – under the Land Transfer Program (PTT) that was part of the 1992 Peace Accords. Thus, in total 401,232 ha were distributed – a fifth of the national territory – to 120,597 beneficiaries and 25 percent of rural households received land under the programs.

The agrarian reform has had four main effects on natural resources. First, at the social level, the land transfer processes opened up the possibility of new forms of tenurial and territorial organization, which in some cases have lead to collective decisions about land, including natural resources management such as forest management, fire control, watershed councils and hunting controls. Local forms of regulation and political lobbying thus emerged from the history of revolutionary and solidarity organizations. Second, some agrarian reform cooperatives and the PTT communities have received technical assistance and funds as part of international cooperation programs. As a result of institutional interactions with NGOs that emphasized environmental ideas and sustainable development projects, such as reforestation, agro-ecology and resource protection, environmental awareness in rural communities has increased.

Third, the redistributive structure itself has created a different agricultural landscape that is much more multi-use and multi-purpose in its focus. In the case of coffee, for example, small farms of less than 7 ha represent 80 percent of the individual farms (about 18,500) and are integrated into mixed production systems that supply coffee as well as other ancillaries, like fruits, artisanal inputs, forage, medicines and fuel-wood for cooking (Méndez *et al.* 2001). These plots have an important role in self provisioning and buffering households and extended families from the economic volatility of the Salvadoran economy and create a highly structurally and ecologically diverse landscape mosaic.

These fragmented holdings mean higher diversification at the plot level, and much more "inertia" for land transformation at the scale of the landscape. This tenurial "patchiness" blocks the large scale clearing more typical of large holdings. This is illustrated by Table 3.4, where the most fragmented holdings in the producer association embrace the greatest diversity since a range of cultivated forest products such as citrus, other fruits, posts, small scale building materials, firewood and artisanal materials are also produced.

Another significant effect of the agrarian reform and the war was to wean El Salvador's economic elites away from the rural economy. The Peace Accords mandated expropriation of many large agro-industrial rural holdings for agrarian

Table 3.4 Diversity characteristics of coffee shade: three small farmer tenurial systems.

Type of cooperative	Tree abundance/ha	Total species diversity/ha	Diversity/ parcel	Management/tenure
Reform Sector*	390	74	12	196 Ha Collectively managed
Traditional Cooperative**	350	51	12	Collective (31 Ha) and private parcels
Producer Association	900	110	22	Private parcels

Source: Méndez *et al.* 2001.

Notes:

* Cooperatives following the 1980 law that ceded holdings to cooperatives formed of their old workers, and are profit sharers in the collective returns.

** Formed from the 1983 FINATA law. Includes collective management and returns to coffee plus private holdings.

reform, while maintaining the security of urban and financial assets of national elites. This, along with the triumph of the conservative Arena party in national politics, helped create an economic and political class with virtually no interest in (and perhaps a political resentment of) rural development questions except as they might effect hydropower and drinking water. As rural areas became less important for economic accumulation, they also were less important in policy and as sites of investment. El Salvador's national economic focus in the 1990s emphasized its financial and industrial policies – in line with structural adjustment strategies that embraced free trade and slashed rural subsidies and credits. After a long agricultural history, El Salvador embarked on an urban based and urban biased development model even though more than half its population remained in the countryside. Within that set of approaches, cheap food policies remained a central strategy to reduce urban unrest, and was largely achieved through trade liberalization with the Central American Common market and not through increasing national production, a situation that sharply undermined small farmers.

Economic integration: global prices and rural commodities

El Salvador is a relatively small country which has always emphasized export led development. It is integrated into the Central American Common Market and the Initiative for the Americas, and partakes of the trade treaties that minimize tariff barriers. The impact of global prices for two key commodities – grains and coffee – have had important effects on rural production and natural resources use. As Figure 3.3 reveals, agriculture's relative prices have been in sharp decline since the 1970s. The dynamics in the grain and coffee economies are different: one has simply been affected by the downward trend in grain prices through the elimination of tariff barriers and cheap food policies, while the other is buffeted by the high volatility and expanded competition in the international coffee market. This global integration significantly undermined the importance of the rural economy in the national GDP.

Grains

During the 1980s agricultural production of all kinds stagnated due to a general downturn in global markets, El Salvador's problems of debt burdens and of course, the war (Conroy *et al.* 1996). Food imports were the norm due to production constraints in the countryside and as a consequence, food import infrastructure and distribution systems became quite well developed. The 1990s witnessed relatively slow annual rates of growth (1.2 percent) overall, because the grain sector was hampered by cheap food imports which were necessary to calm urban political pressures in the post Accord period. The price for grains today is a mere 27 percent of the real value of production in 1978. The impact of these low prices can be seen in the structure of rural incomes which are shown in the next

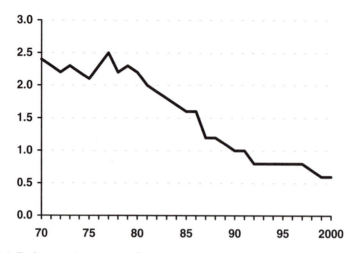

Figure 3.3 Decline in relative prices of the agricultural sector, 1970–2000 (GDP agricultural price index/GDP price index, 1990 = 1) (National accounts base 1990).

Source: PRISMA based on data from the Central Reserve Bank of El Salvador.

table. Sales of grains comprise only 5 percent of rural income even though almost 90 percent of all households produced them (see Table 3.5).

The total *area* in grain decreased. Table 3.6 indicates the magnitude of these crop declines except for sugar which has a protected market and subsidized credit reflecting the political power of the refining owners. Since about 90 percent of rural households grow maize, the contraction of the area by 18 percent indicates a sharp withdrawal from market production in this sector. Overall, the area devoted to producing food crops declined by 15 percent. If grains alone are calculated, the area in production has declined by 28.6 percent. More dramatic has been the contraction of pasture which has reverted to levels of the late 1960s.

Agricultural wages

The key impact of cheap food policies has been to drive down the value of agricultural production and wages. That a decline of this magnitude did not explode in social unrest speaks to the buffering of this sector by the deep semi-proletarianization, access to land for self provisioning as an outcome of agrarian reform, and the impact of remittances. This ensemble however, coupled with its stagnant wages has made the sector undynamic, a factor that works against investment in it, and thus has further constrained its expansion with the exception of sugarcane which receives subsidized credit and a protected market.

This agricultural contraction can be partially explained by price signals and the relentless capital scarcity in this sector, but several other social factors also

Table 3.5 Rural household income sources (%).

Type of activity	%
Agricultural sales	38.6
Grains	5.3
Other agricultural products	3.4
Animals	11.5
Other agricultural income	0.4
Sale of labor in agriculture	18
Non agricultural income	52.3
Work away from farm	44
Other income	9.2

Source: FUSADES (1998).

Table 3.6 Areas planted in agricultural products, 1980–1997 (1000s of hectares).

Crop	1980	1990	% Change 1980–1990	1997	% Change 1990–1997	% Change 1980–1997
Coffee	180	165	−9.1	165	−	−9.1
Sugar cane	57	44	−23	80	+81	+40
Maize	291	261	−11	239	−9	−18
Beans	52	62	+19.2	63	+1.6	+21
Rice	16	14	−13	8.4	−40	−52
Sorghum	119	128	+7.5	101	−22	−16

Source: Central Reserve Bank (1982, 1992, 1998).

played a role. These included the effects of male migration on the gender division of labor, and the declines in availability of family labor. Finally, the changing aspirations and exposure to mass media have made farming less attractive and detracted from its precarious prestige. If one is going to proletarianize oneself, it makes more sense to do so away from the low wage rural zones, and by preference in the US where pay rates are higher. Table 3.7 illustrates the profound rural wage stagnation compared to other employment sectors in the economy. Thus, while grain farming is still important for food security and as a buffer against the vaguaries of the economy (and thus grown by most households), the grain frontier as an *economic* frontier is no longer as significant in deforestation. Grain production might more usefully be seen as domestic horticulture and provisioning rather than a commercial enterprise.

Table 3.7 Real wages by sector (1992 national average = 100).

Sector	1992	1995	1998	2000
Agriculture	53	50	51	50
Industry	98	104	126	108
Construction	101	108	120	123
Commerce	107	112	127	130
Services	108	124	124	129
Transportation/Communication	165	191	169	185
Finance	174	242	206	186
Utilities	182	164	192	226
National average	100	110	124	125

Coffee

Global coffee production has expanded enormously since the late 1970s because virtually every tropical development project in the world now includes coffee in its suite of rural commodities. El Salvador' s coffee sector has declined throughout much of the 1990s reflecting structural change in agriculture (as agrarian reform shifted the sector into many small holdings and cooperatives) global price volatility, a shift from plantations into housing lots for urban development, contraction of credit, and the continuing problems of banditry, extortion and theft of product in rural zones. While Salvadoran coffee can participate in "fair trade" markets, the caliber of most of the product limits its price in elite markets.

One of the historic sources of wealth in Central America, coffee remains critical to the rural economy and environment of El Salvador. Coffee forests, cover about 9 percent of the country, and of these roughly 80 percent are the high biomass, high diversity shaded coffee. Coffee is the largest generator of foreign exchange in agriculture and a significant source of seasonal direct employment (about 30 percent of the wages income in the agricultural sector generating some 120,000 full and seasonal employment, and roughly another 26 percent of the jobs in agroindustries (MAG 1998). About 88 percent of the coffee farms are on approximately 40,000 ha, (about 25 percent of the total area in coffee) and average 2.6 ha. The economic condition of these owners precludes shifts to sun grown coffee, and minimizes the use of biocides in small holdings

The decline in prewar deforestation pressures, the impact of remittances, semi proletarianization and agrarian reform coupled with a downward trend in agricultural returns substantively change the physical, tenurial and structural role of agriculture in the rural economy. These contributed to forest resurgence, a process enhanced by the national policy milieu.

Structural adjustment programs

Structural adjustment programs (SAPs) were put into place with the Peace Accords. The economic package included trade liberalization, reduction of subsidies and decentralization policies. SAPs, in this case had two essentially beneficial effects on environment, although, as we have noted, it was not uniformly wonderful for rural producers. The first was trade liberalization which provided agricultural commodities, like grain and beef, at low costs. This undermined national grain production, reduced rural incomes and transformed the role of agriculture in the national economy. The second was the elimination of fiscal incentives, subsidized credit as part of the Peace Accord adjustment programs. These had the effect of: 1) virtually eliminating high input cotton which had been critical in contaminating mangrove ecosystems and other aquatic systems; and 2) unraveling the highly subsidized livestock sector (which also suffered from rustling during and after the war, and later from cheap beef imports from Honduras). The sharp contraction of these two land uses eliminated the two subsidized systems that had the most impact on pollution and especially on direct and indirect deforestation. While the state was not entirely against subsidized credits to other sectors, such as the Maquiladoras – the manufacture assembly plants – rural areas were under severe austerity programs and sharp import competition. Agrarian debates and fiscal transfers to agriculture were off the agenda whether through policy, indifference or revenge on the part of governing elites after the Peace Accords.

Decentralization and the new environmentalism: formal and social conservation

Decentralization was also an important policy of Structural adjustment programs. The goals of administrative decentralization were to dismantle the power of central state, increase local accountability, and enhance administrative efficiency.[10] While some finances accrued to the municipalities – a mere 6 percent of the national budget – the reality is that decentralization, especially within poor rural areas, often produced an economically impoverished, isolated and thus quite powerless local government. This transfer of governance however, set the framework for the development of new local political arenas where an emerging environmental language and approaches to rural development could be evolved, largely through NGOs and civil organizations.

This greening of national politics was reflected at all levels: from the President Flores five year plan: *Plan de la Naccion* (GOES 1999) which emphasized that any development would have to take on environmental issues in a substantive way, to the emergence of the peasant "Socioambiental" – social/environmental – coalitions such as CACH (Comite Ambiental Chalatenango – the Chalatenango environmental Committee), an outcome of the end of 50 years of authoritarian regimes and decentralization. In this new context, approaches to environmental governance tended in two contradictory directions: one focused on community resource management, informed by the sustainable development approaches, the other on a highly privatized model of conservation.

The idea of sustainable development and increasingly ecosystem services tended to be linked to reconstruction after the civil war and focused on community initiatives. CACH, one of the earliest of such coalitions developed in the mountain areas of Chalatenango, a former rebel strong hold. Their approach is based on collective environmental protection (such as fire control) regional forest management, watershed councils, etc. and emphasizes broader territorial coordination and planning among the municipalities through the development of mancommunidads – associations of adjacent municipalities – that are more appropriate for resource management and strengthens the political powers of the municipalities (CACH 1999). CACH is a model that has inspired other regional programs. Its significance lies in its coalitions that produce regional rather than isolated municipal strategies, and that mobilize the social capital produced by the war and by reconstruction after it. This community and collective approach contrasts with the privatization of conservation embraced by international conservation agencies and the Ministries involved with natural resources, MAG (Ministry of Agriculture) and MARN, (Ministry of Environment and Natural Resources).

The "formal" side of the new El Salvadoran environmentalism, produced a conservation strategy endorsed by the central state that focused regional (in the sense of Central American) conservation and international monies. Emphasizing National Park of El Impossible, (which is not run by the state but rather by a private NGO, Salva Natura, that is funded by subscription, grants and endowments), and on the private coffee forests, this park and its surrounding "corridor" is now the organizing principle for much national natural resource funding (GOES 1999). Financed by the (GEF) Global Environment Facility and CCAD (Central American Commission on Environment and Development) as well as other donors, funds will mainly go to developing technology packages, certification programs and international markets. This however has produced a disjuncture between "social" versus "formal" conservation in the country. By privileging privatized forms of conservation, innovations that could affect a great deal more territory than the 4 percent of El Salvador likely to benefit from the MBC face a funding and policy void. Indeed, the areas where forest resurgence is most marked, and where the social capital is most fully elaborated, the eastern mountain zone, is largely a "policy orphan."

IMPLICATIONS: THEORIES AND MODELS OF FOREST PERSISTENCE AND RESURGENCE

The case of El Salvador has several implications that extend well beyond the boundaries of this small country. The country, unexpectedly, presents phenomena of forest resurgence. In terms of understanding the theories of forest cover change, this study illuminates the strengths and weaknesses of the models on offer. We review these, and how well they explain forest return, and what these might in the end mean both for theory and policy. This next section reviews the cases model by model.

Malthus and markets

First, in spite of the fact that 50 percent of its inhabitants still reside in the countryside, and its rural population densities are over 150/km2, forests are making a comeback except in periurban areas near Metropolitan San Salvador, second housing development on the coast, and subsidized cane farming on the coastal plain. This situation throws into question some of the long held Malthusian ideas about population, population density and resource degradation. While many scholars have argued that a more complex analysis must be applied to resource questions, as we noted early in this paper, El Salvador has historically been the target of exceptionally simplistic discussions of population and resources. What we see now are high rural population densities *and* forest resurgence. A similar process has been noted by Klooster (2003) and by Rudel *et al.* (2001).

Markets

Markets in El Salvador are deeply globalized, and the way they operated was at variance with the way markets have been thought to affect natural resources in most of the models of deforestation. Historically, the action of markets was reflected in the demand for products (that produced deforestation), or alternatively, "greening" as an outcome of substitution either less damaging production techniques or forest products (that stopped deforestation). In most agricultural contexts, as commodity prices drop producers either intensify, extensify or migrate. In this case there is migration (although mostly it was stimulated by the war), but the sectoral volatility and low returns coupled with a national emphasis on cheap food policies produced an agricultural *retraction*. This is possible only in a situation of profound semi proletarianization.

This case suggests that the impact of *labor markets – domestic and international –* may be especially significant in forest resurgence. Labor markets have been largely invisible in discussions of the impacts of markets on clearing. The impacts of globalized markets on land use thus need to be viewed through the lens of economic portfolios. The dynamics of this retraction were reinforced by neoliberal agricultural policies that starved the rural sector, and banished tariff barriers for agricultural products. On its own, however, the collapse in prices would not necessarily reduce agriculture, especially for small farmers, if other elements were not in place. Subsistence agriculture would have continued if farmers had no purchasing power, but, in this case, they did, through wages and remittances.

The dynamics of these markets should be understood as an outcome of globalization and trade liberalization that contributed to agricultural retraction, expanded labor markets and the transnational and national flow of remittances.

Livelihoods

Forest persistence models have emphasized the importance of livelihoods in forest conservation, and this approach was one of the major conservation policy innovations of the 1990s. What are the implications of the livelihood explanation for forest resurgence? Rural populations in El Salvador are highly dependent on anthropogenic forests for firewood, whether coffee plantation prunings, hedge rows, living fences, demarcations, etc., for fodder, fruits medicinals and so on. The high diversity of Salvadoran coffee plantations with their array of native trees provides a regular source of firewood and a wide variety of ancillaries. Studies of plot diversity and households show that anthropogenic forests are important, and that the "livelihood" model does help explain the continuous planting of hedgerows, demarcations and highly wooded patio forests, the "domestic forests" in Table 3.2 (cf. Chazden and Coe 1999; Hecht *et al.* 2002a; Méndez *et al.* 2001). The modest home gardens and hedgerows are an important shadow element of rural livelihoods, a subsidy from nature. And they actively managed elements of the rural landscape. So the livelihood aspect is certainly part of the explanation.

Institutions: property politics and social movements

Property regimes have both stimulated forest clearing and forest persistence. But how might they affect forest resurgence? As mentioned earlier, much of the plantation literature takes private property regimes as the *sine qua non* of forest recovery, and most conservation set asides are use exclusionary property regimes to shape the form of ownership. In El Salvador, the main park is in fact owned by a foundation. So it is certainly the case that private property regimes can produce forest persistence and forest recovery. Collective properties, such as indigenous reserves and *ejidos* have also maintained and managed forests on the basis of livelihood arguments and those of cultural identity.

The emergence of Mancommunidads in El Salvador that specifically emphasize forest recuperation in their management point to the broader impacts that social movements and agrarian reforms can have on landscape recovery. While much is made of the form of property regime, it probably matters less, at the end of the day than the socio-political institutions that surround it. In this case, agrarian reform, which enhanced the equity in rural resource distribution, political decentralization which provided an improved local accountability in political institutions, and the social capital and solidarity developed in the mountain zones during the years of civil war seem to support Torras and Boyce (1998) and the Equity model of Koop and Tole (2001) and which illuminate some of the dynamics of forest resurgence. In terms of conservation politics, a wide variety of property institutions are on offer, but only one – set asides – seems to infuse policy.

Transitions and Kuznets curves

There is no question that structural change in the economy had a significant impact on forests, but this change in the case of El Salvador was not the outcome of endogenous processes of development, but rather an abrupt outcome of being a cold war hotspot. The war simply collapsed the rural economy and forced urban migration. But the Accords also were significant for structuring the new industrial economy. Since they stipulated that urban and financial assets would not be expropriated, and these sources of considerable wealth became the focus of El Salvador's national development policies as economic elites embraced an industrial and financial development agenda, leaving the rural sector starved of capital and state interest, at the mercy, for good or for ill, of new civil institutions. What does this say about environmental Kuznets curves and the forest transition?

Under conditions of globalization and the ubiquitous implementation of structural adjustment policies – especially those pertaining to increased economic openness in a context of regional integration, resource dynamics must be contextualized by understanding how forest trend and land use are affected by household strategies and the politics of localities in the face of global pressures. The El Salvador case differs from most studies of the forest transition in that these rural areas are not forsaken but remain occupied, and often form part of complex transnational and circular migration networks, unlike the history of the US and European transitions. Rural areas have become in many ways a redout.

The environmental Kuznets curve for El Salvador, if one can speak of one, reflects globalized commodity and labor markets, enhanced rural equity, improved political access and accountability at the local level, and adequate social capital. It also reflects another significant global process: the rise of environmental ideologies and projects.

Ideologies of environment

The explosion of international environmental activities by multilateral and international conservations agencies has characterized much of the rural politics of the last decades. These were reflected at the institutional level through tied lending, national activism and greater attention to environmental institutions in national politics. The thrust of rural development as it now evolves is as much an environmental as an agrarian/social project. Landscape recuperation, conservation planning, enhanced agroforestry development, and Agroecological projects increasingly define the "Nature" of rural approaches. Environmental services are thus often seen as viable and valuable rural options than producing agricultural commodities. International investment in national environmental agencies, in conservation planning and in the hordes of NGOs engaged in projects has been one of the main means of transferring funds into rural areas. In Central America this trend increasingly dovetails with the rise of new social institutions as a result of the politics of reconstruction and decentralization.

This was a profound deflection in the rural development discourse away from an agrarian equity to ecological issues, and took on five main characteristics:

- The rise of the rhetoric of sustainable development throughout the 1990s as a product of international post war assistance and recuperation throughout war ravaged Central America. This approach was especially important in El Salvador due to its reputation of ecological ruin, and devastating social impacts of environmental hazards including earthquakes, floods, hurricanes and ENSO (El Niño) related drought whose effects were exacerbated by inappropriate land uses.[11]
- The increasing recognition by the nation of the importance of forest resources in regulating water flow and quality for electrical power and for urban industrial development, and thus the rise in the importance of the idea of environmental services.
- Local social movements began to organize for watershed councils to enhance forests for water flow and to minimize landslides.
- Regional integration in biological terms, through the Meso American Biological Corridor.
- The increased availability of funds for the development of environmental institutions and projects.

Within this post war framework, the agrarian question had been construed increasingly as an environmental one, and one of the few activities where funds would be transferred to rural zones. This "redefinition" of the rural, which also required local organizing for community development has infused virtually all of El Salvador's rural development projects and organizations. Agriculture programs have been displaced to some degree by environmental investment, a process that is occurring more widely (cf. Padoch *et al.*, 1999).

The dynamics of forest resurgence in El Salvador suggest that the directions of forest trend need to be understood in a more complex manner, one that integrates levels of analysis ranging from household strategy, regional to global processes. In this case, one can argue that the impact of globalization – on commodity, labor markets, capital flow (remittances, aid monies) and environmental ideologies produced a general contraction in the agricultural frontier. But national institutional changes such as those associated with agrarian reform that enhanced equity,[12] decentralization, the rise of local environmental groups and the political development of Mancommunidads are also significant.

These kinds of changes – the emergence of substantial areas of anthropogenic forest – need to be incorporated into conservation approaches far more explicitly than they have been. Forest regrowth is occurring elsewhere in Latin America (cf. Klooster 2003; Rudel *et al.* 2002), and yet with the exception of a few case studies, the social ecology, diversity dynamics and political economy of these areas are not well known. These areas, with their substantive contributions to environmental services of all kinds – conservation of biodiversity, protection against natural hazards, carbon sequestration, water provisioning, soil protection and aesthetics are "policy orphans" in today's environmental focus.

Rural inhabitants and the forests they manage, unfortunately, are not viewed as priorities, because these resources are not considered "high value," in spite of relatively little research on the diversity dynamics of anthropogenic forests, in spite of the fact that most of the parks in Central America are in historically dense sites of human occupation and thus not "pristine" in the least (see Denevan 1992).[13] Conservation practice has not really addressed the importance of matrix ecology in anthropogenic environments. As Vandemeer and Perfecto (1997, 2001) among many others argue, this is a major lacuna in research. We argue that it is also a major policy void.

Matrix ecology Mancommunidads and the future of environments and environmental policy

One of the central points of the emerging discipline of fragment ecology is that the structure of the ecologies surrounding conservation ecosystems is central to the biodiversity dynamics of the remaining forests. Forest fragments in the world are found in complex matrices: coffee plantations, diverse tropical fruit orchards, manipulated and secondary successions, wood lots, multicropped milpas, agroforestry systems, abandoned or weedy pastures, door yard gardens, hedgerows. The dynamics and structure of matrices goes far to explain why many highly endangered ecosystems do not experience the kinds of extinction rates that are predicted by theory (Gillespie *et al.* 2000; Whitmore 1997). In these areas, human actions of soil improvement and selection for edible foliage, fruit bearing or seedy vegetation often enhances the value of these anthropogenic ecosystems for animals and can act as nurseries and havens for forest species (Andren 1994; Chokkalingham *et al.* 2001; Estrada *et al.* 1986; Harvey and Haber 1999; Komer 1998; Whitmore 1997). Data from Central America increasingly shows that the diversity in these sites is often significant (cf. Gillespie *et al.* 2000; Greenberg 1996; Harvey and Haber 1999; Perfecto *et al.* 1996; Perfecto and Vandermeer 2001). Thus it may be human impacts on age and heterogeneity of habitats, seral complexity and enrichment of matrices might be on par with "naturalness" in generating diversity at a landscape level. Yet these kinds of forests, from hedgerows and gardens to community successional forests are still largely invisible to conservationists and thus are largely ignored in policy and programs. More critically, this geographic fabric is also a matrix of social processes that have enormous potential to sustain environmental integrity over time.

The larger problem has to do with the conservation discourse and its relentless focus on large parks, devoid of people as the primary conservation emphasis, a discourse that in the end ignores the "secret" forests of El Salvador, by refusing to see them as forests at all. This is not a question of semantics, because in the context of extreme bias against rural economies and no agrarian policy other than neglect, this "invisibility" places peasants and their landscapes in a regressive and short sighted policy vacuum within a political economy of exclusionary environment-alism. As long as anthropogenic and regenerating forests are "invisible" as conservation entities, they are potential sacrifice zones without environmental support. Certainly, these ecologically and socially important landscapes deserve more policy attention that supports their positive impact on biodiversity and environmental services and recognizes the populations that increasingly manage and create these ecosystems.

THE CONTEXT AND CONTRADICTIONS OF CONSERVATION

Central American environmental approaches have historically been overwhel-mingly dominated by conservation ideologies that emerged from ecologists associated with the US. Many influential researchers emphasize "set asides" as the key approach to conservation (See for example Sanderson and Redford 1997). The impact of these researchers on resource policy in Central America is significant because of their prestige, their capacity to shape the international discourse about environment, and to mobilize international economic resources for national ecological projects (such as national parks). El Salvador barely counts in their scheme of things.

Their position has several implications. First, rural Salvadorans remain very poor, and without economic transfers and market products that support peasant incomes, they have increasingly focused on activities to enhance their ability to mobilize environmental funds as parts of international cooperation, green trade and fair trade initiatives, and through payment for environmental services. These efforts in the context of the decentralized state of El Salvador have produced formidable organizations for regional resources management as well as more local efforts to manage water courses, forests and biodiveristy through a range of collective agreements. The mobilization of such social capital creates good perspectives for success in these kinds of environmental enterprises (Gibson *et al.* 2000). This very positive trend in approaches to resource management runs up against serious bias, not just from Salvadoran ruling elites who remain indifferent to the rural poor although concerned about the environmental services they generate, but also from the international environmental community unwilling to invest money, effort and policy development in anthropogenic forest ecosystems other than those associated with coffee or indigenous people. The dynamics of forest trend are much more complicated now than they were a decade ago, and the importance and complexity of different types of secondary, anthropogenic forests, and their role as key elements in conservation remains still largely unrecognized.

The other issue is that while conservationists may have been emphasizing "hotspots" and conservation of selected forest fragments, they have missed, dare one say it, the forests for the trees.

NOTES

1 Consultive Group on International Agricultural Research.
2 Resettlement programs for natural resources development like Dams and for conservation set asides have fallen far short of the promises made to local peoples, and further aggravated the resistance to parks.
3 Again: research from virtually all forest types suggest substantial human impact in the past.
4 This is data that analyses political rights and civil liberties – and are based on 12 different freedom related criteria mainly focused on the electoral process, and 25 indicators of civil rights such as free presses, independent judiciary system, rule of law, freedom of assembly, etc. (See Scully 1992.)
5 Even in the early 1900s El Salvador was thought to only have 10 percent of its "intact" forests (USAID 1985, Daugherty 1969).
6 These studies tend to be overviews based on statistics and compendiums lacking in field work on location, and thus are likely to repeat "received ideas."
7 El Salvador's silvo pastoral systems have not been extensively analyzed, but a detailed survey in Costa Rica showed that 190 of the 360 species that occurred in the area were found in pastures (of which 60 percent were primary forest species). Ninety-four percent of the species in pastures were known to be used by birds, bats and other animals (Harvey and Haber 1999).
8 This dynamic has also been noted elsewhere in Central America by Kaimowitz (1995).
9 Pearse provides an exceptionally moving oral history of the history and war in Chalatenango.
10 The debates over the efficacy of decentralization and natural resource management remain extremely complex, since this form of local governance was implemented throughout the developing world and its results have been mixed. For further discussion see Hecht *et al.* (2002a), Ribot (2001).
11 Many of the most devastating land slides such as that of Sta Tecla on the outskirts of San Salvador were the outcome of faulty road and urban construction.
12 The Gini coefficients for El Salvador are more equitable than all other Meso American countries with the exception of Costa Rica.
13 El Impossible and its adjacent areas was a coffee farm until 1992. Neumann (1999) has noted that many conservation sites were often agricultural landscapes before they were designated "wilderness."

REFERENCES

Aide, T., Zimmerman, J., Pascarella, J., Riveira, L. and Marcano-Vega, H. (2001) "Forest regeneration in a chronosequence of tropical abandoned pastures," *Restoration Ecology*, 8: 328–38.
Andren, H. (1994) "Effects of habitat fragmentation on landscapes with different proportions of suitable habitat: a review," *Oikos*, 71: 355–66.

Anderson, A. (ed.) (1995) *Alternatives to Deforestation*, New York: Colombia Press.

Andrade R. and Andrade, T. (1994) "Sustainable use of the rain forest: evidence from the avifauna in a shifting cultivation habitat mosaic in the Colombian Amazon," *Conservation Biology*, 8 (4): 544–54.

Angleson, A. and Kaimowitz D. (2001) *Agricultural Technologies and Deforestation*, London: CAB.

Barbier, E. (1997) "Introduction to the environmental Kuznets curve," *Environmental and Developmental Economics*, 2 (4): 369–81.

—— (2000) "Links between economic liberalization and rural resource degradation in developing regions," *Agricultural Economics*, 23: 299–310.

Bebbington, A. (1997) "Organizations and intensifications: campesino federations, rural livelihoods and agricultural technology in the Andes and Amazon," *World Development*, 24 (7): 1161–77.

—— (1999) "Capitals and capabilities: a framework for analysing peasant viability, rural livelihoods and poverty," *World Development*, 27 (12): 2021–44.

Berendsohn, W. (1995) "Investigaciones Botanicas en el parque 'El Impossible'," *Jardin Botanico La Laguna*, Informe Tecnico, 5.

Bhattarai, M. and Hammig, M. (2001) "Institutions and environmental Kuznets curve for deforestation: a cross country comparison," *World Development*, 29 (6): 995–1010.

Bush, M., Piperno, D., Colinvaux, P., De Oliveira, P., Kriisek, L., Miller, M. and Rowe, W. (1992) "A 14,300 paleoecological profile of a lowland tropical lake in Panama," *Ecological Monographs*, 62: 251–75.

CACH, Comité Ambiental de Chalatenango, (1999) Plan departamental de manejo ambiental, PADEMA. Bases para el desarrollo sostenible de Chalatenango, El Salvadore: PROCHALATE.

Chazden, R. and Coe, F. (1999) "Ethnobotany of woody Species in secong Graoth, Old Growth and Selectively Logged Forests," *Conservation Biology*, 13 (6): 1312–22.

Chokkalingham, U. de Jong, E., Smith, J. and Sobogal C. (2001) Special issue of *Tropical Forest Science*: "Secondary forests in Asia: their diversity, importance and role in future environmental management," 13 (5).

Coe, S. (1997) *The True History of Chocolate*, New Haven: Yale Press.

Conroy, M., Murray, D. and Rosset, P. (1996) *A Cautionary Tale*, San Francisco: Food First.

Cromes, F. (1999) "Researching forest fragments," in W. Laurance and R. Bierregaard (eds) (1999) *Tropical Forest Remnants*, Chicago: UC Press.

Dale, V., O'Neill, R., Pedlowski, M. and Southworth, F. (1993) "Causes and effects of land use change in Central Rondonia, Brazil," *Photogrammic Engineering and Remote Sensing*, 59 (6): 997–1005.

Daugherty, H. (1969) "Man-induced Ecological Change in El Salvador," Ph.D. thesis, UCLA Department of Geography.

—— (1972) "The impact of man on the zoogeography of El Salvador," *Conservation Biology*, 4 (4) 273–278.

de Jong, W. (2001) "Tropical; secondary forests in Asia 2001: introduction and synthesis," *Journal of Tropical Forest Science*, 13 (4): 536–77.

Deere, C. and Magdalena, L. (2000) *Genero, Propiedad y Empoderamiento: tierra, Estado y mercado en América Latina*, Colombia: Tercer Mundo Editores.

Demarest, A. (1988) "Political evolution in the Maya borderlands: the Salvadoran Frontier," in: E. Boone and G. Willey (eds) *The Southeast Classical Mayan Zone*, Washington DC: Dumbarton Oaks: 335–94.

Denevan, W. (1992). "The pristine myth: the landscape of the Americas in 1492," *Annals of the Association of the American Geographers*, 82: 369–85.

—— (2001) *Cultivated Landscapes of Native Amazonia and the Andes*, Oxford: Oxford University Press.

Didia, D. (1997) "Democracy, political instability and tropical deforestation," *Global Environmental Change*, 7 (1): 63–76.

Diskin, M. (1996) "Distilled conclusions: the disappearance of the agrarian question in El Salvador," *Latin American Research Review*, 3 (2) 111–26.

Downing, T., Hecht, S. and Pearson, R. (1992) *Development or Destruction: the livestock sector in Latin America*, Boulder: Westview Press.

Dull, R. (2001) "El Bosque Perdido: A cultural-ecological history of Holocene environmental change in El Salvador," Ph.D. thesis, UC Berkeley.

Dull, R., Southorn, R. and Sheets, P. (2001) "Volcanism, ecology and culture: a reassessment of the Ilopango eruption in the southern Mayan realm," *Latin American Antiquity*, 12: 25–44.

Durham, W. (1979), *Scarcity and Survival in Central America*, Stanford: Stanford University Press.

Estrada, A., Coates Estrada, R., Meritt, D., Montiel, S. and Curiel D. (1993) "Patterns of frugivore species richness and abundance in forest islands and in agricultural habitats at Los Tuxlas, Mexico," *Vegetatio*, 107/108: 245–57.

Ezzati, M., Singer, B. and Kammen, D. (2001) "Towards an integrated framework for development and environmental policy," *World Development*, 29 (8): 1421–34.

Faber, D. (1993) *Environment under fire: imperialism and the ecological crisis in Central America*, New York Monthly Review.

Fairhead, J. and Leach, M. (1998) "False forest history, complicit social analysis: rethinking some West African environmental narratives," *World Development*, 23 (6): 1023–35.

FAO (Food and Agricultural Organization) (2000) *Forest Assessment*, Rome: FAO.

—— (2001) *Global Forest Inventory*, Rome: FAO.

FESAL (1978) *Encuesta Nacional de Salud Familiar: Informe Final*, El Salvador: Asociación Demográfica Salvadoreña.

—— (1998) *Encuesta Nacional de Salud Familiar: Informe Final*, El Salvadore: Asociación Demográfica Salvadoreña.

Flannery, K. (1982) *Maya Subsistence*, New York: Academic Press.

Fowler, W. (1988) "La poblacion nativa de El Salvador al momemto de la conquista Espanola," *MesoAmerica*, 15: 79–116.

FUSADES (Fundacion Salvadorena de Desarollo) (1998) *Rural Development Study*, San Salvador: FUSADES.

Gibson, C., McKean, M. and Orstrom, E. (2000) *People and Forests: Communities, Institutions, Governance*, Cambridge: MIT Press.

Gillespie, T., Grijalva A. and Farris, C. (2000) "Diversity, composition and structure of tropical dry forests in Central America," *Plant Ecology*, 147: 37–47.

Gobbi, J. (2000) "Is biodiversity friendly coffee economically viable?" *Ecological Economics*, 33 (2): 267–81.

GOES (Gobierno El Salvador) (1999) *Plan de La Naccion*, San Salvador: GOES.

Greenberg, D. (1996) "Managed forests patches and bird diversity in Southern Mexico," in J. Schelas and D. Greenberg (eds) *Forest Patches in Tropical Landscapes*, Washington DC: Island Press.

Haenn, N. (1999) "The power of environmental knowledge: ethnoecology and environmental conflicts in Mexican conservation," *Human Ecology*, 27 (3) 477.

Hall, A. (l996) *Productive Conservation,* Manchester: University of Manchester Press.

Hardin, G. (1967) "The tragedy of the commons," *Science,* 162 (3859): 1243–6.

Harvey, C. (n.d.) "The colonization of windbreaks by forest trees in Costa Rica: implications for regeneration," Ph.D thesis, Cornell University.

Harvey, C and Haber, W. (1999) "Remnant tress and the conservation of biodiversity in Costa Rican pastures," *Agroforestry Systems,* 44: 37–68.

Hecht, S.B. (1985) "Environment, development and politics: capital accumulation and the livestock sector in eastern Amazonia," World Development, 13 (6): 663–84.

—— (1994) "The Logic of Livestock and Deforestation in Amazonia," *Bioscience,* 43 (8): 687–95.

—— (1999) *Forests 2021,* San Salvador: MARN.

—— (2003), "When solutions become drivers: Policy and politics in Bolivian deforestation," *Development and Change,* In press.

Hecht, S.B., Anderson, A. and May, P. (1988) "The subsidy from nature: shifting cultivation, palm forests and rural development," *Human Organization,* 47 (1): 25–35.

Hecht, S.B. and Cockburn, A. (1989) *Fate of the Forest: Developers, Destroyers and Defenders of the Amazon,* London: Verso.

Hecht, S.B. and Saatchi, S. (2003) Landscapes of recovery, Chapman Conference, American Geophysical Union.

Hecht, S.B., Kandel, S., Cuellar, N. and Rosa, H. (2002a) *Globalization, Forest Resurgence and Environmental Politics in El Salvador,* El Salvador: PRISMA.

—— (2002b) *Forms of Decentralization, Governance and the Politics of Natural Resources in El Salvador,* San Salvador: PRISMA.

Holdridge, L. (1975) *Zonas de Vida Ecologicas de El Salvador,* San Salvadore: Ministerio de Agricultura y FAO.

Kaimowitz, D. (1995) *The End of the Hamberger Connection?,* Bogor, Indonesia: CIFOR:.

Kaimowitz, D. and Angelsen, A. (1998) *Economic Models of Tropical Deforestation: A Review,* Bogor, Indonesia: CIFOR.

Kaimowitz, D., Martinez, P. and Vanclay, J. (2002) "Spatial regression analysis of deforestation in Sta Cruz, Bolivia," in C. Wood and R. Porros (eds) *Deforestation and Land Use in the Amazon,* Gainesville: University of Florida Press.

Kandel, S. (2002) Migraciones, medio ambiente y pobreza rural en El Salvador, Documento de Trabajo, San Salvador: PRISMA.

Klooster, D. (2000) "Beyond deforestation: the social context of forest change," *Conference of American Geographers,* 26: 47–59.

—— (2003) "Forest transitions in Mexico," *Professional Geographer,* 55 (2): 227–37.

Klooster, D. and Masera, O. (2000) "Beyond deforestation: the social context of forest change," *Conference of American Geographers,* 26: 47–59.

—— (2000) "Community forest management in Mexico: carbon mitigation and biodiveristy conservation through rural development," *Global Environmental Change,* 10: 259–72.

Komar, O. (1998) "Avian Diversity in El Salvador," *Wilson Bull,* 110 (4): 511–33.

Koop, G. and Tole, L. (1999) "Is there an environmental Kuznets curve for deforestation?," *Journal of Development Economics,* 58: 231–44.

—— (2001) "Deforestation, distribution and Development," *Global Environmental Change,* 11: 193–202.

Kuznets, S. (1957) "Reflections on modern economic growth of nations," *Economic Applications,* 10 (2–3): 211–59.

Laurance, W., Albernaz, A. and de Costa, C. (2001) "Is deforestation accelerating in the Brazilian Amazon?," *Environmental Conservation*, 28 (4): 305–1.

Lentz, D.L., Beaudry-Corbett, P., Reyna de Aguilar, M. and Kaplan, R.L. (1996) "Foodstuffs, forests, fields and shelter: a paleoethnobotanical analysis of vessel contents from the Ceren site," El Salvador: Latin American Antiquity, 7: 247–62.

Leonard, J. (1987) *Natural Resources and Environment in Central America*, New Brunswick: Transaction Press.

Lima, D. (1999) "Equity, sustainable development and biodiversity preservation," in C. Padoch *et al. Varzea*, New York: NYBG, 247–65.

Lopez, R. and Mitra, S. (2000) "Corruption, pollution and the environment," *Ecological Economics:* 137–50.

Lovell, W. (1992) "Heavy shadows and dark night. Disease and depopulation in Colonial Spanish America," *Annals of the Association of American Geographers*, 82: 426–43.

Lungo, M. and Kandel, S. (eds) (1999) *Transformando El Salvador: Migración Internacional Sociedad y Cultura*, San Salvador: FUNDE.

MacArthur, R. and Wilson, E.O. (1967) *The Theory of Island Biogeography*, Princeton: Princeton University Press.

MAG (Ministerio de Agriculture) (1998) *Informe de Conyunctura*, San Salvador: MAG.

—— (1999) *Informe de Conyunctura*, San Salvador: MAG.

MARN (Ministerio de Medio Ambiente y Recursos Naturales) (1999) *El Salvador: Medio ambiente y recursos naturales*, San Salvador.

—— (2000), *Política Nacional de Medio Ambiente y Lineamientos Estratégicos, Ministerio de Medio Ambiente y Recursos Naturales*, San Salvador: MARN.

Mather, A and Needle C. (1998) "The forest transition," *Area*, 30 (2): 117–24.

Méndez, E. and 15 others (2001), Café con sombra y Pago por Servicios Ambientales: Riesgos y Oportunidades para impulsar mecanismos con pequeños agricultores de El Salvador, case study for the "Pago por Servicios Ambientales en Latinoamérica" Project, San Salvador: PRISMA – Ford Foundation.

Moran, E., Brondizo, E., Tucker, J., Falesi, I. and McCracken S. (2000) "Strategies for Amazonian afforestation," in A. Hall (ed.) *Amazonia at the Crossroads*, London: Institute for Latin American Studies.

Murray, D. (1994) *Cultivating Crisis: The Human Cost of Pesticide Use in Latin America*, Austin: University of Texas Press.

Myers, N., Mittermeier, R., Mittermeier, C., Fonseca, G. and Kent, J. (2000) "Biodiversity hotspots for conservation priorities," *Nature*, 403: 853–58.

Neumann, R. (1999) *Imposing Wilderness*, Berkeley: UC Press.

Padoch, C. Ayers, M., Vasquez, P. and Henderson, R. (1999) *Varzea*, New York: NYBG

Paige, J. (1997) *Coffee and Power*, Cambridge: Harvard University Press.

Pearce, J. (1986) *Promised Land: Peasant Rebellion in Chalaltenango*, El Salavdor, London: Latin American Bureau.

Perfecto, I and Vandermeer, J. (2001) "Quality of agroecological matrix in a tropical Montane landscape," *Conservation Biology*, 16 (1): 174–82.

Perfecto, I, Rice, R., Greenberg, R. and van der Voolt, M. (1996) "Shade coffee is a refuge for biodiversity," *Bioscience*, 46: 598–608.

Pimentel, D., Stachow, U., Tackacs, D., Brubaker, H., Dumas, A., Meany, J., O'Neill, J. and Onsi, D. (1992) "Conserving biodiversity in agriculture/forest systems," *Bioscience*, 42 (5) 354–362.

Pohl, M., Pope, K., Jones, J., Jacob, S., Piperno, D., de France, S., Lentz, D., Gifford, J.,

Danforth, M. and Josserand, J. (1996) "Early agriculture in the Maya lowlands," *Latin American Antiquity*, 7: 355–72.

Posey, D. and Balee, W. (eds) (1989) "Natural resource management in Amazonia," *Advances in Economic Botany*, 7, New York: New York Botanical Garden.

Primack, R., Bray, D., Galletti, H. and Ponciano, I. (1998) *Timber Tourists and Temples*, Washington, DC: Island Press.

PRISMA (2002) *National Level Environment Analysis. Report to World Wildlife Fund*, San Salvador: PRISMA.

PRISMA/UCLA (n.d.) *Surveys. Materiales Primarios de La Montanona, Tacuba, Barra-Santiago*, San Salvador: PRISMA.

Procafé (1998) *Estatisticas del Economia de Café*, San Salvador: MAG.

Ramirez, C. (2001) "Vegetation of a subtropical pre-montane moist forest in Central America," Ph.D thesis, CUNY.

Repetto, R. and Gillis, M. (1988) *Public Policies and the Misuse of Forest Resources*, Washington: World Resource Institute.

Ribot, J. (2001) "Integral local development," *Agricultural Resources, Governance and Ecology*, 1 (3/4): 327–51.

—— (2002), *Democratic Decentralization of Natural Resources: Institutionalizing Popular Participation*, Washington: World Resources Institute.

Rudel, T., Bates, D. and Machinguiashi, R. (2002) "A tropical forest transition? Agricultural change, out-migration in the Ecuadorian Amazon," *Annals of the AAG*, 92 (1): 87–102.

Rudel, T., Perez-Lugo, M. and Zichal, H. (2000) "When fields revert to forest," *Professional Geographer*, 52 (3): 386–97.

Sanderson, S. and Redford, K. (1997) "Biodiversity politics and the contest for the ownership of the World's biodiversity," in R. Kramer, C. van Schaik and J. Johnson (eds) *The Last Stand*, Oxford: Oxford Press: 115–33.

Schortman, E. and Urban, P. (1991) "Patterns of pre-classic interaction and the formation of complex societies in the southeastern Maya Periphery," in W. Fowler (ed.) *The Formation of Complex Society in Southeastern MesoAmerica*, Boca Raton: CRC Press: 121–42.

Schmink, M. and Woods, C. (1992) *Contested Frontiers in Amazonia*, New York: Colombia University Press.

Schuuman, D. and Partridge W. (1989) *Human Ecology in Tropical Forests*, Boulder: Westview.

Schwartzman, S., Moreiera, A. and Nepstad, D. (2000) "Rethinking tropical forest conservation," *Conservation Biology*, 14 (5): 1351–7.

Scott, J. (1997) *Seeing Like State*, New Haven: Yale University Press.

Shanley, P., Perice, A., Laird, S. and Guillen, A. (eds) (2003) *Tapping the Green Market: Management and certification of non-timber products*, London: Earthscan.

Sharer, R. (1978) *The Prehistory of Chaluapa, El Salvador*, Philadelphia: University of Pennsylvania Press.

Sheets, P. (1979) "Environmental and cultural effects of the Ilopango eruption in Central America," in P. Sheets and D. Grayson (eds) *Volcanic Activity and Human Ecology*, New York: Academic Press: 525–64.

—— (1982) "Prehistorical Agricultural systems in El Salvador," in K. Flannery (ed.) *Maya Subsistence*, New York: Academic Press: 99–116.

—— (1984) "The prehistory of El Salvador: an interpretive summary," in F. Lange and D. Stone (eds) *The Archeology of Lower Central America*, Albuquerque: University of New Mexico Press: 85–112.

Spense, M. (1999) *Dispossessing the Wilderness*, New York: Oxford University Press.

Steininger, C., Tucker, C., Ersts, P., Killen, T.O. and Hecht, S.B. (2001) "Clearance and fragmentation of tropical semi-deciduous forest, in the Tierras Bajas," *Conservation Biology*, 15 (4): 856–66.

Stern, D., Common, I. and Barbier, E. (1996) "Economic growth and environmental degradation: the environmental Kuznets curve and sustainable development," *World Development*, 24 (7): 1151–69.

Storper, M. (1997) *The Regional World*, New York: Blackwell.

Sunderlin, W., Angelson, A., Dermawan, A. and Rianto, E. (2001) "Economic crises, small farmer well being and forest cover change in Indonesia," *World Development*, 29 (5): 767–82.

Templeton, S. and Scheer, S. (1999) "Effects of demographic and related macroeconomic change on land quality in hills and mountains of developing countries," *World Development*, 27 (6): 903–18.

Terborg, J. (1999) *Requiem for Nature*, Washington: Island Press.

Torras, M. and Boyce, J. (1998) "Income inequality and pollution: assessment of the environmental Kuznets Curve," *Ecological Economics*, 25 (2): 147–69.

Utting, P. (1993) *Trees People and Power*, London: Earthscan.

Uusivuori, J., Lehto, E. and Pallo, M. (2002) "Population, income, and ecological conditions as determinants of forest area in the tropics," *Global Environmental Change*, 12: 313–23.

van Shaik C., Terborgh, J. and Dugelby, B. (1997) "The silent crisis: the state of rain forest nature," in R. Kramer, C. van Shaik and J. Johnson (eds) *The Last Stand*, Washington: Island Press: 64–89.

Vandermeer, J. and Perfecto, I. (1997) "The agroecosystem: the need for the conservationists lens," *Conservation Biology*, 11: 1–3.

Whitmore, T. (1997) "Tropical forest disturbance, disappearance and species loss," in L. and R. Bierregaard (eds) *Tropical Forest Remnants*, Chicago: UC Press.

Williams, R. (1986) *Export Agriculture and the Crisis in Central America*, Chapel Hill: University of North Carolina Press.

Zier, C. (1983) "The Ceren site: a classic period Maya residence and agricultural field in the Zapotitan Valley," in P. Sheets (ed.) *Archeology and Volcanism in Central America*, Austin: University of Texas Press.

Zimmerman, J. (2001) "Puerto Rico as a model for understanding Tropical reforestation," *Tropinet*, 12 (2): 1–2.

ACKNOWLEDGMENTS

The field research for this project was developed over several years with wonderful colleagues from the research and policy NGO PRISMA (Program on Environmental Research in El Salvador) through a grant from the MacArthur Foundation to Hecht/UCLA and PRISMA. More detailed expositions of the field work can be found in the monographs of Hecht *et al.* 2002a and b. Hecht also received research assistance from the UCLA Academic Senate Grants and the Latin American Center. Special acknowledgements go to Herman Rosa and Susan Kandel from PRISMA. The theoretical issues raised by the paper reflect more recent research.

Part II

DISCOURSE AND PRACTICE

4

ENVIRONMENTAL DISCOURSES ON SOIL DEGRADATION IN BOLIVIA

Sustainability and the search for
socioenvironmental "middle ground"

Karl S. Zimmerer

THE EROSION PROBLEM IN BOLIVIA

Beginning in the late 1970s various reports were sounding the alarm about worsening soil erosion in Bolivia, a landlocked and mostly mountainous republic of over 4 million people in central South America. Books such as *Bolivia: The Despoiled Country* by Walter Terrazas Urquidi (1974) and *The Wasted Country: The Ecological Crisis in Bolivia* by Mariano Baptista Gumucio (1977) alerted many Bolivians and Latin Americans to the country's grave dilemma. The widely read *Losing Ground* by North American Erik Eckholm (1976) introduced it to a still larger audience in the United States and Western Europe. Academic and governmental studies spelled out some of the serious consequences of Bolivia's erosion crisis (Grover 1974; LeBaron *et al.* 1979; Preston 1969). Accelerating erosion was degrading farm and range-land, forcing floods downstream, and leading to destructive desertification and dust storms. Bolivia's major newspapers regularly raised concerns that the country now suffers moderate to extreme loss of soils (*Los Tiempos* 1991; *Presencia* 1990).

Widespread alarm about soil erosion cannot be attributed solely to the new gravity of the problem. A variety of historical sources from the 1570s to the 1920s allude to severe soil loss resulting from grazing and farming and to deforestation carried out in the absence of conservation measures (Larson 1988; Zimmerer 1993b). The predominant base of erosion-prone sedimentary rock, steep terrain, and semi-arid climate and vegetation in the Bolivian Andes have long rendered its mountainous landscape vulnerable to soil loss (de Morales 1990; Montes de Oca 1989). While the Inca overlords probably enforced soil conservation in the fourteenth and fifteenth centuries, the colonial rulers (1530s–1825) and republican governments (1825–present) did not instill soil conservation or grant official notice to the erosion problem. The countryside's environmental destruction almost disappeared from view for the broader Bolivian public between the, 1530s and 1970s. Even advisors from the United States Department of Agriculture, present from 1943, failed to bring attention to the country's catastrophic erosion (USDA 1962).

A heightened awareness of erosion in Bolivia during the 1970s was inspired by international conferences and publications coming from other Latin American countries, the United States, and the United Nations. Their environmental ideas, together with the downright worsening of erosion, led Bolivian agronomists Walter Terrazas Urquidi and Mariano Baptista Gumucio to publicize the country's environmental crisis. Yet the widely held concern about soil loss in Bolivia has not cemented a public consensus about the nature of the erosion problem or the prospects for conservation. In fact individuals and institutions living and working in Bolivia hold a variety of divergent views on the causes, as well as the preferred solutions, for the problem of soil degradation. Together their articulated perceptions make up a regional discursive formation (see the beginning of this chapter), which is actively forged at sites throughout the mountainous agricultural areas of Bolivia (and often beyond). It takes the form of several distinct-and-intersecting dialogues on the nature of these mountainous soils environments. Recent discourses on soils degradation in Bolivia show the persistent yet changing place of environmental scientists and reasoning derived from their quarters. Certain environmental science has tended toward a downplaying of the problems of soils erosion. For example, an influential environmental scientific overview, published in *The Environment* (generally a pro-nature journal), makes a claim for the new unimportance of soil degradation problems in developing countries (Crosson 1997). This claim is derived from accounts of successful soil conservation and, equally central to the argument, the perceived substitutability of adequate soils quality on the scale of the global economy. In my view, the relative drop-off in the environmental importance of soils issues is indicative of a shift in conservation and environmental scientific circles to the biodiversity phase of environmentalism and, concomitantly, toward the so-called ecological phase of capital (Escobar 1995; Robertson 2000). That shift underlines the prior preeminence of modern science's concern for soil erosion and conservation in developing countries, at least when cast at the national scale such as in the aforementioned works of Terazas (a socialist analysis) and Baptista (framed in a pro-capitalist position). This earlier concern appears to have been filtered through an environment-related discourse of modernity, with an emphasis on nationalist control over resources and economic class. (On class in environmental discourse see Forsyth 2001; Peet and Watts, this volume.) Subsequently environmental scientific concerns tended to shift toward an emphasis on biodiversity that is emblematic of current environmental and conservation science (Zimmerer 2000c; on land use cover change and modernity see Robbins 2001).

My interest in the predicament of Bolivian soil erosion and conservation prospects led me to study the diverse perceptions of these issues in relation to the political ecology of the changing environment (Zimmerer 1993a, 1993b, 1994,

2000a, 2000b, 2003). I chose to focus on the period of the recent past (1950s–present) in the Cochabamba region, a geographical area of abrupt ecological transitions between the mountainous Andes and the rainforest lowlands of the upper Amazon. Often claimed to be the "heartland" or "breadbasket" of Bolivia, the productivity of Cochabamba is threatened due to soil degradation, misguided water privatization, and a volatile mix of economic and political reforms that have disadvantaged the rural sector (Healy and Paulson 2000; Laurie and Marvin1999). Site specific approximations of annual erosion range from 50 to 150 tons per hectare, rates exceeding those of soil formation (CORDECO 1980; Zimmerer 1991). The erosion threat has become a pressing issue for many institutions and inhabitants – including more than 100,000 peasant farmers in the region whose livelihoods depend on small-scale cropping, livestock raising, and a wide variety of non-farm work. Adding to their voices, development institutions of the government and non-governmental organizations (NGOs) have voiced alarm and launched analyses and programs to address the erosion problem. Focusing on soil degradation in Cochabamba enabled me to study regional discursive formations and the political ecology of the changing environment in a well-ground historical and regional setting. In all, three primary perspectives have taken shape on Cochabamba's erosion crisis and proposals for conservation-with-development (i.e. sustainable development): (i) government and non-government institutions; (ii) peasants in their personal perspectives; and (iii) rural trade unions.

SOIL DEGRADATION AND DISCOURSE IN POLITICAL ECOLOGY

To assess the perspectives on soil degradation and sustainable development I combine a framework of concepts from political ecology and the analysis of articulated perceptions or discourse (Blaikie 1985; Blaikie and Brookfield 1987; Eagleton 1983; Emel and Peet 1989; Gupta and Ferguson 1997; Peet and Watts 1996; Watts 1983, 1985). From political ecology I adopt a socio-economic and political analysis of environmental change. A process of uneven economic development has in effect cornered peasant farmers in the Cochabamba countryside, pressuring them to modify land use in erosion-inducing ways – such as curtailing once common conservation measures as a consequence of labor-time constraints – while not offering economic alternatives sufficiently remunerative to permit land-use alteration in a more environmentally sound manner (Zimmerer 1993b; see also de Janvry *et al.* 1989; Storper 1991). The analysis shows that the so-called "scissors effect" has led peasants to intensify production without the use of conservation techniques.

I also pursue a pair of other themes recently being brought to bear in the political ecology approach. These themes are the analysis of civil society, especially so-called "new social movements" distinct from government institutions and the analysis of environmental discourses (Peet and Watts 1996; Zimmerer 1993a).

Rural trade unions in Cochabamba resemble other "new social movements" that seek to revitalize existing institutions at the grassroots or popular level. The rural trade unions gained a growing prominence in public debate and discussion on environmental issues, among which soil erosion looms large. Interestingly, the recent renewal of Cochabamba trade unions as social movements – as well as the region's proliferation of NGOs returns us to a main theme of political ecology: how government policies in many developing countries enforce the extremely biased processes of uneven development. Amid the biases of distorted development policies, non-state institutions spread pell-mell in certain regions and economic sectors, including those of peasants (Slater 1985). Applying a political ecology approach to the language-rich realm of articulated perceptions about erosion requires a focus on discourses representing the ideas and ideologies held by groups of involved individuals and institutions (Peet and Watts 1996; Zimmerer 1993a). As environmental discourses, their views represent diverse dimensions that are directly related to soils degradation. Through my fieldwork I realized that a set of three facets tend to be most central: (i) the understanding of the nature of soil degradation; (ii) the view (if any) of the problem associated with soils degradation; and (iii) the perceptions of the proper solution or response to the problem(s). These facets are characteristic of global discourses and national Bolivian ones, and, as seen in the next section, they also distinguish those at regional and local sites within Cochabamba. The natural basis of soils degradation is rife with varied interpretations and often much-debated meanings. This type of debate tends to privilege the contribution of environmental and agronomic science although, as seen in the following sections, it is frequently engaged by the articulated perceptions of non-scientists as well. One central question is concerned with the processes and rate of soil erosion and whether it is accelerating. Environmental scientific and political ecological investigation of these questions in Bolivia has increasingly relied upon historicizing the interpretation of soils changes and the key elements of human-soils interaction (Pestalozzi 2000; Preston et al. 1997; Zimmerer 1993a). Similar to the interpretive scientific discourses aimed at other types of human-environmental change, this emphasis on historicization is entwined with an analytical appreciation of site-specific differences (Fairhead and Leach 1996; Forsyth 1998). A related focal point, specific to the environmental scientific discourse on soil erosion, is framed to apportion causation to "natural" factors (climate, topography, vegetation) and/or "cultural" factors (changing land use, adoption or abandonment of conservation measures). Diverse meanings are likewise abundant in the interpretation (if any) of problems associated with changes in soils environment. The previously mentioned position of Crosson (1997), for example, exemplifies the view that overall there exists little problem if any of soil degradation. That view is supported through the growing evidence of environmental resilience, the reversibility of degradation through soil conservation, and the economic prospect of global-scale substitutability (especially the relocation elsewhere of agriculture). In the Bolivian Andes, the soil erosion problem also invokes the significance of the long-term nature of erosion and the part-time character of farming. Such reasoning may tend to overlook that soil degradation can actively incur adverse impacts to local inhabitants and land users who are already poor, even if their poverty is only partly a problem of farming. This environmental justice perspective has been inherent in the longstanding contribution of political ecology to interpreting the meaning of soils change (Blaikie 1985; Blaikie and Brookfield 1987; Grossman 1997; Zimmerer 1993a, 2000c). As shown in the following sections, land users themselves, as well as environment-and-development institutions, are active participants in present-day discussions of whether there exists a soils problem (erosion and/or fertility loss). What to do about the soils problem is an aspect of equal significance. Environmental discourses on solutions tend to be closely related to the perceptions of a soil problem. Indeed the following sections provide several examples of the close discursive connection of how soil degradation is understood and the perceived solutions to it. In addition to being closely connected, these views of soils issues tend to be etched in more general cultural discourses, particularly those of locality/place and community (see Gupta and Ferguson 1997). The relation of environmental discourses (such as soils) to those on locality/place and community is often powerful in the immediate political spheres and in their national- and global-scale articulations. The political ecology of such multi-scale discursive relations is being richly demonstrated in the works of Juanita Sundberg on environmental conservation projects in Guatemala's Maya Biosphere Reserve (Sundberg 1998; nd.). In Bolivia, the more general cultural discourses of relevance to environmental issues are increasingly cast amid the enlarged importance and general fluidity of identity politics (Healy and Paulson 2000; Lagos 1997; Zimmerer 2000c). The rise of identity politics in Bolivia is being combined with market-led, neoliberal reforms and structural adjustment packages (SAPs) in powerful yet previously unpredicted ways (Laurie and Marvin 1999).

Also through my fieldwork I realized that people and institutions there did not form their environmental discourses in absolute isolation or, figuratively speaking, as discursive islands of self-contained dialogue (cf. the narrowly poststructuralist interpretation of Orlove 1991). Instead, they expressed and gave distinctiveness to their viewpoints through interaction within and among the groups. Processes of resistance and contestation as well as accommodation and agreement guided their elaboration of environmental ideas. A historical approach and sensitivity to discursive alterations over time were crucial to my gaining insights on their views of changing nature and efforts to conserve it. Finally, I have found over the course of the past decade that their environmental discourses are regularly articulated in diverse ways, sometimes subtle though often not, with global environmental policy and environmental science. These articulations underscore the concrete connections between works introduced at the outset of this chapter's introduction and the materials that follow. It also highlights a perspective of political ecology that takes global environmental policies or prevailing scientific interpretations as a point of departure (Bassett and Zueli 2000; Bryant 2001; Nesbitt and Wiener

2001; Neumann 1995, 1998; Turner 1999). This political ecological approach to environmental discourse is distinctively cross-scalar, rather than singularly local-scale, in that typically it follows linkages of ideas, policies, and resources that are North–South or rural–urban within advanced industrial countries.

In general, I seek also to renew consideration of local knowledges and peasants' personal or "everyday" perspectives on the soil erosion dilemma as part of an effort to invigorate political ecology through the analysis of discourse: i.e. as part of what Peet and Watts call liberation ecology in this book. Such projects include the study of conservation-related knowledges and institutions (Zimmerer 1993b, 1994; 2003). Focusing on the situating of erosion-related knowledge in peasants' discourses, the present study offers an example of a broadened political ecology and the impetus for a more open "liberation ecology" (Peet and Watts, Chapter 1 in this volume). Indeed the local knowledge apparent in the everyday discourses of Cochabamba peasants about erosion has rarely been manifest as a strictly self-contained and self-referential dialogue. Their common-place observations suggest strongly that such expressions have influenced, and been influenced by, ongoing efforts to negotiate erosion-related meanings, policies, and resource access with other social groups and institutions in the region.

SOIL EROSION DISCOURSES IN COCHABAMBA

In undertaking the present study I relied on written and sometimes published materials and open-ended interviews with thirty-four Cochabamba peasants taped and transcribed with their consent. Our conversations usually combined Quechua, the first language of Cochabamba peasants, and Spanish, a widely used second language.

Development institutions: blaming the peasants

> The land users have not developed any awareness about the problems of soil erosion.... Overgrazing and trampling by livestock, together with the removal of shrub cover for fuel in the Altiplano and the Mesothermic Valleys, are the most important causes of soil erosion.
>
> (IIDE and USAID 1986)

Government institutions in Bolivia paid little heed to soil erosion and failed to support any sizeable effort at conservation despite the accumulating accounts of a dramatic erosion dilemma. Throughout the 1980s national governments refused to establish a policy or program on soil conservation (II DE and USAID 1986). When agencies in the Bolivian government did address the erosion problem, they placed the blame squarely on the shoulders of peasant farmers and herders. A 1977 report on "'Renewable natural resources" by its Ministry of Peasant Agriculture and Ranching, for instance, claimed that the primary cause of soil erosion could be found in the land-users' failure to employ modern techniques (MACA 1977). Such reports reasoned that the transfer of proper tools and techniques to ill-equipped peasants would stem erosion. Market signals and the articulation of the peasants' economy with agricultural businesses would induce the necessary innovations and transfer of land-use techniques including modern soil conservation (Adams 1980).

But the capacity of Cochabamba's peasant sector to generate market demand for modern technologies fell during the 1980s under an onslaught of sectoral, social, and spatial inequalities enforced by national economic policies. Agribusiness integration, meanwhile, was confined to small segments of the total peasant economy (Weil 1983). Restructuring the national economy since August 1985 in accord with a neo-liberal model imposed by the International Monetary Fund and the World Bank dashed most remaining hopes that market-induced technological change would reduce erosion. To be sure, some agencies in the Bolivian government have recently urged assistance programs to aid in transferring modern farm tools and techniques for conservation goals. In a 1987 "National Meeting on Natural Renewable Resources," governmental institutions, together with major international aid agencies, proposed establishing a national soil conservation program (MACA 1977). Yet such programs faced scant chance of government financing in the aftermath of 1985 restructuring and Decree 21060 that officially abdicated the nation's responsibility for rural development in highland regions (Pérez Crespo 1991).

As the fiscal and administrative capacity of the Bolivian government stagnated, soil conservation became the hallmark mission of a proliferating pool of international aid agencies and NGOs. Beginning in the 1970s, these institutions advocated technical assistance to peasant producers. USAID, for instance, sponsored a number of soil erosion and conservation studies in peasant communities in the guise of larger projects aimed at modernizing Bolivian agriculture (LeBaron et al. 1979; Wennergren and Whittaker 1975). Another international aid organization, the Swiss Technical Corporation, supported a number of soil conservation program's, most notably in the form of small-scale forestry projects. Overall NGOs accounted for the bulk of the new erosion prevention and conservation programs. By the late 1980s, more than 300 NGOs had initiated assistance programs in Bolivia. At least eighty clustered in the Cochabamba region, where many sponsored studies and small projects designed to abate soil erosion.

The statements on erosion causes by international aid agencies and most NGOs coincided in large part with assessments by the government agencies. Most non-Bolivian development institutions concurred with the mainstream belief that peasant ignorance was culpable for the erosion crisis. Consider for example the "Environmental profile of Bolivia" in which NGO and USAID authors allege that "land users were not at all aware of the soil erosion problem" (IIDE and USAID 1986: 99). Other USAID reports went further in concluding that worsened erosion originated in the "cultural backwardness" of rural inhabitants (LeBaron *et al.* 1979; Wennergren and Whittaker 1975). The director of the Center for Forestry Development (CFD) in Cochabamba held that "men cause soil erosion where they do not know better" (Estrada 1991). Rooting their logic in the perceived bane of cultural backwardness, the modernizers deduced that the techniques and technologies of peasant land use were the chief inducers of erosion (de Morales 1990: 52; Estrada 1991; IIDE and USAID 1986; MACA 1977).

Curiously an indictment of ill-suited techniques and inadequate knowledge similar to this modernizing logic was shared by an anti-modernizing group of mostly NGOs. They held that environmentally damaging land use and the lack of necessary knowledge are consequences of cultural degradation, in effect too much modernity rather than too little (Eckholm 1976; van den Berg 1991). The erosion-inducing peasant farmers of Cochabamba and other Bolivian regions are thought to have fallen from an earlier, near-Edenic state of advanced culture and sophisticated environmental knowledge. Current residents represent, it is said, the culturally deprived descendants of the Inca and Aymara empires that ruled before the onset of Spanish colonialism. Despite the radical difference in interpretation, the anti-modernizing analysts arrived at a conclusion not much different to their mainstream counterparts: cultural inadequacy and lack of knowledge cause the current erosion crisis.

Hence, many soil conservation projects in Cochabamba were designed to address the perceived ignorance of peasant farmers (e.g. IIDE and USAID 1986). Characteristically the projects disseminated educational pamphlets and organized conservation seminars for local farmers. They funded posters, supported demonstration plots, set up farmer-to-farmer forms of knowledge transfer, and arranged for speakers at meetings of peasant communities or unions known as *sindicatos*. Though the project methods were not fundamentally wrong, their analysis was. The projects frequently prescribed new conservation strategies without really assessing present practices and the rationales behind them. Circumstances in peasant farmers' livelihoods shaping their land use were largely ignored. A number of soil conservation projects found themselves promoting measures such as the large-scale construction of contour terraces that were impractical given the reality of peasant resources (Zimmerer 1993b).

The peasants: diverging perspectives

It was not like this before, the hills weren't barren nor were there many erosion gullies. Look, I'm only 27 years old but I've seen it deteriorate bit by bit ... the soil has lost its productive force, each year it no longer produces as before. Soil from the slopes is being swept downwards-leaving bare rock, subsoil, and gullies – due to the heavy rains ... the development institutions claim that they know the solutions, but when we look at it, we recognize that we know as a result of our experience, we know how to take care of the earth.

(Interview, Ubaldina Mejia, Aiquile, Cochabamba, 14 Oct. 1991)

Peasant farmers in Cochabamba blamed themselves for soil erosion. But their viewpoints are distinct from the development institutions in two principal ways. First, most peasant farmers express a sophisticated knowledge of soil erosion, utilizing a complex lexicon from Quechua and Spanish to discuss diverse erosional landforms and their management, while relating erosion to soil types and farming practices (Zimmerer 1994). Second, they invoke the supernatural world of religious beliefs and customs in explaining the causes of erosion (see also Staedel 1989). In both regards, a distinguishing characteristic of everyday peasant perceptions of soil erosion is the sense – at times vivid and quite personal – of prolonged historical time and a close relation to place.

Historical illustrations of peasants' environmental discourse are revealed in their judicial depositions lodged in efforts to defend the right to land, water, and forest resources. Records filed during the nineteenth century, currently housed in the Municipal Archive of Cochabamba (*Archivo Municipal de Cochabamba*, AMC) reveal the twin features of prolonged time and a close relation to place. For example, in 1832 Isidoro Ayllita who, as the Cacique of Colcapirgua, was a local indigenous authority in the Cochabamba countryside, defended his right to irrigate with waters of the Collpa River on the basis of traditional and hence long-term historical use: "we have possessed these waters of the Collpa since time immemorial ... and we have used them continuously ... since the creation of the world." Similar in form to concepts embraced by many present-day environmentalists, Ayllita evoked the rights conferred by sustained use over the long-term past. By referring to various local places, he offered a detailed as well as highly personal and familiar knowledge of resources and landscapes.

A personalized and long-term view of resources continues to infuse the everyday accounts of Cochabamba peasants. In their commonplace discourse on soil erosion most attribute the worsening problem to an increased frequency and intensity of torrential downpours referred to as "crazy rains" or *loco paras*, an amalgam of Spanish (*loco* for "crazy") and Quechua (*para* for "rain"). This may seem to blame nature. But the ultimate responsibility for "crazy rains" is seen as personal. A neglect of ritual obligations toward the main non-Christian deity – the climate-controlling "Earth Mother" – has brought on recent worsening of "crazy rains." As Leocardia Gonzalez said:

> When I was a child my parents made offerings to the "Earth Mother" *{Pachamama}*. They cooked special foods which they buried in the soil, along with maize beer *{aqha}*. They did all this so that they would be looked on favorably by her. But today these practices aren't common although we still make offerings on Carnival and on Saint John's Day, and when we start to plant. But it's less than before; perhaps for this reason she's angry with us and maybe that's why there are so many "crazy rains" *{loco paras}*.
>
> (Leocardia Gonzalez, Tiraque, Cochabamba, 2 March 1991)

This account attests to how most peasants envision personalized reciprocity as the basis for obligations to the "Earth Mother." Such a customary reciprocity may form the basis of a peasant ethic on environmental conservation according to some indigenist anthropologists in Cochabamba (Rocha 1990; van den Berg 1991). Yet, while soil erosion is often attributed to ritual neglect, the region's peasants do not see this transgression as solely responsible for the divine thought to wreak dire environmental consequences. Many people reason that transgressions in the realm of social reciprocity also incite punishment from the earth deity who orders the heightened onset of "crazy rains" and ensuing soil erosion. Following their style of cause–effect thinking, erosion is born in the breakdown of customary social rights and obligations. In our conversations, numerous persons commented on a social world being undermined by disrespect, animosity, inequality, and violence.

Although the ideal of social reciprocity continues to infuse life and livelihood in the Cochabamba countryside, this belief often veils outright domination and growing differentiation of groups defined by wealth, status, age, and gender (Mallon 1983; Orlove 1974; Weismantel 1988). Among the Cochabamba peasantry, the elderly and the young adults differ in amount of schooling and non-farm work experience. Discourses on soil erosion correspond in an indirect, albeit important, way to this rift in experience. Elderly peasants are most likely to voice the explanations of soil erosion similar to those outlined above. Some elderly peasants find young people in their communities inviting divine wrath saying, for example, that "some people rebel against their parents," "children do not respect us," and, in one conversation, "parents are being killed." Numerous young peasants in Cochabamba, by contrast, cast soil erosion less in terms of the earth deity's wrath and more in terms of human-induced causes. The young people commonly blame their elders, those from whom they are inheriting degraded fields and pasture of obviously diminished value. More schooled, more likely to speak Spanish as well as Quechua, and more experienced in off-farm work, many young peasants admit skepticism about the earth deity, although few deny her existence outright. Conversational accounts of two 23-year-olds, a man and a woman, illustrate the generational shift:

> It's true that "crazy rains" have increased, the thunder too is greater than before. They have increased the problem of erosion, but the problem of erosion is due also to the fact that the ground is "naked." It no longer has grass or trees. These were depleted by our parents and the others [elders].
> (Casimiro Vargas, Tarata, Cochabamba, 15 June 1991)

> Due to erosion the fields that we [young adults] inherit are infertile. Seeing this some of us migrate to the Chapare. Furthermore, there's not much land left, and all of it is pure rock or at least rocky. There aren't good agricultural lands available for inheritance or partitioning. Look up there, for example, it's bedrock, along with some other rock-filled

fields. It looks as though the rain or perhaps the wind has removed the soil.

(Ninfa Salazar, Tarata, Cochabamba, 11 Oct. 1991)

During the 1980s, the discontent of young Cochabamba peasants about economic, political, and environmental dilemmas led to their increased involvement in peasant unions or trade unions. An increasingly common perspective on soil erosion formed as several rural trade unions initiated critiques that combined local perspectives with a broader consideration of related national and international issues. By the decade's end, young peasant voices were mingling with, and eventually adding to, prior explanations on environment and development in a revitalized branch of the traditional trade unions.

Rural trade unions

The peasants will no longer tollarate ... the exploitation of our natural resources by the oligarchy and the imperialists.

(*Resolutions of the Third Congress*, FSUTCC
(The Sole Trade Union Federation of the Peasant Workers of
Cochabamba), Cochabamba, 1986)

Rural trade unions were first organized in Bolivia after national defeat in the Chaco War with Paraguay (1932–5) and the ensuing weakening of its small governing elite. Through the mid-1980s, the rural trade unions – which belonged initially to the main miner-led Confederation of Bolivian Workers (COB) – did not state official positions on soil erosion or other degradation of rural environments, although they did advance criticisms and social analyses of water pollution in mining centers. The apparent absence of critique persisted even as leadership in the COB shifted in 1977 from mining centers to an axis combining city with countryside under the growing ethnic and social movement known as *katarismo*. *Katarismo* quickly gained much popularity among Bolivia's indigenous peasants for its commitment to cultural autonomy, social justice, and economic betterment. But during its early years the popular *katarismo* movement did not address the erosion dilemma. Even after 1979, when *katarismo* activists founded the first national trade union for peasants, known as "The Sole Trade Union Confederation for the Peasant Workers of, Bolivia" or CSUTCB, the problem of erosion could not be counted among its concerns (Albó 1987; Flores 1984; Healy 1989).

The absence of a discourse on soil loss became more conspicuous and perplexing since the *katarismo* leaders and their rural followers rose to power amid a growing awareness of worsening erosion in the late 1970s. That discursive silence did not result from mere coincidence, nor simple oversight, for at the same time the trade union movement advanced its critique of other environmental problems, especially water pollution and lowland deforestation. In fact, the official resolutions drafted at the national and regional meetings of rural trade unions and other union groups through the mid-1980s repeatedly detailed these environmental dilemmas (Calla *et al.* 1989; COB 1985; CSUTCB 1989; FSUTCC 1986). Yet the notable absence of erosion in this early environmental discourse coincided with the epistemology implicit in the trade union analysis of resource-related problems. Economic and political domination by transnational corporations and imperialist countries could be held accountable for water pollution in the mining centers and large-scale deforestation. Soil erosion, on the other hand, was not obviously extra-local in origin. Instead, for most trade union members, the causes of this problem were contained in local settings, among local people, including themselves.

Prevailing accounts in other social sectors and settings reinforced the otherwise perplexing omission of soil erosion from syndicalist discourse. Neither the vocal discourse of government and non-governmental institutions that blamed peasants' land use nor the personal perspectives that singled out ritual neglect offered much of a model that could be absorbed into a union position. But as erosion in Cochabamba worsened and a clamor began about its effects, the rural trade unions began to articulate their concern for the first time. Their efforts coincided with many young adult peasants gaining leadership posts in the late 1970s and the 1980s. The new trade union discourse on soil erosion stressed land-use practices as the main cause, extending to unfavorable economic policies affecting the peasant farm sector. One leader in the Campero province of Cochabamba said the following:

> The national government maintains a contradictory position ["thinks two times"]; on the one hand, they want us [peasants] to conserve the environment but on the other hand they pressure us to exploit the environment because we keep having to produce more to earn a livelihood.

The success of soil conservation projects, he continued, hinges on favorable policies for peasant farming.

This leader and others in Campero, who became especially involved in discussing soil erosion and conservation, referred to a lengthy historical past and personalized views of the environment in their trade union discourse. With a broad base of popular support and participation, Campero unions resembled other social movements in Latin America (Slater 1985). The Campero *sindicatos* selectively adopted several views and beliefs about resources from earlier generations of peasants, such as the Cacique Isidoro Ayllita mentioned above, which they used to defend their resource rights. Yet the environmental traditions were reinvented under the inspiration of a growing politics of cultural revindication launched by the ethnically charged *katarismo* movement. Reinvention of earlier ideas led one Campero peasant leader to thread environmental deterioration into the much-debated 1992 quincentennial of the Spanish invasion:

Throughout the last 500 years we peasants have been stepped on by the wealthy, the mestizos, and the Spaniards; the trees and animals similarly have been abused and are being extinguished, and thus we share much suffering along with the environment.

(Victor Flores, Aiquile, Cochabamba, 30 March 1991)

SUSTAINABILITY AND THE SEARCH FOR A "MIDDLE GROUND"

Discourses on soil loss in the Bolivian "heartland" of Cochabamba conjure distinct visions not only of the causes of degradation but also of scenarios for, conservation and development or sustainability. The majority of aid agencies – national, international, and NGO – used their backward-peasant argument to justify conservation programs emphasizing on-farm technical assistance and training. By contrast, the analyses by rural trade union leaders combined assessments of local and extra-local conditions. Some union officials in Cochabamba's Campero province, for instance, pressed the NGOs to design intermediate-scale technologies – such as small dams for irrigation that would improve existing land-use patterns – while also urging regional and national leaders to protest against unfavorable farm policies that worsen erosion. Clearly these discourses on soil loss were not simply mirrors of experience. They also constituted differing efforts to shape conservation and development through engagement with the existing policy narratives (Giddens 1979; Roe 1991).

Notwithstanding contrasting prescriptions for sustainable development, the major groups found themselves motivated by a similar concern: the impacts of soil loss. In a general way the groups came close to a broadly defined "middle ground." This concept has recently been applied by ethnohistorians to places and issues where different groups, cultural, social, political, economic, with quite distinct self-interests, were able to negotiate shared understandings and solutions to everyday problems (e.g. Merrell 1989; White 1991). Characterizing the erosion crisis in Cochabamba as a broadly defined "middle ground" resembles other environmental dilemmas where diverse groups pursue linked goals of sustainability. Indeed the idea of a "common future" was claimed by now classic statements on sustainability (Mathews 1989: World Commission on Environment and Development 1987). Political analysis and individual case studies indicate that such assumptions need to be tempered by the realization that environmental dilemmas may divide as much as they unite (Denevan 1973; Hecht and Cockburn 1989; Schmink and Wood 1992; White 1966).

The dilemma of soil erosion in Cochabamba did in fact divide people. There was little evidence of a neatly defined "common future" or "middle ground" at work in the major discourses on soil erosion. Yet, taking a broader view, it is possible to see some efforts at creating shared understandings. First, as individuals and groups altered standpoints through contending or accommodating other discourses, they created at least partially commensurate realms of meaning. Consider the changing perspective on erosion of many young chochabamba peasants and the leaders of their unions' social movement. Reminded daily of their degraded landscape, and in many cases frustrated by the abysmal failure of solely technical solutions to soil conservation, they sought sustainability in terms different than their parents and the region's development agencies. Yet even as the young people's discourse opposed others, it nonetheless drew them closer together in debate. Similarly some Cochabamba NGOs reacted to peasant discourse by advocating political and economic reform, thus extending beyond the counsel of technical assistance alone (Rist and San Martin 1991). Though their dialogues did not self-consciously seek "middle ground," an awareness and engagement with other positions enacted a preliminary sort of negotiation on the soil erosion issue.

Another semblance of "middle ground," albeit modest, was evident in the processes of justifying, or challenging, the government's role in conservation. At first the rural trade unions, development institutions, and peasants reinforced one another by indirectly affirming the Bolivian state's denial of an erosion problem. Beginning in the early 1980s, when this shared account of soil loss first started to fall apart in Cochabamba, the contestors' goal was not just accurate diagnosis but rather challenging and changing the government's policy. For its part, the Bolivian government moved to marginalize peasant and union discourses on erosion, typically by omitting them. Even after national officials acknowledged the erosion problem, they undermined the prospect of a mediated "middle ground" by excluding peasants and their organizations. But when the national government hastened its withdrawal from rural aid under neo-liberal reforms in the mid-1980s, the prospect for "middle ground" processes improved. The Cochabamba peasants and their local unions stood a better chance of negotiating conservation measures with the new development institutions.

The small semblances of a "middle ground" do not of course grant an arena in which sustainability can be easily attained. The activity of the new development agencies, dominated by NGOs, is rife with potential difficulties and possible cross-purposes. The NGOs are not held directly responsible to a citizenry or in some cases even to the Bolivian government. Thus far most environmental NGOs tend to offer partial models for facilitating the sort of public debate and democratic participation necessary for a consolidation of "middle ground" approaches toward sustainability. Though the role of the Bolivian government in discourses on soil loss diminished in the late 1980s and early 1990s it is likely that national government will continue to shape whether a fairly negotiated "middle ground" can be approached. Without a duly and democratically established arena for debate on sustainability, Cochabamba and other complex, developing societies are unlikely to find a sufficient area of "middle ground," leaving little hope for resolving environmental dilemmas.

ACKNOWLEDGMENT

A Post-Doctoral Fellowship from the Social Science Research Council and a research grant from the National Science Foundation funded the field project. The Graduate School at the University of Wisconsin – Madison, supported the analysis and interpretation of fieldwork findings. I am grateful for the collaboration and co-operation of numerous peasant farmers, rural trade unions, government agencies, and NGOs in Cochabamba, Bolivia. I have chosen to use pseudonyms for living farmers in order to maintain their anonymity. Their lack of secure civil and human rights makes this precaution a necessity since even a seeming innocuous discussion might conceivably jeopardize them at some later day.

REFERENCES

Adams, K. (1980) "Agribusiness integration as an alternative small farm strategy," Report to the Consortium for International Development (CID), Cochabamba.

Albó, X. (1987) "From MNRistas to kataristas to katari," in S.J. Stern (ed.) *Resistance, Rebellion, and Consciousness in the Andean Peasant World, Eighteenth to Twentieth Centuries*. Madison: University of Wisconsin Press, pp. 379–420.

Baptista Gumucio, M. (1977) *El pais erial: la crisis ecológica boliviana*. La Paz: Los Amigos del Libro.

Bassett, T. and Zueli, K.B. (2000) "Environmental discourses and the Ivorian savanna," *Annals of the Association of American Geographers*, 89 (3): 377–401.

Blaikie, P. (1985) *The Political Economy of Soil Erosion in Developing Countries*. Essex: Longman.

Blaikie, P. and Brookfield, H. (1987) *Land Degradation and Society*. London: Methuen.

Bryant, R. (2001) "Politicized moral geographies: Debating biodiversity conservation and ancestral domain," *Political Geography*, 19: 673–705.

Calla, R., Pinelo, N. and Urioste, M. (1989) *CSUTCB: debate sobre documentos politicos y asamblea de nacionalidades*. Talleres CEDLA Number 8. La Paz: EDOBOL.

Central Obrera Boliviana (COB) (1985) *VI Congreso de la COB: protocolos y tésis de la discussion politica*. La Paz: HISBOL.

Confederacion Sindical Unica de Trabajadores Campesinos de Bolivia (CSUTCB) (1989) *IV Congreso Nacional Ordinario: Comision de recursos naturals y tenencia de tierras*. Tarija. CSUTCB.

Corporación de Desarrollo de Cochabamba (CORDECO) (1980) *Análisis de problemas y potenciales en el desarrollo regional de Cochabamba*. Cochabamba: CORDECO.

Crosson, P. (1997) "Will erosion threaten agricultural productivity?," *Environment*, 39 (8): 4–31.

de Janvry, E. Sadoulet and L.W. Young (1989) "Land and labor in Latin American agriculture from the 1950s to the 1980s," *Journal of Peasant Studies*, 16, 3: 396–424.

de Morales, C.B. (1990) *Bolivia: medio ambiente y ecologia aplicada*. La Paz: Universidad Nacional Mayor de San Andrés.

Denevan. W.M. (1973) "Development and the imminent demise of the Amazon rain forest," *Professional Geographer*, 25 (2): 130–5.

Eagleton, T. (1983) *Literary Theory*. Minneapolis: University of Minnesota Press.

Eckholm, E. (1976) *Losing Ground: Environmental Stress and World Food Prospects*. New York: Norton.

Emel, J. and Peet, R. (1989) "Resource management and natural hazards," in R. Peet and N. Thrift (eds) *New Models in Geography*. London: Unwin Hyman, pp. 49–76.

Escobar, A. (1995) *Encountering Development: The Making and Unmaking of the Third World.* Princeton: Princeton University Press.

Estrada, V.J.G. (1991) "El fenómeno erosivo en Bolivia," *Pro Campo* (Chohabamba), 22: 29–30.

Fairhead, J. and Leach, M. (1996) *Misreading the African Landscape: Society and Ecology in a Forest–Savanna Mosaic.* Cambridge: Cambridge University Press.

Federación Sindical Unica de Trabajadores Campesinos de Cochabamba (FSUTCC) (1986) *Resoluciones del III Congreso Campesino, 17–19 de Julio 1986.* Cochabamba: FSUTCC.

Flores, G. (1984) "Estado, políticas agrarias y luchas campesinas: revision de una década en Bolivia," in F. Calderón and J. Dandler (eds) *Bolivia: la fuerza histórica del campesinado.* Cochabamba: CERES, pp. 445–545.

Forsyth, T. (1998) "Mountain myths revisited: integrating natural and social environmental science." *Mountain Research and Development* 18 (2): 107–16.

—— (2001) "Environmental social movements in Thailand: How important is class?" *Asian Journal of Social Science*, 29 (1): 35–51.

Giddens, A. (1979) *Central Problems in Social Theory: Action, Structure, and Contradiction in Social Analysis.* Berkeley: University of California Press.

Grossman, L. (1997) "Soil conservation, political ecology, and technological change on Saint Vincent," *Geographical Review*, 87/3: 353–74.

Grover, B. (1974) *Erosion and Bolivia's Future.* Utah State University Series 31/74, USAID Contract 511-56T. La Paz: USAID.

Gumucio, M.B. (1977) *The Wasted Country.* La Paz: Los Amicos del Libro.

Gupta, A. and Ferguson, J. (eds) (1997) *Culture, Power, Place: Explorations in Critical Anthropology.* Durham: Duke University Press.

Healy, K. (1989) *Sindicatos campesinos y desarrollo rural, 1978–1985.* La Paz: HISBOL.

Healy, K. and Paulson, S. (2000) "Political economies of identity in Bolivia, 1952–1998," *Journal of Latin American Anthropology*, 5 (2): 2–29.

Hecht, S. and Cockburn, A. (1990) *The Fate of the Forest: Developers, Destroyers, and Defenders of the Amazon.* New York: Harper Collins.

IIDE and USAID. (1986) *Perfil ambiental de Bolivia.* La Paz: USAID.

Lagos, M. (1997) "'Bolivia La Nueva': Constructing New Citizens." Paper presented at the Latin American Studies Association International Congress, Guadalajara, Mexico.

Larson, B. (1988) *Colonialism and Agrarian Transformation in Bolivia: Cochabamba, 1550–1990.* Princeton, NJ: Princeton University Press.

Laurie, N. and Marvin, S. (1999) "Globalisation, neoliberalism, and negotiated development in the Andes: Water projects and regional identity in Cochabamba, Bolivia," *Environment and Planning*, A31 (8): 1401–15.

LeBaron, A., Bond, L.K., Aitken, P.S. and Michaelsen, L. (1979) "Explanation of the Bolivian highlands grazing-erosion syndrome," *Journal of Range Management* 32 (3): 201–8.

Los Tiempos (Cochabamba) (1991) "Más del 35% del territorio nacional está sufriendo un proceso de erosion," 12 May 1991.

Mallon, F.E. (1983) *The Defense of Community in Peru's Central Highlands: Peasant Struggle and Capitalist Transition, 1860–1940.* Princeton, NJ: Princeton University Press.

Mathews, J.T. (1989) "Redefining security," *Foreign Affairs*, 68 (2): 162–77.

Merrell, J. (1989) *The Indians' New World.* Chapel Hill: University of North Carolina Press.

Ministerio de Asuntos Campesinos y Agropecuarios (MACA) (1977) "Recursos naturals renovables," in M. Baptista Gamucio (ed.) *El pais erial: la crisis ecoló boliviana.* La Paz: Los Amigos del Libro, pp. 43–56.

—— (1987) *National Meeting on Natural Renewable Resources.* Tarija: MACA.

Montes de Oca, I. (1989) *Geografía y recursos naturals de Bolivia.* La Paz: Editorial Educacional.

Nesbitt, J.T. and Weiner, D. (2001) "Conflicting environmental imaginaries and the politics of nature in Central Appalachia," *Geoforum*, 32: 333–49.

Neumann, R. (1995). "Ways of seeing Africa: colonial recasting of African society and landscape in Serengeti National Park," *Ecumene*, 2 (2): 149–69.

—— (1998) *Imposing Wilderness: Struggles over Livelihood and Nature Preservation in Africa.* Los Angeles and Berkeley: University of California Press.

Orlove, B.S. (1974) "Reciprocidad, desigualdad, y dominación," in G. Alberti and E. Mayer (eds) *Reciprocidad e intercambio en los Andes peruanos.* Lima: Instituto de Estudios Peruanos, pp. 290–321.

—— (1991) "Mapping reeds and reading maps: the politics of representation in Lake Titicaca, *American Ethnologist*, 18: 3–38.

Peet, R. and Watts, M. (eds) (1996) *Liberation Ecologies.* London: Routledge.

Pérez Crespo, C. (1991) "Why do people migrate? Internal migration and the pattern of capital accumulation in Bolivia," IDA Working Paper no. 74. Binghamton, NY: IDA.

Pestalozzi, H. (2000) "Sectoral fallow systems and the management of soil fertility: Indigenous knowledge in the Andes of Bolivia." *Mountain Research and Development*, 20 (1): 64–71.

Presencia (La Paz) (1990) "El problema de la erosion de suelos," 16 May.

Preston, D.A. (1969) "The revolutionary landscape of highland Bolivia," *Geographical Journal*, 135 (1): 1–16.

Preston, D., Macklin M. and Warburton, J. (1997) "Fewer people, less erosion: The twentieth century in southern Bolivia," *Geographical Journal*, 163 (2): 198–205.

Rist, S. and San Martin, J. (1991) *Agroecología y saber campesino en la conservacíon de suelos.* Cochabamba: Runa.

Robbins, P. (2001) "Tracking invasive land covers in India, or why our landscapes have never been modern," *Annals of the Association of American Geographers*, 91 (4): 637–59.

Robertson, M. (2000) "No net loss: Wetland restoration and the incomplete capitalization of nature," *Antipode*, 32 (4): 463–93.

Rocha, J.A. (1990) *Sociedad agrarian y religion: cambio social e identidad en los valles de Cochabamba.* La Paz: HISBOL.

Roe, E.M. (1991) "Development narratives, or making the best of blueprint development," *World Development*, 19 (4): 287–300.

Schmink, M. and Wood, C.H. (1992) *Contested Frontiers in Amazonia.* New York: Columbia University.

Slater, D. (1985) "Social movements and a recasting of the political," in D. Slater (ed.) *New Social Movements and the State in Latin America.* Dordrecht: CEDLA, pp. 1–26.

Staedel, C. (1989) "The perception of stress by campesinos: a profile from the Ecuadorian sierra," *Mountain Research and Development*, 6 (1): 35–49.

Storper, M. (1991) *Industrialization, Economic Development, and the Regional Question in the Third World: From Import Substitution to Flexible Production.* London: Pion.

Sundberg, J. (1998) "Strategies for authenticity, space, and place in the Maya Biosphere Reserve, Petén, Guatemala," *Yearbook, Conference of Latin Americanist Geographers*, 24: 85–96.

—— (n.d.) "Conservation, globalization, and democratization: Exploring the contradictions in the Maya Biosphere Reserve, Guatemala," In K.S. Zimmerer (ed.) *Geographies of Conservation and Globalization: New Dimensions of Environmental Management.*

Terrazas Urquidi, W. (1974) *Bolivia: País saqueado* [*The Despoiled Country*]. La Paz: Ediciones Camarlinghi.

Turner, M. (1999) "Merging local and regional analyses of land-use change: the case of livestock in the Sahel. *Annals of the Association of American Geographers*, 89 (2): 191–219.

Universidad Mayor de San Simón (UMSS). (1963) *Mesa redonda sobre desarrollo agroeconómico del valle de Cochabamba, 19–21 Marzo 1963*. Cochabamba: Imprenta Universitaria.

Urioste Fernández de Córdova, M. (1984) *El estado anticampesino*. Cochambamba: El Buitre.

USDA Mission to Bolivia (USDA) (1962) *Bolivian Agriculture: Its Problems, Programs, Priorities, and Possibilities*. Washington, DC: USDA

Van den Berg, H. (1991) "Conviven con la tierra," *Cuarto Intermedio* (Cochabamba) 18: 64–83.

Watts, M. (1983) "On the poverty of theory: natural hazards research in context," in K. Hewitt (ed.) *Interpretations of Calamity: From the Viewpoint of Human Ecology*. Boston: Allen & Unwin, pp. 231–62.

—— (1985) "Social theory and environmental degradation," in Y. Gradis (ed.) *Desert Development: Man and Technology in Sparse Lands*. Dordrecht: Reidel, pp. 14–23.

Weil, C. (1983) "Migration among landholdings by Bolivian campesinos," *Geographical Review* 73: 182–97.

Weismantel, M.J. (1988) *Food, Gender, and Poverty in the Ecuadorian Andes*. Philadelphia University of Pennsylvania Press.

Wennegren, E.B. and Whittaker, M.D. (1975) *The Status of Bolivian Agriculture*. New York: Praeger

White, G.F. (1966) "Formation and role of public attitudes," in G.F. White (ed.) *Environmental Quality in a Growing Economy: Essays from the Sixth RFF Forum*. Baltimore: Johns Hopkins University Press, pp. 105–27.

White, R. (1991) *The Middle Ground: Indians, Empires, and Republics in the Great Lakes Region*, 1650–1815. Cambridge: Cambridge University Press.

World Commission on Environment and Development. 1987. *Our Common Future*. Oxford: Oxford University Press.

Zimmerer, K.S. (1991) *Informe diagnóstico: uso de la tierra y la erosion de suelos en la cuenca del Rio Calicanto*. Report, Cochabamba: Centro para la Investigación de Desarrolle Rural (CIDRE).

—— (1993a) "Soil erosion and social discourses: perceiving the nature of environmental degradation," *Economic Geography*, 69 (3): 312–27.

—— (1993b) "Soil erosion and labor shortages in the Andes with special reference to Bolivia, 1953–91: implications for conservation-with-development," *World Development*, 21 (10): 1659–75.

—— (1994) "Local soil knowledge and development: answers to some basic questions from highland Bolivia," *Journal of Soil and Water Conservation*, 49 (1): 29–34.

—— (2000a) "Re-scaling irrigation in Latin America: The cultural images and political ecology of water resources," *Ecumene*, 7 (2): 150–75.

—— (2000b) "The reworking of conservation geographies: Nonequilibrium landscapes and nature-society hybrids," *Annals of the Association of American Geographers*, 90 (2): 356–69.

—— (2000c) "Social science intellectuals, sustainable development, and the political economies of Bolivia," *Journal of Latin American Anthropology*, 5 (2): 2–12.

—— (2003) "Environmental zonation and mountain agriculture in Peru and Bolivia: Toward a model of overlapping patchworks and agrobiodiversity conservation," in K.S. Zimmerer and T.J. Bassett (eds) *Political Ecology: An Integrative Approach to Geography and Environment-Development Studies*. New York: Guilford Publications, pp. 141–69.

5

PURITY AND POLLUTION

Racial degradation and environmental anxieties

Jake Kosek

It is not just the wood and soil or other resources that we are interested in preserving and protecting – it is something more, something deeper . . . it is the integrity, vitality, and purity of the wilderness that we want to maintain.

(Bryan Bird, Forest Guardians[1])

Wilderness is something that is entirely a white man's invention; it is not something I relate to . . . it is something that I have a deep reaction against. We have a close tie to the land – I have lived on and worked on and lived off this land my whole life. They just don't get it.

(Ike DeVargas, land grant activist[2])

ENVIRONMENTALISM'S TROUBLED (AND TROUBLING) "HEART OF WHITENESS"

One spring morning in Santa Fe, while conducting his morning tasks at the office of the Forest Guardians, a staff member went outside to retrieve the mail. Inside the mailbox, he found more than membership applications: carefully placed atop the letters was a large pipe bomb packed with ball bearings. The bomb's fuse had been inserted into one end of a filterless cigarette; it was evident that the cigarette had been lit, but had gone out a quarter of an inch before its embers would have reached the tip of the fuse. Sergeant Tom Stolee of the Santa Fe Police Department's bomb squad said that had the bomb exploded, it would have blown the Guardians' door off its hinges and killed any pedestrians within twenty feet. Two days later, the Guardians found an envelope in the mail: on the enclosed sheet was a drawing of a rifle scope's cross hairs over the words "Forest Guardians" and "see-ya" written underneath. It was signed "MM – the Minute Men" (Neary 1999: C3).

Sam Hitt, President of the Forest Guardians, considered it another case of what he termed "Green hate," and vowed that their mission to ensure the "protection and restoration of wild places will not be compromised by such cowardly acts" (Hitt 1999). The Guardians' then-executive director John Talberth said he was "not surprised," noting that "it's one small step from killing old growth forests and spotted owls to killing people" (Lezon 1999a: A2). Then-board member Charlotte Talberth pointed an accusing finger at Chicano activists Ike DeVargas, Santiago Juarez, and their supporters for "fomenting the hatred" that led to the bombing. She pointed to an all-day meeting the week before, held by officials and activists from Northern New Mexico's rural counties. They had come together to discuss their opposition to the Forest Guardians' regional plan for "rewilding" the southern Rockies, from southern Colorado to Northern New Mexico. The Minute Men were never identified, and neither were the parties involved in the attempted bombing. But this did not mark the first threat of violence to the Forest Guardians; in fact, they had received numerous threats before this event, and received more afterwards. One activist told me: "The only surprising thing about the bomb attempt on the Guardians is that it has not happened earlier."[3]

What is most interesting about the incident is that the potential culprits spanned the spectrum from radical Chicano activists to conservative property rights' advocates. In fact, many environmentalists theorized – without a shred of evidence – that the two factions had colluded in the coordination of the attack (Talberth, C. 1999). The dim view of local Chicanos was nothing new. Many members of the environmental community, including members of the Sierra Club, Forest Watch, the Forest Council, and the Forest Protection Campaign, had repeatedly expressed confusion and frustration over why they could not forge any significant alliances with Hispanos from Northern New Mexico. George Grossman of the Sierra Club put it this way: "I am not sure why we [environmentalists] get the brunt of so much hatred – we really should have a lot in common [with Chicanos]" (Grossman 1999). Others such as John Talberth felt that "the people of Northern New Mexico have been manipulated by a few extremists; in reality we (Forest Guardians) are their real allies; we have the same interests as they do" (Talberth and Talberth 1999). Talberth went on to write in a newspaper editorial, that "[t]he protesters [against the Forest Guardians] are tragically deluded as to who their real enemies are – the advocates for big industry and the Forest Service, who have consistently ignored the needs of small communities" (Talberth and Talberth 1999). Sam Hitt concurred: "They don't have the right enemy.... They are just throwing punches and not knowing where they are landing.... There are no real conflicts between the needs of rural communities and the goals of environmentalists" (Hitt 1999).

In what follows, I explore the notion of wilderness, the bitter responses elicited by its proponents, and its relationship to historical forms of whiteness. More specifically, I examine how notions of wilderness have been infused with racialized notions of purity and pollution.[4] Using links between contemporary New Mexico and the rise of particular racialized notions of nature around the turn of the twentieth century, I investigate how the movement to protect forests from degradation and pollution in that region draws on national metaphors regarding the contamination of pure white bodies and unsoiled bloodlines. I trace the entanglement of eugenicist conceptions of bodily purity with wilderness protection, and demonstrate how past formations of whiteness articulate with current struggles over wilderness in New Mexico. And finally, I argue that local Hispano activists' animosity towards environmental groups espousing strict preservation of forests is not so mysterious as it may seem to some environmentalists: it has very much to do with the ways in which forest preservation activities are haunted by exclusionary rhetoric of purity and entrenched fears of racial pollution.[5]

Most often, the history of the environmental movement is traced to either abusive land practices at the turn of the twentieth century, greater scientific understanding of "natural" processes, or the rise and expansion of modern enlightenment thinking into nonhuman realms (Hays 1959; Nash 1967). Even more progressive critiques of capitalism have become part of some wilderness advocates' rationale for the protection of "wild" spaces. These histories have clearly contributed to the development of the wilderness movement, but current battles within it point to a still greater diversity of origins.[6] From among those, I want to call into view an estranged ancestor: the movement for white racial purity, a specter of environmentalism's past that is hardly acknowledged, yet rarely, if ever, entirely absent. For as others have pointed out, while wilderness is a concept that by definition runs counter to modernity and politics, it is, in truth, a product of both (Cronon 1996; Neumann 1997; Spence 1999).[7] It carries with it complicated inheritances that counter its own claims to timelessness and universality. One need only look at the evictions of Native Americans from such icons of wild America as Yosemite, Yellowstone, and Glacier National Parks (among many others) to understand the deep and material contradictions of claims to pure, untouched nature (Solnit 1994; Spence 1999).

These aspects of wilderness have been well explored by others; it is not my intent to rehearse here what William Cronon calls the "trouble with wilderness," nor what his critics call "the trouble with Cronon" (Cronon 1996; *Wild Earth* 1992). Nor do I want to rework the ground that has been so fruitfully cultivated by political ecologists around questions of parks and people, though both are related. I do, however, hope to shed light on the complex relationships between forms of nature and forms of difference, and more practically, to illuminate tensions permeating the environmental movement in New Mexico. Because so much is at stake in these debates I want to be clear: I am not speaking generally about the current environmental movement or all environmentalists; neither am I denying that the wilderness movement has many different origins beyond what I discuss here, including many which are quite progressive. Instead, I mean to unearth some of the wilderness movement's deep and troubling roots, and to invite critical examination of the ways in which the movement – both in its past and in the present – is implicated in the reproduction of racial difference and class privilege.[8]

On another level, my work here is an attempt to broaden the capacity of political ecology to critically engage with, or even place at its analytical core, a cultural politics of difference. This shift does not simply call for the inclusion of racialized bodies as another variable or factor in a pre-existing frame of analysis. Rather, it proposes that the practices, politics, and effects of racial formation be examined as sites central to the politics of nature. This form of engagement requires three critical moves. First, it entails a treatment of nature as more than a physical environment filled with external landscapes. More than this, nature needs to be understood as a broader and deeper terrain that incorporates internal essences, evolutionary imperatives, and would-be universal Truths. Second, it entails a rejection of assumptions about fixed forms of difference, and instead an exploration of the ways in which nature and the environment are complicit in making and remaking forms of difference, including race. Finally, I maintain that these moves are important precisely because this routing of race through notions of nature has made for dangerous and tenacious couplings, securing technologies of unjust governance, stabilizing relationships between subjects and their assumed, disabling essences, and shaping landscapes, both material and emotional.

The history of the environmental movement stretches far beyond debates between Muir and Pinchot to encompass multiple, deeply political sites wherein nature is produced outside of the realm of the environment. For conceptions of nature are slippery. From their formation in one site, they travel through metaphor and material practice, crossing seamlessly between bodies, souls, universal laws and forests processes, between the species and the individual, all with troubling invisibility and stunning audacity. This approach raises new questions, both about the links between nature and forms of difference, and about the nature of the environment itself.[9] By learning from and integrating insights from critical race theory, political ecology will be better equipped to untangle forms of social difference from biology and treat nature as more than a fixed set of environmental objects, thus allowing its theorists and practitioners to better illuminate the symbolic and material ways in which formations of nature and difference are made and manifest in resource struggles. This reconfigured analytic can do more than enrich political ecology's approach to environmental politics; it can also help to destabilize the universal and timeless ground of nature upon which essentialist ideas of race have been and continue to be built.

This more critical political ecology engages both ethnographic research and critical race theory to address pressing questions of injustice, drawing attention to the fine-grained practices and politics through which racial and class identities are formed, naturalized, and contested within the arena of environmental politics. I hope to show how these entanglements of race, class and nature are manifest both in abstract ideas, such as "wilderness," and in material forms, such as the gunpowder, ball bearings and lead pipes found in the Forest Guardians' mailbox. The divisions between various progressive ideas of the environmental movement are clearly manifest in New Mexico, where longtime environmental advocates typically line up on very different sides of the fence – often reaching across only to grab at one another's throats. The struggle over the environmental movement there is, in large part, a struggle over these different roots.[10]

OF BLOOD AND POWER: "OVERLAPPINGS, INHERITANCES AND ECHOES"[11]

[The population of Northern Mexico is] a sad compound of Spanish, English, Indian and Negro bloods ... resulting in the production of a slothful, indolent, ignorant race of beings.

Columbus Delono, congressman from Ohio, 1846
(quoted in Horsman 1981: 240)

I will begin by looking in greater depth at historical notions of purity in relation to race, first in a broader sense, and then more specifically as these relationships have played out in the wilderness movement over time in New Mexico. I do not intend this to be a comprehensive genealogy of race in the United States, nor do I claim that all forms of race are inherently the same. My intent here is to outline how racial discourses – especially notions of racial purity and improvement – articulate with formations of nature. And I hope to show how fears of the dilution and degradation of race – in particular, of forms of whiteness – became entangled with fears of the degradation of New Mexico's "pristine" forest landscapes. My claim here is that discourses of purity placed diluted racial subjects and degraded landscapes into the same "grid of intelligibility," wherein understandings of and fears surrounding race at the turn of the twentieth century became the raw substance out of which wilderness as an idea and a landscape was forged (Foucault 1978). Moreover, these fears of bodily pollution folded into new formations of "wild" landscapes at a particularly tense moment in American history: former slaves were emancipated and migrating, immigration was rapidly rising, and the protection of a white, masculine notion of nationality had become a central preoccupation.[12] Notions of wilderness and its importance to the nation must be understood within this temporal and spatial context.

The notion of protecting or maintaining the purity of a racially exclusive national body politic has long been central to American nationalism. From the first naturalization laws in 1790 limiting the privilege of citizenship to "free white persons," to the nineteenth century's Chinese Exclusion Act, to California's Proposition 187 in the late twentieth century, this country's history is riddled with legislated racial exclusion and definition. Regardless of contemporary mythmaking about the nation's longstanding multiracial identity, numerous battles have been fought – some, ongoing to this day – to preserve and reproduce this nation's white racial "character." When President Theodore Roosevelt considered the weakening of whites' "strong racial qualities" and the declining population among whites amid rising immigration to be "race suicide," and when President Coolidge, upon signing the 1924 Immigration Act drastically limiting immigration into the United States, stated that "America must remain American," each echoed deep-seated fears of racial degradation (Bederman 1995).

Many scholars have noted that racial discourses have hidden attachments (Almaguer 1994; Cosgrove 1995; Hall 1986a; Roediger 1991; Ware 1992). But these fears of bodily pollution in the United States in the mid-1800s reached significant proportions. Discourses of pollution became deeply imbedded in the formation of the nation, including its narratives of improvement and progress, its selective construction of its own "common" national history, and its desired national future. Indeed, the rationale of American expansionism was embedded within a racial logic substantiating the expansion of the social and political principles of the American Anglo-Saxon offshoot of the Caucasian race (Horsman 1981). It was posited that the superiority of the white race not only enabled its conquest of other races and the spread of "good" government, commercial prosperity and Christianity throughout the world but, in fact, this undertaking was the manifestation of its destiny. This pungent mixing of paternalism and colonialism became, as Kipling's oft-quoted poem proclaimed later in the century, "the white man's burden." (Kipling 1899). This mixture also became one of the driving and legitimating forces in western expansion within the United States.

Another equally troubling consequence of the racial logic of western expansionism was the conviction that progress – as seen in this rubric – was inevitable in the war between races. This was particularly true in relationship to the Spanish and Mexican Southwest in the mid- to late-1800s. Class distinctions had been a prominent feature in that region, forging links between the elites of the Spanish blood caste system and wealthy white capitalists. Yet the tension between American and Mexican/Spanish elites grew as racial tension in the United States became more entrenched. As renowned racial scholar Josiah Nott outlined in the mid 1850s, important distinctions existed even between those of European descent:

> The Ancient German may be regarded as the parent stock from which the highest modern civilization has sprung. The best blood of France and England is German; the ruling caste of Russia is German; and look at the United States, and contrast our people with the dark-skinned Spaniards. It is clear that the dark-skinned Celts are fading away before the superior race, and that they must eventually be absorbed.
>
> (Nott, as quoted in Horsman 1981: 131).

Contempt for "mixed blood" Mexicans was even greater. In 1846, US Representative Columbus Delono from Ohio described the population of Northern Mexico as "a sad compound of Spanish, English, Indian and Negro bloods ... resulting in the production of a slothful, indolent, ignorant race of beings" (Horsman 1981: 240). At the center of this discourse – one in which races were set off as different and simultaneously assigned to a singular, evolutionary hierarchy – was the need to legitimate the expansionism dictated in manifest destiny (Horsman 1981).[13]

Furthermore, many scholars have pointed out how the individual body and the social body have been deployed as metaphors and metonymys for each other (Thongchai 1994). In this way, the bodily health of the individual citizen and the well-being of the collective nation become culturally intelligible through commonly deployed metaphors of blood, vitality, and race.[14] At times, fears of pollution and contagion in the colonies, for example, became a central concern, spawning efforts to control colonial officers' sexuality for fear of diluting the potency and purity of the European race (Arnold 1996; Grove 1995). The nation is also often seen as embodied in individuals – athletes, cultural icons, and political leaders, among others; their success or failure often linked implicitly to patriotic notions of the strength and well-being of the national character. Whether seen as virile or viral, the body has served as both metaphor and metonymy for processes that occur well beyond the boundaries of the skin.

Particularly powerful has been the symbol of blood as a means of defining the boundaries of difference. Blood has served to link identity to the body, present generations to past and future, and individual characteristics to the vitality of species. Foucault called attention to blood as a mechanism of power:

> For a society in which the system of alliance, the political form of the sovereign, the differentiation into orders and castes, and the value of descent lines were predominate; for a society in which famine, epidemics and violence made death imminent, blood constituted one of the fundamental values.... Power spoke through blood: the honor of war, the fear of famine, the triumph of death, the sovereign with his sword, executioners, and tortures; blood was a reality with a symbolic function.
>
> (Foucault 1978: 123).

Foucault traces the course by which power operates in society from a "symbolics of blood" to "an analytic of sexuality," which becomes the basis for his understanding of modern power (Foucault 1978). Of course, control of the purity of blood has much to do with sexuality, and Foucault does not assume a complete break in this transition. He notes that "[w]hile it is true that the analytics of sexuality and the symbolics of blood were grounded at first in two very distinct regimes of power, in actual fact the passage from one to the other did not come about without overlappings, interactions and echoes" (Foucault 1978: 149).

The bio-political shift that occurred in relation to blood and its purity is significant. Foucault suggests that the move from a preoccupation with a royal elite, whose blood purity must be protected from a contaminative society, to the defense of an implicitly racially pure society from the biological dangers of another race, represents a shift not only in formations of race but also in the operations of power. As Ann Stoler points out, the key elements in this calculus are still "society," "enemies," and "defense," but their arrangement is different. What must be defended – and what must be defended against – significantly changes; thereafter the role of the state is transformed from that of an unjust state to a state that "is and must be the protector of the integrity, the superiority, the purity of the race" (Stoler 1995: 71). Racial and blood purity no longer pit one social group against another, or against the state, but instead serve as mechanisms by which to sustain the health and life of both the individual and the entire population. As Foucault argued, in this new formulation, the war of the races changes shape to become a racism that "society will practice against itself, against its own elements, against its own products; it's an internal racism – that of constant purification – which will be one of the fundamental dimensions of social normalization" (as quoted in Stoler 1996: 67). Stoler points out that this understanding makes racism:

> more than an ad hoc response to crisis; it is a manifestation of preserved possibilities, the expression of an underlying discourse of permanent social war, nurtured by the biopolitical technologies of "incessant purification." Racism does not merely arise in moments of crisis, in sporadic cleansings. It is internal to the biological state, woven into the weft of the social body, threaded through its social fabric.
>
> (Stoler 1996: 69)

Foucault sees these new forms of racism as rebuilding the previous symbolics of blood, spawning new, biologizing forms of racism (Foucault 1978). Foucault takes some of the most significant of these to be the modern forms of racism that arose in the second half of the nineteenth century: the science of eugenics, and one form of its state expression, Nazism, which attempted to cleanse the German national body by exterminating individuals and populations that it understood as pollutive threats. Foucault traces this modern racism back to seventeenth-century beliefs that the social body was divided into two separate, warring races. He posits that nineteenth-century bourgeois class anxieties were constructed according to this racial grammar, spawning the call to cleanse and purify the social body of these threats. Efforts were made to differentiate the social into natural or biological orders of race, caste and descent lines. The rise of new forms of intervention surrounding the body and everyday life found expression at the level of health and hygiene which, he notes, indicates another effort to protect the vitality and purity of race (Foucault 1976).

WILD NATURES: THE MAKING OF A TRUE-BLOODED AMERICAN

Like the links between nation, blood, and body, the connections between nation and "wild" nature in America are anything but arbitrary, simple, or benign. Perhaps the most influential origin story of American nationalism grows out of these persistent connections. In 1893, Fredrick Jackson Turner delivered his famous paper "The Significance of the Frontier in the Building of American National Identity" (Turner [1898] 1994: 61). His basic premise, in this paper, is that the confrontation between civilization and the wild, demanding frontier, transformed the fundamental character of Americans as a people, transforming them into strong individuals with a propensity for democratic principles of governance. It was not just any immigrant that Turner had in mind; implicit in his frontier thesis is the transformation of English and German "stock" into a new, Anglo-Saxon, fundamentally masculine, American stock.

Speaking of the frontiersman, Turner states that "[l]ittle by little, he transforms wilderness, but the outcome is not the old Europe, not simply the development of Germanic germs . . . here is a new product that is American" (Turner [1898] 1994: 35). What drove these white explorers? Turner, directly echoing the rhetoric of manifest destiny, quotes Grund's famous essay on America, positing that "it appears then that the universal disposition of Americans to emigrate to the western wilderness, in order to enlarge their dominion over inanimate nature, is the actual result of an expansive power which is inherent in them" (Turner [1898] 1994). According to Turner's treatise, the "Americanization" of the European, or at least a particular class of European, takes place in the western "wilderness," which is itself made "American" by free white men – not by former slaves migrating West after abolition, nor by Chinese laborers, nor Mexican sheep herders, all of whom significantly transformed western landscapes (Turner [1898] 1994).

Turner imagined that the nation's strength came from its wilderness and argued that "the existence of an area of free land, its continuous recession, and the advance of American settlement westward, explain American development" (Turner [1898] 1994). Turner also claimed that "the frontier is gone, and with its going has closed the first period of American history," and "a great historical movement" (Turner [1898] 1994: 56). The closing of the frontier, Turner feared, signaled an end to Americans' conquering spirit. What, he worried, was going to test and distinguish Americans as a people if the very material and forces that formed American identity ceased to exist? The anxiety over the closing of the frontier came at a moment of great transition in American society, and Turner's words resonated with an anxiety about the character and boundaries of racial dominance in America. The closing of the frontier meant the loss of wilderness, which in turn implied the loss of the site in which white American masculinity had been produced – and with it, the "superior" institutions and civility through which the nation had been constituted.

This anxiety over the protection of national and racial superiority is especially visible in the context of immigration. From the late 1880s through 1914, the United States experienced one of its largest influxes of immigration, reaching almost 1.3 million people in 1907 alone. Between 1870 and 1920, over 26 million people migrated to the United States (Jacobson 2000). Not until the 1980s would an equal number of people enter the nation in one year (Cosgrove 1995: 34). So deep were anti-immigrant fears that President Theodore Roosevelt campaigned against birth control among Anglo-Saxons, believing that the overwhelming numbers of non-Anglo-Saxons would diminish the quality and quantity of the superior "native American stock" (Graves 2001: 129). Similarly, the young Woodrow Wilson commented on biological threats to Anglo institutions that stemmed directly from an increasing influx of immigrants, whom he described as hailing from "the ranks where there was neither skill nor energy nor any initiative of quick intelligence" (Graves 2001: 131).

Fear of contamination by immigrants led to direct conflict with the desire to create an immigrant "army of surplus labor." This tension resulted in a paradox: immigration took place, but so, too did the segregation of those same immigrants. Laws preventing Chinese from testifying in court, the exploitation of sharecroppers in the South, and later the *bracero* program, which imported Mexican immigrant laborers without offering them basic human rights – all these became means by which to contain racial difference within the national body, while at the same extracting labor and profits from immigrant bodies. If immigrants had to be part of the means by which the national body could extract profit, then internal forms of differentiation and a means of protecting the nation had to be developed. Many tensions are at play here; at this point it is enough to note that during the early twentieth-century wave of immigration, many Anglo-Saxons were as concerned about the mixing of the races – something they believed would lead to a less pure nation – as they were of the immigrants themselves.[15]

THE SCIENCE OF DESTINY AND THE "GREAT WHITE MISSION"

These racialized fears found some of their strongest articulation and legitimation within the language of science. Theories of polygenesis – which posited that different races had, in fact, different origins – were the most widely accepted theories of racial difference at the time. Indeed, tensions over theories of polygenesis revolved not around the argument that non-Anglo races were inferior but rather their potential challenges to the biblical genesis story.[16]

Nineteenth-century race theorists Dr. Josiah Nott and Egyptologist and professional lecturer George Gliddon drew off the work of prominent scientists, ethnologists, evolutionary biologists and phrenologists, asserting that "a long series of well-conceived experiments has established the fact that the capacity of the crania of the Mongol, Indian and Negro, and all dark-skinned races, is smaller than that of the pure white man. And this deficiency seems to be especially well-marked in those parts of the brain which have been assigned to the moral and intellectual faculties" (Horsman 1981: 131). Nott goes on to claim that: "everything in the history of the Bee shows a reasoning power little short of that of a Mexican" (Horsman 1981: 131).[17] His sentiments about racial purity reflected ones that were becoming deeply entrenched in the mid- to late-1800s. Fears abounded that pure strains of Aryan blood would be polluted, thus weakening the nation. Nott explained that "[t]he adulteration of blood is the reason why Egypt and the Barbary states never can again rise, until the present races are exterminated and the Caucasian substituted" (Horsman 1981: 130). This scientific naturalization of racial difference helped to create not just the idea of a hierarchy among races, but something on the order of "natural" distinctions among races that could not be changed. Dr. S. Kneeland wrote, in an introduction to the 1852 English version of Darwin's *The Natural History of the Human Species*: "the dark races are inferiorly organized, and cannot, to the same extent as the white races, understand the laws of nature" (Horsman 1981: 134).

Debates about Darwin's ideas of natural selection and species diversity and their relationship to race are far too involved to engage in detail here, but they warrant brief comment. More than any other theory, Darwin's theory of evolution (written in 1860) became deeply intertwined with racial debates, and anchored even more firmly in the popular imagination this conception of race as a subset of naturalized hierarchies of difference. Intentionally or not, Darwin both drew from and contributed to debates about race. And while his position directly countered ideas about the polygenesis of the human race, it opened the door for new ways of understanding racial difference. Herbert Spencer, Ernst Haeckel and members of the rising eugenics movement in the United States were deeply influenced by Darwinian concepts, as well as by Mendelian theories of heredity in farm animals. Both were harnessed to explain a wide variety of moral, intellectual and social traits in humans, including poverty, patriotism, and of course racial difference.[18]

EUGENICS: PURIFYING AND PROTECTING NATURE

Francis Galton, preeminent British scientist and cousin of Charles Darwin, coined the term "eugenics" – meaning "good in birth" – in 1883 (Selden 1999). Galton believed in the genetic superiority of the British ruling class – thus, he reasoned, their leadership and economic position – and he became a popular advocate of selective breeding in the late 1860s, long before the term eugenics appeared. Though the tenets of eugenics had their roots in earlier ideas of race, the rise in production of "scientific" knowledge regarding racial difference found its traction at this junction where new theories of evolution mixed with the burgeoning field of genetics and deepening anxieties concerning racial degradation. Galton's notions borrowed from and contributed to work in the United States, and, by the turn of the century, eugenic theories of social behavior underpinned the "common sense" understanding of racial difference and provided the legitimating authority for a whole host of new policies and social programs. In fact, at the turn of the century, eugenic theories found transpolitical support – from conservatives to progressives to libertarians – and were deployed in immigration reform, sterilization programs, marriage laws, health policies and segregation programs.

Organizations such as the American Eugenic Society, the Galton Society, the American Breeders Society, and the Immigration Restriction League were formed to guide and implement immigration and population control policies in the United States. Prominent eugenicist and avid naturalist Charles Davenport was recruited to lobby Congress on immigration issues.[19] With the help of the Carnegie Institute and the Rockefeller Foundation, Davenport founded the prestigious research institute at Cold Spring Harbor to "investigate and report on heredity in the human race, and emphasize the value of superior blood and the menace to society of inferior blood" (Spence 1999: 4). Davenport was extremely successful at persuading Congress, the Surgeon General and other officers within the US Public Health Service and the Department of Education to align with the eugenics movement. He actively published articles on the importance of eugenics, using it to support immigration restriction and population controls. The results helped make eugenics a key public health issue and brought it to the center of education policies. As L.K. Sadler declared to the Third International Congress of Eugenics: "The stocks which carry the germ plasm of leadership, talent and ability must be nurtured and increased; better babies must be the watchword ... the race must be purified" (Selden 2001: 22). Explicit in Sadler's and others' arguments are fears of contagion and pollution of blood purity, the rise of "social inadequacies" due to improper breeding, and the increased social burden on a nation yoked into supporting genetically inferior races. Eugenicists were able to exploit historically resonant fears of impurity and convince Congress that the "American" gene pool, originating with the Puritans themselves, was being polluted by defective germ plasm and creating a growing number of a genetically inferior American "stock." As a direct result, Congress passed the 1921 Immigration Act.[20] President Calvin Coolidge made the law's premise explicit when he signed the 1924 Act into law: "Biological Law shows that Nordics deteriorate when mixed with other races."[21]

Throughout the mid- to late-eighteenth century, notions of whiteness and superiority relied deeply on formations of nature. From the natural "destiny" of whites to "manifest" their "innate" tendencies toward western expansion, to the basis of racial difference in the eugenics movement, nature has been central to concepts of racial purity in the United States. It is no coincidence that in this context – one filled with obsession over the purity of bloodlines and the nation's body politic – the wilderness movement was born. It was at the very moment when immigrants were "flooding" the cities, when new epidemics were "infecting" the population, and when the frontier that had supposedly both tested and made white men and their institutions of governance was believed to be "closing" that the early "fathers" of environmentalism, such as John Muir, George Perkins Marsh, Gifford Pinchot and Aldo Leopold, developed and began to propagate concerns over degradation of the natural integrity of pure wilderness.

"HIDDEN ATTACHMENTS:"
THE PURITY OF BLOOD AND SOIL[22]

I wish now to suggest that Muir, Leopold, Marsh and other early environmentalists, though they may seem so, were not overtly racist so much as they were creatures of their historical moment. Though many of their writings have troubling, often explicitly racist overtones, these and other men drew from prevailing understandings of and anxieties around race to make environmental issues intelligible, and their impulse to create and protect national wilderness areas flowed directly from the perceived need to differentiate and protect the "pure" from the "polluted," the "natural," from the "unnatural." The result was that racial and class fears surrounding purity and degradation became a primary means through which wilderness and the environment became discernible. By feeding on the prevailing fears of that particular moment in American history, they galvanized support for wilderness preservation; the importance of maintaining in perpetuity the purity of the nation's environment – the very environment that was said to embody white nationalism and help forge the nation's individual character and institutions – resonated with popular understandings and fears of the nature of race.

When John Muir went into the Sierra to, as he put it, "get their good tidings," he did not just discover the forest through his wanderings; he brought with him his life history as an immigrant Scot who had worked as a laborer and had developed a deep distrust of things modern. On the wanderings that took him to California, he brought the New Testament, Robert Burns' poems, Milton's *Paradise Lost*, and the writings of Darwin, Whitman, Emerson and Thoreau (Strong [1971] 1988). By the time he embarked on his first summer in the mountains, Muir already carried with him idealized notions of the West, deeply held Judeo-Christian beliefs, and perceptions of the changing condition of the working class, all of which were part of the means through which Muir came to understand landscapes. Muir also packed in with him contemporary fears and attitudes about race that led him to conclude that not everyone belonged in his beloved mountain cathedrals. Muir wrote disdainfully about the "Chinaman" and "Digger" Indian who first set off with him into the Sierra, and about the lack of enlightened appreciation on the part of the Hispanic herders for the majestic grandeur of the mountains. Along with scorn for the "filthy," "lazy" habits and perpetual "dirtiness" of the herders, he also deplored the sheep themselves, calling them "wooly locusts," "dirty," "wretched," "miserably misshapen and misbegotten." (Spence 1999: 23) He saw both the sheep and these men as out of place in the mountains, and placed them all – sheep, Hispanos, "Chinamen," fallen Indians – in opposition to the purity and grandeur of "Nature." He complained that he could not find the "solemn calm" when they were present and described the Indians in Yosemite as "mostly ugly, and some of them altogether hideous." He argued that they had "no right place in the landscape."[23]

The wilderness sanctuaries Muir held so dear were not, as he believed, simply "created by god"; they were created by the US Cavalry, armed with the nineteenth-century authority of manifest destiny. In fact, it was in pursuit of Indians that whites first discovered Yosemite Valley. And it was that same pursuing battalion that finally captured Chief Tenaya on the shores of Pyweack Lake and marched him and his band to a reservation in the flat, hot San Joaquin Valley. Upon the group's capture, the United States soldiers told the Chief that they were going to rename the lake after him "because it was upon these shores of the lake that we had found his people, who would never return to it to live. . . . His countenance," one soldier wrote, "indicated that he thought the naming of the lake no equivalent for his loss of territory" (Solnit 1994: 220). As Rebecca Solnit points out in her essay on Yosemite, it is on this same site that, 25 years later, John Muir camped and wrote of the purity and wildness of the valley: "[Lake Tenaya] with its rocky bays and promontories well-defined, its depth pictured with the reflected mountain, its surface just sufficiently tremulous to make the mirrored stars swarm like water-lilies in a woodland pond. This is my old haunt where I began my studies.... No foot seems to have neared it" (Solnit 1994: 220).

Muir was not opposed to the US Army's presence in Yosemite; in fact, he continued to promote its presence in the valley to keep out perceived undesirables – especially Hispanics and Native American grazers. Muir declared "blessings on Uncle Sam's soldiers! They have done their job well, and every pine tree is waving its arms for joy" (Meyerson 2001: ix). It is likely that he knew of at least some of the occupying Army's past, though he wrote very little about it. Though he depicted it otherwise, John Muir's unblemished wilderness was, in fact, a space of violent, racially driven dispossession, one of a series of removals, massacres and impoverishments that had reduced the Native American population in California from 250,000 to 16,000 within half a century (Ehrlich 2000: 85). These brutal acts created the conditions not only for the "wild" Sierra that Muir and others exalted over so passionately, but also the "solemn calm" they unapologetically experienced there. Indeed, this type of "pure," "natural" space, created by the elimination of Native Americans and others who were deemed to have "no right place in the landscape," became the basis for the National Park system in the United States (Spence 1999: 133).

Muir and many others thus helped create an external nature shaped by internal lines and boundaries that separate pure wilderness from sullied society. Parks and wilderness areas are, in fact, monuments to the ideological separation between nature and society. This is not just an abstract separation of nature and culture; this is a particular form of separation reflecting the anxieties, politics and relationships – human and inhuman – of a particular time. Parks and wilderness areas have served as material, naturalized reaffirmations of this spatial separation and those relationships. They are, of course, not fixed; their meanings are the site and source of constantly changing politics. But the meanings themselves are not easily changed. The density with which the social relations of race and class are embedded within these spaces of "pure" wilderness has helped reproduce attitudes about the nature of race and perpetuate the racialization of nature.[24]

WILDING SUBJECTS: THE "PURIFICATION MACHINE"

Nature served as a purification machine, a place where people "became white." ... The journey into nature [for purification] was just as much a journey away from something else, and that something else was race."

(Braun 2003: 197)

Nature's external purity was also celebrated as a catalyst for internal purity. While society degraded the human spirit, and modernity and its trappings polluted both nature and the human soul, the solution, many thought, was to be cleansed by a return to that which is timeless, to nature as it was before humanity's fall, to the "true," pre-social world of wilderness. This process of purification merits more attention, for the creation of such wilderness did more than make nature divinely and racially pure, spatially separate and materially expressed in trees, mountains and rivers. It also created subjects of wilderness. It was, as Braun describes, the "purification machine" that took polluted individuals and helped make them pure again (Braun 2003). The act of going out into wilderness also was and continues to be an act of looking inward. This is perhaps one of the most recurring themes in the argument for wilderness. The formation of individual subjects has also served as one of the central themes of nation building, from as early as Fredrick Jackson Turner's white pioneers creating both country and character. And so the intertwined formation of a nation and its people continues to serve as a central logic for preserving wilderness. Wallace Stegner, one of the most eloquent supporters of the wilderness, wrote this in support of the passage of the Wilderness Act of 1964: "We need wilderness preserved – as much of it as is still left, and as many kinds – *because it was the challenge against which our character as a people was formed*" (Stegner 1961; emphasis added). This implicit grouping, this trinity of body, nature, and nation, is not accidental or insignificant; rather, it has its origins in the belief in racial salvation through a return to nature.

A tacit assumption of many of these early arguments was that nature's healing capacities, or rather, the ability of whites to benefit from nature's curative powers, depended on the absence of, and distance from, those with darker skin. Braun addresses this rather large caveat: "nature served as a purification machine, a place where people became white." In fact, he argues, "the journey into nature [for purification] was just as much a journey away from something else, and that something else was race" (Braun 2003). This myth of white purification was made more persuasive and insidious by its suggestion that what the wilderness adventurer had to learn was internal and eternal. Because wilderness has been created as a space beyond the social, the wilderness traveler believes he/she is experiencing the essence of nature, pure nature, unpolluted by the social, cultural aspects of society. It is this myth that makes the search for our inner selves so compelling, something to "get back to," a place that serves as a mirror to our own true nature. Of course, wilderness does not underlie our true being any more than nature determines culture. As Donna Haraway observes in *Primate Visions*: "Nature [serves as] the raw material of culture, appropriated, preserved, enslaved, exalted or otherwise made flexible for disposal by culture." She claims that "the appropriation of nature [serves] the production of culture" and acts as a means for the "construction of the self from the raw material of the other" (Haraway 1989: 16).

Muir is just one of many advocates for this kind of natural transformation of the inner self, of finding the soul through the exploration of nature. From Ralph Waldo Emerson to Charles Darwin, Theodore Roosevelt to Edward Abby, Aldo

Leopold to Gary Snyder, the discovery of the self in the supposedly timeless material of nature has served as one of the most dominant themes in western environmentalism. People go to nature to find their "true selves," to "remember" the basis of life. The "call of the wild," is, in truth, nature's hailing. It is a green version of Althusser's famous "Hey, you there," but in this case the interpellating agent is not a state official but a social and political history that is vested in and bound up in the materials of mountain, rivers and forests (Althusser 1971). But because the hailing is outside of humanity, because it is from a "pure" source, the calling goes unexamined and its political histories remain obscured. Thoreau exclaimed:

> Give me the ocean, the desert, the wilderness! ... When I would recreate myself, I seek the darkest wood, the thickest and most interminable and ... the most dismal swamp. I enter a swamp as a sacred place, a *sanctum sanctorum*. There is the strength, the marrow, of Nature. The wildwood covers the virgin mould, and the same soil is good for men and trees. ... In such soil [civilization] arose and out of such wilderness comes the reformer eating locusts and wild honey.
>
> (Oelschlaeger 1991: 165)

The same theme is present in Aldo Leopold when he exhorts us to "think like a mountain,"[25] or when Muir "discovers" himself in Yosemite, or when hikers come to "find" themselves through the timeless wisdom of nature. Acts of self-discovery are, of course, not unique to western subjects; transformations of the self through nature occupy many different traditions far beyond those of western environmentalism. Even in the West, it can be argued that acts of self-discovery by white environmentalists have different purposes and effects; subjectivization, like nature, is contingent and uneven. Many dynamics are at play here; for now it is enough to simply acknowledge the persistent links between individual subjectivity and nature forged through these acts of self-discovery.

Walt Whitman was another believer in nature's role in forming individuals. In *Leaves in the Grass*, he wrote: "Now I see the secret of the making of the best persons. It is to grow in the open air, and to eat and sleep with the earth" (Whitman [1860] 1961: 319). But, like Muir, Whitman did not extend this character-building ability to non-Anglo-Saxons. When Whitman was editor of the *Brooklyn Daily Eagle* in the 1840s, he argued that American expansion and manifest destiny would be good for the whole world. He wrote: "[w]hat has miserable, inefficient Mexico ... to do with the great mission of peopling the New World with a noble race" (Horsman 1981: 235)? He celebrated General Taylor's capture of Mexican territory as "another clinching proof of the indomitable energy of the Anglo-Saxon character" (Horsman 1981: 235). What Whitman and others like him wanted to preserve, evidently, were not just the leaves of grass, but also the physiological and psychological milieu out of which the individual white male was formed in America.

The link between race and nature was even more direct in the work of George Perkins Marsh.[26] In a frightening foreshadowing of Turner, Marsh believed that American government was the product of this mixing of a potent strain of Germanic–Anglo tradition with the wilds of America. In 1868 he wrote: "The Goths are the noblest branch of the Caucasian race. We are their children. It was the spirit of the Goth that guided the May-Flower across the trackless ocean; the blood of the Goth that flowed at Bunker Hill" (quoted in Horsman 1981: 181). For Marsh, nature – both human and environmental – was something that could be controlled, that needed protection and the proper management. It followed, then, that a love of liberty and effective governance were exclusive attributes of the Germanic people (Horsman 1981). Marsh argued that "they [California and New Mexico] are inhabited by a mixed population, of habits, opinions, and characters incapable of sympathy or assimilation with our own; a race, whom the experience of an entire generation has proved to be unfitted for self-government, and unprepared to appreciate, sustain, or enjoy free institutions" (Horsman 1981: 183). At stake for him in these debates is no less than a loss of purity, the decline of the race and the consequent corrosive effect on the white nation.

But Marsh also recognized that the return to nature was not without peril. In words eerily reminiscent of Hitler's, Marsh argued that "if man is indeed above nature, wherever he fails to make himself master [of nature], he can be but her slave" (Marsh [1864] 1973: xxvi). In this formulation, there is a balance: the potential destruction of nature – leading to the further decline of civilization and ultimately to barbarism – is tempered by the fact that nature is manageable by "man." So it follows that we must govern "her," nature, both for the good of nature and of "man." Such arguments allowed Marsh's work to feed directly into the eugenics movement after the Civil War. This need to manage nature fits well with eugenicists' arguments and was widely used by them. They wanted to take nature's evolutionary process, as described by Darwin, and make improvements on it. Those who claimed some knowledge of or control over nature demonstrated, by their own logic, their superiority over those who did not. Thus while the "lesser races" were subject to nature's whims, the "higher races" were able to bend nature and its subjects to their will, for their own good.

Francis Galton, the father of the eugenics movement, made this explicit in his landmark paper in *The American Journal of Sociology*: "What nature does so blindly, slowly and ruthlessly, man may do providently, quickly and kindly" (Galton 1904: 2). Galton, along with a growing group of scientists, politicians and popular supporters, sought to "introduce [eugenics] into the national conscience, like a new religion. It has, indeed, strong claims to become an orthodox tenet of the future, for eugenics co-operates with the workings of nature by securing that humanity shall be represented by the fittest races" (Galton 1904: 2). Darwinian conceptions of nature here are combined with Marsh's vision of a nature that needs to be both protected and managed for the well-being of civilization.[27] The answer to the dilemma was to manage nature more efficiently, more benignly; to protect nature's purity while at the same time developing better subjects through closer interaction with it. Many pick up on these insights, most notably Gifford Pinchot, who was himself actively supportive of both Marsh's ideas and the eugenics movement. He compared the managing of people's nature to the managing of forest nature, claiming that "only in this way could the forest, like the race, live on" (Guha 2000: 30; Kevles 1985). His models for managing the nature of the forest and the nature of the race both called, at their core, for the proper governance of nature's purity.

"DANIEL BOONEING" AMERICAN HISTORY: THE "DARK AND BLOODY REALITIES OF THE PRESENT"

Fears of contagion were expressed by environmental leaders from Muir to Roosevelt to Pinchot and others; all saw immigration restriction as vital to the protection of nature's purity. But these fears are not limited to the late nineteenth and early twentieth century; these issues of purity and perceived national threat continue to be at the forefront of contemporary debates around the protection of nature – whether the contagion is national, racial, or environmental.

The Sierra Club's proposed initiative to support California's Proposition 187, which would have defined the Club's position as actively anti-immigration, was a clear relic of these turn-of-the-century fears.[28] Though the Sierra Club measure lost, the massive support it received, including from a number of prominent environmentalists, was very telling. Stewart Udall, Gary Snyder, Dave Foreman, David Brower, Farely Mowat, Herman Daly and Lester Brown were just a few of the well-known environmentalists who publicly supported the measure.[29]

Edward Abbey, prominent author and modern-day environmental renegade and hero, was probably the most often invoked in the Sierra Club debates over the issue. Abbey wrote: "I certainly do not wish to live in a society dominated by blacks or Mexicans, or Orientals. Look at Africa, Mexico and Asia" (Petersen 1998). He even invoked Garrett Hardin, a neo-Malthusian biologist who developed the infamous "tragedy of the commons" theory: "Garrett Hardin compares our situation to an over-crowded lifeboat in a sea of drowning bodies. If we take more aboard, the boat will be swamped and we'll go under. [We must] militarize our borders [against illegal immigration]. The lifeboat is listing" (Petersen 1998). He went on to even more directly echo turn-of-the-century eugenicists, stating that "it might be wise for us, as American citizens, to consider calling a halt to the mass influx of even more millions of hungry, ignorant, unskilled, and culturally-morally-genetically impoverished people.... Why not [support immigration]? Because we prefer democratic government, for one thing; because we still hope for an open, spacious, uncrowded, and beautiful – yes beautiful! – society, for another. The alternative, in the squalor, cruelty and corruption of Latin America, is plain for all to see" (Petersen 1998). Abbey's views, as well as those of many others engaged in recent immigration debates, clearly reflect long-standing conceptions of a pure nature threatened by various forms of racial difference.[30]

Aldo Leopold was still another environmental movement founder indebted to bodily metaphors and to a rhetoric lamenting the degrading national health and its consequences to nature. Like George Perkins Marsh and other predecessors, Leopold believed that nature had to be properly managed for the "good of man" and for its own "well-being" (Leopold [1949] 1987: xxiii). Indeed, Leopold considered wilderness to be the purest and "most perfect norm" and, as such, he believed it "assumes unexpected importance as a laboratory for the study of land-health" (Leopold [1949] 1987: xxiii). We are lost without it, he wrote: "we literally do not know how good a performance to expect of healthy land unless we have a wild area for comparison with sick ones" (Leopold [1949] 1987: 196–7).

Leopold also agreed with Muir that human use of wilderness involved "direct dilution" that "destroys" the "pure essence of outdoor America" (Leopold [1949] 1987: 172–3). However, he allowed himself one caveat. Like Muir and Marsh, Leopold conceived of "wild" nature as central to the formation and/or regeneration of the citizen – or at least, the white male citizen (Leopold [1949] 1987).[31] He argued: "Wilderness areas are . . . a means of perpetuating . . . the more virile and primitive skills in pioneering travel and subsistence" (Leopold [1949] 1987: 192). The experience of wilderness, he insisted

> reminds us of our distinctive national origin and evolution, *i.e.* it stimulates awareness of history. . . . For example, when a boy scout has tanned a coonskin cap, and goes Daniel Booneing in the willow thicket below the tracks, he is reenacting American history. He is, to that extent, culturally prepared to face the dark and bloody realities of the present.
>
> (Leopold [1949] 1987: 177)

Of course, the "American history" reenacted by the boy scout and revered by Leopold overlooks the "dark and bloody realities" of the past – as well, I believe it is safe to argue, as those of the present. In the boy scout's performance, the theater of wilderness bears no traces of land dispossession, immigrant labor, or slavery. Rather, the celebration of his "Daniel Booneing" reinforces a "purified" white national history, one that relies on nature to bind national citizenship to gender and race.

Indeed, as Robert Finch points out in the 1987 introduction to the reprinting of Leopold's classic compilation, *The Sand County Almanac*: "[N]o idea of Leopold's has been more important . . . than his assertion that our encounters with wild nature can reveal, not only interesting and useful observations about natural history, but important truths about human nature" (Leopold [1949] 1987: xxiii). This claim must have seemed almost self-evident to Leopold, given his belief in Americans' "wild rootage" (Leopold [1949] 1987: 177). However, despite his attempts at deciphering these natural "truths," Leopold neglected to grasp the profoundly political nature of these roots, particularly the fears of the loss, degradation and infirmity caused by social degradation, which was largely defined as a mixing of upper and middle-class whites with those of another ilk.

Ultimately for Leopold and many other conservationists, a healthy landscape, like a healthy body, is the purest one. As I hope the foregoing discussion has suggested, this equation of purity and health, of both land and body, is closely linked to the history of racial struggles over the purity of white bodies as they battle against contamination by unhealthy, impure peoples and nations. Particularly telling are Leopold's metaphors of the human body, metaphors that are no less deeply immersed in national and regional discourses of race than they had been for thinkers a generation or so earlier. When Leopold says that "the evidence indicates that in land, just as in the human body, the symptoms may lie in one organ and the cause in another," and "the practices we now call conservation are, to a large extent, local alleviation of biotic pain," he is tacking back and forth between metaphors – of the nature of the body and the nature of the landscape – that are necessarily grounded within historical and contemporary notions of the bodily health of the individual and the nation (Leopold [1949] 1987: 195). These metaphors are also, interestingly enough, grounded in the debates around race in turn-of-the-century New Mexico, where Leopold, as a young Forest Service ranger, developed the germs of many of his ideas and the country's first wilderness area (Leopold [1949] 1987: 175).

NEW MEXICO: GUARDING THE FOREST, PROTECTING THE PURE

The Southwest has been a seedbed for such great visionaries of the environmental movement. I mean, Aldo Leopold, Dave Foreman, Ed Abbey. Look at the people that have come out of this blasted landscape. There is a clarity of vision; there is a singleness of purpose that instills people in the Southwest, and I don't know where that comes from, but it's absolutely part of the landscape.

(Sam Hitt, Forest Guardians 2000)

The idea of wilderness we have used is flawed. This flaw is never acknowledged when "white" or urban environmentalists gather because the concept has been driven into us so completely.

(Chelise Glendening 1996)

On 20 April 1999 – Earth Day – I attended a public presentation in Santa Fe organized by the Forest Guardians "to educate and inform people about health and threats to Northern New Mexico Forests" (Hitt 2000). The presentation was one of many events going on that day in Santa Fe, including a tree-planting ceremony, a kids' educational fair – the usual events one might expect to mark the occasion.

The talk was held in Santa Fe's public library, and about 60 people were in attendance, all of whom were white, well-dressed, and seemingly genuinely concerned about forest issues. Three staff members of the Forest Guardians introduced themselves and explained that the Forest Guardians was the most active and most uncompromising of the groups engaged in the protection of the forests. They then began a slide show with an opening image of a plantation: a large white house stood in the background, surrounded by a green, manicured garden.

Bryan Bird, one of the presenters, told the crowd: "These are the true roots of the environmental movement. . . . When people tell us we must compromise, that we must lower our standards and commitment to the integrity and health of wild forest, we remember that compromise did not end slavery." He added: "it was the Civil Rights movement activists in the 1960s and their discovering the words of Muir and Leopold that launched the modern-day environmental movement." The presentation went on to address broader struggles over the forests in Oregon and Washington, and the civil disobedience techniques that people were using in the struggle for "what is left of the pure and pristine wild spaces of the West." The audience was reminded of the importance of the national campaign for "zero cut" and "zero grazing" on federal lands and the work that the Forest Guardians were engaged in, locally, nationally, and with other groups, to forward these agendas both in public opinion and within federal agencies.[32]

We were then told that in nearby Vallecitos the Guardians had lost a recent battle to stop logging in the area. "The dangers to this area are real; what lies in the balance is the last best hope for the preservation of one of the few remaining unspoiled areas of forest in the region." We were left with a sense that a small island of pure wilderness stood alone against a rising tide of human imposition. The talk ended with a call to not compromise the last free and wild places in the West, and a commitment to "rewilding the West" through the creation of zones and corridors and more open, untouchable areas.[33]

The audience had a few questions. The first came from an elderly woman whom I recognized from a Sierra Club meeting a few weeks before. She asked about how the group was dealing with the "real problems" that underlie the "threat to wild spaces," which she defined as issues of "population control." The speakers nodded as she spoke and did their best to answer the unwieldy question, pointing to the loss of the Sierra Club initiative as a loss for the environment. Bird reassured her that many people were continuing to work on that issue, and that the fight had not been lost. He also pointed out that while population was part of the problem, another factor was also our over-consumption of resources. The woman conceded that "yes, that is true," but reasserted that, "to save our resources we need to protect both our borders and our wild lands."[34]

Another question, raised by an elderly man, was a simple one: "Why are the environmentalists so disliked in Northern New Mexico?" A tense moment followed, but the matter was something that almost anyone involved in New Mexico public life, or who even regularly read the newspaper or listened to the radio, knew to be true. Bird referred back to a controversy surrounding fuelwood collection in the wake of a 1995 injunction to protect the endangered spotted owl,[35] but claimed that the Forest Service had "manufactured the tension" in a "divide and conquer" move "that fractured the possibility of alliances between the community and environmentalists." He also claimed that a few "radicals" like Ike DeVargas and Max Cordova helped stir up the problem, which he claimed was "in fact not as widespread as it seemed." Soon after, the presentation came to a close and small knots of people gathered to talk more individually with the speakers. The event itself was not at all surprising. I had been to many such meetings before, but this was the first time I heard environmentalists claim a direct lineage to the civil rights movement. Indeed, after this event, it became a much more common refrain among environmentalists.

When I recounted environmentalism's newly claimed civil rights heritage to Santiago Juarez, a longtime radical Chicano activist and organizer in the region, the 180-pound man responded: "Yeah ... and I am Snow White." Then he turned serious: "Where the fuck were they during the 1960s and the *La Raza* movement? I didn't see any of them at the marches in Albuquerque. They weren't canvassing for *La Raza Unida* or being arrested by the cops with Corky [Gonzales]. Where were they when we marched against the racist policies and unfair hiring and wage practices at Los Alamos? Where the fuck are they when some white cop pulls me over for nothing?" He finished with an outright dismissal: "They're as tied to the civil rights movement as much as I am part of the Klan." As I sat with him, he got even angrier. His sentiment was not unique; before and since, I heard similar reactions from many others throughout the region (Juarez 1999).[36]

Kay Matthews, one of the editors of the region's radical community newspaper, *La Jicarita*, frames the conflict this way: "You have two of the most progressive environmental justice groups in the West just down in Albuquerque. When was the last time they went down and marched or organized for Chicanos?"[37] Matthews referred to Forest Guardian leader Sam Hitt's claims that the small-time irrigators are the culprits in the water wars in New Mexico, that one major problem is that agriculture uses 90 percent of the water in New Mexico and only about one-third actually reaches the fields. Hitt's position is that locals are using technology from the nineteenth century, a flood irrigation system called the *acequia*, and this is what is causing the degradation of local rivers (Matthews 1999: 3). But Matthews argues correctly that nowhere near that amount of water goes to agriculture and that the water used is central to growing food and maintaining the livelihoods of the people who live in the area and hold legal title to that water – access rights dating back hundreds of years. Matthews asserts: "Sam is fighting to take waters away from the *acequia* for the silvery minnow, calling the *acequia* inefficient. But when was the last time he went down and fought with Intel in Albuquerque over their water use – which is massive – in the production of computer chips? Or when was the last time they addressed white urban sprawl?" (Matthews 1999).[38] Matthews sums it up this way: "You need to judge the Guardians both by what they do and also what they don't....

It is very telling which struggles they are involved in, who they blame, and which ones they avoid in the region" (Schiller and Matthews 2000).[39]

Though there are many examples of the tensions in New Mexico between commitments to social justice and environmental concerns, probably none make this tension clearer than a public letter entitled "A Letter to Environmentalists" written by Chelise Glendening, who describes herself as a "recovering environmentalist."[40] The letter was published in the *Santa Fe Reporter* in April 1996, after the Federal fuelwood injunction that halted all logging on Forest Service lands. The letter was a small but telling part of the intense controversies that followed in its wake. Glendening knew many people involved in that struggle, including local Chicano activists Ike DeVargas and Max Cordova. Her time with them helped convince her that there was a serious tension within the environmental movement that needed to be addressed. She started the letter by "asking environmentalists to stand behind the politics of indigenous Chicano people 100 percent," claiming that "the idea of wilderness we have used is flawed" (Glendening 1996). She pointed out that "this flaw is never acknowledged when 'white' or urban environmentalists gather because the concept has been driven into us so completely." But, she added, "whenever indigenous people join us, the flaw is quickly pointed out." She writes eloquently about the need to lift the veil and see what is behind what is unquestioningly referred to as "our" efforts at preservation. "Put most simply," she says, "this veil concerns our unthinking use of the word 'we.' 'We' must save the forests! 'We' must build a better world!" She then asks: "How different are these statements from the outlandish manifest destiny rationales used to conquer these lands in the first place?" The rewards of this internal examination, she claims, would be to have "new ideas, new strength, new comrades and, best of all, to understand that 'we' ... is something entirely different from saying it inside the empire" (Glendening 1996).[41]

The letter resonated powerfully with many environmentalists, including members of the Forest Trust and the New Mexico Green Party, and it spawned a number of meetings and teach-ins at the *Oñate* Community Center in Española. Here, at long last, was a respected environmentalist siding with Chicano loggers. Her position was, of course, not surprising to DeVargas or Cordova or many other local loggers who knew her. But because of the position from which she spoke, her letter had powerful repercussions within the environmental community in Northern New Mexico. It also identified a legacy that would later become more apparent, a fracture in the environmental community that would only become wider in the coming years. The initial letter was followed by another, entitled "Inhabited Wilderness." Written by Glendening, along with fellow "recovering environmentalists" Marc Schiller and Kay Matthews, and signed by 75 others who supported an alternative vision of wilderness, the letter characterized the tension in the following way: "The most recent tragedy to emerge from this injustice is a conflict that is tearing the people of New Mexico apart. On the one side stand the advocates of pure wilderness, working to halt a toxic civilization by isolating areas away from human use; on the other, the Indio-Hispanic communities of the north, fighting for their lands, livelihood and culture" (Matthews 1999.[42] The letter goes on to pronounce that "we support their [natives' and Indio-Hispanos'] right to sustainable forestry, including community-based logging and restoration, as well as hunting, fishing, herb-gathering, firewood collection, and water. And we honor their right to make decisions about the lands that, according to the Treaty of Guadalupe Hidalgo, are theirs" (Matthews 1999).

The letter and the subsequent meetings only led to greater animosity toward the Forest Guardians' position. Contrary to what Bryan Bird had claimed at the 1999 Earth Day meeting, the animosity toward environmentalists in Northern New Mexico was very widespread. As a matter of fact, it was one of the few topics on which almost all Hispano locals I met seemed to agree. The shared anti-environmentalist sentiment assumed the status of the weather as an assuredly universal topic of conversation at the post office, at food counters, in parking lots and along the shaded sides of adobe buildings where people gather. It provided the safest conversational gambit because, just as everyone suffers the same deprivations from the weather, everyone shared frustrations about the most recent lawsuit, or overheard statement, or letter to the editor about environmentalists. Though many locals may not have agreed with everything that DeVargas or Cordova or other leaders in the forest struggle did or said, these activists became ever more popular for their vocal opposition to environmentalists in general, and to the Forest Guardians in particular.

One especially contentious encounter took place in August 1996, when members of the Forest Guardians and the Forest Conservation Council hosted a camp-out at the timber sale site in the Vallecitos Sustained Yield Unit.[43] Over the course of three days, the groups organized workshops on bird identification, tree sitting and nonviolent protest techniques. After hearing of the event, DeVargas, other members of *La Compañía* (a community based logging organization), and members of the community of Vallecitos all chose to stage their own counter-event to protest against the presence of the environmentalists. After all, although 75 percent of the timber sale had been guaranteed for *La Compañía*, the Forest Guardians had forced an injunction to stop the logging. *La Compañía* could not begin the logging, and as a result many people had no work (Wilmsen 1997).

Locals claimed that since the Guardians and *La Compañía* were involved in litigation over the area, it was "provocative" for the Guardians to have staged the workshops at the timber sale site. Local counter-demonstrators began by hanging John Talberth and Sam Hitt in effigy, as they had in an earlier demonstration in Santa Fe; this time the effigies hung from the trees along Forest Road 274, which led to the timber sale site and the site of the Guardians' workshops. They posted signs that read: "It's not the owl . . . it's a way of life that's at stake." When asked if a compromise was possible, Sam Hitt responded: "This stand must be protected," and "my bottom line is that these old growth pines will not be cut" (quoted in Matthews 1999). He claimed that it was simply "culturally irresponsible" to log in the area (Matthews 1999).[44] Eventually, DeVargas and Cordova and others met to discuss their differences with the Guardians, an event which devolved to a great deal of finger-pointing, followed by heated threats and further discussions. After the incident, DeVargas commented that Guardians' leader Sam Hitt "can no longer portray himself as David fighting Goliath, out to save the poor people against the corporate giant. He has now become Goliath" (quoted in Matthews 1999).

WILD NATURES AND A NEW WAVE OF COLONISTS

This vision of environmentalists as the new Goliath, a new "wave of colonists," was widely shared by many involved in New Mexico forest politics (DeVargas quoted in Goldberg 1997: 19). The Guardians' choice to use a litigious, one-size-fits-all approach to environmental issues only exacerbated these feelings. As Chelise Glendening claims, there is "a lot of wisdom in the environmental movement, but it has been a mistake to try to disseminate it through litigation" (DeVargas quoted in Goldberg 1997: 21). The legal tactic was effective in stopping large-scale logging nationally, but in this region it seemed to exacerbate existing antagonism between the environmentalists and the Forest Service. Yet the community's antagonism toward environmental groups runs far deeper than that of the Forest Service; and in most of my interviews, it did not appear so much in discussions of the Vallecitos Sustained Yield Unit, or even in the specifics of the firewood controversy. Rather, it was clearly articulated in disbelief at the arrogance of the Forest Guardians.

Truchas resident and ex-Reverend Alfredo Padilla said during one firewood gathering trip: "Who the heck do they think they are?... They act like they can send down commandments and that we should all get out of the way or get on our knees for them." (Padilla 1999).[45] He later asked rhetorically: "Who put them in charge of these woods? The Forest *Guardians*? I did not ask them to *guard* these forests. Who are they guarding them for? Not for the people who live here. They are guarding [the forest] so that they can have their own playground" (Padilla 1999).[46] Others, like Sam Cordova, who worked thinning and selling firewood, said: "I do not feel the woods are going to be safer because they are guarding them.... I cannot think of any bigger threat to these forests than the Guardians ... all they are doing is making it safe for people to develop here [in Northern New Mexico]" (Cordova 1999).[47]

In a letter to the *Albuquerque Journal* entitled "Green Vision Blind to Native Hardship," three Chicano county commissioners asked: "What right other than conquest do these people [environmentalists] have to develop a vision for our communities or for the lands stolen from us?" (Montoya *et al*. 1997: A15). Referring to the injunction against logging on federal lands, the three added: "The courts have been used to rid the United States of our kind for too long. We will do whatever is required, as individuals and as elected public servants, to defend our country and our people from a sophisticated, treacherous and deceptive attempt at cultural extermination" (Montoya *et al*. 1997: A15).

Both the intent and tone of the editorial are very clear; so, too, are its racial undertones. "Our kind" is vaguely defined here. At times it might be a class reference; it can also be tied to culture and place – but it is most certainly racial. In the preceding paragraph, the letter asks, in reference to the Southwest Forest Alliance (which is made up of 16 environmental groups including the Forest Guardians), whether "there is a Hispanic, a Native American, or even an individual raised in Northern New Mexico among them" (Montoya *et al.* 1997: A15)? The answer is, not surprisingly, no. The letter articulates one of the biggest concerns expressed in innumerable meetings, interviews and conversations with local Hispanos. What they find most objectionable is often not the details of environmentalists' claims but rather the claim of the almost all-white, largely male contingent to be the singular, rightful voice for nature, and for its protection.

In fact, the Forest Guardians assume this mantle with little equivocation. As Sam Hitt said during an interview: "We might not always be popular, but if we did not look after the forest, who would?" He broadens the rationale to a campaign whose bounds are as noble and inevitable as those expounded by the environmentalists' founding thinkers: "We are doing something bigger than ourselves; we are working to preserve the forest for people who will be living beyond our lifetimes. [We work] to maintain its health and protect its integrity." The best way to do this, according to Hitt and others, is to "keep as much of it as wild and free of degradation as possible" (Hitt 1999).[48]

When I asked Hitt what was at stake for him personally in the preserving wilderness, he said:

> When I go out deeper into it [wilderness], I end up going deeper into myself. It does not happen all the time; most of the time I go to the forest and I see problems. I see cows in the wilderness. I see roads that are polluting sediment into streams. I don't see the creatures that should be there.... If you're not sad, you have no right to be alive in the twenty-first century. You're living inside a cocoon. You're numb. You've lost connection with the wild and you're blind to the incredible ecospasm that's going on, on the planet. It's global suicide, this greatest extinction in 60 million years.... There is something about it [wilderness] which makes us stronger, physically and mentally; it recharges our batteries; it restores our souls.... It is these trips, both the problems and the beauty, that reaffirm my commitment to what I am doing and remind me of why it is so important.

> (Hitt 1999)

Bryan Bird expressed a similar sentiment: "Yes, it is about preserving endangered species; yes, it is about protecting old growth forests and maintaining biodiversity. But it is also about reminding ourselves of who we are. The fact is, our inner nature is connected to our outer nature" (Bird 1999).[49] These sentiments express the selfsame notions of pure wilderness espoused by Muir, Marsh and Leopold. It is

150

this understanding of pristine, non-human nature, as well as these deeply personal, sentimental, and political connections to it, that are at stake in struggles over the forests in New Mexico. It is also this understanding of nature that makes the preservation of its purity and the commitment to its improvement so sacrosanct. This is the key point, the fulcrum on which my argument turns; I am arguing that if it is through this connection to nature that contemporary environmental citizens are formed, and if we accept that nature has deeply racialized roots, then it must follow that the ways that environmental subjects form themselves and their ideologies through nature must be examined more carefully.

This ideological heritage was clearly demonstrated in *The State of the Southern Rockies*, a report authored by Bird, along with then-Forest Guardians member John Talberth, and published by the Forest Guardians. Sam Hitt traced the genealogical connection himself, claiming that the report "was an offspring of Aldo Leopold's vision of land health and John Muir's vision of wilderness" (Hitt 1999). The report grew out of a 1996 meeting in which 23 environmental organizations in the region agreed to collaborate under an umbrella organization they would call the Southwestern Wildlands Initiative. It was part of a larger set of initiatives of the Wildlands Project, which hoped to establish an "audacious plan" because, in their words, "North America is at risk" (Talberth and Bird 1998: 4). According to them, this plan is central to the region's survival and recovery. The intent is to create a "vast, interconnected area of true wilderness" by means of a connected system of reserves that span from Panama and the Caribbean to Alaska and Greenland. The plan was most clearly articulated in a special issue of *Wild Earth* dedicated to "Plotting a North American Wilderness Recovery Strategy."[50] The magazine featured articles by EarthFirst! activist David Foreman, poet Gary Snyder, conservation biologist Michael Soulé and many others, all in support of the plan. Foreman goes so far as to call the Wildlands Project plan "one of the most important documents in conservation history," claiming that what its creators "seek is a path that leads to beauty, abundance, wholeness and wildness" (Foreman 1992).

The *State of the Southern Rockies* report claims that, if the region is "managed properly," it will be possible to restore much of the area to its wild state (Talberth and Bird 1998: 9). The authors continually invoke metaphors of a sick and imperiled patient in need of a recovery strategy; they propose to restore natural health to the forest through scientific management and rational planning.[51] Needless to say, given the environment of conflict over the forests in Northern New Mexico, the Forest Guardians' role in producing the plan did little to help it receive favorable reviews.

Among the reviews it did receive was an eloquent letter from land grant activist Max Cordova (Cordova 1999).[52] He stated that he was drawn to write a response because, though "the plan is an abstraction, disconnected from the day-to-day lives of people living in the area ... the Forest Guardians' lawsuits have themselves demonstrated [that] these abstractions are based on objects and issues as real and concrete as the wood in my backyard, the temperature of our homes in winter and the sovereignty of our lands" (Cordova 1999). Cordova stated that he was concerned that:

> the report describes a plan to build a "wilderness," a "land where the earth and its community of life are untrammeled by man." The plan's prescription to create a "pristine nature" out of a landscape that is deeply related to our history – from the births and deaths of family and friends, to the sweat and labor of our ancestors as herders, hunters, farmers, firewood gatherers, community loggers, acequia members, miners – is deeply disturbing.... These "wildlands" are not wild; they are the products of intensive use dating back hundreds, if not thousands, of years.... The Forest Guardians are not the first to use the notions of wilderness in this way. The concept of an open, unoccupied, "wild" frontier has been the myth that has fueled the dispossession of lands in America for a long time. Whether the planners of this report are conscious of it or not, the report carries on this legacy that empties the landscape and erases our history ... which disingenuously dismisses the past, with disturbing implications for the future.
>
> (Cordova 1999)

Cordova's recollection of racialized dispossession from his ancestors' point of view is powerful. But it, too, needs to be complicated, lest this story appear to be a simple one in which a traditional, rural Hispano group is pitted unfairly against an overwhelming force for national/natural purification, itself driven by a racially haunted past. It is important to interrogate diverse notions of blood, and notions of nature, as much those of local Hispanos as those of environmentalists. Hispanos claim that their land was stolen by the Forest Service, and they invoke claims to blood purity that often seamlessly cross centuries, eliding the brutal histories of Spanish colonialism in America as well as centuries of cross-racial intermarriage. Such claims staked upon the purity of bloodlines enable the possibility for land title restoration, according to the terms of the Treaty of Guadalupe Hidalgo, and yet also implicate the title recipient directly in the legacy of colonial violence. Indeed, blood politics in the region are deeply complicated by histories and contradictions that make simple ideas of Hispano opposition to environmental claims of wilderness and whiteness – while powerfully compelling – less than straightforward. Moreover, still more troubled variations on the narrative of purity and pollution haunt these bodily and natural landscapes. Among the most compelling are the very real fears of radiation pollution from Los Alamos, one of the biggest employers in the region, and the skyrocketing deaths from heroin overdoses that have earned the region the ignominious distinction as the nation's rural heroin death capital. All of which is to say that any investigation of the intersections of environmental politics and blood politics should not end with a critique of the eugenicist roots of environmentalist thought; instead, it might begin there, and move outward.

CONCLUSIONS

> We seek the purity of our absence [in nature], but everywhere we find our own fingerprints.
>
> (Richard White[53])

My argument here has been simple. Tensions that exist around the nature of wilderness in New Mexico (and elsewhere) are deeply rooted in very particular formations of nature – formations that owe much of their shape, size, and even soil structure, to anxieties over the loss of bodily and national purity in mid- to late-nineteenth-century North America. Nature, race, nation, have been intimately and insidiously bound together for well over a century, from Darwin's theories of natural improvement and progression, to Turner's warnings about the closing of the American frontier, to the invention of polygenesis and eugenics to ensure the integrity and health of the middle- to upper-class white populace. Wilderness advocates and other proponents of the early conservation movement, including Muir, Marsh, Whitman, Leopold, and Abbey, were deeply influenced by these intersecting notions and are equally implicated in their disturbing effects. As with Turner, Calhoun, Galton, Davenport, Haeckel, Grant and so many others before (and after) them, the efforts of these men to protect the purity of nature were intertwined – whether explicitly or implicitly – with their desire to ensure the strength of their nation, their fellow citizenry, and themselves. Nature thus became a social template that needed to be "guarded" – kept or made pure – not only for its own sake, but for the good of the nation and select, deserving individuals within it.

The conception of nature as always-already pure and yet in continual need of purification – in need of protection from the ever-threatening elements which "have no right place in the landscape" – continues to trouble the contemporary environmental movement. This is not to say that every reference to wilderness is bound to historical formations of race, class, and nation in the same way; wilderness draws off of many forms of knowledge, from Judeo-Christian traditions to Enlightenment thought. However, as long as racial histories remain hidden, racist and racialist practices will continue to find some form of expression; and efforts at environmental protection will continue to be cast as attempts to guard and restore a natural, God-given purity, by the pure, for the pure. I hope to have demonstrated that such efforts must be recognized as more than that. By looking at forest politics in a contentious corner of the Southwest, I have tried to illustrate some of the dangers of regarding nature as a pure template for moral guidance. Nature itself has a social history that is anything but pure. Efforts to preserve and restore "wilderness," to create "healthy" forests, and to treat "sick" and "degraded" landscapes are not as simple as they may at first seem.

This article has traced a history of the production and conservation of wilderness to a history of the defense of middle and upper class whiteness in America. My central claim is that the environmental movement, particularly as it pertains to the protection of forest wilderness, is haunted by the specters of its own racist creation. In part, the very meaning and impact of the environmental movement in the United States is at stake. If environmental groups continue to conceive of the debate so narrowly around the question of wilderness as traditionally defined, they will do so at their own peril.

In the battles over the forest in Northern New Mexico, many environmentalists blame local Chicano activists and the "recovering environmentalists" who roused the internal debate within area environmentalist groups for "fomenting the hatred" that led directly to the escalation of tensions there.[54] But the tensions in the region run still deeper. At issue are historically sedimented fears and understandings regarding nature, race, and class, and they are made manifest in material, often violent struggles over the forest. New Mexico and the racially charged forest landscapes that populate it demonstrate that these tensions and their lengthy historical lineages are inescapable and deserve closer, more careful attention.

These spectral pasts are powerfully expressed in contemporary struggles over the forests – in debates over wilderness preservation and in the zero-cut/zero-grazing initiatives of the environmental movement in New Mexico. I do not mean to claim that this is the only history that infuses wilderness. Wilderness is, of course, vested with all kinds of anxieties, aspirations and politics. In fact, a part of the wilderness debate has been about how to conceive of wilderness as anything other than simply a landscape of resource production. Next to this, however, there is a reactionary, conservative aspect of wilderness as well, which reasserts itself through such figures as Edward Abbey, David Foreman, Sam Hitt and others.

Some conceptions of wilderness protection have echoed substantial critiques of capitalism. However, these critiques seem to have quieted amid the advent of "green capitalism," which implies that we can "save the environment" while simultaneously saving corporate interests and profit margins, thus maintaining the inequitable distribution of resources and the security of suburban white enclaves. A more critical political ecology would cultivate an awareness of the production of nature and the construction of wilderness and draw out the hidden labors and constitutive silences implicit in the making of wilderness (Cronon 1996; Smith 1984; White 1995). Yet one of the biggest disappointments of the environmental movement has been its stubborn inability to critically examine the politics involved in its own contribution to the formation of the environment itself, as well as the social legacies imbedded and reproduced within the movement's understandings of nature (Braun and Castree 1998). Though recent debates about wilderness and environmental justice have become more widespread, a radical rethinking of wilderness has yet to occur. Most notably, leftist and conservative environmentalists alike continue to deal ineffectively – or not at all – with issues of race as they intersect with questions of wilderness.

I have raised concerns here about spaces of whiteness in federal forest lands in the United States, and challenged what is being protected and perpetuated through these spaces. I hope, however, to have done more than merely point to the problematic ways in which race, class, gender are linked to environmental politics; I want also to have opened the door for a reconceptualization of wilderness areas and public lands more generally. The fact is, public spaces in the West have too long been defined as white; too few people and ideas have contributed their reconceptualization in broader, more politically engaged ways. What does it mean to remake the notion of wilderness in the United States? What does it mean if nature is not something to be protected, but something that is continually produced? How do "we" begin to remake spaces of nature in ways that make clear the histories present within them, while also forging new ways to openly engage these spaces as alternatively raced, classed and gendered? Northern New Mexico forest politics demonstrate that this process of radically remaking forests landscapes in the United States is an intensely complicated, contentious one – but one that can, and does, indeed happen.

Ultimately, much more than the environmental movement or the 15 percent of the country that lives in "wilderness areas" are at issue here. More centrally at stake are the notions of nature and its purity that continue to work as a reservoir for "common sense" conceptions of race and for the reproduction of exclusionary logics of racial difference. What is at stake are lived experiences of difference that are naturalized and reproduced through those notions of nature. Given these stakes, the responsibilities and possibilities of environmental politics are even greater than we have yet imagined.

NOTES

1 Bird, Bryan (1999). Conversation with the author, Santa Fe, NM, 12 March.

2 DeVargas, Ike, (2000). Interview by the author, Cervieta Plaza, NM, 12 March.

3 Anonymous New Mexico activist, 1999. Interview by the author, Española, NM, 5 April.

4 This is not to say that the history of wilderness is racist, *per se*, or that wilderness does not have other histories. I intend here merely to map these hidden genealogies of wilderness and the grounded implications of this history of the movement in Northern New Mexico, and to point to the fact that this history is very much still part of many conceptions of wilderness. "Purity and Pollution" comes from Douglass (1996), but I am using her conception in a slightly different formulation here.

5 I am not saying in any way that wilderness works the same in all places. Neither do I want to imply that all efforts at wilderness preservation in all places carry the same meanings. Ideas of wilderness have many genealogies and articulate differently with the particularities of places, practices and histories. Likewise I am not saying that to work for the preservation of the forest is inherently racist. What I do want to claim is that notions of wilderness and, more broadly, nature, articulate with historical tensions in New Mexico, and that some of these meanings have long, entangled histories with racial formations and anxieties.

6 Though the wilderness movement is only one part of the environmental movement, it was a particularly important one at the turn of the century. I do not mean to conflate the two but rather discuss the wilderness movement as an important component of the environmental movement.

7 See Cronon (1996), Neumann (1997), and Spence (1999) for a discussion of the implications of notions of wilderness. For the embeddedness of the notion of wilderness in social relations in relation to labor, see Smith (1984), White (1995) and Williams (1980).

8 The connections between racial purity and nature's purity have a long history, with many and sometimes contradictory paths, more than I can do justice to here. I attempt only to trace the necessary links between them, not to map the comprehensive, entwined epistemology of nature, race and purity.

9 How does knowledge about the human body – its health, contamination, and virility – become the means for understanding the health and well-being of the forests? To understand this claim in relation to the forest, we need to begin with a simple reiteration of the following postulate: the cultural history of the forest is inextricable from the forest itself, from the very material fibers of the wood. That wilderness areas have this social history is the starting point. Second, we must remember that the discourses of forest wildernesses are not produced from the forest alone; rather, they are woven together by the iteration and reiteration of established norms, meanings and understandings. This iterative process is what makes forests intelligible as wilderness areas. The epistemology of wilderness purity may be constructed without specific reference to race, but it is bound to turn on references to notions of purity that conform to established norms regarding race. These meanings of wilderness, of course, can drift, be contested and remade. Likewise they can serve to reproduce racialized ideas of difference without intention and without direct reference to forms of difference. As a result, notions of wilderness that arose in the late 1800s and early 1900s were governed by regulatory norms and anxieties that function even in the absence of their explicit articulation. I'm grateful to Bruce Braun for conversations related to the development of this point.

10 The environmental justice movement is the only aspect of the contemporary environmental movement to truly engage difference. The environmental justice movement has shaken the foundations of the "old school" environmental movement, forcing some of its adherents to reexamine their own practices and assumptions. To date, the environmental justice movement has concentrated almost exclusively on the inequitable access to resources or disproportionate exposure to hazardous pollution based on race or class difference. This work has radically changed approaches to environmentalism, especially in relation to pollution and health. But the roots of race questions lie still deeper. Scholars have scrutinized the racially charged statements of individuals such as Muir, Thoreau, Pinchot and others, but those critics too have stopped short of exploring the *origins* of the ideas behind these statements and the ways in which these origins continue to shape environmental agendas. The few who have tried to understand the colonial or Judeo-Christian traditions as they manifest in the notion of wilderness have indeed opened new ways of understanding the familiar logics of these claims. However they have continued to treat wilderness as a coherent, homogeneous, universal concept in which lived and contested formations are rarely – if ever – situated in specific times and places. My work here seeks to build off these insights by exploring the roots of notions of wilderness while also examining how they articulate and are lived within a particular time and place.

11 Foucault (1978).

12 Of course, the threat here is to the formation of a white masculinized notion of nationality. These developments were not threatening to others, except in that the reproductions of the fear led to some of the darkest and most violent incidents in American history.

13 Nowhere was the notion of "manifest destiny" more explicitly expressed than in the Mexican-American War. Given the widely accepted belief that it was the destiny of this white nation to reach "from sea to shining sea," the idea of annexing a territory with a large Mexican population was deeply troubling to many in the United States (Horsman 1981). Ironically, the debate over whether or not to get involved in the Mexican-American War was fought largely between those who thought that it was "our" mission, a nationally shared burden, to civilize the Mexican race, and those who feared what the mixing of races would do to the national character. Senator John C. Calhoun put it this way: "Can we incorporate a people so dissimilar from us in every respect – so little qualified for free and

popular government – without certain destruction to our political institutions?" Calhoun directed his words to those who that felt it was Americans' duty to spread civil and religious practices across the continent, stating that "[w]e have never dreamt of incorporating into our Union any but the Caucasian race – the free white race." (as quoted in Horsman 1981, 241). The Mexican-American War at mid-century, coupled with growing tensions over slavery, placed the racial question at the heart of scholarly discussion; nineteenth-century American theorists and popular writers were deeply engaged in defending the innate differences between races and warned of the dangers of mixing blood between races – both at the level of the individual body and within the body of the nation.

14 See McClintock (1995), Stoler (1995); Young (1995). For notions of cultural intelligibility, see Butler (1993). For a discussion in relationship to nature, see Braun (2000).

15 Deep fears of "the enemy within" resonate strongly with contemporary fears of domestic terrorism in the wake of the 11 September 2001 attacks on the Pentagon and the World Trade Center, a time in which anxieties regarding not only the external "other" but imagined internal threats to the national body have lead to calls for the purging of "alien elements" and re-affirmation of the narrow boundaries of the nation. This has been clearly illustrated by the increase in violence towards Muslims in the wake of the attacks. It is also important to note that, even in efforts to include Muslims into the national body, the terms on which Muslims are included – that is, what constitutes acceptable behavior and what is suspect – are tightly bound within the liberal norms of western national rationalities.

16 For example, W.G. Ramsay from South Carolina wrote two articles in the *Southern Agriculturalist* in 1839 in which he argued, "We are almost tempted to believe that there must have been more Adams than one, each variety of colour having its own original parent." (As quoted in Horsman 1981, 141). Abetted by the popularizing zeal of Josiah Nott and George Gliddon, Agassiz and Morton helped the scientific postulations behind theories of polygenesis become widely accepted. In an influential book first published in 1852, *Types of Mankind*, Nott and Gliddon took these theories to one of their seemingly natural conclusions, proclaiming whites as the carriers of civilization (Horsman 1981). They wrote: "The creator had implanted in this group of races an instinct that, in spite of themselves, drives them through all difficulties, to carry out their great white mission of civilizing the earth. It is not reason, or philanthropy, which urges them on; but it is destiny" (Horsman 1981). By conquering the globe and, in particular, expanding westward, they reasoned, Caucasians were "fulfilling a law of nature" (Horsman 1981, 136). The overriding message of *Types of Mankind* was that superior races would make the world a better place by exterminating, or at least governing, the inferior races that stood in their way.

17 Nott drew specifically off the work of the influential ethnologists Dr. Samuel Morton and Dr. George Combe, phrenologist Dr. Charles Caldwell, and many others. (Horsman 1981)

18 Spencer's earlier work also influenced Darwin's own thinking. See Spencer (1855), *The Principles of Psychology*, for his early ideas of evolution. He and nine other well-known British intellectuals formed Club X to discuss Darwin's ideas.

19 He later became president of the American Breeder Association, and director of Cold Springs Harbor.

20 These efforts echo German eugenic projects taking place at the same time. In fact, Germany directly mirrored American sterilization programs, the Immigration Act, and the research at Cold Springs, which the American eugenics movement had developed for the creation of its own "racial hygiene movement." In fact, Davenport and leading German eugenicist Eugene Fisher (whom Hitler relied directly upon in *Mein Kampf*) were such close colleagues that Davenport asked Fisher to take over as chair of the International Federation of Eugenics organization when he stepped down. Nazi and US notions of race are in fact far more closely linked than is often acknowledged.

21 These were President Calvin Coolidge's comments as he signed the 1924 Immigration Act. See "A Science Odyssey: People and Discoveries: Eugenics Movement Reaches Its Height." By 1915, the rise of the eugenics movement had helped spawn anti-miscegenation laws in 28 states, invalidating marriages between "Negroes and white persons;" six of those states went so far as to write this prohibition into their constitutions. For example, Virginia's Racial Integrity Act of 1924 warned of the "dysgenic" dangers of mixing the blood of different races. The law declared that: "it shall thereafter be unlawful for any white person in this State to marry any save a white person, or a person with no other admixture of blood than white." (Source: http://www.pbs.org/wgbh/aso/databank/entries/dh23eu.html.) These very sentiments are frighteningly parallel to those echoed in Nazi Germany 20 years later. Dr. Gerhard Frey, founder of the German People's Union and Nazi activist, similarly stated in 1933 that: "Germany should remain German."

22 I borrow the notion of "hidden attachments" from an essay by Dennis Cosgrove (1995). Cosgrove's argument is similar to mine here, in that he also finds race a hidden attachment to contemporary environmentalism.

23 Indeed, modern notions of wilderness were not created in a vacuum, and did not simply emerge self-evidently by virtue of John Muir's or others' wanderings and discovering what was "really" there. Instead, stories of the discovery of pristine wilderness gained relevance and support because they emerged from and addressed prevailing anxieties around the need to protect the purity of the body and of the nation – and not just any bodies, but white bodies, and their associated pure blood. In other words, the term "wilderness" and the meanings conferred to it emerged not through objective observation of a "real," timeless nature, but rather through historical sedimentation of discourses that incorporate notions of race and class. The point here is that meanings of wilderness have not come into the world fully formed; neither have they been simply induced from dispassionate observation of socially disconnected material objects. But these meanings of wilderness are also not produced by the intention of the subject who makes the observations. Rather, these understandings of wilderness connect present acts to prior ones in ways that conform to the iterable norms, fears, and understandings of the social and political context in which they are created.

24 The opposition of wilderness and modernity was not a gesture of Muir's invention, but his life reaffirmed it and made him a passionate preacher of this divide. While he worked making parts for carriages in his late twenties, Muir stopped to untie the belt of a machine with a file. The file flipped up and hit him directly in the eye. He was convinced that his eye was lost, though it was only partially damaged. However, Muir had to spend four weeks in a dark room to enable it to heal; when he got out he was left with an even greater disdain for machines and factories. The incident helped spur Muir's trip west in search of Yosemite, which he had read about in a small pamphlet. He set out in the "direction by the wildest, leafiest and least trodden way [he] could find." Immediately upon his arrival in San Francisco, he asked a fellow traveler (who was British): "What is the quickest way out of the city?" The traveler asked him: "Where do you want to go?" "Anywhere that is wild," Muir responded. Muir had already found what he wanted well before he arrived in Yosemite, well before he "discovered" true wilderness. What he sought was an Other against which to pit modernity, with which to measure the fall of man from grace. Like transcendentalists before him, Muir found God in nature, and with God on his side, he drew ever more clearly the line between that which was "pure," "cleansing," "light-filled" and true, and that which was "fallen," "degraded," polluted and impure. The "grandeur" of the mountains was on one side, and the "squalor" of the cities and their inhabitants on the other; "God's wild gardens" and their protectors set against the "temple destroyers."

25 Muir also wrote an essay about forests called "Thinking Like a Forest."

26 Marsh is widely considered the founder of the environmental movement in the United States. He was more than this; he was also a lawyer, a manufacturer, a philosopher, a

congressman, a diplomat and one of the founders and earliest supporters of the Smithsonian Institute. He was a broad and influential thinker of his day. His most celebrated work, *Man and Nature*, is considered to be the greatest contribution of his life. It is widely seen as the standard-bearer of ecological thinking in this country. Luis Mumford and others considered it to be the "fountainhead of the conservation movement." Gifford Pinchot called it "epoch-making," and more recently, Stewart Udall claimed it to be the "beginning of land wisdom in this country." (Strong 1988, 27, 36)

27 Support for Galton's position was broad, spanning the social spectrum from John D. Rockefeller to Emma Goldman. Even noted leftist writer George Bernard Shaw said, in response to Galton's paper "Eugenics: Its Definition, Scope and Aims": "I agree with the paper, and I go so far as to say that there is now no reasonable excuse for refusing to face the fact that nothing but a eugenic religion can save our civilization from the fate that has overtaken all previous civilizations." (See Galton 1904, 1996.) With these words, Shaw echoed Marsh and Galton – and amplified – common fears of the decline of civilization in the wake of man's folly against nature.

28 The proposition was a central debate within the Sierra Club for over a year, starting in 1998.

29 See *The Wild Duck Review* 4, no. 1 (Winter 1998).

30 Here again, articulations of difference become tightly wrapped around the body of nature, both as national landscape and internal marker of an essential identity. As such, the protection and improvement of nature deeply link blood and landscape; the threat of pollution of this body necessitates the proper governance and management of nature for its and humanity's own good. Virile white males step up to protect and improve the body of nature in the face of foreign threats to its purity. This management requires masculinity, science and proper governance.

31 Leopold's answer to concerns regarding nature's sickness is proper "husbandry" and a recognition of the value of America's "wild rootage." This "wild rootage" is similar to Muir's notions, as are his ideas of wilderness as the fountain and purity from which humanity has emerged and which humanity must now protect. His attention to the managerial notion of "husbandry" also owes a profound debt to George Perkins Marsh.

32 Forest Guardian Earth Day meeting in Santa Fe Public Library, April 2000. I should note here that the zero-cut campaign became renamed the National Forest Protection Campaign.

33 Forest Guardians, Earth Day meeting 2000.

34 Forest Guardians, Earth Day meeting 2000.

35 Locals flouted a federal injunction and "trespassed" on Forest Service land to continue harvesting the fuelwood on which many base their livelihoods. I examine this conflict in more detail below and in a chapter of my dissertation, *Understories: The Political Life of Forests in Northern New Mexico.*

36 Juarez, Santiago. 1999. Conversation with the author, Española, NM, 16 February.

37 The groups to which Matthews refers are SWOP – Southwest Organizing Project and Southwest Network for Environmental and Economic Justice.

38 The Intel Corporation in Albuquerque uses over 7 million gallons of water a day in the middle of a desert to produce Pentium chips, and urban sprawl is probably the most threatening process facing the Southwest today. See *Southwest Organizing Project* Web site, http://www.swop.net/intelinside.htm.

39 Schiller and Matthews (2000). Conversation with the author, El Valle, NM, 3 April.

40 Chelise Glendening, Kay Matthews, her partner Marc Schiller, and a handful of other activists have all lived in Northern New Mexico for years, and all of them have at one time or another been intimately involved in environmental struggles. In fact, some of them have even worked with Sam Hitt in the past. But the impact of living in this area, coupled with their commitment to other issues such as labor rights, racial justice and local sovereignty, have changed how they understand and engage in environmental struggles.

41 Glendening (1996). For a complete copy of the letter, see HTTP: <http:/www.lajicarita. org/justice.htm>.

42 The letter "Inhabited Wilderness" was printed in different newspapers; see *La Jicarita News* 1996.

43 One of the most recent, intensely public, and long-fought battles over New Mexico's Carson National Forest is the battle over *La Manga* and *Agua/Caballos* timber sales – a part of the Vallecitos Sustained Yield Unit, or VSYU. The VSYU was established by the Forest Service in 1948 as one of a number of test sites on federal lands in which the logging and processing of timber are guaranteed for local communities (Goldberg 1997, 15–21; Wilmsen 1997). In most cases, the Forest Service policy is to present a timber sale for competitive bid; in the case of the VSYU, the Forest Service awards a bid to a lumber company at the timber's appraised value in exchange for the company's promise to employ only local loggers, provide a sawmill, and conduct primary manufacturing on-site, as well as provide Hispanic residents with a supply of wood for domestic use. However, logging companies and the Forest Service have continually tried to dissolve the VSYU, only to have community members challenge these efforts.

44 Matthews (1999). In my interviews with Sam Hitt, John Talberth, Bryan Bird and many others, none wanted anything to do with compromises. After *La Compañía Ocho* won the right to log the *La Mango* timber sale in court, Bird declared: "We will appeal and ... we will litigate" in order to stop the logging in the area. Both Bird and John Talberth were most centrally interested in "ecosystem health and integrity," "wildlands restoration and preservation," and they were committed to the new mantra – "to protect and restore native biological diversity." They were, in their own words, "uninterested in making concessions." As Talberth stated, in relation to the sale: "There are places you just cannot compromise.... Letting old growth be slaughtered for commercial gain is clearly one of them."

45 Padilla, Alfredo (1999). Discussion with author, Truchas, NM, 13 December.

46 Padilla, Alfredo (1999). Interview by the author, Truchas, NM, 13 December.

47 Cordova, Sam (1999). Interview by the author, Truchas NM, 12 November.

48 Hitt, Sam (1999). Interview by the author, Santa Fe, NM, 16 April.

49 Bird, Bryan (1999) Conversation with the author, Santa Fe, NM, 12 March.

50 See the Wildlands Project mission statement in *The Wild Earth* special issue (1992).

51 This approach to "wilderness preservation" is not unique to the Forest Guardians. The Sierra Club's broader national zero-cut campaign – which proposes to end all commercial logging on federal lands – as well as the more recent zero-grazing campaign, express this same hubris. Indeed, the zero-cut campaign has been at the center of national forest debates since 1997. However, some environmentalists from Northern New Mexico dissented. Most notably, long time award winning member, George Grossman, a well-respected member of the local Sierra Club chapter, came out in favor of the cut. His position, however, led to serious tensions not only between the local chapter and the state chapter, but also within the national policy of the Sierra Club. Members of the Guardians, many of them members of the Sierra Club, complained to the state chapter and members of the board of national Sierra Club, voicing the complaint that the local chapter was at odds with the national policy. They claimed that Grossman "violated Club policy, misrepresented the Sierra Club, and misused the Sierra Club name." They went on to call for Grossman to "step down from his position." As one member of the state chapter and a supporter of the Guardians explained at a regional Sierra Club meeting: "There is a tendency for this group [the Rio Grande chapter of the Sierra Club and Grossman in particular] to wander away from the pure environmental focus to the sociological ... [Hispanos] seem to think they have a right to live rurally and they can take it off the taxpayers anyway they want." The censorship of the Rio Grande chapter of the Sierra Club from both within and without, coupled with pressure from the Forest Guardians, further divided environmentalists in Santa Fe and Albuquerque, forcing people to choose between two very different strains of the environmental movement.

52 Max Cordova's letter was widely distributed to politicians and activists throughout Northern New Mexico.
53 White (1995: 173). I am indebted to Anand Pandian for bringing this quotation to my attention.
54 See email from Charlotte Talberth to *Earth First Journal* and the subsequent lawsuit.

REFERENCES

Abbey, E. (1996) *Confessions of a Barbarian*, Arizona: Little Brown & Company.

Almaguer, T. (1994) *Racial Fault Lines: The Historical Origins of White Supremacy in California*, Berkeley: University of California Press.

Alonso, A.M. (1995) *Thread of Blood*, Tucson: University of Arizona Press.

Althusser, L. (1971) "Ideology and ideological state apparatus," in *Lenin and Philosophy and Other Essays*, London: New Left Books.

Arnold, D. (1993) *Colonizing the Body: State Medicine and Epidemic Disease in Nineteenth-Century India*, Berkeley: University of California Press.

—— (1996) *The Problem of Nature: Environment, Culture, and European Expansion*, Oxford, England: Blackwell.

Bederman, G. (1995) *Manliness and Civilization: A Cultural History of Gender and Race in the United States, 1880–1917*, Chicago: University of Chicago Press.

Bird, B. (1999) Conversation with the author, Santa Fe, NM 12 March.

Braun, B. (2000) *The Intemperate Rainforest: Nature, Culture and Power on Canada's West Coast*, Minneapolis: University of Minnesota Press.

—— (2003) "On the raggedy edge of risk: articulations of race and nature after biology," in D. Moore, J. Kosek and A. Pandian (eds) *Race, Nature and The Politics of Difference*, Durham: Duke University Press.

Braun and Castree, N. (1998) *Remaking Reality: Nature at the Millennium*, London: Routledge.

Cohen, M.P. (1984) *The Pathless Way: John Muir and American Wilderness*, Madison: University of Wisconsin Press: 23.

Cordova, D. (1997) "Enviropression," *The Santa Fe New Mexican*: Op-ed.

Cordova, S. (1999) Interview with the author. Truchas, NM, 12 November.

Cosgrove, D. (1984) *Social Formation and Symbolic Landscape*, Madison: University of Wisconsin Press.

—— (1995) "Habitable earth: wilderness, empire and race in America," in D. Rothenberg (ed.) *Wild Ideas*, Minneapolis: University of Minnesota Press.

Cronon, W. (1996) "The trouble with wilderness; or, getting back to the wrong nature," in W. Cronon (ed.) *Uncommon Ground: Rethinking the Human Place in Nature*, New York: W. W. Norton and Company.

Darwin, C. [1871] (1981) *The Descent of Man, and Selection in Relation to Sex*, Princeton: Princeton University Press.

DeLuca, K. (1999) "In the shadows of whiteness: the consequences of constructions of nature in environmental politics," in T.K. Nakayama and J.N. Martin (eds) *Whiteness: The Communication of Social Identity*, London: Sage Publications.

Douglas, M. (1966) *Purity and Danger: An Analysis of the Concepts of Pollution and Taboo*, London: Routledge.

Editorial (1995) "Conversations that kill." *Albuquerque Tribune*. Albuquerque: A6.

Ehrlich, G. (2000) *John Muir Natures Visionary*, Washington DC: National Geographic.

Foreman, D. (1992) "Around the campfire," *The Wild Earth* Special Issue. The Wildlands Project: Plotting A North American Wilderness Recovery Strategy.

Foreman, D., Davis, J., Johns, D., Noss, R. and Soule, M. (1992) "The Wildlands Project Mission Statement." *The Wild Earth* Special Issue. The Wildlands Project: Plotting A North American Wilderness Recovery Strategy.

Foucault, M. (1978) *The History of Sexuality*, vol. 1, trans. Robert Hurley (1990) New York: Random House.

—— (2000) *Power*, James Faubion (ed.) New York: New Press.

Galton, F. (1904) "Eugenics: its definition, scope and aims," *The American Journal of Sociology*, vol. X; July, no. 1.

—— (1996) *Essays in Eugenics*, Washington DC: Scott-Townsend Publishers.

Gasman, D. (1971) *The Scientific Origins of National Socialism: Social Darwinism in Ernst Haeckel and the German Monist League*, New York: P. Lang.

—— (1998) *Haeckel's Monism and the Birth of Fascist Ideology*, New York: P. Lang.

George M. (1964) *The Crisis of German Ideology: Intellectual Origins of the Third Reich*, New York.

Gilroy, P. (2000) *Against Race: Imagining Political Culture Beyond the Color Line*, Cambridge, Mass: Harvard University Press.

Glendening, C. (1996) "A letter to environmentalists," *Santa Fe Reporter*. April 3–9: 21. See HTTP:<http:/www.lajicarita.org/justice.htm>.

Goldberg, D.T. (ed.) (1990) *Anatomy of Racism*, Minneapolis: University of Minnesota Press.

—— (1997) *Racial Subjects*, New York: Routledge.

Goldberg, J. (1997) "Finding the people among the trees," *The Reporter*, Santa Fe: 17–23.

Grant, M. [1918] (1970) *The Passing of the Great Race*, New York: Arno Press and The New York Times.

Graves Jr., J.L. (2001) *The Emperor's New Clothes: Biological Theories of Race at the Millennium*, New Brunswick: Rutgers University Press.

Grossman, G. (1999) Interview by the author, Santa Fe, NM 17 March.

Grove, R. (1995) *Green Imperialism*, Cambridge, England: Cambridge University Press.

Guha, R. (2000) *Environmentalism: A Global History*, Boston: Addison-Wesley.

Hall, S. (1980) "Race, articulation, and societies structured in dominance," in UNESCO (ed.) *Sociological Theories: Racism and Colonialism*. Paris: UNESCO.

—— (1986a) "Gramsci's relevance for the study of race and ethnicity," *Journal of Communication Inquiry*, 10 (2): 5–27.

—— (1986b) "The problem of ideology: Marxism without guarantees," *Journal of Communication Inquiry*, 10 (2): 28–43.

—— (2002) "Reflections on 'race, articulation, and societies structured in dominance'," in P. Essed and D.T. Goldberg (eds) *Race Critical Theories*, Oxford, England: Blackwell.

Haraway, D. (1989) *Primate Visions: Gender, Race, and Nature in the World of Modern Science*, New York: Routledge.

Hardin, G. (1968) "The tragedy of the commons," *Science*, 162 (13 December): 1243–48.

Hays, S. (1959) *Conservation and the Gospel of Efficiency: the Progressive Conservation Movement 1890–1920*, Cambridge, MA: Harvard University Press.

Hitler, A. [1927] *Mein Kampf*, trans. Ralph Manheim (1947) Boston: Houghton Mifflin.

Hitt, S. (1996) "Green Hate in the Land of Enchantment," Press release.

—— (1999) Interview with author. Santa Fe, NM 16 April.

Hitt, S. and Talberth, J. (1995) "Forest service demonizing activists," *Albuquerque Journal*: Op-ed.

Horsman, R. (1981) *Race and Manifest Destiny: The Origins of American Racial Anglo-Saxonism*, Cambridge, MA: Harvard University Press.

Hough, F.B. (1897) "On the duty of government in the protection of forests," Paper presented to American Forestry Association, Portland, ME.

Jacobson, M.F. (2000) *Barbarian Virtues: The United States Encounters Foreign People at Home and Abroad, 1876–1917*, Hill and Wang, New York.

Juarez, S. (1999) Interview with author. Española, NM, 3 May.

Kevles, D.J. (1985) *In the Name of Eugenics: Genetics and the Uses of Human Heredity*, New York: Knopf.

Kipling, R. [1899] (1971) "The white man's burden," in T.S. Eliot (ed.) *A Choice of Kipling's Verse*, London: Faber and Faber.

Leopold, A. [1949] (1987) *A Sand County Almanac; Sketches Here and There*, New York: Oxford University Press.

Lezon, D. (1999a) "Forest guardians blame bomb on 'Minuteman' radicals," *The Albuquerque Journal*, 20 March: A2.

—— (1999b) "Police explode pipe bomb," *The Santa Fe New Mexican (Journal North)* 20 March 1999: B12.

McClellan, D. (1995) "Protestors hang environmentalist in effigy," *Albuquerque Journal*, Albuquerque, NM: D3.

McClintock, A. (1995) *Imperial Leather: Race, Gender and Sexuality in the Colonial Contest*, New York: Routledge.

Malthus, T.R. [1798] (1970) "*An Essay on the Principle of Population and a Summary View of the Principle of Population*," A. Flew (ed.) London: Pelican.

Marsh, G.P. [1864] (1973) *Man and Nature*, New York: Charles Scribner and Company.

Matthews, K. (1999) "Sierra Club hears from minorities locally and nationally," *La Jicarita News*, August IV (7).

—— (2000) Interview with author. El Valle, NM, 7 July.

Matthews, K. and Shiller, M. (1997) "Environmentalists must work with us," *Santa Fe New Mexican (Journal North)*: 4.

Meine, C. (1988) *Aldo Leopold: His Life and Work*, Madison: University of Wisconsin Press.

Meyerson, H. (2001) *Nature's Army: When Soldiers Fought For Yosemite*, Lawrence, KS: University of Kansas Press.

Montoya A., Morales, M. and Tafoya, R. (1997) "Green vision blind to native hardship," *The Albuquerque Journal*, 5 November: A15.

Moore, D.S. (1998a) "Subaltern struggles and the politics of place: remapping resistance in Zimbabwe's eastern highlands," *Cultural Anthropology*, 13 (3): 344–81.

—— (1998b) "Clear waters and muddied histories: environmental history and the politics of community in Zimbabwe's eastern highlands," *Journal of Southern African Studies*, 24 (2): 377–403.

Moore, D.S., Kosek, J. and Pandian, A. (2002) *Race, Nature, and the Cultural Politics of Difference*, Durham: Duke University Press.

Muir, J. (1915) *Travels in Alaska*, Boston: Houghton Mifflin.

Nash, R. (1967) *Wilderness and the American Mind*, New Haven: Yale University Press.

National Forest Protection Campaign (1997) *Mission Statement of Zero-Cut: The Campaign Too End Logging on Public Lands*. Online. Available HTTP: <http://ef.enviroweb.org/zerocut/packet/solution.html>.

Neary, B. (1999) "Bomber targets forest guardians – police defuse deadly pipe bomb," *The Santa Fe New Mexican*, 20 March: C3.

Neumann, R. (1997) "Primitive ideas: protected area buffer zones and the politics of land in Africa," *Development and Change*, 28 (3): 559–82.

—— (1999) *Imposing Wilderness*, Berkeley: University of California Press.

Obsatz, S. (1995) "S.F County, Española decry firewood ban," *The Santa Fe New Mexican*: B12.

Oelschlaeger, M. (1991) *The Idea of Wilderness: The Prehistory to the Age of Ecology*, New Haven: Yale University Press.

Orr, D. (1997) "Zero-cut on public lands," *Earth Island Journal*, Summer 12 (3).

Padilla, A. (1999) Interview with author, Truchas, NM, 13 December.

Peet, R. and Watts, M. (eds) (1996) *Liberation Ecologies: Environment, Development, Social Movements*, New York: Routledge.

Peluso, N. (1992) *Rich Forest Poor People*, Berkeley. University of California Press.

Peluso, N.L. and Watts, M. (eds) (2001) *Violent Environments*, Ithaca, NY: Cornell University Press.

Petersen, D. (1998) "Immigration and Liberal Taboos' Redux: Reflections on Racism, Truth . . . and Edward Abbey," *Wild Duck Review*, winter IV (1): 21.

Pinchot, G. (1901) *Forest Destruction*, Washington D.C., Smithsonian: 401–4.

—— (1905) *The Use of The National Forest Reserves*, Washington D.C.: US Department of Agriculture.

—— (1909) "The address of the Hon. Gifford Pinchot," *The Santa Fe New Mexican*, Santa Fe, 15 March: 4.

—— (1949) *Breaking New Ground*, Washington D.C.: Island Press.

Prince, B.L. (1910) *New Mexico's Struggle for Statehood: Sixty Years of Effort to Obtain Self Government*, Santa Fe: The New Mexico Printing Company.

Raymond D. (1987) "The Nazis and the nature conservationists," *The Historian*, XLIX (4) (August).

Roediger, D. (1991) The Wages of Whiteness, New York: Verso.

Rosales, A. (1996) *¡Chicano!: The History of the Mexican American Civil Rights Movement*, Houston: Arte Publico Press.

Rose, M.R. (1998) *Darwin's Spectre*, Princeton: Princeton University Press.

Schiller M. and Matthews, K. (2000) Interview with author, el Valle NM, 7 July.

Selden, S. (1999) *Inheriting Shame: The Story of Eugenics and Racism in America*, New York: Teachers College Press.

Smith, N. (1984) *Uneven Development: Nature, Capital and the Production of Space*, Oxford: Blackwell.

Solnit, R. (1994) *Savage Dreams: A Journey into the Landscape Wars of the American West*, Berkeley: University of California Press.

Spence, M.D. (1999) *Dispossessing the Wilderness: Indian Removal and the Making of the National Parks*, New York: Oxford University Press.

Spencer, H. (1855) *The Principles of Psychology*, London: Thoemmes Press.

—— [1892] (1965) *The Man Versus the State*, Caldwell, ID: Caxton Printers.

Staudenmaier, P. (1995) "Fascist ideology: The 'Green Wing' of the Nazi Party and its historical antecedents," in J. Beihl and P. Staudenmaier (eds) *EcoFacism: Lessons from The German Experience*, San Francisco: AK Press.

Stegner, W. (1961) "The idea of wilderness," in M. Nelson (ed.) (1998) *The Great New Wilderness Debate*, Athens, GA: The University of Georgia Press.

Stoler, A.L. (1995) *Race and the Education of Desire: Foucault's History of Sexuality and the Colonial Order of Things*, Durham, N.C.: Duke University Press.

Strong, D. (1988) *Dreamers and Defenders: American Conservationists*, Lincoln, NB: University of Nebraska Press.

Talberth, C. (1997) "Don't be pawns of forest service," *Albuquerque Journal*, Op-ed.

—— (1999) "Bombers target forest guardians," *Earth First Journal*, 25 March.

Talberth, J. and Bird, B. (1998) *State of the Southern Rockies: Greater San Juan-Sangre de Cristo Bioregion*, Santa Fe: The Forest Guardians and the Wildlands Project.

Talberth, J. and Hitt, S. (1995) "Environmentalists' views of firewood restrictions," *The New Mexican*, Santa Fe: 12.

Talberth, J. and Talberth, C. (1999) Interview with author, Santa Fe, NM, 13 January.

Thongchai Winichakul (1994) *Siam Mapped: A history of the Geo-body of a Nation*, Honolulu: University of Hawaii Press.

Toppo, G. (1995) Striking an emotional cord: environmentalists hanged in effigy in logging ban, *The Santa Fe New Mexican*, Santa Fe: A1.

Tuan, Y.F. (1977) *Space and Place: The Perspective of Experience*, Minneapolis: University of Minnesota.

Tucker, E.A. and Fitzpatrick, G. (1972) *Men Who Matched The Mountains: The Forest Service in the Southwest*, Washington D.C.: United States Department of Agriculture.

Turner, F.J. [1898] (1994) "The Significance of the Frontier in American History," in J.M. Faragher (ed.) *Rereading Frederick Jackson Turner*, New York: Henry Holt and Company.

USDA (1905) *The Use of the National Forest Reserves: Regulations and Instructions*, Forest Service: Washington D.C.: USDA.

—— (1922) *The National Forests of New Mexico*, National Forest Service, Department Circular 240, Washington D.C.: USDA.

—— (1989) *Landscape Character Types of the National Forests in Arizona and New Mexico*, Washington D.C.: The Forest Service.

Wade, P. (1993) "Race, nature, and culture." *Man N. S.* 28 (1): 17–34.

Walker, F. (1896) "Restriction of immigration." *Atlantic Monthly*, 77 (June): 828.

Ware, V. (1992) *Beyond the Pale: White Women, Racism, and History*, London: Verso.

Watts, M. (1983) *Silent Violence*, Berkeley: University of California Press.

Weber, D. (1994) *The Spanish Frontier in North America*, New Haven, CN: Yale University Press.

Weiss, D. (1993) "John Muir and the Wilderness Ideal", in Sally Miller (ed.) *John Muir: Life and Work*, Albuquerque: University of New Mexico Press.

White, R. (1995) *The Organic Machine*, New York: Hill & Wang.

Whitman, W. [1860] (1961) *Leaves in the Grass*, Ithaca, New York: Cornell University Press:.

Wild Earth, (1992) "The Wildlands Project: plotting a North American wilderness recovery strategy," Special Issue, no date.

Willems-Braun, B. (1997) "Buried epistemologies: the politics of nature in (post) colonial British Columbia," *Annals of the Association of American Geographers*, 87: 3–31.

Williams, R. [1972] (1980) "Ideas of nature," in *Problems in Materialism and Culture*, London: Verso.

Wilmsen, C. (1997) *Fighting for the Forest: Sustainability and Social Justice in Vallecitos, New Mexico*, Dissertation Thesis, Clark University, Department of Geography.

Young, R.J.C. (1995) *Colonial Desire: Hybridity, Culture and Race*, 1st edn, London and New York: Routledge.

Zorrilla, L.G. (1965) *Historia de las relaciones entre Mexico y Los Estados Unidos de America 1800–1958*, 2 vols, Mexico: University of Mexico.

6

ECO-GOVERNMENTALITY AND OTHER TRANSNATIONAL PRACTICES OF A "GREEN" WORLD BANK*

Michael Goldman

INTRODUCTION

In a 1996 report written by a prominent environmental organization for the World Bank sits a hand-drawn map of Lao People's Democratic Republic (Laos). This map does not demarcate the nation's capital, its towns or villages; the only cartographic markings are round, oblong, and kidney shapes, each labeled with initials such as WB, SIDA, WCS, and IUCN. That these splotches cover almost one-fifth of the territory of Laos, that they represent newly classified zones for environmental protection and conservation, and that the symbols translate to the World Bank, Swedish International Development Agency, Wildlife Conservation Society, and IUCN–World Conservation Union, tell an important political story about new efforts to classify, colonize, and transnationalize territory – in the name of "eco-governance."

In this chapter, I present the case that debates on the specific value of, and effects produced by, the recent shift toward "environmentalism" by our largest international development and finance institutions (Darier 1999; Fox and Brown 1998; Pincus and Winters 2002; Sachs 1993; Wade 1997; Young 2003) could be enhanced by looking at the recent phenomenon of new global regulatory regimes for the environment. Indeed, despite disparate interpretations of what is "good" environmentalism, there has been a change in the character and increase in intensity of activity around constructing global *truth* and *rights* regimes on the environment and natural resource use. My focus is on the relationship among regimes of rights within and among states and the interstatal system, with special attention to the World Bank's new green regulatory policies for its borrowing countries. I argue below that in spite of the worldwide movement challenging the World Bank and its development and finance partners, the Bank has strengthened itself in its reinvention as a catalyst for green institutional change. Indeed, through the tussle with its social-movement critics, the Bank has been able to enlist scores of social actors and institutions to help generate a new development regime that is coherently *green* as well as *neoliberal*, one that is as inclusive of ministries of environment, natural resources, and finance, as it is of some of its best-funded international environmental organizations such as IUCN and WWF. While activists and academics successfully build their case for a World Bank that is ecologically and socially destructive, the Bank industriously plugs away at greening its works in more countries engaging more ecosystems and populations, expanding its portfolio and including more "partners" from the private and public sectors. In other words, since the late 1980s, when the World Bank was forced to "reform or die" because of the success of campaigns documenting the Bank's horrendously destructive projects (e.g., Indonesia's Transmigration and Brazil's Polonoroeste projects), twenty years later the Bank has been at the center of constituting a hegemonic, albeit fragile, project of "green neoliberalism."

For an illustration of the making of what I call *green neoliberalism*, as well as a description of how the Bank's power and authority gets strengthened and dispersed, we turn to the Mekong, where the World Bank represents its recent interventions as reflecting its new *modus operandi*: "environmentally sustainable development." Over the next two decades, the multilateral banks and the Lao government plan to build more than a dozen hydro-electric dams on the Mekong River, converting Laos into the Tennessee Valley Authority (TVA) of Southeast Asia. Unlike the TVA, however, these plans are being implemented *through* new ideas and tools of conservation, preservation, and sustainability. When a whole range of actors, from World Bank lawyers to international conservation scientists, are commissioned to rewrite national property rights laws, redesign state agencies, and redefine localized production practices based on new global norms, they transform conventional forms of state power, agency, and sovereignty.

I argue below that these new "green" practices are impacting the production of, first, national and global truth regimes on nature; second, rights regimes to more effectively control (and increase the market value of) environments, natural resources, and resource-dependent populations; and third, new state authorities within national boundaries and in the world system. Hence, the World Bank's practices are facilitating the birth of *environmental states* in the South, but not in the way that ecological modernization theorists suggest, i.e., that states are unified, rational actors and eventually graduate into eco-rational modernity (Frank, *et al.* 2000; Mol and Sonnenfeld 2000; Schofer, *et al.* 2000; Spaargaren and Mol 1992). Instead, the environmental states that are emerging around the world today are marked by new global forms of legality and eco-rationality that have *fragmented*, *stratified*, and unevenly *transnationalized* Southern states, state actors, and state power.

These changes affect what Foucault called the "art of government" (Dean 1994; Foucault 1991), a concept he deployed to decenter the traditional notion of the state as the main site of modern societal power ("the transcendent singularity of Machiavelli's prince"). He preferred to emphasize the multiplicity and widely dispersed "forms of government and their immanence to the state" (Foucault

1991: 91) that had been left undertheorized in political theory. Foucault argued there were three basic types of government and each was connected to a particular science or discipline: self-government or morality/ethics, the proper way to govern the family from which emerges the modern science of economy, and the science of state rule or politics. For our purposes, the art of government also includes, on the one hand, the *making* of the modern rational subject and the efficient state which s/he would help build, and, on the other hand, the intensified regulation of the relation of these subjects to their natural territory. I call these productive relations of government – with their emphasis on "knowing" and "clarifying" one's relationship to nature and the environment as mediated through new institutions – *eco-governmentality*. That is, in the process of analyzing this new type of global green governmentality, I engage Foucault's question: "What rules of right are implemented by the relations of power in the production of discourses of truth?" (Foucault 1994). We can learn a lot about relations of power through an inquiry into the co-production of regimes of territorial rights and discourses of environmental truth.

My emphasis, however, differs from the recent literature on governmentality in that here the contested terrain is the arena that Foucault and his interlocutors have overlooked and rendered undifferentiated: nature, qualities of territory, and the political–epistemic rationalities that give meaning, order, and value to them (Braun 2000, Kuehls 1996; Moore 2001; Sivaramakrishnan 1997). It is through what I call the "green neoliberal project" in which *neocolonial* conservationist ideas of enclosure and preservation and *neoliberal* notions of market value and optimal resource allocation find common cause – that institutions such as the World Bank have made particular natures and natural resource-dependent communities legible and accountable. Confronted with what Foucault called the "problem of government," unevenly transnationalized state and non-state actors have sought to "improve" conditions of nature and populations by introducing new cultural/scientific logics for interpreting qualities of the state's territory. In doing so, a hegemonic discourse of ecological difference rooted in neoliberal market ideology emerges, defining some "qualities of territory" as degraded, and others as necessary instruments for the improvement of populations, states, and natures. In this way, new domains of political–economic calculation are forged that facilitate the disciplinary (normalizing) practices and legitimating devices for *trans*nationalizing access to the Mekong.

My emphasis also differs from the perspective of economic geographers who effectively argue that nature is socially produced and, under a capitalist regime, produced specifically for commodification (Harvey 1996; Smith 1990). Instead of seeing this process as *fait accompli*, and hesitant to gloss over the contested terrain from which such transformations occur, I choose to emphasize here the heated productive relations out of which new political, economic, and scientific rationalities are borne, and become institutionalized, resisted, and everything in-between. That is, I find it useful to interrogate the process of production from which new hegemonic forms emerge, to better understand the actual routes from which the adjective "green" and the noun "neoliberalism" may congeal as naturalized artifact, becoming part of the "There Is No Alternative" (TINA) state of mind circulating professional–epistemic communities across the globe.

In what follows, I begin by drawing some parallels between imperial science in the colonial period and the neoliberal global environmentalism being practiced today by the world's leading multilateral institutions. I then turn to my case study of the World Bank in Laos, first describing its projects in the Mekong region, and then analyzing more fully the character, meaning, and wide dispersion of its interventions. I divide this analysis into two parts. The first focuses on what are perhaps the more obvious exercises of World Bank power as it seeks to "green" or ecologically neoliberalize its borrowing country clients, including its efforts to restructure the state itself and related non-state local institutions (most importantly, those governing property rights). This part of my analysis also reveals the extent to which the Lao state has become transnationalized, albeit highly unevenly and traveling particular neoliberal routes. The second part focuses on the more subtle, yet perhaps more significant, exercises of power that operate through the practice of a new global green science. In exploring the concept of eco-governmentality, I try to make concrete the connections between the different modalities of government (in the Foucauldian sense) being deployed and their effects on the social whole. This chapter attempts to make more transparent and explain the routine practices of the construction of a specific type of environmentalism – neoliberal, transnational, capitalist – that is becoming hegemonic and multi-scalar, and being advanced in localized/globalized sites across the "development" map (Crush 1995; Escobar 1995; Ferguson 1990; Hart 2001; Watts 1995).

KNOWLEDGE PRODUCTION AND INSTITUTION BUILDING: PAST AND PRESENT

A robust literature on the role of imperial science in empire- and state-building attests to how scientific and administrative missions into colonies facilitated the crafting of state power, population management and exploitation, and resource expropriation in both the colonies *and* the metropole (Comaroff 1985; Rabinow 1989; Said 1978; Stoler 1995). Unwittingly, according to Bernard Cohn and others, the British not only conquered territorial but epistemological space as well; since "the facts" of the colonializing space did not reflect what the British knew, the British embarked upon a project of translation of its colonies into knowledge it could understand ("establishing correspondence could make the unknown and the strange knowable") (Cohn 1996: 3). In this way, the metropole's elite could explore and conquer *through* translation. Cohn argued that the British took a number of steps in the colonial project that could be understood most generally as "investigative modalities" – the command of local languages, the construction of historiography and museology – and more specifically, the survey and census which helped to define the colonial subject. Some modalities were eventually transformed into sciences, such as tropical medicine, ethnology, and economics. In the end, colonial scientists and investigators, from botanists to anthropologists, played crucial roles in the colonial project of "civilizing" through translation. More than creating power through knowledge for the imperial masters based in the colonies, these practices helped foster the idea of modernity through state- and institution-building in both the colonies *and* the metropole.

In a related vein, Michel Foucault wrote on early modern relations between institutional practices and the rationality of "seeking to organize, codify, direct such practices" (Foucault 1994: 169) emphasizing the tactical productivity and reciprocal effects of power and knowledge. Foucault extended his analytic inquiry to the ways in which new political technologies, which he called "biopower," brought "life and its mechanisms into a realm of explicit calculation and made knowledge/power an agent of transformation of human life" (Foucault 1990). Foucault's most innovative ideas in this regard connected "practices of government" and "practices of self," viewing the fields of politics and ethics as irreducible and linked to modernity projects. Similarly, Arturo Escobar (1995) argues that post-World War II development practices have very much been a discursive production process of new forms of human behavior, conduct, and ethics in regards to ideas of modernity, e.g. civilization, progress, rationality, poverty alleviation, and now – environmental sustainability. State-building and subject-creating exercises, therefore, become mutually constitutive. As Ann Stoler (1995) argues of the colonies and the metropole, and Escobar contends of postcolonial institutions and elite classes working in the "developing" South, biopower produces subjectivities across national boundaries. This argument is not intended to elide or sidestep the brutal force of colonial power or capitalist imperialism; rather it is to focus more sharply on the legitimating technologies that get constituted, dispersed, and influence the shaping of norms and behaviors of colonialized and capitalized societies.

Although this perspective of the legacy of colonialism offers a foundation for exploring the twin themes of territorial and epistemological conquest and transformation, the colonial and imperial projects in Laos and beyond were substantially different from present-day activities. Today, under multilateral pressure to integrate into the regional and global economies, Laos is being re-envisioned as the next Switzerland or Kuwait of Southeast Asia, a prospective engine for borderless commerce and energy-driven capital accumulation. In particular, the newly identified populations of Laos (the hill tribes, rice growers, technocrats, and entrepreneurs) are being called upon to awake from their "sleepy" state to become global market-oriented, scientifically based, and ecologically sustainable – to be eco-scientific, transnationally rational, *and* to accumulate. In this construction, the good of the nation and the citizen hinge on the globalization project, which in turn, hinges on people's and nature's *intelligibility* to global experts, managers, and investors (cf. Braun 2000, Goldman 1998).

Different from colonial projects, as well as from more recent national-development projects (McMichael 2000) however, what we find in Laos and in many other World Bank borrowing countries is a form of knowledge and set of practices that are shaped by the discourse of neoliberalism and their proprietary rights-based orientation, on the one hand, and the disciplinary practices of globalized environmentalism, on the other. Specifically, the most remote populations of the mountainous jungle of the Mekong are being made accountable for their ecologically destructive conduct and are at the core of the massive development schemes responsible for bringing exportable hydro-electric production to their mountainous terrain, in exchange for an improved, civilized lifestyle. Unlike colonial conservationists who acknowledged remote resource-dependent communities in only the most rudimentary terms, the World Bank and its partners have initiated and financed a process that *targets* resource-based populations, accounting for them and the qualities of their environments in new discourses of ecological improvement, and compelling them to participate in the new neoliberal process of eco-government. The science of judging their needs and deficiencies becomes critical to the Bank's intervention, and gets refracted through the new environmental state institutions being designed for Laos. In this way, the art of eco-government circulates and expands through multiple sites of encounter (e.g., beyond and below the national and the state) and leads to new modalities of power/knowledge.

In rhetoric and in practice, infrastructure investment in today's "green stage" of capitalist development (O'Connor 1994; O'Connor 1998) requires a tremendous public architecture of scientific discursive practices. Coterminous with the rapid expansion of green scientific practices are powerful effects that get dispersed over a wide range of social spaces, influencing the formation of governing agencies that oversee scientifically constructed objects of study such as fisheries, watersheds, wetlands, and indigenous peoples' extractive reserves. With these new governing bodies come new social and natural bodies. For example, for the first time in parts of Southeast Asia, there are watersheds, national biodiversity conservation areas, indigenous peoples' extractive reserves, and sustainable logging zones, managed by international experts, action plans, surveys, and non-governmental organizations (NGOs). Lao professionals charged with studying and working on these newly defined "eco-zones" receive accreditation (and training program certificates) from attending short courses in Geneva, Washington, D.C., and Bangkok. Moreover, the subjectivities of the people dependent upon these natural resources also get reconstituted. "Unknown" peoples living in the mountains, river valleys, and forests are being categorized as distinct and accountable populations – non-timber forest users, wetlands managers, sustainable forest laborers, as well as trespassers, poachers, and slash-and-burn cultivators.

As the case of Laos will show, when we add up all the differentiated populations and natures, we find that much of a borrowing country – its nature, populations, governance, and knowledge – has the potential of becoming reconstituted as subjects of new forms of government according to new cultural logics of eco-rationality, enabling new and old frontiers of World Bank-fostered capital accumulation.

GREENING LAOS

Transnational development boosters trumpet Laos as the future "crown jewel" (Traisawasdichai 1997; Usher 1996a; Usher 1996b) of Southeast Asia, hoping to offer abundant energy resources and services to the economic fireballs in the more developed regions of the Mekong, such as Bangkok. The Lao state, the World Bank, the Asian Development Bank (ADB), and private consortia of foreign investors have all been keen on building hydro-electric dams in the Annamite mountains on the major tributaries of the Mekong River. In the next two decades, these actors hope to direct the transboundary plan (of the newly created "Greater Mekong Sub-region") to relocate millions of mountain inhabitants in six neighboring countries (Laos, Cambodia, Vietnam, Burma, Thailand, southern China) in order to construct dozens of hydro-electric dams. In the process, hill dwellers would become the agro-industrial workforce in the newly irrigated and electrified plains *and* a new population of eco-rational natural resource managers. According to an Asian Development Bank (ADB) director, Noritada Morita: "We may need to reduce the population of people in mountainous areas and bring them to normal life (*sic*). They will have to settle in one place . . . but don't call it resettlement. It is just migration" (*The Nation* 1996). Morita estimated that 60 million people lived in the hill areas in these six neighboring countries, and noted that the ADB has targeted them because "these people are not a part of their national economies." The ADB depicts this project as environmental because its goal is to stop forest destruction through "developing" the hill tribe populations whom the ADB blames for engaging in "slash-and-burn" cultivation, encroachment, illegal logging, and too-rapid reproduction. The plan calls for scientists, governments, and (select) NGOs to join the ADB in this $50 *billion* engineering project.

Somewhere in the middle of this desire spectrum is Nam Theun 2, the dam, hydro-electric power, and forestry project that is considered to be the test case for these larger transnational dreams. Nam Theun 2 is currently the biggest investment in Laos, with an estimated cost of $1.5 billion, slightly smaller than the country's GDP, and almost four times the national budget (GOL 1997; World Bank 1997, 1999b, 2001a). The financial consortium (Nam Theun Electricity Consortium or NTEC) of French, Thai, and US investors that will own and operate the project claims that when the dam is built, the annual revenues from its electricity sales to Thailand will generate up to $233 million annually for the Lao government, which is equivalent to 43 percent of the country's current income from exports (World Bank 2001b).

The site of the Nam Theun 2 dam, watershed, and reservoir happens to house one of the most biologically diverse forests in the world, and was once listed for global protection by the World Bank's Global Environmental Facility (GEF). Aside from being a propitious site for a dam, the region (especially the newly designated NNTCA or catchment area) hosts an amazing array of diverse and apparently rare animal and plant species, some of which are, according to the latest commissioned studies, at risk of global extinction. On the one hand, according to some scientists' reports, local elephant, tiger, bird, deer, frog, and fish populations will be seriously threatened once the ecological and social landscape is transformed through the damming and rerouting of the Theun river and the inflow of infrastructure-maintenance activities. Anthropologists have raised similar concerns about the human populations that currently occupy the forested plateau and catchment, whose communities and cultures – especially the indigenous tribes (two of whose languages represent "newly discovered" linguistic groups) – could be destroyed by resettlement and the influx of other populations attracted to the formal and informal opportunities arising from these development/conservation projects (Chamberlain and Alton 1997; Chamberlain *et al*. 1995).

On the other hand, other scientists' studies have argued that within the current political–economic situation (i.e. Laos has been designated by the World Bank as a Heavily Indebted Poor Country or HIPC), these natural resources would be better protected if large-scale capital projects are approved, because the revenues generated for the state could be spent on "much needed" conservation, preservation, and sustainable development (IAG 1997; Scudder, *et al*. 1998; World Bank 1999b). Without immediate action, this global environmental "hot spot" will deteriorate under the destructive weight of "over-" population, hunting, slash-and-burn cultivating, fishing, poaching, and tree felling. As professional experts hired to help Laos have explained to me, for the future of Laos, there is no alternative (TINA).

Proponents of "immediate action" believe that only a massive intervention can stem the rising tide of ecological destruction and human poverty, and that a large capital project would be the best, and perhaps the only, vehicle for changing the way that the environment has been managed. As such, hydro-electric dam builders, the World Bank, the Lao state leadership, and international conservation groups have joined together in support of a hydro-dam solution to the problem, and hope to use the intervention as a means of introducing the state and society to new techniques and conduct of eco-rationality. Although this unified front encompasses plenty of dissenters, the net effect is to move ahead with this myriad of structural-adjustment, infrastructural, institutional, and capacity-building projects. The most crucial intervention, for which there is much less dissension, is the scientific assessment work that has served as a legitimating technology for many other interventions.

Because the project region is described as so ecologically and socially fragile – and because similar large dam projects by the World Bank have been so severely criticized for their disastrous effects on people and nature in the past (McCully 1996; Rich 1994, 2002; World Commission on Dams 2000) – Nam Theun 2 is being painstakingly designed to be environmentally and socially "sustainable." Teaming up with international mega-fauna and biodiversity conservationists, such as World Conservation Union (IUCN), Worldwide Fund for Nature (WWF), and Wildlife Conservation Society (WCS), the World Bank and its partners are linking hydro-electric dam and transmission financing to an ambitious string of conservation and protected areas, mega-fauna running corridors, watershed conservation sites, eco-tourism projects, biodiversity research and development sites, and indigenous peoples' extractive reserves. Roads, markets, experimental farms, Lao-language schools, health clinics, and workshops in agronomy, resource management, family hygiene and birth control, comprise the proliferating list of social projects that are part of the package. Initially, Northern development NGOs (e.g., CARE) will be contracted to run many of these social/ecological modernization projects, until a cadre of domestic professionals can be trained to indigenize these efforts. Because of its sophistication, the project has become a prototype for the World Bank (Interviews, 1998–2000; World Bank 1999a).

Whether these efforts at mitigation, rehabilitation, and modernization could ever really "work" as planned is beside the point (Ferguson 1990); the effect of such an effort to become more environmentally and socially pro-active is that the World Bank's interventions also become much more inclusive, authoritative, and disciplinary. The World Bank has successfully engaged the nascent professional class in the neoliberal discourse of "how to govern oneself, how to be governed, how to govern others, by whom the people will accept being governed, how to become the best possible governor" (Foucault 1991: 87). It has also begun the process of converting the previously inconsequential forest, hill, and river communities into visible, communicative, and accountable populations. In short, the Bank has instigated a proliferating domain of human activity – the activity of government and subject creation – that *works* to make sites and populations more compatible for these large capital investments, even as these investments evolve to include new ways to improve biodiversity, mountain populations, and the professional class.

At the same time that the World Bank is seeking to appease those concerned with improving the conditions of populations and ecosystems, it is even more concerned about its ability to interest private capital in its new ventures. Large fixed-capital investments are not like speculative capital; they require secure and unambiguous property rights and minimal political risks over a substantial period of time to ensure high profit rates. Yet to achieve this requires a number of fundamental changes in countries such as Laos, starting with the state and its institutions of regulation and law governing usufruct and ownership rights to rivers, mountains, forests, and the natural resources within these ecosystems. In other words, before the World Bank can expect to persuade Northern investors to invest in "hardware" in Laos, it must invest in the "software" of state restructuring – including new state regulations, new state authorities, and new state actors (or what some Bank officials refer to as "cultivating champions").

Rewriting laws, restructuring state agencies, and financing green projects

Under the rubric of state restructuring, there are three types of interventions in which the World Bank is engaging: rewriting laws (particularly related to the regulation of natural resources, the environment, and property rights); restructuring state agencies that regulate environments (broadly defined to include many state ministries); and funding large-scale "green" infrastructural projects. All three interventions are inextricably linked: the development of fixed capital infrastructure (in this case, a joint-ventured hydro-electric facility) requires laws that establish certain property rights, which can only occur through the restructuring of state institutions. The environmental projects are the legitimizing vehicle for the dam: Without such a strong public commitment to environmentally sustainable development, the World Bank and its counterparts would not be able to proceed without incurring robust resistance from the highly effective campaigns to stop "traditional" World Bank-style developmentalism. In effect, the Bank's pro-active response to transnational environmental organizations, networks, and movements are new strategies of global environmentalism that have become institutionalized (with greater and lesser effectiveness) throughout the world.

Before 1975, the colonial French created the Lao legal, juridical, and administrative systems to maximize social control, resource taxation, and forced labor for their empire. Upon taking power, the socialist Pathet Lao abolished the French system and replaced it with a general declaration that all land and resources would belong to the people and held in a public trust (Evans 1995). By the late 1980s, as its main source of foreign aid from the USSR dried up and its foreign debt ballooned out of control, the Pathet Lao introduced a market-oriented set of economic reforms that were in part in response to pressure from its main creditors, the World Bank and the Asian Development Bank, as well as to dramatic shifts already occurring in Vietnam and China. Foreign fiscal advisors, natural resource planners, and lawyers soon moved into the capital city, Vientiane, to facilitate the policy shift. Subsequently, the Prime Minister's office passed a number of important decrees relating to property rights and natural resource use, especially forest, water, and land. Each was motivated and largely written by foreign consultants to international finance institutions (IFIs), donor "trust funds," or international NGOs. Each was followed up with Northern loans, aid, and foreign direct investments, leading to larger and more permanent offices and staff in Vientiane for Northern aid and development agencies. With each legal change came institutional restructuring of the Lao government, and in distinction from the colonial era, the *greening* and *neoliberalization* of the state, as the following sections show.

In 1989, the country's first national forestry conference produced the Tropical Forest Action Plan (TFAP), which was drafted and funded by UN agencies. It was a "boilerplate" plan that received criticism as being pro-timber industry by international environmental groups who were leading successful campaigns against TFAPs in other borrowing countries (Hirsch and Warren 1998; Lohmann and Colchester 1990; Parnwell and Bryant 1996). Undeterred, the Lao government and the World Bank began an extensive campaign to document the social and ecological processes occurring in the fairly inaccessible but populated forests of Laos. The Prime Minister even ordered a ban on all logging operations until a national audit could be properly conducted with international support (Decree No. 67). The government acknowledged it lacked sufficient or reliable data to fulfill the demands of both international development institutions and their detractors. (In fact, it even lacked the capacity to enforce the ban; as Lao military generals depend upon logging revenues for their unit's operating costs and their own income, logging of the Nakai Plateau has expanded since the ban (Southavilay and Castren 2000; Tropical Rainforest Programme 2000; Walker 1996; Watershed 1999). The World Bank, at the very least, wanted to establish a coherent analysis of the supply of ecological resources, dynamics of ecosystems, and the utilization patterns of different forest users. The Bank and Northern aid agencies filled this void by commissioning stacks of scientific studies, a process on which both the government's growing professional staff and transnational consultants have been cutting their teeth over the past decade, inventing and implementing techniques and tools for rapid appraisals and diagnostics of extremely complex and unstudied terrains.

As these scientific studies were being conducted, the government passed more decrees to incorporate their findings, imposing classificatory systems imported from abroad by the development banks' environmental consultants. Prime Minister's Decree No. 169, established in 1993, created a classification system for the nation's forests: *protection forests* for watershed catchment areas as well as for the supply of timber and non-wood products; *conservation forests* for biological diversity and the promotion of scientific and cultural values; *village forests* for subsistence production only; and *degraded forests* for sedentary agriculture. Three years later, the National Assembly approved this decree as forestry law, thereby legalizing state control over forests, a law imprinted atop of hundreds of localized customary resource-use institutions. Decree No. 164, passed in the same year, further classified nearly three million hectares (or one-seventh of Lao's total land mass and one-fourth of Lao's forested land) as conservation and protection forests. This decree also established eighteen (now expanded to twenty) National Biodiversity Conservation Areas (NBCAs), a concept promoted by the World Bank's Global Environmental Facility and the largest international conservation NGOs. As I note below, emerging from these Northern-financed initiatives is a new classificatory system and knowledge regime for land relations: socially diverse, semi-nomadic, shifting, kinship-based, interdependent processes of production and management are "out" in the new framework, while biodiversity conservation, sustainable timber production, and watershed management are "in."

These forestry decrees and laws systematically re-constitute administrative and cultural boundaries into rationalized *eco-zones* delineated by new attributions of the value of the forest and of different groups of forestry users. Every user group – from the timber industry to semi-nomadic forest users, pastoralists, nature preservers, pharmaceutical producers, the global energy and eco-tourism industries – receives its plate of rights to and regulations of one portion of the nation's forests. These plans seek to clarify property rights and resource use rules through the transnational environmental science of tropical forestry management, matching newly collected data on ecological resources and capacities (i.e., degradation and recovery rates) with the demands of diverse new markets for these natural goods and services, from hardwoods to biodiversity aesthetics to electricity.

New forestry laws have also authorized the shift of the fiscal and taxation dimensions of forestry from the provincial government to the central government, and centralized all these new undertakings under the Ministry of Agriculture and Forestry at the local, district, provincial, and national levels. The Ministry is growing rapidly as a result of the new influx of millions of dollars, and is creating additional branches and divisions every year, most of which are skeletal units in which transnationally funded projects and programs are housed. Even those UN and bilateral agencies that have spearheaded forestry projects that try to *decentralize* authority over localized resources and land, contribute to strengthening the central authority of the state through the rents that the central bureaucracy demands from the dollar-based aid money that flows into the provincial and district government agencies, and the villages. Hence, these decentralizing investments are also fortifying central state power, as well as institutionalizing aid-based corruption/rent-seeking.

The 1992 Land Decree and 1997 Land Law have had the joint effect of establishing a land market and new standards for land use. New land titling projects in pilot villages are taking cadastres, and drawing up state-sanctioned land titles to replace a decentralized system of customary property rights. These new titles guarantee rights of usufruct, transfer, and inheritance to their owners, and allow for land to be bought and sold. According to the new laws, any land left fallow for more than three years can be claimed by the state, and any land can be expropriated for development projects as long as the users receive compensation. This last law, when enforced, is devastating to upland cultivators who rotate cultivation in eight to 20 year cycles to maintain the land's long-term fertility. Land use that results in "degradation" or "neglect" based on new criteria and priorities (e.g., land left fallow for three years), can be confiscated – even if such lands will regenerate fruitfully in these long fallow cycles.

A unique characteristic of these new green laws is the authority of a new network of transnational actors – particularly within the World Bank – in their creation. Evidence of the World Bank's authority and influence is apparent in its confidential Staff Appraisal Report for the Forest Management and Conservation Project (FOMACOP), which actually named a deadline by which the Lao National Assembly was required to pass into law certain policy changes as a pre-condition for the project (World Bank 1993, 1994). Regulatory reforms and state restructuring have always been pre-conditions for Bank loans and private investments (George and Sabelli 1994; Kapur *et al.* 1997); but these green preconditions are unique for being so encompassing, disciplinary, and neoliberal.

The threat of being denied a large loan makes it difficult for states to say "no." Such financial withdrawals would be devastating to the workings of borrowing states, many of which are dependent upon development grants, loans, and resources for their operating budgets, and are already carrying high levels of debt. Indeed, by 1993–94, fully half of Laos' domestic revenue came from foreign grants, 80 percent of the state's Public Investment program came from foreign aid (GOL 1997; UNDP 1997), and its per capita debt load stripped its per capita GDP by $140 per person ($500 vs. $360). These are important features of the environmental Lao state.

By 1999, more new laws had passed the national assembly that effectively created new state authorities and regulatory mechanisms over natural resources. One of the most significant is the Environmental Protection Law. The United Nations Development Program (UNDP) helped write and push this law through the assembly and, along with the Swedish aid agency SIDA, is providing substantial support for the fledgling environmental ministry it helped create, called STEA (recently changed from STENO). Northern aid and finance institutions have helped establish the government agency that oversees all protected areas and wildlife activities (CPAWM), and strengthened the Ministry of Agriculture and Forestry, which receives most of its budget as grants and loans from these aid agencies. The budget of the Hydropower Division of the Ministry of Industry has grown exponentially due to foreign contributions directly related to big dam investments.

The amount and breadth of environmental programming within the state – instituted by transnational actors and introduced within the latest wave of dam initiatives – is impressive. The Forestry Department alone is buckling under the ballast of more than fifty separate projects, named with English acronyms such as FOMACOP and NAWACOP and official titles that reveal their origins and nationality. In 1999, the Lao Department of Forestry found itself responsible for the Lao Swedish Forestry Programme, the Lao ADB Commercial Tree Plantation Project, the Lao-WB-Finnida (Finnish) Forest and Management and Conservation project, the UN-FAO Benzoin Improvement Project, and the Lower Nam Ngum Catchment JICA-FORCAP (Japanese) Project. Some of the larger projects, such as the World Bank's FOMACOP or Forestry Management and Conservation Programme, represent a major wing of the organizational structure of the Forestry Department. Together, foreign donors and creditors finance almost all of the department's annual budget, which goes to implementing these transnational projects (and paying the high cost of their foreign staff), managing the forests in a huge expanse of Lao territory, supporting the Forest Training School and Training Centre, collecting and analyzing the data required by these new projects, and implementing the laws and decrees described above (Forestry Dept. 1997).

As agencies such as the Ministry of Agriculture and Forestry and the Ministry of Industry and Handicrafts become more involved in the receipt and management of foreign capital inflows, their staff have the opportunity to become vertically integrated into the transnational professional class, which increases their relative power and influence at the domestic level. At the same time, this flow of foreign money reshapes these agencies' domestic priorities to focus on the large-scale investment projects they are now financed to implement and regulate. As the Lao government readily admits, because of the region's currency crash in the late 1990s and the fiscal austerity programs demanded of it by the World Bank, there is a growing disparity between public expenditures on health, education, and public services, and expenditures on the (increasingly transnational) energy, forestry, construction (to house these new actors and agencies), and transport sectors (GOL 1997). Indeed, 84 percent of total state investment was in the latter group of sectors (UNDP 1997; World Bank 1997). In short, the new Lao state, like so many other states today, must bleed the social sectors to nourish the newly capitalized ones.

Hybrid actors

In these upgraded state agencies, the traditional work of state actors is now being dispersed across a new array of *hybridized* "state" actors. The most common is the Lao civil servant – the privileged of whom have been retooled and professionalized in new skills and norms, and if lucky, sent abroad for special training. The second type is epitomized by the Northern (semi-nomadic) transplant who works *inside* the Lao state as a consultant (invariably wielding enormous power relative to his or her Lao counterpart). The third is the Northern expert who sets up or staffs a shadow organization that conducts the work of an existing state agency, but without the obvious representational or bureaucratic constraints. Both the work and the actor of the old regime are paradoxically under-funded and yet judged ill-equipped for the highly valued and transnational state work of the new regime.

A striking feature of government offices in Vientiane is the sharp degree of contrast among them. Some are dusty, hot, and slow-paced, while others are air-conditioned, computerized, and run on international clocks – typically European. At one forestry department office, for example, Finnish consultants work with Lao assistants in the redesigned top floor of a dilapidated government building behind sliding glass doors in arctic air conditioned offices; they are busy reorganizing the "subsistence" forestry sector. The Lao counterparts to the Finns work on the ground floor in pre-gold rush style and calm.

In an effort to provide them with highly specialized technical training, the World Bank and Lao government send civil servants to workshops and conferences abroad, and to short courses in environmental assessment and management designed by the World Bank Training Institute (which has also recently grown in size and influence). These international excursions are quite a departure for a country that is still unable to afford to send high-level delegates to international meetings of the United Nations and ASEAN. Indeed, this process of "building up Lao human capital" takes more than just time, as many civil servants are put in the position of being paid a tiny fraction of what their Northern counterparts earn ($20 per month salary versus $3,000–$5,000 per month). Scientists with the country's lead environmental agency, STEA, have told me that they are often treated as second-class citizens by foreign consultants and staff, regarded more as translators than scientists, assisting the consultants as they quickly traverse the countryside and computer data bases gathering evidence, yet left with little but the authoritative, English-written report.

STEA officials explained that although their budget has grown substantially, almost all of the funds coming from Northern agencies funnel directly to Northern scientists, consultants, and engineering firms shuttling back and forth between Laos and their home countries. As one senior official described the situation: "The problem is that our whole agenda and budget is project driven. We can grow and do the job of an environmental agency *only* if Nam Theun 2 gets funded. Meanwhile the donors demand so much from us. Even though a lot of foreign money flows through our agency, most of it goes to pay for foreign consultants" (interview, Vientiane 1998).

Besides writing the new environmental laws and regulations, these traveling hybrid actors are critical conduits for the transnationalization of ideas. They are also the ones who are paid to construct environmental data, without which STEA (and the World Bank's loan and guarantees packages) cannot move forward. In short, these foreign "state" actors are designing and carrying out the mandates of Lao's new laws and regulations.

With World Bank funding and initiative, and an eye toward readying Laos for large-scale capital investments, Finnish, Swedish, and German government aid agencies have *de facto* taken over major wings of the Forestry Department to thoroughly restructure it. The budget of each Northern agency includes a portion for training Lao civil servants in environmental technocracy and management as well as in English. That most of Laos' public investment funds are voted upon and allocated at a meeting in Geneva, Switzerland suggests the power of these transnational actors. That the skills, worldviews, and conduct of the Northerners in Laos are being indigenized by savvy Lao staff is progressively more apparent.

When listening to some Northern officials and consultants portray the Lao people and their "lacks," it is clear that within the discourses of development, progress, and sustainability lie some very neo-colonial attitudes and practices on the part of those "doing the development" toward those "being developed." For example, Northern conservationists work under the premise that the Lao people know nothing about conservation. Their funding proposals and projects reflect the view that it is best to start from scratch with Laos: first teach proto-professionals English, then send them abroad to learn how to identify endogenous (flora, fauna, fish) species, and then return them to Laos to staff the newly designed wetlands, watershed, and conservation agencies (Chape

1996a, 1996b; IUCN 1993, 1998; McNeely 1987; WCS 1995, 1996). Yet such a perspective ignores, among other things, the fact that the actors being "produced" are the ones who made this knowledge possible to generate in the first place. As one Thai forestry specialist argues, "It's not as if only people with [the Northern experts'] kind of knowledge are equipped to conserve forests . . . if it were not for those villagers with their knowledge of the terrain and animals, Northern wildlife scientists . . . would not be able to make those amazing 'discoveries' of rare or newly found animal species" (Watershed 1996: 40). Some Lao government scientists expressed frustration that they had already undergone scientific and linguistic training abroad during the 1980s, under different world-systems imperatives, in Bulgaria, Hungary, and Czechoslovakia. Why retool again?

Reflecting this powerful neo-colonial attitude, the Australian public relations specialist with the Nam Theun Electricity Consortium (NTEC), upon being asked about the possible negative effects of the dam, listed the issues raised by the anti-dam critics, and then remarked, "Remember, these people on the Plateau are primitive and anything is better than what they have" (interview, Vientiane 1998). This representation does not seem to be lost on the people who are "being professionalized" through the development process. For example, an environmental specialist for the Laos environmental agency explained the asymmetry of his relationship with Northern scientists hired to conduct work for his agency: "We want to learn the consultants' trade, but we are pretty much left carrying their bags" (interview, with an STEA staff scientist in Vientiane, 1998). While this "modernization" process is supposed to reduce uncertainty and risk, the nascent class of Lao state actors has expressed a sense of *increased* uncertainty and risk in engaging in a process over which it has little control. These Lao actors describe what Harvey (1982) and some geographers call the time–space compression of having to get on or off this fast-moving vehicle called "environmentally sustainable" capitalist development.

The discursive field in which these development actors work is so powerful that it is hardly possible to speak with expertise of Laos today *except* by deploying these eco-rationalities. Indeed, the proliferating body of scientific research on the Mekong is produced either as a condition of multilateral bank loans and development interventions, or as a reaction to them. Fish biologists, cultural anthropologists, environmental economists, development professionals, and Lao citizens participating in World Bank public consultations have spoken mostly within this discursive field. Indeed, international conservation groups have framed their mega-fauna discoveries and their mechanisms to protect biodiversity exclusively in terms of a commensurable trade-off between dams and conservation. "The old," as David Scott (1995) argues of colonial governmentality, "(is now only) imaginable along paths that belong to new, always already transformed sets of coordinates, concepts, and assumptions." In this way, localized forms of scientific production proceed along highly pre-figured political fields of development (Bourdieu and Wacquant 1992).

ENVIRONMENTALLY SUSTAINABLE DEVELOPMENT AS TECHNOLOGY OF GOVERNMENT

Which theoretical-political *avant-garde* do you want to enthrone in order to isolate it from all the discontinuous forms of knowledge that circulate about it?

(Foucault 1994: 205)

The World Bank's influence in Laos is not limited to the visible means of restructuring and environmentalizing the state described in the preceding sections; indeed, the very same analytical and methodological tools that the World Bank and its partners invent and use, and the classificatory systems they establish in pursuit of environmentally sustainable development, represent an exercise of power. These tools, methodologies, and classification systems serve to create a new cognitive mapping of Lao nature and society, state and citizen, through new forms of knowledge production and institutional collaborations. They are a powerful set of discourses of norms, rights, and truths of global eco-rationality that seeks to build upon and replace prior formations that have dealt with the "subjects" of these new policies: hill tribes, forest dwellers, scientists, and development officials.

As noted above, for example, the new forestry law deploys a classification scheme that rezones Lao forests into distinct administrative categories. *Protected forests* are designated to protect *ecological services* such as watersheds and soil from erosion, and *national security*, such as porous international borders. *Conservation forests* protect high-valued biodiversity while permitting limited uses, such as non-timber forest production, tourism, and hydro-electric dams (Forestry Dept. 1997). These new environmental zoning classifications carve up territory *and* sovereignty through scientific distinctions of forest use (e.g. sustainable timber production, transboundary protection parks, subsistence production). They also have been catalysts for resettlement plans of populations, requiring numerous public consultations and scientific studies that seek to demonstrate the feasibility of transplanting people who do not fit the required social characteristics for eco-zone habitation (Franklin 1997; GOL 1999; Sparkes 1998).

In addition to this forest classification system, Lao's new policies and regulatory schema also identify corridor zones, controlled use zones, total protection zones, and National Biodiversity and Conservation Areas (NBCAs). Within these various eco-zones, there are clear regulations as to what people are and are *not* allowed to do: in NBCAs, for example, no firearms are allowed, entry by motor vehicle is limited, and commercial logging is banned. In total protected zones, there is no hunting, fishing, or collection of non-timber forest products; no entry without permission, agriculture, and no overnight stays. In controlled use zones, there is no in-migration, and only limited village rights to grazing, fishing, and fire wood collection. The only requests that are allowed "by special permission" are eco-tourism and hydro-dam reservoirs (Forestry Dept. 1997). In other words, dams and tourism receive greater weight than forest dwellers' right to hunt, gather and sleep in their forests. Indeed, the implementation of these newly ascribed eco-zones is shifting the rights and access to the vast natural resources of the forests, mountains, and rivers, from the forest dwelling populations to the energy, conservation, and tourism industries. In the case of the NBCAs, moreover, responsibility and rights over a significant amount of national territory (now an estimated 15 percent of total land mass) is being handed over to international environmental organizations, soon to be empowered to manage these enormous areas (through off-shore accounts overseen by an international board of directors), albeit through the support of "responsibilized" individuals from the mountains trained to become productive citizens, guides, rangers, and park police. Key areas are planned to be off-limits to the majority of the population, except for those who work at the dams, ranger stations, and tourist sites. With the shifting of rights and access comes the ontological transformation of the forest dwellers, hunters, gatherers, fishers, and swidden cultivators of the Vietic, Brou, Tai, and Hmong ethnic groups, among many others.

Ten years ago the World Bank was severely criticized for *only* doing engineering and economic assessments of its proposed projects and dismissing the significance or reliability of assessing the social or environmental; today the Bank has established itself as the world's leading expert in environmental impact assessments (EIAs), social assessments (SIAs), and green cost-benefit analyses (CBAs). EIAs, SIAs, and CBAs are notable not only for the legitimacy they give the Bank as it seeks to restructure states, peoples, and environments, but also for the *particular* knowledges they produce and privilege, as they implicitly and explicitly assign values to groups of people and parcels of environment. Typically, these values get expressed in narrow economic terms, due to the urgent need to make them "commensurable." How else can one assess the cost of relocating a cluster of villages, destroying customary rights to forest resources, and threatening the fate of a fish species or an indigenous language? The World Bank's brand of assessment science has become virtually hegemonic, as scores of trainers, consultants, and engineering firms apply these strategic tools of economic rationality. In Laos, as elsewhere, scores of impact assessments are conducted that help clarify who and what is at stake, socially and environmentally, in big infrastructural projects, and the role of large capital investments in society as a whole. They effectively frame the debates, so that critics must engage not only these reports and their knowledge, but the particular scientific practices, and their institutionalization in national agencies, laws, and norms. This must happen if they are to enter the political debate on the future of these projects of Laos and, more generally the struggle over the "global commons."

These impact assessments also act as important levers to open up the countryside for investigation, and intervention. Since so little data has been collected on Laos, any data that is generated becomes an important element in the struggle over the particular meanings and values of nature and the environment. No matter how much controversy erupts over the reliability of the knowledge-production process, these assessments have enabled a seemingly reliable platform for claims-making for state and transnational development officials.

Through the use of these tools, the World Bank makes its objects of study accountable in two senses: first, in being counted and hence made visible locally and transnationally; and second, in reference to new environmentalist norms and responsibilities with new institutional policing and extractive capacities. As the unknown gets explored and translated, and as the language of translation gets concretized in this discovery-classification-capitalization process, newly identified citizens gain responsibility to act in specific ways. Hence, people categorized as slash-and-burn cultivators must eschew the life of an eco-outlaw, and try to take up a different livelihood (Dove 1983; Fisher 1996; Forsyth 1999; Watershed 2000). In this way, new subjects are born and new subjectivities are created, however targeted individuals choose to act. These new environmental norms adjudicate between the problematic categories of the traditional and the modern, the ecologically irrational and rational, and the ways in which people do, and should, interact with nature(s). The cumulative effect of these mechanisms has been to scientize, de-politicize (Ferguson 1990), and institutionalize certain notions of global environmentalism and citizenship, regulation and subjectivity, while de-legitimizing and destabilizing others. Not withstanding the more conventional and brutal tactics of national militaries, the Mekong's decade of capitalization could not have occurred without this wholesale normative shift.

In "sleepy" Vientiane, real estate prices have skyrocketed and urban resources have become expensive and relatively scarce for the urban denizen. The new "eco-development" business boom has had the immediate and inadvertent consequence of depriving some locals of basic resources while increasing the city's consumption of goods such as petrol, water, and public funds, to service the transnational class in their new settlement enclaves. This chapter, however, has emphasized some of the more political effects – for example, the focusing of the investigative lens on rural peoples and environments in which they inevitably find degradation, mismanagement, poverty, and backwardness. Ironically, these human populations are now being thrust onto center stage as their conduct becomes the scrutinized subject of new global technologies of government, which has become routinely known as environmentally sustainable development.

In sum, the new legal, institutional, and investment modalities mentioned above are all buttressed upon corresponding forms of knowledge, such that state-building is not just about gaining new forms of control over territorial but also epistemological space. Moreover, the newly emerging art of government of Laos is being framed by a global environmental scientific discourse that includes a politics of ethics, i.e., you're now accountable to the global community (Goldman 1998; Rose 1999). Percolating up from the investigative modalities in Laos is a classification system that has helped facilitate the proliferation of a new set of disciplinary technologies running through Laos, as well as through an emergent transnational professional class. EIAs and SIAs, NBCAs and the "best practices" of Nam Theun 2, FOMACOP and STEA, are all critical effects of power circulating through Laos, constituting the double movements of state- and global institution-building, on the one hand, and the science and art of government, on the other. In addition to constructing an environmental state, the World Bank is instigating the rise of an inclusive global environmentalism that is not based on mere rhetoric but on powerfully dispersed regimes of science, regulation, and capital investment. These power/knowledge interventions come at a time when the World Bank has had to wrestle with a demand crisis for its services and goods largely due to trenchant social activism and borrowers' increased access to foreign direct investment from private sources. That is, the Bank's response has been to create new demands for its services with renewed rigor, gaining access to new populations and environments to the benefit of its main clients, Northern-based capital goods and financial sectors.

CONCLUSION: THE RISE OF ENVIRONMENTAL STATES AND ECO-GOVERNMENTALITY

As the Laos case demonstrates, even with the expansionary powers of financial and speculative capital, multilateral development banks, and transnational regulatory regimes, state power is not vanishing. Instead states are rapidly changing. Some state functions are being created or strengthened, while others are being devalued. States are being reconfigured with new regulatory regimes and hybridized transnational state actors. The World Bank leads the vanguard of this state restructuring, helping states to better respond to forces of globalization.

The emphasis in this chapter is not only on these changes, but also on the regimes of power, truths, and rights on which these institutional practices are based. These knowledge/power relations run through the scientific and legal practices of the World Bank's new green work and become concretized through loan conditionalities, environmental assessments, scientific reporting, methodologies, classifications, policy papers, decrees, legislation, and large-scale foreign investments. Newly transnationalized state agencies, staffed with new hybrid actors, emerge with the strengthened mandate to oversee the re-territorialization (Brenner 1999) and re-evaluation of borrowing country landscapes, resulting in a radical alteration in the ways in which people interact with people and nature. In analyzing this process, I stress the making of hegemonic forms of rationality that translate into effects of government: constructing the environmental science and art of targeting populations, production practices, and behaviors *vis-à-vis* nature that are judged as guilty or innocent of ecological degradation. In this way, the modern eco-rational subject and the environmental state are being mutually constituted.

Different from the prevailing debates on governmentality, however, I have emphasized that national territory is not just an unchanging stage on which new political rationalities are exercised (Burchell *et al.*1991; Darier 1999); in fact, the problems of government hinges on the contest over defining "nature's intelligibility" (Braun 2000; Moore 2001). In the case of resource-rich and capital-poor borrowing countries such as Laos, natural wealth and natural-social relations are being transformed through proliferating scientific and political processes under the mantle of *environmentally sustainable development*. Based on actual practices, however, it should be renamed *green neoliberalism*, a political rationality that has fostered the scientization, governmentalization, and capitalization of some very hotly contested eco-zones (e.g. the Mekong, the Amazon).

Interestingly, this strategy was not anticipated by the leading actors. The fiercely nationalistic Lao state and rigidly economistic World Bank did not originally intend to hinge new capitalistic investments on transnationalized eco-zones of biodiversity and conservation. At the start, international conservation NGOs did not want to sign on to large-scale projects that dam rivers, submerge forests, and threaten the health of rare animal, fish, and reptile species. These conservation NGOs were part of the larger NGO community questioning the motives and outcomes of the World Bank. Indeed, the most effective activist networks and movements forced the Bank, by the late 1980s, to "green" itself. Yet, within a decade, in the process of establishing the epistemic and ethical differences of territory and nature (e.g. the ways in which some species have become more important than others, some knowledge privileged over others), the cognitive mappings of this strange collection of bedfellows have, remarkably, converged. These epistemic and legal interventions have triggered new political rationalities in this transnational "scramble for the Mekong."

New environmental regimes being indigenized in Laos and elsewhere do not roll quietly into town on the train of progress, but rather storm in on the wild bull of global economic integration. Any resources that might be harnessed for environmental protection and improvement are concentrated on the natures that will support capital investments – such as forested mountainous terrain that can act as a watershed buffer for the hydro-electric dams. Other natures are differentially defined based on the needs of development-related rationalities: Some are retooled for rice cultivation through privatization, while others are judged to be best for export logging. This Laos story is an extreme case of a set of broader trends found in larger borrowing countries with more robust state institutions and more active and autonomous civil societies. With more heterogeneous localized participation, influence, and invention, we find greater variation and difference. Nonetheless, from Mexico to India, these practices strengthen neoliberal and environmental authorities *vis-à-vis* other aspects of state and non-state institutions.

These hegemonic processes and Bank-related activities are not, needless to say, going unchallenged. In post-socialist countries, people raised on the politics of resisting "scientific socialism" are quick to respond to the Bank's "scientific capitalism," and its large infrastructure projects (e.g. power plants, highways, toxic waste incinerators) smoothed over by public hearings, environmental economics, and scientific eco-rationalities. In countries where public space for protest is extremely limited (e.g. Laos, Chad), environmental and human rights activists have entered on (and arisen from) the coat tails of large Bank projects which now require some sort of civil-society participation and "safeguard policies." In these places where civil protest is dangerous, regional and international activists, working quietly with locals, have tried to force the Bank to become more scientific, transparent, and participatory, but based on criteria which they believe reflect the needs of the social majority, in marked contrast to the Bank's. Activist networks have grown that connect social activists in Brazil, Mexico, South Africa, and Thailand (where the World Bank and its partners have carved out new environmental states) with people in Cameroon and Pakistan, (where green neoliberalism is just underway), and with others in London and Washington D.C.

In the short period of anti-globalization protest from 1999 to 2003, tens of thousands of people demonstrated in the media spotlights of WTO, ADB, and WB/IMF meetings. But, this same short time span has seen hundreds of other localized strikes, rallies, and battles with police protesting World Bank and IMF policies and projects around the world. In the month of May 2000 alone, half of South Africa's workforce went on general strike against World Bank-instigated job-cutting policies; 20 million workers in India went on general strike protesting Bank/IMF neoliberal policies; 80,000 people participated in anti-IMF demonstrations in Argentina; and thousands in Turkey, Haiti, Paraguay, and Thailand hit the streets protesting Bank/IMF policies (Bond 2000; Goldman 2002).

In the summer of 2000, Thailand's Assembly of the Poor organized thousands of demonstrators (for months) and scores of hunger strikers to shut down the government and force the multilateral banks to publicly discuss the possibility of closing down the disastrous Pak Mun dam project, allowing for the river to flow again. Transnational organizing is bringing the "voices of the rural poor" together with the "voices of scientists." Practically speaking, Thai and Northern organizations have funded activists, scientists, and activist-scientists to shadow the ADB and World Bank throughout the Mekong region, producing their own "just-in-time" scientific reports, and using them strategically. As these counter-discourses multiply, they incite *contre histoires* that historicize and politicize hegemonic practices, challenge state and World Bank authority, and articulate alternative globalization-from-below politics.

ACKNOWLEDGMENTS

I gratefully acknowledge the intellectual support of Jonathan Fox, Serife Genis, Philip Hirsch, Patrick McCully, Donald Moore, Rachel Schurman, James Scott, Yildirim Senturk, David Smith, Tuba Ustuner, and the numerous people who were gracious enough to allow me to interview and/or hang out with them. This chapter is a revised version of an article from *Social Problems*, 48: 4, 2001.

NOTE

* Parts of this chapter previously appeared in M. Goldman, "Constructing an environmental
state: eco-governmentality and other transnational practices of a Green World," *Social
Problems* 48, 4, 2001, and are reprinted by permission from the copyright holder, The
Society for the Study of Social Problems and the publisher, University of California Press.

REFERENCES

Berger, Inc., Louis (1997) *Economic Impact Study of Nam Theun Dam Project.* Washington D.C.,
28 July.

Bond, Patrick (2000) "African grassroots and the global movement." Znet commentary, a service
of Z magazine, 19 October. Online. Available HTTP: <http://www.lbbs.org/weluser.htm>.

Bourdieu, Pierre and Loic Wacquant (1992) *An Invitation to Reflexive Sociology,* Chicago:
University of Chicago Press.

Braun, Bruce (2000) "Producing vertical territory: geology and governmentality in late
Victorian Canada." *Ecumene,* 7: 1.

Brenner, Neil (1999) "Beyond state-centrism? Space, territoriality, and geographical scale in
globalization studies." *Theory and Society,* 28: 39–78.

Burchell, Graham, Gordon, Colin and Miller, Peter (eds) (1991) *The Foucault Effect: Studies in
Governmentality,* Chicago: University of Chicago.

Buttel, Frederick (2000) "World Society, the nation-state, and environmental protection."
American Sociological Review, 65: 117–21.

Chamberlain, James R. and Alton, Charles (1997) "Environmental and Social Action Plan for
Nakai-Nam Theun Catchment and Corridor Areas." Vientiane: IUCN.

Chamberlain, James R., Alton, Charles and Crisfield, Arthur G. (1995) *Indigenous People's
Profile: Lao People's Democratic Republic,* Vientiane: CARE International.

Chape, Stuart (1996a) *Biodiversity Conservation, Protected Areas and the Development Imperative in
Lao PDR: Forging the Links,* Bangkok: IUCN.

—— (1996b) "IUCN Programme Focus and Development in Relation to the Forestry Sector."
Donor's Meeting, Vientiane, April.

Cohn, Bernard (1996) *Colonialism and its Forms of Knowledge,* Princeton: NJ: Princeton
University Press.

Comaroff, Jean (1985) *Body of Power, Spirit of Resistance: The Culture and History of a South
African People,* Chicago: University of Chicago Press.

Crush, Jonathan (ed.) (1995) *Power of Development,* London and New York: Routledge.

Darier, Eric (ed.) (1999) *Discourses of the Environment,* Malden, Ma: Blackwell Publishers.

Dean, Mitchell (1994) *Critical and Effective Histories: Foucault's Methods and Historical Sociology,*
London and New York: Routledge.

Dove, Michael (1983) "Swidden agriculture and the political economy of ignorance."
Agroforestry Systems 1, 1: 85–99.

Escobar, Arturo (1995) *Encountering Development: The Making and Unmaking of the Third World,*
Princeton, NJ: Princeton University Press.

Evans, Grant (1995) *Lao Peasants under Socialism and Post-Socialism,* Bangkok: Silkworm Books.

Ferguson, James (1990) *The Anti-Politics Machine: "Development," Depoliticization, Bureaucratic
Power in Lesotho,* Minneapolis: University of Minnesota Press.

Fisher, R.J. (1996) "Shifting cultivation in Laos: Is the government's policy realistic?," in Bob Stensholt (ed.) *Development Dilemmas in the Mekong Sub-region*, Workshop Proceedings of Monash Asia Institute.

Forestry Department (1997) *Development of Policy and Regulations under the Forestry Law for Protected Area Management in Lao PDR*, Vientiane.

Forsyth, Timothy (1999) "Questioning the impact of shifting cultivation." *Watershed*, 5: 1, July–October. Bangkok.

Foucault, Michel (1990) *The History of Sexuality: An Introduction*, New York: Vintage Books.

—— (1991) "Governmentality," in Graham Burchell, Colin Gordon and Peter Miller (eds) *The Foucault Effect*, Chicago: The University of Chicago Press.

—— (1994) "Two Lectures," in Nicholas B. Dirks, Geoff Eley and Sherry B. Ortner (eds) *Culture/Power/History: A reader in Contemporary Social Theory*, Princeton, N.J: Princeton University Press.

Fox, Jonathan and Brown, David (eds) (1998) *The Struggle for Accountability: The World Bank, NGOs and Grassroots Movements*, Cambridge: MIT Press.

Frank, David John, Hironaka, Ann and Schofer, Evan (2000) "The nation-state and the natural environment over the twentieth century." *American Sociological Review*, 65: 96–116.

Franklin, Barbara (1997) *A Review of Local Public Consultations for the Nam Theun 2 Hydroelectric Project*. Vientiane.

George, Susan and Sabelli, Fabrizio (1994) *Faith and Credit: The World Bank's Secular Empire*, London and New York: Penguin Books.

Goldman, Michael (1998) *Privatizing Nature: Political Struggles for the Global Commons*, New Brunswick: Rutgers University Press, and London: Pluto Press.

—— (2001) "The birth of a discipline: Producing authoritative green knowledge, World Bank-style." *Ethnography*, 2, 2, July.

—— (2002) "Notes from the World Summit in Johannesburg: 'history in the making?'" *Capitalism, Nature, Socialism*, 13: 4.

Government of Lao PDR (GOL) (1997) *Socio-Economic Development and Investment Requirements, 1997–2000*, Government Report, Sixth Round Table Meeting, Geneva, Switzerland.

—— (1998a) *Project to Stop Shifting Cultivation, Allocate Stabilized Livelihoods and Protect the Environment 1998–2000*, Vientiane.

—— (1998b) *The Rural Development Programme 1998–2002: The Focal Site Strategy*, Vientiane.

—— (1999) "Resettlement Action Plan Seminar Notes," taken at Lane Xang Hotel, Vientiane.

Hart, Gillian (2001) "Development Critiques in the 1990s: Culs de sac and Promising paths." *Progress in Human Geography*, 25: 649–58.

Harvey, David (1982) *Limits to Capital*, Oxford: Oxford University Press.

—— (1996) *Justice, Nature and the Geography of Difference*, Oxford: Oxford University Press.

Hill, Mark (1995) "Fisheries ecology of the Lower Mekong River: Myanmar to Tonle Sap River." *Natural History Bulletin of Siam Society*, 43: 263–88.

Hirsch, Philip and Warren, Carol (eds) (1998) *The Politics of the Environment in Southeast Asia: Resources and Resistance*, London: Routledge.

International Advisory Group (IAG) (1997) *World Bank's Handling of Social and Environmental Issues in the Proposed Nam Theun 2 Hydropower Project in Lao PDR*, Vientiane.

International Rivers Network (1999) *Power Struggle: The Impact of Hydro-development in Laos*, Berkeley: International Rivers Network.

IUCN (1993) *Improving the Capacity of the Lao PDR for Sustainable Management of Wetlands Benefits*, Vientiane, May.

189

—— (1998) *Environmental and Social Plan for Nakai-Nam Theun Catchment and Corridor Area*, Vientiane, July.

Kapur, Devesh, Lewis, John and Webb, Richard (eds) (1997) *The World Bank: Its First Half-Century*, Washington, D.C.: Brookings Institution.

Kuehls, Tom (1996) *Beyond Sovereign Territory*, Minneapolis: University of Minnesota.

Lohmann, Larry and Colchester, Marcus (1990) "Paved with Good Intentions: TFAP's Road to Oblivion," *The Ecologist*, 2: 3, May/June.

McCully, Patrick (1996) *Silenced Rivers: The Ecology and Politics of Large Dams*, London: Zed Books.

McMichael, Philip (2000) *Development and Social Change*, Thousand Oaks, Ca: Pine Forge Press.

McNeely, John (1987) "How dams and wildlife can coexist: Natural habitats, agriculture, and major water resource development projects in tropical Asia." *Conservation Biology* 1, 3: 228–38.

Meyer, John, Boli, John, Thomas, George and Ramirez, Francisco (1997) "World society and the nation-state." *American Journal of Sociology*, 103 (July), 1: 144–81.

Mol Arthur P.J. and Sonnenfeld, David (eds) (2000) Ecological Modernisation Around the World: Perspectives and Critical Debates. Ilford (UK) and Portland: Frank Cass & Co.

Moore, Donald (2001) "The ethnic spatial fix: geographical imaginaries and the routes of identity in Zimbabwe's eastern highlands." Yale University, Program in Agrarian Studies, April.

O'Connor, James (1998) *Natural Causes: Essays in Ecological Marxism*. New York: Guilford Press.

O'Connor, Martin (ed.) (1994) *Is Capitalism Sustainable? Political Economy and the Politics of Ecology*, New York, NY: Guilford Press.

Ong, Aihwa (1999) *Flexible Citizenship*, Durham: Duke University Press.

Parnwell, Michael and Bryant, Raymond (eds) (1996) *Environmental Change in South-East Asia*, London: Routledge.

Pincus, Jonathan and Winters, Jeffrey (2002) *Reinventing the World Bank*, Ithaca: Cornell University Press.

Rabinow, Paul (1989) *French Modern: Norms and Forms of the Social Environment*, Cambridge: MIT Press.

Rich, Bruce (1994) *Mortgaging the Earth*, Boston: Beacon Press.

—— (2002) "The Smile on a Child's Face." Manuscript, Environmental Defense. Washington, D.C.

Roberts, Tyson (1995) "Mekong Mainstream Hydropower Dams: Run-of-the-River or Ruin-of-the-River?" *Natural History Bulletin of Siam Society*, 43: 9–19.

—— (1996) "Fluvicide: An Independent Environmental Assessment of the Nam Theun 2 Hydropower Project in Laos, with Particular Reference to Aquatic Biology and fishes." Consultant's report, Bangkok.

—— (1999) "A Plea for Proenvironment EIA." *Natural History Bulletin of Siam Society*, 47: 13–22.

—— and Baird, Ian (1995) "Traditional Fisheries and Fish Ecology on the Mekong River at Khone Waterfalls in Southern Laos." *Natural History Bulletin of Siam Society*, 43: 219–62.

Rose, Nikolas (1999) *Powers of Freedom: Reframing Political Thought*, Cambridge, UK: Cambridge University Press.

Rotberg, Eugene (1994) "The Financial Operations of the World Bank, *Bretton Woods Commission Background Papers*, Washington, D.C.: Bretton Woods Committee.

Ryder, Grainne (1996) "The Political Ecology of Hydropower Development in the Lao People's Democratic Republic." Environmental Studies Program. Toronto: York University.

Sachs, Wolfgang (ed.) (1993) *Global Ecology: A new arena of political conflict*, London: Zed Press.

Said, Edward (1978) *Orientalism*, New York: Vintage.

Schofer, Evan, Ramirez, Francisco and Meyer, John (2000) "The effects of science on national economic development, 1970 to 1990." *American Sociological Review*, 65, 6 (December): 866–87.

Scott, David (1995) "Colonial governmentality." *Social Text*, 42: 191–220.

Scudder, Thayer, Lee Talbot, and Ted Whitmore (1998) *Third report of the International Environmental Social Panel of Experts*, Vientiane: GOL, 21 January.

Sivaramakrishnan, K. (1997) "A limited forest conservancy in Southwest Bengal, 1864–1912," *Journal of Asian Studies*, 56: 75–112.

Smith, Neil (1990) *Uneven Development: Nature, Capital and the Production of Space*, Oxford: Oxford University Press.

Southavilay, Thongleua and Tuukka Castren (2000) *Timber Trade and Wood Flow-Study, Lao PDR*, Vientiane: Mekong River Commission.

Spaargaren, Gert and Mol, Arthur (1992) "Sociology, environment, and modernity: Ecological modernization as a theory of change," *Society of Natural Resources*, 5: 323–44.

Sparkes, Stephen (1998) *Public Consultation and Participation on the Nakai Plateau*, Vientiane: NTEC.

Stoler, Ann Laura (1995) *Race and the Education of Desire: Foucault's History of Sexuality and the Colonial Order of Things*, Durham: Duke University Press.

The Nation (1996) "Relocation in sight for hill people." Bangkok, 4 August.

Traisawasdichai, M. (1997) *Rivers for Sale*, Reuters Foundation, Oxford University.

—— Tropical Rainforest Programme (2000) *Aspects of Forestry Management in the Kao PDR*, Amsterdam.

UNDP (1997) "Report of the Sixth Round Table Meeting for Lao PDR." The Roundtable Meeting for the Lao PDR, 19–20 June.

US Treasury Department (1994) *The Multilateral Banks: Increasing US Exports and Creating U.S. Jobs*, Washington, D.C.

Usher, Ann Danaiya (1996a) "The Race for Power in Laos," in Michael Parnwell and Raymond Bryant (eds) *Environmental Change in South-east Asia*, London: Routledge.

—— (1996b) *Dams as Aid* London: Routledge.

Wade, Robert (1997) "Greening the Bank: the struggle over the environment, 1970–1995," in D. Kapur *et al. The World Bank: its First Half Century*, vol. 2, Washington D.C.: Brookings Institution.

Walker, Andrew (1996) "The timberland industry in Northwestern Laos: A new regional economy?" *Development Dilemmas in Mekong Region*, Conference Proceedings, Melbourne.

Watershed (1996) "Why the Nam Theun 2 Dan won't save wildlife." 1: 3 March–June. Bangkok.

—— (1999) Special Forum on Swidden Cultivation, 5: 1 July–October. Bangkok.

Watts, Michael (1995) "A new deal in emotions: Theory and practice and the crisis of development," in Jonathan Crush (ed.) *Power of Development*, London and New York: Routledge.

Wegner, David (1997) *Review of the Nam Theun 2 Environmental Assessment and Management Plan*, Ecosystem Management International, Arizona.

Wildlife Conservation Society (WCS) (1995) *Preliminary Management Plan for Nakai Nam Theum National Biodiversity Conservation Area*. Vientiane.

World Bank (1993) *Lao PDR Forest Management and Conservation Project, EA Category B*. East Asia and Pacific Regional Office. Washington D.C.

191

—— (1994) *Lao PDR Forest and Conservation Management Project*, East Asia and Pacific Regional Office. Washington D.C.

—— (1995) *Annual Report*. Washington D.C.

—— (1997) *Lao PDR Public Expenditure Review*, East Asia and Pacific Regional Office. Washington D.C.

—— (1999a) *Annual Report*, Washington D.C.

—— (1999b) *Country Assistance Strategy for Lao PDR*, East Asia and Pacific Regional Office. Washington D.C.

—— (2001a) *Lao PDR at a glance*, World Bank website. Online Available HTTP: <http://www.worldbank.org/data>.

—— (2001b) *Dams for Development in Laos*, Press release, external affairs department, Washington D.C., 15 May.

World Commission on Dams (2000) *Dams and Development*, London: Earthscan.

World Rainforest Movement (1993) "Notes from a NGO/Bank consultation." London, April.

World Rainforest Movement Bulletin (2001) "Strangling the life-source of millions: China dams the Mekong." 46, May.

Young, Zoe (2003) *A New Green Order: The World Bank and the Politics of the Global Environmental Facility*, London: Zed Press.

Part III

INSTITUTIONS AND GOVERNANCE

7

NATURE–STATE–TERRITORY

Toward a critical theorization of conservation enclosures*

Roderick P. Neumann

A proper game policy is an important item of the general statecraft of African territories, a more important item than seems to be realized by any otherwise well-informed people

(Julian Huxley, *Africa View*)

In August 2002 I happened to be in Yosemite National Park during the celebration of the seventy-fifth anniversary of the opening of the Ahwahnee Hotel, one of the "classic" park lodges of the western United States. Deciding this was a unique opportunity, I signed up for a guided historical and architectural tour of the hotel. The tour was lighthearted, entertaining, and full of fascinating if disconnected bits of trivia. At the end of the tour, the guide, a long-time employee of Yosemite National Park and the hotel, abruptly shifted from her playful tone and focus on hotel lore to offer an impassioned tribute to and defense of national parks. "National parks are one of the United States' greatest contributions to the world," she exclaimed. "They serve to protect wild nature and have never caused harm to anyone." What was this about? Scholars widely accept the fact that the national park was invented in the postbellum US and later adopted as a model for the rest of the world. The idea that parks and other protected areas are the principal means of biodiversity protection is also well established in policy. But what of her assertion that the establishment of national parks "never caused harm to anyone?" None of the questions posed by her audience even remotely suggested that they did, so why did she find it necessary to rise to defend the national park concept as harmless?

Perhaps the answer can be found in a brief examination of the intellectual trajectory of scholarship on US national parks, beginning with one of the earliest historical accounts of the first national park, Yellowstone. Hiram Chittenden, a US Army Corps of Engineers captain, published the first edition of his descriptive survey, *The Yellowstone National Park*, in 1895. Chittenden (1903) noted that there were Native American tribes in the region, but none seemed to have knowledge of the area and only one had any history of occupation of park lands and that was sporadic, scattered, and not of recent origin. In a manner not unusual for the period, he essentialized the various tribes of the region in terms of their moral, social, and intellectual inferiority. The Crow's "tribal characteristics were an insatiable love of horse stealing and a wandering and predatory habit" (Chittenden 1903: 6). The Blackfeet were "perpetual fighters, justly characterized as the Ishmaelites of their race" (p. 7). "The Shoshones as a family were an inferior race ... destitute of even savage comforts ... [and] ... feeble in mind" (p. 8–9). Fast-forwarding to the latter half of the twentieth century we find that scholars have expunged not only the racist language, but also any mention of Native Americans altogether. In Runte's (1979) highly regarded *National Parks: the American Experience*, there is no entry in the index for "Native American" or any variation thereof. When Native Americans are mentioned in the text, they appear as a point of trivia, as when the author notes the Sierra redwoods (*Sequoia gigantea*) are named "after the Indian chief Sequoyah" (p. 27). Well into the 1990s there was "almost nothing" published that linked the history of US national parks with the history of Native Americans (Keller and Turek, 1997: xii). The long silence on the historic relationship between Native Americans and national parks is now being resoundingly filled.

As the millennium turned, the scholarship on US national parks turned with it. An important part of the context for this shift was the maturation of the field of environmental history, which in recent years began to critically examine the moral certitude of the mainstream conservation movement. As the past editor of the field's flagship journal, *Environmental History*, recently explained "[t]hese comfortable ideas began to come apart in the 1980s, and the genuine critique took a decade to evolve" (Rothman 2002: 18). This critical turn has led, among other explorations, to a reexamination of the process of park establishment in the western US and its relationship to the process of Native American removal and dispossession. New histories challenge the notion that iconic natural landscapes, such as Glacier National Park were "virgin wilderness," and instead emphasize a record of centuries of occupation by Blackfeet and other tribes (Keller and Turek 1997; Spence 1999; Warren 1997). These narratives suggest that "uninhabited wilderness had to be created before it could be preserved" (Spence 1999: 4). Recent studies by cultural geographers and environmental historians have emphasized that the open, "park-like" landscapes that early conservationists so admired were not shaped by nature, but by the agency of Native American tribes (Olwig 2002; Warren 1997). In campaigns to eliminate their access to wildlife, Native American hunting practices were characterized as morally reprehensible and ecologically destructive. At Yellowstone, Shonshone and Banock hunters were accused of "wanton slaughter of game" (quoted in Spence 1999: 65) and at Glacier the park superintendent claimed that elk leaving the boundaries were "recklessly slaughtered" by Blackfeet hunters (quoted in Warren 1997: 148). Through a detailed study of the "crown jewels" of the national park system,

Yosemite, Yellowstone, and Glacier, Spence (1999) details how the removal of Native Americans from parks and the establishment of a system of Indian reservations coincided in the west following the Civil War. Challenging the characterization of the conservation movement as a moral triumph, Germic points to baser impulses, such as the desire for individual profit, the genocidal aspects of Manifest Destiny, and the fear of economic ruin that led nineteenth-century bankers and railroad magnates to treat parks as "objects of venture capital" (Germic 2001: 105). Given the emergence of something like a revisionist history of US national parks and wilderness preservation, it is understandable that those who find the park ideal challenged, such as perhaps the Ahwanhee tour guide, take a defensive stance.

That scholars have begun to link US national parks to one of the most regrettable chapters in the nation's history – the dispossession and near genocide of hundreds of thousands of Native American peoples – is truly a sea change. Earlier "definitive" histories, such as Runte's (1979) and Nash's (1982 [1967]) *Wilderness and the American Mind* treated the preservation of national parks and wilderness as a story of moral and political triumph. Both, in their own way, sought to demonstrate the importance of wild areas, particularly in the west, to defining a uniquely American national identity. They can be read as morality tales, where high-minded ideals of public access and national heritage preservation triumph over plunder and profit-seeking. The new narratives turn these tales on their heads. They offer a different morality tale, one in which local commons are enclosed by the state, sovereign nations are disposed of their territory and written out of history, and the ecological consequences of park establishment are not altogether positive (Germic 2001; Spence 1999; Warren 1997). These alternative narratives of dislocation and enclosure offer new possibilities for geographical and historical comparative analysis that can provide the empirical foundation for theorizing the importance of national parks and conservation to the establishment of the modern territorial state and the creation of a corresponding national citizenry.

POSSIBILITIES FOR GEOGRAPHICAL AND HISTORICAL COMPARISON

Wilderness preservation and national park establishment are now standardized practices in virtually every Third World state, largely based on the US model and experience. The prevailing understanding of the effects of transferring this model "offshore" has been that it does not readily fit the circumstances of developing countries where protected area establishment can result in conflicts with resident peasant and tribal populations over access to land and resources (West and Brechin 1991). The recent studies previously cited suggest that the circumstances may not be so different as has been assumed, and that thus the US, rather than representing a model and an exception, may be, for theoretical purposes, treated as another example of a European settler state in conflict with indigenous populations. Furthermore, as several of these studies indicate, the conflicts and negotiations between Native American nations and the Anglo-European state over access to park land and resources are not "ancient history" but are ongoing and unresolved (Keller and Turek 1997; Spence 1999; Warren 1997). In all likelihood the maturation of a global indigenous peoples movement (Hodgson 2002) will further blur the boundaries between First and Third World as Blackfeet and Shoshone find common cause with San and Maasai in their efforts to come to terms with state conservation enclosures.

While I will not attempt a full comparative analysis here, the new US national park histories suggest that analogous and possibly generalizable processes and relationships among state policies and practices are experienced in other regions of the world. Focusing on examples and an extended case study from sub-Saharan Africa, I will highlight some of these in the pages that follow. Though there are many new forms of conservation territories that feature "nature–society hybrids" (Zimmerer 2000: 356), my analysis will concentrate on the "traditional" forms of state conservation territories – national parks, game reserves, and officially designated wildernesses. These are the "fortress conservation" models that require the absence of human occupation and impact and "reflect the priorities of national conservation agencies and international organizations" (Adams 2001: 272–3). The case study focuses on an area of roughly 200,000 square kilometers in the southeastern corner of Tanzania, particularly two features of the state's administrative geography, Liwale District and the Selous Game Reserve (GR). The study is largely based on primary colonial documents housed in the Tanzania National Archive, supported by reviews of secondary literature and field visits that included interviews conducted in communities bordering the Selous and with state officials and international conservation organization officials. During the German and British colonial periods, Liwale was the site of a series of state interventions designed to rationalize social and spatial relations and thereby create the necessary conditions for development. The location of these mid-twentieth century interventions is today the Selous GR, which encompasses about half of Liwale District's land base and is "popularly known as 'Africa's last wilderness area'" (Tibanyenda 1997: 38). Combined with the revisionist park histories from the US, the examples and case study from Africa will provide the basis for a preliminary theorization of the role of conservation territories in the construction of the modern state. Let us begin, first, with an examination of some of the patterns in the establishment of parks common across continents and historical periods.

I have identified four themes in the histories of national parks and other reserves in Africa that have parallels in the new US park histories. First, we can identify a pattern wherein park advocates deny or fail to recognize the historic human occupation of an area and the role of human use and management on the ecology and landscape that is targeted for preservation. In the case of Omo National Park in Ethiopia, for example, the main report advocating its establishment greatly underestimated the existing population of Mursi agro-pastoralists and characterized it as "Ethiopia's 'most unspoiled wilderness' which has 'retained its primeval character from ages past'" (quoted in Turton 1987: 179). Careful ecological study, however, would reveal that "virtually every square inch of this country bears the imprint of human activity" (Turton 1987: 180). Indeed, research has determined that much of East Africa's savanna grasslands, the predominant ecological community of the most renowned game parks in the region, are not wild landscapes but rather were shaped by centuries of pastoralists' herding and burning activities (Homewood and Rodgers 1984; Moe *et al.* 1990). Areas in West and Southern Africa that were also long settled, cultivated, and grazed, have been evacuated to create national parks in the name of wilderness preservation (Ranger 1999; Zuppan 2000). Where occupation is acknowledged, moral and western scientific discourse constructs African land use practices as inferior and destructive, thereby making populations subject to eviction in the name of wilderness preservation (e.g., Brockington and Homewood 2001; Collett 1987; Giles-Vernick 2002; Homewood and Rodgers 1987; Neumann 1995).

A second theme, closely related to the first, is the lack of empirically sound ecological justification for evictions of resident populations. This work is part of a general trend among researchers from a variety of disciplinary backgrounds who are critically examining "consensus views" and "received wisdom" of environmental degradation narratives expounded by expert western scientists (Beinart 2000; Stott and Sullivan 2000). Ranger's study of the creation of Matopos National Park revealed that numerous environmental interventions, including the ultimate eviction in the 1960s of the remaining residents, "were vainly attempted despite the fact that successive investigations found that the Matopo Reserve was not, after all, in a state of [ecological] crisis" (1999: 80). In an investigation of one of the more recent mass evictions from a protected area, Mkomazi Game Reserve in Tanzania in 1988–1989, researchers examined the empirical basis for evacuation advocates' claims that the area was of exceptional ecological importance and was threatened by human use (Brockington and Homewood 2001; Homewood and Brockington 1999). They found a general lack of ecological data and an absence of studies on human–environment interactions that could substantiate the claims. Nonetheless, once the reserve was cleared, eviction advocates wrote of an immediate, almost miraculous recovery of the ecosystem, raising the suggestion that either the "degradation of the reserve has been remarkably ephemeral" or that the environment is highly resilient and not characterized by "the fragility the writers thought they had saved" (Brockington and Homewood 2001: 462). Outside of national parks and protected areas, recent studies by geographers, historians, and anthropologists demonstrate that conservation scientists' and resource managers' unexamined acceptance of the premise that African land use practices are destructive has resulted in a misunderstanding of human-environment relations and ecological change and in misguided state environmental interventions (Bassett and Zueli 2000; Fairhead and Leach 1996; Sullivan 2000).

Third, conservation enclosures have been linked with other colonial and post-colonial state policies of social control and spatial segregation (Beinart 2000). In colonial Central African Republic, French administrators employed the technique of *regroupment* in an effort to corral Africans into "civilized" settlements in strict distinction from "wild" forest areas and thereby reorder the landscape. The postcolonial state, in cooperation with the Worldwide Fund for Nature (WWF) implemented similar settlement policies in the 1990s as part of their program to manage the Dzanga-Sangha National Park (Giles-Vernick 2002). In South Africa, the evacuation of part of Kruger National Park was associated with the creation of "homelands" under the former Apartheid regime (Carruthers 1994). In the first half of the twentieth century, San "Bushman" resident in Namibia's Etosha National Park were rounded up and relocated to labor on white farms or concentrated in "rest camps" to provide labor for government projects (Gordon 1992). More recently, the Botswana government began a three-year campaign in 1998 to remove nearly 3,000 resident Basarwa San from the Central Kalahari Game Reserve and relocate them in resettlement villages to facilitate their assimilation as Botswana citizens (Survival International 2001). In colonial East Africa, British authorities linked policies to sedentarize nomadic pastoralists to conservation enclosures in Tanzania and Kenya (Collett 1987; Igoe and Brockington 1999; Neumann 1998). The colonial vision for the Maasai and other nomadic pastoralists was to concentrate them into group ranches to facilitate market-oriented livestock production while enclosing their former pastures in national parks and reserves. In essence, colonial and postcolonial evacuations of protected areas have been characterized as part of an overall policy to "civilize" and "develop" the citizens of the state.

Fourth, parks and protected areas have functioned as enclosures (see *The Ecologist* 1993) that have curtailed access to local commons and a variety of communal resources such as wildlife, pasture, water, and fuelwood. East African pastoralists, for example, have lost over 20,000 square kilometers of grazing commons to national parks and game reserves in Kenya and 3, 234 square kilometers in Tanzania's Mkomazi Game Reserve alone (Igoe and Brockington 1999). It is difficult to assess the full impact of the enclosures for other resources, but it is well-established that access to wild meat protein was a critical element in many African agrarian economies across the continent, especially in times of dearth (see Neumann 2002). African hunting on forest and savanna commons is usually the first activity to be outlawed upon the declaration of new conservation enclosures. Often wildlife immediately adjacent to protected areas are considered "park animals" or "park property," which raises conflicts and debates over communal property rights in wildlife conservation (Naughton-Treves and Sanderson 1995) More generally, people's customary rights of movement through the commons are curtailed by the creation protected area boundaries (Giles-Vernick 2002; Neumann 1998). The overall effects of enclosure have been far-reaching, restructuring not only human–environment relations and livelihood strategies, but also even basic social relations within communities (e.g. Giles-Vernick 2001; Gordon 1992).

THEORIZING STATE-BUILDING AND
CONSERVATION ENCLOSURES

In this section I want to explore the question of why the patterns and processes detailed in the previous section recur in the conservation geography and history of country after country, from the late nineteenth century to the early twenty-first century. The answer, I argue, lies with the origins and characteristics of the modern state, particularly the assumption of territorial sovereignty, and, more generally, with the expression of modernity as a project of rational ordering to achieve "progress" and "development." Under modernity, a new and large category of people emerges – typically rural residents and forest dwellers – who are "backward" rather than "progressive," "traditional" rather than "modern." These are people "in the way of history, of progress, of development" (Berman 1982: 67). The transformation of backward citizens in the name of progress commonly has been referred to as the state's "civilizing" mission (Scott 1998), which requires the fundamental reordering of society in space. The types of projects mentioned previously (sedentarization, concentration, reservation) are common manifestations of the civilizing mission of colonial and postcolonial states around the world. Where Scott (1998) has emphasized the role these sorts of projects play in the simplification of forestry, agriculture, and the social life of the citizenry, we can see that they also play a role in clarifying boundaries and removing ambiguities. We can think of this aspect of modernity's rational ordering as a process of "enframing" that fixes "distinction between outside and inside" (Mitchell 1991: 55), most notably for the purposes at hand, the critical distinction between nature and culture.

The conventional understanding of what makes a modern state different from other forms of political and social organization is the requirement of clearly bounded territory and a recognition of the individual states' "claim to total sovereignty" over that territory (Agnew 1999: 503). The fulfillment of this claim is an assertion of ownership through the process of dismantling and enclosing the commons (*The Ecologist* 1993). Enclosure originated historically from the fifteenth to the nineteenth centuries in England as a process of transferring land and resources from communal ownership to private ownership and was later exported around the globe through colonialism. In the process of constructing empire, European colonial powers made claims of ownership to all lands they considered "wastelands" or "uncultivated," thereby seizing control of forest fallows, grazing lands, water sources, and other commons (*The Ecologist* 1993). Blaut (1993) labeled the justification for European appropriation of land in far flung regions as the "myth of emptiness," the assertion that the lands that Europe colonized were either vacant or, if occupied, were not subject to any legitimate claim of ownership beyond that land which was under cultivation. The myth of emptiness takes the form of specific policy and legal doctrine in different regions, such as *terra nullius* in Australia (Rangan and Lane 2001).

At the same time that the colonial state makes claims on all "vacant" lands, it also claims sole ownership over the natural resources of the territory, exclusive of those on recognized private land. For example, in a comparative study of forestry in Malaysia, Thailand, and Indonesia, Peluso and Vandergeest (2001) found that under varying historical and political circumstances, all of the states took similar actions, declaring the state to be sovereign owners of all land and resources within their territorial boundaries. "They all enacted laws enabling the demarcation of permanent state forests, causing forests to be defined subsequently in terms of state property regimes rather than ecologically ..." (Peluso and Vandergeest 2001: 768). They term these designated forests, "political forests" to emphasize their origins in the state rather than in nature. The designation and bounding of political forests was part of a greater state effort to reorder the landscape "into new categories of 'forests,' 'agricultural enterprise,' and 'settlement'" as a means of controlling nature and citizenry within its territorial boundaries (Peluso and Vandergeest 2001: 763). The colonial state made similar claims of ownership for wildlife resources as well. In British Africa, all wildlife was declared the property of the Crown and the state claimed sole authority over access and control (Neumann 1998, 2002). Postcolonial states by and large kept the land and resource laws and institutions in place, in some cases merely substituting the "President" for "Crown" (e.g. United Republic of Tanzania 1992).

The point I wish to make here is that these proprietary claims and the process of mapping, bounding, containing and controlling nature and citizenry are what make a state a state. States come into *being* through these claims and the assertion of control over territory, resources, and people. Julian Huxley, in the epigraph of this chapter, was clearly aware of the central role of natural resource regulation in the practice of statecraft (see also Scott 1998). The Colonial Office in London had sent Huxley to Africa – several years before he became the first director of UNESCO – to provide advice on "native education." In writing about his experiences there, he explicitly linked wildlife control to African education and social progress through the rational ordering of the territorial space of the colonies.

> Meanwhile it would, I think, be quite legitimate for the Government to plan ahead, work out what would be the best boundaries for game and native areas under a hypothetical scheme of division, and then concentrate all their work for native development – water-borings, dams, schools, demonstration farms, and the rest – in the "native" area, while deliberately keeping the other as untouched as possible.
>
> (Huxley 1931: 241)

Huxley's vision for Africa demonstrates the centrality of natural resource regulation and management to the origins, meaning, and functioning of the modern state and its citizens. It is a modernist vision, characterized by a faith in the mastery of science over nature and in the ability of the state to rationally reorder its territory. As Peluso and Vandergeest observed in the case of modern state forestry, "the science and the politics of territorial control go hand in hand" (2001: 781).

The claims of modern western science serve to support the proprietary claims of the state by asserting its superiority over other, more localized forms of knowledge. The claims and transformative ambitions of science are in the end claims "about ownership, not just of discourse but of nature" (Smith 1998: 274). Colonial science, with little exception, condemned African land and natural resource use and practices as wasteful, environmentally destructive, and inefficient thereby supporting the state's moral justification for its proprietary claims. The presumptions of the superiority of western science and of Africans as poor land stewards provided the basis for the construction of a set of "degradation narratives" that were initiated in the early colonial period and persist to the present (Bassett and Zueli 2000; Brockington and Homewood 2001; Fairhead and Leach 1996; Sullivan 2000). In some cases, there is only the authoritative claim and presumption of degradation without the support of scientific study or empirical evidence (e.g., Brockington and Homewood 2001; Sullivan 2000). Degradation narratives provide the rationale for state interventions into Africans' interactions with their environments, including the complete severing and reordering of those interactions.

The Selous Game Reserve in Tanzania provides an exemplary case for further theorizing and exploring the relationship between the establishment of conservation territories and grander civilizing/modernizing/development initiatives of the state. British colonial conservation and development plans in Tanzania (then Tanganyika) required a fundamental and geographically extensive spatial reorganization of wildlife populations, land uses, and African settlements. The colonial state planned and initiated these relocations against a background of evolving development strategies between 1919 and 1961, the period of formal British rule. Central to this process was the spatial segregation of human and wildlife populations that required the displacement and relocation of both. The motivations behind the plans to reorder the Tanzanian landscape were encompassed by the "civilizing" mission of British colonialism, which in turn was embedded in ideologies of racial and cultural superiority and a faith in western scientific achievements and mastery over Nature. The colonizers of Tanzania placed great faith in this plan. A provincial commissioner (PC) confidently predicted in 1947 that the "creation of National Parks for game and the creation of Closer Settlements for people ... would appear to be a solution which would serve most situations for very many years to come in Tanganyika."[1]

In the end, the fulfillment of the plan produced the landscape of the modern Tanzanian state, including its officially designated wilderness. For every settlement concentration, there were new spaces created to contain wild nature, spaces that I call "artifactual wilderness." It is artifactual in the dual sense that it is wilderness produced not by nature, but by human hand and that is an artifact of the state's assertion of territorial ownership and control. Containment and control of nature in conservation territories was inseparable from the colonizing state's efforts to control its African subjects and ultimately create a new kind of person: civilized, productive, and observable.

THE CASE OF THE SELOUS GAME RESERVE

The region of the case study is comprised of two administrative units of the Tanzanian state, the Selous GR and the Liwale District (see Figure 7.1). The terrain of the region is mostly rolling, forested hills – called *miombo* (*Brachystegia* spp.) woodland after the dominant tree species – and is heavily bisected by frequent streams and rivers. The larger valleys have deep alluvial soils. Most of the land falls within the range 300 to 700 meters elevation and receives an average of 600 to 800 millimeters of rainfall annually. The German colonial administration initially established two smaller game reserves in the northern area in 1905, which the British later incorporated into the Selous GR. During most of the British colonial period, the Liwale District fell within the Southern Province, now called the Lindi Region, and was administered at various times from district administrative offices in Liwale, Kilwa, and Nachingwea. The colonial government recognized Liwale as the ancestral home of the Ngindo people and in 1926 created the Ngindo Native Authority, whose boundaries more or less overlapped with those of the district.[2]

In the late nineteenth century, prior to the imposition of German colonial control, Liwale and the territory surrounding it functioned as the economic hinterland of the Indian Ocean trade in African commodities funneled through Zanzibar and Kilwa (Wright 1985). Slaves, ivory, rubber, and, to a lesser extent, various non-timber forest products such as gum-copal, beeswax, and honey, flowed through and from Liwale to the coast. By the late 1870s, rubber, from vine-grown rubber collected from the forest and bush by Ngindo, Makonde, and Ndonde inhabitants of the hinterland, had surpassed both slaves and ivory as Kilwa's principal export. Later, under German occupation in the early 1900s, Ngindo collectors experienced an economic boom period based on income derived from collecting and marketing wild rubber (Iliffe 1979). Beginning in 1906 another bush product, beeswax, was on its way to becoming an important cash earner for Liwale and cotton and tobacco were established as peasant cash crops.

Throughout Tanzania, wide-ranging negative demographic, ecological, and economic effects accompanied the imposition of German colonial rule in 1885 (Ford 1971; Iliffe 1969; Iliffe 1979; Kjekshus 1977; Turshen 1984; Wright 1985). In the Kilwa hinterland, human and animal disease epidemics, followed by German military actions, took a huge toll on the economy and population. Following the suppression of the 1905 Maji Maji rebellion German estimates of population in Songea District, part of which would later be included in the Selous, declined from 166,000 in 1902/03 to 20,000 in 1907 (Turshen 1984: 113). It is estimated that Liwale and surrounding areas suffered a loss of one-third

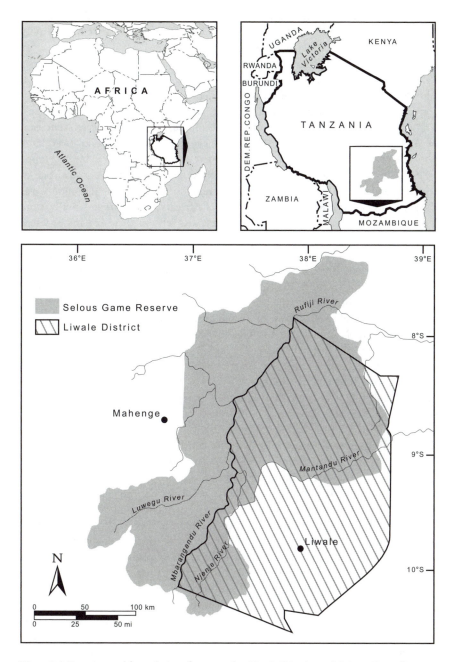

Figure 7.1 Location and boundaries of present-day Liwale District and Selous Game Reserve.

Source: Adapted from GTZ/Selous Conservation Programme (1995).

of their population in the aftermath of Maji Maji (Iliffe 1979: 200). African peasants and their livestock have never reoccupied many areas that were heavily cultivated in the late nineteenth century.

One of the main factors in the early twentieth century inhibiting the reoccupation of territory in southeastern Tanzania, and throughout East Africa, was the spreading presence of tsetse fly (*Glossina* spp.), which is the vector for trypanosomiasis in livestock and wildlife and sleeping sickness in humans. Through the combined effects of conquest, ecological crisis, and the reorientation of African labor to European enterprises, tsetse fly began to take over large portions of East Africa beginning in Uganda about 1900 (Ford 1971; Hoppe 1997; Iliffe 1995; Langlands 1967). In effect, an unintended consequence of conquest and the early incorporation of the region into the colonial economy was the expansion of wild nature at the expense of African settlement and civilization. Unlike their Belgian and the French counterparts who medicalized the problem of tsetse fly, the British emphasized a spatial strategy of population evacuation of affected areas and settlement concentration elsewhere (Ford 1971; Hoppe 1997). British authorities first ordered sleeping sickness evacuations in Uganda in 1906 and followed suit in the 1920s in Tanzania after taking control of the territory from Germany. Many of the major sleeping sickness evacuation areas formed the core of East Africa's well-known protected areas, including Queen Elizabeth and Murchison Falls National Parks in Uganda (Kinloch 1972; Langlands 1967).

Beginning in the early 1930s the Tanganyika Game Department implemented wildlife conservation and control practices that would transform the political ecology of wildlife and peasant agriculture in Liwale. Early colonial wildlife policies and practices were not designed merely to conserve wildlife from overhunting, but also to control it in terms of numbers, variety, and location. During most of the British colonial occupation, game policies were an attempt to balance wildlife conservation with wildlife control, always with an eye toward increasing peasant agricultural surplus. In the 1920s, the Game Preservation Department scouts were killing over 800 elephants a year in Tanzania in an effort to limit crop predation. Throughout the colonial period, the Game Department spent about 75 percent of its energies trying to control raiders, particularly elephant. Nevertheless, the government estimated that animals consistently consumed one-quarter to one-third of the Territory's food crop production. In 1932–33, as the elephant population climbed, the government inaugurated six elephant control schemes across the Territory in response to food shortages and excessive crop losses.[3] In Liwale, an ambitious and geographically extensive scheme was devised after the Governor toured the district in 1933 and was alerted to the serious damage being done to crops by elephant and other wildlife.[4] The scheme, in essence, was an attempt to corral elephants toward the west and eradicate them in the east.[5] In the first year of the scheme, European game rangers accounted for the shooting of 1,304 elephants in the Southern Province alone.[6]

As elephant herds were driven westward, the government strategy required the creation and expansion of the game reserves in the Liwale District to contain them. In the early 1930s the government approved a southward extension of the extant Selous GR to accommodate the exiled elephant herds. This lengthened the reserve to about 241 kilometers along a north-south axis, but in 1937 the GW declared that it was too narrow for elephant movement and recommended a westward expansion that would include an area where Game Department hunting bans and the in-migration of exiled elephant herds had already created a *de facto* reserve.

While the effort to herd elephants into what would eventually become the Selous GR progressed, the Game Department's opposition to peasant occupation of Liwale fertile western valleys increased. The government's strategy was to have elephant "control be intensified to the east and abandoned to the west, to try and force the natives in the west to come into country which could be protected."[7] The resident Ngindo, who were cultivating some of the most fertile soils in the district, resisted relocation and continued to exercise their legal rights to defend their crops. Game Department officials believed this was foiling their efforts to relocate the elephant herds. Initially, there were no compulsory relocations, though the government supported the Game Department's campaign of neglect with regard to crop protection in the reserve. The strategy was to coerce people into "voluntarily" vacating the game reserve, as the Game Department would provide neither protection within the reserve nor allow residents to arm and defend themselves.[8]

As a result of the state's control efforts, the population of elephants throughout the Territory grew. The Southern Province had the largest concentration and as these were driven westward, their densities in the *de facto* reserve increased. Those villages unfortunate enough to be in the path of the drive came under intense pressure from crop raiding elephants. With the passage of the 1940 Game Ordinance, those pressures increased. The boundaries of the Selous were expanded by the ordinance and it now encompassed the most fertile and productive valleys in the district. Describing the implications of the new boundaries, the Liwale district commissioner (DC) wrote that the Ngindo "are valley cultivators and a cursory glance at a map will show the West and South of the District [now in the reserve] offer at once the greatest number of valleys."[9] As had been the case, the 1940 Game Ordinance did not require forced evictions, but "discouraged" settlement by withholding crop protection.

The effect of these developments on the valley dwellers in the reserve is illustrated by the experience of the villages under one *Mwenye*, Msham Mbamba. From 1935, Mbamba's people found themselves living in the proposed Njenje-Mbarangandu-Luwegu Reserve, later to become the Selous GR. Most of them lived and farmed in "an extremely fertile area" along the Njenje and Mbarangandu Rivers.[10] In 1936, the district office reported that Mbamba was not particularly concerned about elephants, but *was* anxious that they not be forced to relocate. According to one DC, these areas were "probably the most well-watered and fertile in the District and the people are living there for cultivation, not for hunting purposes." However, the "pressure of elephant," he noted, "is already very great and the natives are finding it difficult to maintain their cultivations."[11] By 1941, most of Mbamba's people left the most fertile valleys inside the reserve for less desirable sites because of the problem of rising numbers of elephants and the lack of protection from the Game Department scouts.[12] The situation was no longer one of scaring off the stray bull at night, but "a perpetual day and night warfare."[13] Consequently, the population of the Njenje-Mbarangandu-Luwegu area began to decline, from 378 taxpayers in 1935 to 277 six years later.[14] The evacuation of western Liwale and its construction as a wilderness landscape had begun.

As people began leaving the more fertile valleys they helped to fashion an image of western Liwale as a sparsely populated, remote, and tsetse fly-infested landscape. Some local administrators realized that the Game Department's twin strategy of driving elephants westward while withholding crop protection was promoting an invasion of wildlife and the spread of tsetse fly into the most agriculturally productive areas of the district. By early 1936 Liwale District officials began to note that the number of elephants had increased inside the *de facto* reserves and that the best valley lands subsequently had been given over to wildlife.[15] Later the district records note that the reoccupation of once prosperous settlements on the Mbarangandu River had been curtailed by the 1940 extension of the Selous and that elephants had become "a great menace to the fertile Ndapata valley and to Mbindera which borders with the reserve."[16] Some Liwale district officials objected to the inclusion in the Selous of "so large an area so much of it fertile valley land."[17] Officials depicted other valleys, which the government later included in subsequent expansions of the Selous, as having tremendous agricultural potential. One DC described part of the Matandu River to the north of Liwale town on the Selous boundary as

> an area of some 200 square miles of alluvial valley bottom soil of the highest fertility suitable for growing of millet rice and cotton on an intensive scale. This soil is in general of almost inexhaustible fertility and great depth.[18]

In the Ngarambi-Mihumo area, also north of Liwale town, the sleeping sickness surveyor observed that the "land is cultivated year after year, yielding good crops and very little new land is broken each year."[19] Others noted that the valleys supported a system of double cropping and "can be cultivated for generations without respite or erosion."[20]

While the GW lobbied the local administration to order the Ngindo Native Authority to evacuate the expanding Selous,[21] officials in Dar es Salaam were pondering the larger question of how to proceed in "civilizing" and "developing" the residents of Liwale. In 1943, Tanganyika's administrative secretary, J.E.S. Lamb, revealed that he had "for some time had in mind the need for 'doing something' about the Liwale district."[22] His overriding concern was the same as that of his official predecessors; that Liwale and its inhabitants were just too cumbersome to administer. The only viable solution for future economic and social development was "resettlement of the bulk of the population." Once the area was depopulated, it "should be declared a game reserve," the use for which it was best suited.[23] A subsequent minute by the Governor justifies compulsory relocation as a necessary first step in bringing Africans "the 'civilizing' influences" of colonial occupation.[24]

Since the 1930s, the colonial government had, as part of its civilizing mission to "induce natives to settle in productive areas and develop them," endorsed an overall "concentration policy."[25] According to British authorities, Liwale, where concentration would be in the "the natives" own interest,[26] was an ideal target. For one thing, the administration viewed settlement concentration as a means to strengthen a politically weak Ngindo Native Authority by bringing their subjects under closer supervision. For another, Liwale was in all regards difficult to administer, "especially with regard to tax collection," and concentrating populations closer to large towns and administrative centers would relieve this problem.[27] Following the governor's approval of Lamb's evacuation plan, events progressed swiftly in Liwale and it quickly became the largest single settlement concentration in Tanganyika, with the goal of relocating 30,000 people.[28] The Game Department, no doubt pleased with the government's new proactive policy of concentrating settlement in Liwale, took the opportunity to target communities that had so far resisted their efforts to coerce relocation from the Selous.[29]

The resettlement scheme created three concentration centers in the northeast, central east, and southeast of the district, evacuating everything to the west. This would leave only Liwale town, which would serve as a local headquarters for the game reserve on its new eastern boundary. The operation's records indicate that nearly as many people fled the concentration schemes as were actually relocated by the government. As the news of the evacuation spread through the district, there was a general exodus from areas of western Liwale by people who had only recently been welcomed there after dodging other concentrations. In Madaba, a village north of Liwale, when a game scout was sent to escort the evacuees through the game reserve and into the Eastern Province, they "vanished in the night" before he arrived, leaving the settlement abandoned.[30] Madaba was thus an immediate "success" but progress in the rest of the district was slow so that only 1,200 people were moved in 1945. The greatest movement was in 1947 and much of central Liwale was evacuated. As district administrators made plans to evacuate another 3,195 families in 1948 an even more ambitious plan for Liwale's development appeared on the horizon. Compared glowingly in the press to the Tennessee Valley Project and the settlement of the western frontier of North America, the Overseas Food Corporation's (OFC) enormous groundnut scheme dwarfed and ultimately halted the evacuation plans in the interest of maintaining an in situ labor force. As a consequence, the plan of completely evacuating Liwale District never came to full fruition, but the resettlement project did effectively clear the fertile valleys of the Selous GR of settlements thereby establishing a vast new wilderness.

Elephant control policies and sleeping sickness concentrations helped to transform Liwale inhabitants and their relationship to the environment. First, the pressure from the increasing numbers and density of elephants reduced peasant production in two ways; by increasing crop losses and by displacing cultivation from the more productive soils. Second, elephant control schemes included a general African peasant disarmament policy and most wildlife control efforts were taken over by the state. In the case of the game reserve, the state provided no assistance at all while simultaneously denying the right of farmers to defend fields. The records make clear that colonial officials were aware of increasing elephant populations and crop losses and that wildlife managers and advocates of closer settlement used the knowledge effectively to drive peasants off their lands. Third, being forced into less fertile lands, the displaced Ngindo populations had no choice but to employ more extensive land use practices. Drier, less fertile valleys and uplands supported fewer numbers and were more easily exhausted. Shifting cultivation in the *miombo* woodlands spread as the productive capacity of the remaining valleys was reached. Displacements had the twin effects of reducing peasant production and reinforcing the image of Liwale as unproductive and Ngindo cultivation as backward and inefficient, which together helped to justify extreme measures for achieving development. Fourth, the sleeping sickness concentrations further diminished agricultural production by disrupting planting and harvesting activities. Upon hearing of the government's resettlement plans, some residents quickly fled, "many of them leaving the food which they had just harvested."[31] In the upheaval and uncertainty of the resettlements, fields were abandoned, harvests left to rot, and "hunger and hardship" accompanied each move.[32] In some communities, the government had threatened evacuation for over three years, leaving farmers reluctant to make land improvements or plant extensively and thereby fulfilling the British officials' ideas of lazy, backward farmers in need of transformation.

The removal of valley settlements and their replacement with elephant and other wildlife populations transformed western Liwale into a wilderness "teeming with game." Earlier studies of the Selous concur that very little wildlife was found in the area prior to the 1930s (Kjekshus 1977; Matzke 1972, Rodgers 1976). Ngindo elders interviewed in the 1970s unanimously remembered that no elephants were in the area prior to the 1920s and associated their occurrence with the imposition of British rule (Rodgers 1976: 23).

Each successive state intervention into human–nature relations in Liwale created the ecological and economic conditions for the next. By the early 1930s, it was widely recognized by game officers in the field that elephant numbers were increasing throughout the territory, "occupying great tracts of land where they have not been seen for years."[33] From 1931 to 1950, Game Department officers during control operations shot 33,462 elephants in Tanzania (Tanganyika Territory 1951: 3). The Game Department was killing ever-larger numbers of elephants around cultivation areas, from 800 in the 1920s to over 3,000 annually by the 1940s in a futile effort to control the population. Year after year the Game

Department reported that "[i]n spite of so many beasts being killed, it is estimated that the elephant ... is still on the increase" (Tanganyika Territory 1953: 10).

In the Selous GR, there was virtually no check on the population from Game Department control efforts and the valley cultivation was overrun. The elephant control policies made the most fertile valleys uninhabitable, promoted the advance of bush at the expense of cultivation, and thus encouraged the spread of tsetse fly.[34] When, in 1944, it came time to "do something about Liwale," the environment and human geography had been transformed by two decades of "civilizing" efforts and "elephant control" to an area unfit for habitation. Evacuation was the "rational" solution and the sleeping sickness concentrations that followed merely served to spatially extend what the Game Department had set in motion. Sleeping sickness concentrations eliminated what remained of peasant cultivation and resource extraction and the expansion of the game reserve curtailed any possibility of reoccupation and recovery.

The various colonial plans for Liwale, from the first elephant control scheme to the sleeping sickness evacuation, to the groundnut scheme, never mention a desire to preserve wilderness, a need to protect wildlife populations, or any other significant conservation motivation. Wildlife control policies were, however, inextricably linked to the general policy of settlement concentration of the 1930s, which was driven by overriding concerns for the political control and economic development of the territory. The twin spatial strategy of park and reserve creation and "closer settlement" became the foundation upon which to construct a colonial economic development strategy in the 1940s. This "modernization" strategy failed, however, to translate into "advancement" in Liwale though it did produce a vast wilderness area. As district administrators in the 1950s concluded in hindsight, "it would be untrue to say that Development to any appreciable extent has taken place."[35] That "[s]leeping sickness was used only as an excuse" to evacuate Liwale was no secret to succeeding district administrators.[36] There was, in fact, no sudden outbreak of sleeping sickness in 1943 when the government initiated its evacuation plan. The archival documents make clear that the main problem that the state had with Liwale was one of insufficient political and economic control – Liwale peasants were too easily evading taxes, law enforcement was difficult at best, and the Ngindo Native Authority too weak to be effective.

CONCLUSION

British colonizers sought to re-order the landscape of Tanzania, to establish a new spatial organization of nature and culture in the name of a "civilizing" mission. Behind the colonial plans of civilization and improvement were proprietary claims over the territory and its resources and an overriding faith in the state's power to rationalize space and transform its citizen-subjects. Both the elephant control scheme and the sleeping sickness evacuations were based on what was considered sound scientific calculation and justified by high-minded goals for the improvement of Africans' lives. What this ultimately meant was that the state would have to destroy Liwale in order to save it, but also in the process, would create a new space for wild nature, an artifactual wilderness. The Selous's wilderness, it should be clear, is not a vestigial premodern landscape, but rather is a product of colonizers' modernization efforts and thus a critical component of the landscape of modernity, not of antiquity. Wilderness, in this case, is not "outside" the rational ordering of state territory, but is in fact an integral part of that order.

We can theorize the British colonial administration's plans to reorder nature and human society within its East African territory in terms of a generalizable process of state-building. That is, the colonial administration's property claims combined with its reordering of space through the relocation and containment of wildlife and the movement of people into concentrated settlements were fundamental to this region of Africa becoming a modern state. It is a process that can be documented in every region of European expansion. Bantustands, Indian reservations, *regroupment*, homelands, native reserves, group ranches, villagization, and all of the many other spatial strategies that state employs to make its far-flung citizens visible, contained, and *ordered* are one side of the same process of containing and bounding wild nature in parks and reserves. Both wilderness and concentrated settlement are products of a single process, the creation of the modern territorial state.

The wild areas of national parks and reserves, as products of the creation of the modern nation state, are as much an expression of modernism as skyscrapers, though their establishment does not offer the drama and promise of great social engineering plans like Brasilia or Soviet collectivization. They are, nevertheless, an integral part of the practice of modern statecraft and a result of numerous plans to divide and contain the central antinomies of modernity: nature and culture, consumption and production, wilderness and civilization. The idea of bounding and managing a landscape the size of several mid-sized European countries as a preserve of wild nature has all the hubris and faith in science and progress as any high modernist project. The establishment of protected areas represents an historic transition under modernity wherein human civilization becomes the caretaker of a wild nature that poses no threat beyond the threat of disappearing. Nature becomes a ward of the state.

What links the case of the Selous to the establishment of the first national parks in the US is the fact that wilderness had to be created before it could be protected. In both cases wilderness was created through the assertion of the state's proprietary claims on land and nature and subsequent enclosure of existing commons. It is this historical dispossession in the name of nature protection and wildlife conservation that helped fuel anti-colonial movements in Africa (Beinart 2000). Struggles over property rights continue to fuel the conflicts and negotiations between national parks and Native Americans in the US and displaced peasant and pastoralist communities in Africa. As Goldman has insisted, for many ecological conflicts, whether involving states, local NGOs, or indigenous peoples' movements, the common focus is "the question of property rights – rights to land, forest, yields, burial grounds, seeds, intellectual property rights, ground and surface water . . . the village well and pastoral grazing" (Goldman 1998: 2).

Where the roots of resistance to commons enclosure can be traced to the nationalist movements in 1950s Africa, in the US it can be traced to the "Red Power" movement of the 1960s. The movement featured attempts of various Native American tribes to seek to redress historic dispossessions of lands through, among other demonstrations, the occupation of federal lands such as Mount Rushmore and Alcatraz Island. The creation of national parks on federal lands claimed by Native Americans was a central point of contention at both of these occupations. As one Sioux occupier of Rushmore explained:

> These were the reasons we invaded Mount Rushmore. The Federal Government took two hundred thousand acres from us to build a gunnery range. They didn't return it as promised. We've been waiting twenty-six years. And now the Secretary of the Interior has made a statement that he's going to turn it into a national park. That's a violation of the damn agreement they made with us
>
> (Lee Brighman, quoted p. 88 in
> Lame Deer/John Fire and Erdoes, 1972).

Similarly in the case of the Alcatraz occupation, the US National Park Service attempted to rush through the designation of the island as part of the Golden Gate National Recreation Area. The occupiers responded to that idea: "Our answer to the U.S. government that this island be turned into a park . . . at this time or at any other time is an emphatic NO" (quoted in Germic 2001: 67). What is clear from these two cases and from all of the cases presented in the preceding sections is that the creation of conservation territories is a key focus in the continuing struggles over the commons and proprietary rights of nature in the First and Third Worlds alike.

NOTES

* Parts of this chapter previously appeared in R. Neumann, "'Africa's last wilderness': reordering space for political and economic control in Colonial Africa," *Africa* 71, 4, 2001 and are printed by permission from the copyright holder, Edinburgh University Press.

1 Provincial Commissioner (PC), Southern to Chief Secretary (CS), 3 Jan., 1947. Tanzania National Archive (TNA Acc. No. 16, 22/4).

2 TNA, Nachingwea District Book, sheet no. 8.

3 Minute, illegible, 4 Aug., 1933. TNA 21726.

4 Minute, illegible, 4 Aug., 1933. TNA 21726.

5 Game Ranger (GR) to Acting Game Warden (GW), 30 Jan., 1935, TNA Acc. No. 16, 22/13, vol. II; and "Elephant Control in Tanganyika Territory," Memorandum by Commander D.E. Blunt, 1933, TNA 21773.

6 Acting GW to CS, 20 Jan., 1934. TNA 21726.

7 "Elephant Control in Tanganyika Territory," Memorandum by Commander D.E. Blunt, 1933. TNA 21773.

8 This strategy essentially followed a policy put in place by the first Director of Game Preservation, C.F. Swynnerton. In 1927, he instructed his cultivation protectors to give no aid to sparsely populated or distantly-located settlements. From: Copy of letter of instructions from the Director of Game Preservation (C.F. Swynnerton) to all Cultivation Protectors, Lindi District, 14 Sep., 1927. TNA Acc. No. 16, 22/4.

9 "Report on revision required in the boundaries of the Selous Game Reserve as extended to the Liwale District by the Game Ordinance of 1940," by DO, Liwale, Jan., 1942. TNA Acc. No. 19, 22/13, Vol II.

10 Act. DO, Liwale to PC, Southern, 20 Jan., 1936. TNA Acc. No. 19, 22/13, Vol II.

11 DO, Liwale to PC, Southern, 19 Mar., 1941. TNA Acc. No. 19, 22/13, Vol II.

12 For example, the Nachingwea district reported that "he and his people moved in 1941 from the most fertile valley . . . being driven out by elephant." TNA, Nachingwea District Book.

13 "Report on revision required in the boundaries of the Selous Game Reserve as extended to the Liwale District by the Game Ordinance of 1940," by DO, Liwale, Jan., 1942. TNA Acc. No. 19, 22/13, Vol II.

14 "Report on revision required in the boundaries of the Selous Game Reserve as extended to the Liwale District by the Game Ordinance of 1940," by DO, Liwale, Jan., 1942. TNA Acc. No. 19, 22/13, Vol II.

15 Acting DO, Liwale to PC, Southern, 20 Jan., 1936. TNA Acc. No. 16, 22/13 and District Commissioner (DC), Liwale to PC, Southern, 13 May, 1941. TNA Acc. No. 16, 22/13, vol. II

16 "The Native Authorities of Liwale District," TNA, Nachingwea District Book.

17 "Report on revision required in the boundaries of the Selous Game Reserve as extended to the Liwale Distric by the Game Ordinance of 1940," by DO, Liwale, Jan., 1942. TNA Acc. No. 19, 22/13, Vol II.

18 DC, Liwale to PC, Southern, 21 Nov. 1944. TNA Acc. No. 16, 19/70, Vol I.

19 Sleeping Sickness Surveyor to SSO, 28 Jan., 1943. TNA Acc. No. 16, 19/70, Vol I.

20 "Memorandum on the Concentration of Population in the Liwale District," 11 Sep., 1943 TNA Acc. No. 16, 11/204.

21 CS to PC, Southern, 25 Feb. 1941. TNA Acc. No. 16, 22/13, vol. II

22 Minute by administrative secretary (AS), Lamb, 22 Dec. 1943. TNA 31796.

23 Minute by AS, Lamb, 22 Dec. 1943. TNA 31796.

24 Minute by governor, Tanganyika Territory, 27 Dec. 1943. TNA 31796.

25 Government Circular No. 40 of 1934. TNA Acc. No. 16, 1/102.

26 ADO, Liwale to PC, Southern, 22 Jan., 1935. TNA Acc. No. 16, 1/102. In general, the administration was concerned that Liwale peasants could too easily avoid paying taxes by marketing their products in Kilwa, a longstanding practice. TNA, Nachingwea District Book, sheet no. 6.

27 ADO, Liwale to PC, Southern, 22 Jan., 1935. TNA Acc. No. 16, 1/102.

28 Senior Agricultural Officer, Southern Province to Director of Agricultural Production, 6 Sep. 1944. TNA Acc. No. 16, 19/70, vol. I.

29 GR, Liwale to PC, Southern, 20 Apr. 1945. TNA Acc. No. 16, 19/70, vol III.

30 DC, Liwale to PC, Southern, 23 Nov. 1944. TNA Acc. No. 16, 19/70, vol. I. PC, Southern to AS, 9 Jan., 1945 and DC, Liwale to PC, Southern 7 June, 1944. TNA Acc. No. 16, 19/70.

31 PC, Southern to CS, 21 Nov., 1945. TNA Acc. No. 11/234.

32 "Liwale Closer Settlement Scheme," by DC, Liwale, TNA, Nachingwea District Book.

33 "Elephant Control in Tanganyika Territory," Memorandum by Commander D.E. Blunt, 1933, p. 3, TNA 21773 and Game Ranger (GR) to Acting Director of Game Preservation Department, 1933. TNA 21726.

34 Kjekshus (1977) has argued that the initial effects of European colonialism on Africans in Tanzania was a loss of control over their environment and thus greater vulnerability to

natural hazards and a consequent drop in population. Iliffe (1979: 201–2) points out that, in the early colonial era, the creation of game reserves combined with human depopulation as a result of disease and European military conquest resulted in the spread of tsetse fly and the invasion of wildlife into previously cultivated areas.

35 "Excerpts from the Annual Report of District Commissioners," TNA, Nachingwea District Book.
36 "Liwale closer settlement scheme," TNA, Nachingwea District Book.

REFERENCES

Adams, W. (2001) *Green development: environment and sustainability in the Third World*, 2nd edn, London: Routledge.

Agnew, J. (1999) "Mapping political power beyond state boundaries: territory, identity, and movement in world politics," *Millennium*. 28, 3: 499–521.

Bassett, T. and Koli, B.Z. (2000) "Environmental Discourses and the Ivorian Savanna," *Annals of the Association of American Geographers* 90, 1: 67–95.

Beinart, W. (2000) "African history and environmental history," *African Affairs* 99: 269–302.

Berman, M. (1982) *All that is solid melts into air: the experience of modernity*, New York: Simon and Schuster.

Blaut, J. (1993) *The colonizer's model of the world: geographical diffusionism and Eurocentric history*, New York: Guilford Press.

Brockington, D. and Homewood, K. (2001) "Degradation debates and data deficiencies: the Mkomazi Game Reserve, Tanzania," *Africa*, 71, 3: 449–80.

Carruthers, J. (1994) "Dissecting the myth: Paul Kruger and the Kruger National Park," *Journal of Southern African Studies*, 20, 2: 263–83.

Chittenden, H. (1903) *The Yellowstone National Park*, 4th edn, Cincinnati: The Robert Clark Company.

Collett, D. (1987) "Pastoralists and wildlife: image and reality in Kenya Maasailand," pp. 129–48 in D. Anderson and R. Grove (eds) *Conservation in Africa: people, policies and practice*, Cambridge: Cambridge University Press.

Fairhead, J. and Leach, M. (1996) *Misreading the African landscape: society and ecology in a forest savanna land*, Cambridge: Cambridge University Press.

Ford, J. (1971) *The role of the trypanosomiases in African ecology: a study of the tsetse fly problem*, Oxford: Clarendon Press.

Germic, S. (2001) *American green: class, crisis, and the deployment of nature in Central Park, Yosemite, and Yellowstone*, Lanham, Maryland: Lexington Books.

Giles-Vernick, T. (2002) *Cutting the vines of the past: environmental histories of the Central African rain forest*, Charlottesville: University of Virginia Press.

Goldman, M. (1998) "Introduction: the political resurgence of the commons," pp. 1–19 in M. Goldman (ed.) *Privatizing nature: political struggles for the global commons*, New Brunswick: Rutgers University Press.

Gordon, R. (1992) *The Bushman myth: the making of a Namibian underclass*, Boulder: Westview.

GTZ/Selous Conservation Program. (1995) *Selous Game Reserve General Management Plan*, Dar es Salaam: GTZ/Selous Conservation Program.

Hodgson, D. (2002) "Introduction: comparative perspectives on the indigenous rights movement in Africa and the Americas," *American Anthropologist* 104, 4: 1037–49.

Homewood, K. and Rodgers W. (1984) "Pastoralism and conservation," *Human Ecology*, 12, 4: 431–41.

—— (1984) "Pastoralism, conservation and the overgrazing controversy," pp. 111–28 in A., D. and R. Grove (eds) *Conservation in Africa: people, policies and practice*, Cambridge: Cambridge University Press.

Homewood, K. and Brockington, D. (1999) "Biodiversity, conservation, and development in Mkomazi Game Reserve, Tanzania," *Global Ecology and Biogeography*, 8: 301–13.

Hoppe, K. (1997) *Lords of the fly: environmental images, colonial science and social engineering in British East African sleeping sickness control, 1903–1963*, Doctoral Dissertation: Boston University.

Huxley, J. (1931) *Africa view*, London: Chatto and Windus.

Igoe, J. and Brockington, D. (1999) *Pastoral land tenure and community conservation: a case study from north-east Tanzania*, Pastoral Land Tenure Series No. 11. London: International Institute for Environment and Development.

Iliffe, J. (1969) *Tanganyika under German rule 1905–1912*, Cambridge: Cambridge University Press.

—— (1979) *A modern history of Tanganyika*, Cambridge: Cambridge University Press.

—— (1995) *African: the history of a continent*, Cambridge: Cambridge University Press.

Keller, R. and Turek, M. (1997) *American Indians and national parks*, Tucson: University of Arizona Press.

Kinloch, B. (1972) *The shamba raiders: memories of a game warden*, London: Collins and Harvill Press.

Kjekshus, H. (1977) *Ecology control and economic development in East African history: the case of Tanganyika, 1850–1950*, Berkeley: University of California Press.

Lame Deer/John Fire and Erdoes, R. (1972) *Lame Deer seeker of visions*, New York: Simon and Schuster.

Langlands, B.W. (1967) *Sleeping sickness in Uganda 1900–1920*, Makerere University College, Department of Geography Occasional Paper No. 1. Kampala: Makerere University College.

Matzke, G. (1972) "Settlement reorganization for the production of African wildlife in miombo forest lands: a spatial analysis," *Rocky Mountain Social Science Journal*, 9, 3: 21–33.

Mitchell, T. (1991) *Colonising Egypt*, Berkeley: University of California Press.

Moe, S.R., Wegge, P. and Kapela, E.B. (1990) "The influence of man-made fires on large wild herbivores in Lake Burungi area in northern Tanzania, *The African Journal of Ecology*, 28: 35–45.

Mustafa, K. (1993) "Eviction of pastoralists from the Mkomazi Game Reserve in Tanzania," Typescript, International Institute for Environment and Development.

Nash, R. (1982) *Wilderness and the American mind*, (3rd edn), New Haven: Yale University Press.

Naughton-Treves, L. and Sanderson S. (1995) "Property, politics, and wildlife conservation," *World Development*, 23, 8: 1265–75.

Neumann, R. (1995) "Ways of seeing Africa: colonial recasting of African society and landscape in Serengeti National Park," *Ecumene* 2, 2: 149–69.

—— (1998) *Imposing wilderness: struggles over livelihoods and nature preservation in Africa*, Berkeley: University of California Press.

—— (2002) "The postwar conservation boom in British colonial Africa," *Environmental History* 7, 1: 22–47.

Olwig, K. (2002) *Landscape, nature and the body politic: from Britain's Renaissance to America's new world*, Madison: University of Wisconsin Press.

Peluso, N.L. and Vandergeest, P. (2001) "Genealogies of the political forest and customary rights in Indonesia, Malaysia, and Thailand," *The Journal of Asian Studies*, 60, 3: 761–812.

Rangan, H. and Lane, M. (2001) "Indigenous peoples and forest management: comparative approaches in Australia and India," *Society and Natural Resources*, 14, 2: 145–60.

Ranger, T. (1999) *Voices from the rocks: nature, culture and history in the Matopos Hills of Zimbabwe*, Oxford: James Currey.

Rodgers, W.A. (1976) "Past Wangindo settlement in the eastern Selous Game Reserve," *Tanzania Notes and Records*, 77–8, June: 21–5.

Rothman, H. (2002) "A decade in the saddle: confessions of a recalcitrant editor," *Environmental History*, 7, 1: 9–21.

Runte, A. (1979) *National parks: the American experience*, Lincoln: University of Nebraska Press.

Scott, J. (1998) *Seeing like a state: how certain schemes to improve the human condition have failed*, New Haven: Yale University Press.

Smith, N. (1998) "Nature at the millennium: production and re-enchantment," pp. 271–85 in B. Braun and N. Castre (eds) *Remaking reality: nature at the millennium*, London: Routledge.

Spence, M. (1999) *Dispossessing the wilderness: Indian removal and the making of the national parks*, Oxford: Oxford University Press.

Stott, P. and Sullivan, S. (2000) "Introduction," pp. 1–11 in P. Stott and S. Sullivan (eds) *Political ecology: science, myth and power*, London: Arnold; New York: Oxford University Press.

Sullivan, S. (2000) "Getting the science right, or introducing science in the first place? Local 'facts,' global discourse – 'desertification' in north-west Namibia," pp. 15–44 in P. Stott. and S. Sullivan (eds) *Political ecology: science, myth and power*, London: Arnold.

Survival International. (2001) *Survival International Website*. Online. Available HTTP: http://www.survival-international.org.

Tanganyika Territory. (1951) *Annual report of the Game Division, 1950*, Dar es Salaam: Government Printers.

——(1953) *Annual report of the Game Division, 1951–1952*, Dar es Salaam: Government Printers.

The Ecologist (1993). *Whose common future? Reclaiming the commons*, Philadelphia: New Society Publishers.

Tibanyenda, R. (1997) "Walking safaris: sampling nature in Tanzania's wilderness," *Tanzania Wildlife*, 6, June–Aug.: 38–9.

Turshen, M. (1984) *The political ecology of disease in Tanzania*, New Brunswick: Rutgers University Press.

Turton, D. (1987) "The Mursi and national park development in the lower Omo Valley," pp. 169–86 in A., D. and R. Grove (eds) *Conservation in Africa: people, policies and practice*, Cambridge: Cambridge University Press.

United Republic of Tanzania. (1992) *Report of the Presidential Committee of Inquiry in land matters*, Dar es Salaam: URT, Government Printers.

Warren, L. (1997) *The hunter's game: poachers and conservationists in twentieth-century America*, New Haven: Yale University Press.

West, P. and Brechin S. (1991) *Resident peoples and national parks: social dilemmas and strategies in international conservation*, Tucson: University of Arizona Press.

Wright, M. (1985) "East Africa 1870–1905," pp. 539–91 in R. Olive and G. Sanderson (eds) *The Cambridge history of Africa*, vol. 6, from 1870–1905, Cambridge: Cambridge University Press.

Zimmerer, K. (2000) "The reworking of conservation geographies: nonequilibrium landscapes and nature–society hybrids," *Annals of the Association of American Geographers* 90, 2: 356–69.

Zuppan, M.E. (2000) "Including herders in conservation management: reflections on both sides," *Development Anthropologist*, 18, 1–2: 3–17.

8

WATER, MARKETS, AND EMBEDDED
INSTITUTIONS IN WESTERN INDIA[*]

Navroz K. Dubash

A process of state withdrawal from water control and provision, ceding this space
to the market, characterized global water politics in the 1990s. A series of global
pronouncements during this period lauded the market as the best means of
dealing with the harsh reality of allocating ever more scarce water resources. This
development mirrored a broader political shift, as the dominant "Washington
Consensus" of the 1980s and 1990s urged a relentless unshackling of economic
forces from the constraints of social conditions and political pressures.[1] This
dominance has not gone unchallenged. The implementation of market-approaches
to water has spurred an outbreak of protest, particularly at the municipal level
and in reaction to large water projects. Local communities and their supporters
struggle to articulate an alternative framing of water management, around
community rights, social risks, and local control rather than around supply,
demand, and corporate control.

At the root of global water politics today, then, is a Polanyian tension between
the commoditization of water and the counter-pressures arising from its
embeddedness in social practice and institutions (Polanyi 1944). To this
discussion, political ecology brings a tradition of analysis that focuses on the local
and the particular that provides a basis for questioning the abstract market as a
uniformly valid mechanism of coordination. Starting with Blaikie and
Brookfield's (1987) early concerns with the "land manager," political ecology
serves to re-embed water politics in daily practice through its focus on
management questions and the articulation of these questions with the changing
context of state and market roles. In addition, by exploring the "institutional and
regulatory spaces in which ... knowledges and practices are encoded, negotiated
and contested" (Chapter 1), political ecology helps counter the abstraction of the
apolitical and self-regulating market. Instead, it argues for detailed empirical
attention to the local micro-politics of institutional change. A third theme from
political ecology relevant to water politics is its attention to the environmental
context in which history is made (Guha 1989). Attention to the natural resource
characteristics of water as a commodity requires an appreciation of how nature

and society interact in shaping and constraining markets. When connected with a view of exchange as an "instituted process" through which economic life is re-embedded in social life through the political process (Polanyi 1957), political ecology provides a powerful counter-narrative to the dis-embedded market.

In this chapter, I begin by showing how contemporary water politics is shaped by a tension between market expansion and efforts at social protection. This brief survey sets the stage for a more detailed examination of one case – rural groundwater use and systems of exchange for groundwater in western India – which I use to illustrate the ways in which water and social practice are intertwined. These so-called "water markets" in the state of Gujarat in Western India are far from the abstract price controlled institutions of economic theory. Instead, they are complex systems of exchange shaped by historical and natural context, and regulated at the local level using social practices. By exploring the micro-politics through which local regulatory institutions are forged, I show how groundwater exchange flourishes because, rather than in spite of, local customary practice.

GLOBAL WATER POLITICS: MARKET TRIUMPHALISM AND ITS CRITICS

Water debates have not been immune to the market triumphalism that Peet and Watts (Chapter 1) describe as a distinctive feature of the intellectual and political atmosphere at the turn of the century. But at the same time, the water arena has resounded with calls for greater political space for users of water, in what was often a reaction against a history of state dominance in control over water. A sketch of water politics through the 1990s shows markets in the ascendant, although increasingly in conflict with civil society, and a retreat of the state.

The increasingly uneasy accommodation between market and civil society is aptly captured in a rash of global pronouncements around environment and development in the last decade of the twentieth century. These efforts should by no means be read as unproblematic statements of a global consensus, despite strenuous efforts by organizers to portray them as such. Instead, I use them here as a window onto contending political positions.

Of these, the Dublin Statement on Water and Sustainable Development of early 1992 arguably stirred the most debate by recognizing water as an economic good (ICWE 1992). This principle was one of four – the others called for attention to the ecological dimension of water, endorsed a "participatory approach" to water management, and reiterated the central role of women in water use and control – intended to frame an agenda around water issues going into the 1992 Rio "Earth Summit." Barely beneath the surface of agreement over these principles were fissures mirroring deep ambivalence on leaving water control and management to market forces. For example, as part of the principle on the economic value of water the Dublin Statement prominently included a caveat demanding that "it is vital to recognize first the basic right of all human beings to have access to clean water and sanitation at an affordable price." Within its very formulation the statement contradictorily sought to create space for encroachment of the market, even while containing the seeds of a protective impulse by identifying a basic right to water.

By the end of the decade, a commission established to formulate a "World Water Vision" more emphatically reiterated a market-led approach to water: "the single most immediate and important measure we can recommend is the systematic adoption of full-cost pricing for water services." Economic pricing, it was argued, would provide the basis for mobilizing needed investment from the private sector. To ensure access to water as a "basic need" (but significantly not a right, a retreat from Dublin) it called on governments to mobilize the financial resources necessary to subsidize access for the poor, while not shrinking the space for the market: "it is essential to separate the welfare task (the task of government) from the business task" (ICWE 1992: 3). While the Dublin statement left unresolved potentially conflicting agendas, the World Water Vision was much clearer about prioritizing the market.

Specific arenas of water control reflect the tension in these global pronouncements. Here I use two examples, the privatization of municipal water supply and protest against large water projects, to illustrate both the march of the market and the emergence of defensive efforts to limit its scope.

Defensive action against an ascendant market: privatization of municipal water supply

Municipal water privatization is among the most contested of all water arenas. Countries have been spurred down this path by a perception that the capital necessary to sustain and enhance water services can only come from the private sector, a perception reinforced by an explicit World Bank policy promoting privatization (World Bank 1993). One global estimate, for example, projects that 95 percent of the estimated $105 billion additional funds needed for water needs will have to come from the private sector (GWP 2000). Faced with these pressures, ninety-three countries had partially privatized water or wastewater services by the year 2000 (Brubaker 2001). The water privatization industry that has emerged is extraordinarily concentrated. For example, two French firms together control an estimated 70 percent of the total market, serving between them 215 million customers (Hall 2002).

The privatization of water services raises familiar questions about the fate of social and public concerns in the rush to remove obstacles to expansion of the market (Gleick *et al.* 2002). What would be the fate of under-represented and under-served communities? Would the profit motive lead to an inequitable pricing structure to replace a principle of uniform or progressive pricing? How would water rights be protected and future needs accounted for? Concerns such as these have spurred a counter-movement of considerable vigor.

Among the most celebrated of these cases is that of Cochabamba, Bolivia (Finnegan 2002). Starting from the weak position of seeking bids for an ill-functioning system, the city received only a single offer for its water system from a British-led consortium. The deal included not only the water system but also all the rights to the water in the district including underground water, and guaranteed a 15 percent return on investment in dollar terms. The consortium quickly raised – in some cases doubled – water bills, sparking mass demonstrations in Cochabamba and beyond, which escalated into dramatic confrontation with armed police. As unrest grew, the government eventually revoked the contract (on the grounds that the company had abandoned its concession) and rapidly implemented a new national water law that promised public consultation on rates and provided protection to small-scale water systems. The water story in Cochabamba, however, has no happy resolution – although restored to public hands the municipal supply system remains inadequate and cash-starved.

Advocates of privatization respond to their critics by pointing out that many public water systems are indeed deeply flawed, and by highlighting short-term gains, including to poor populations, that have been achieved by some privatized utilities. They also point out that the experience after the public sector re-takes control is not a happy one either, as Cochabamba bears out. However, even if the record of public control is not encouraging, it is also clear that Cochabamba is not an isolated case of dissatisfaction with private control. Privatization of municipal water has run into a buzzsaw of protest around the world, including a high profile cases in Buenos Aries, Argentina which began with some success, but then led to accusations of embezzlement and a default by the private company (International Consortium of Investigative Journalists 2003), a program of "cost recovery" following privatization in South Africa which contributed to a cholera epidemic (International Consortium of Investigative Journalists 2003), and a surprising return to public control following health scares under private control in Atlanta (Jehl 2003). In many cases the state has been forced into the breach to re-establish control over water, and to mediate economic motives and social concerns.

From our point of view the issue is less one of the inherent superiority of public versus private modes of operation, or of ideological predispositions toward each, as of the tangible sense, validated by experience, that unchecked expansion of the market interest in the water arena does threaten the public interest and spur a counter-movement aimed at restricting market excess. Indeed, this struggle in specific places is being replicated at a global scale, with efforts at the World Trade Organization to continue negotiations on a "General Agreement on Trade in Services," which would severely limit the ability of states to regulate private investors, including in the water arena, and impose severe penalties on governments that seek to undo privatization (Sinclair 2000).

A re-assertion of social life: the struggle over large dams

While water privatization illustrates an effort at defending the growing encroachment of the market on social space, the related history of large water projects, particularly large dams, offers a concrete example of an effort to re-embed economic decisions in social life. The progress toward re-assertion of social life over economic decisions has come only after a decade of intensely political struggle.

The market is a relatively recent entrant to the debate over large dams, which has been characterized by longstanding social struggle against state control over water. Large state-led water control projects have been a primary locus for national, and increasingly transnational, civil society organization. Proclaimed for a half century as monuments to the aspirations of newly independent states, a wave of criticism in the 1980s instead labeled large dams monuments to social, cultural and environmental short-sightedness and misguided economics (McCully 2001). Large dams, charge the critics, displace large populations, typically among the most impoverished, cause long-term ecological harm, and often fail to meet their deliver the expected economic gains. Because the World Bank has provided both financial support and legitimacy for state-led construction of large water projects, national protest movements have built bridges with international social and environmental advocates to single out the World Bank as a strategically important target (Khagram 2000). Over time, the movement against large water control projects became a central component of a larger campaign for reform of the World Bank, and a particularly potent example of what Keck and Sikkink (1997) call "transnational advocacy networks." For example, a milestone declaration by 326 social organizations from forty-four countries signed in 1994 at Manibeli in India, called for a moratorium on World Bank funding of large dams (Manibeli Declaration 1994). These opponents also insisted that efforts at managing water include a voice for affected communities as a check on state power.

By the late 1990s, however, the state was no longer all-powerful in the water arena. As part of a much larger shrinking of the public sphere, states acting at the behest of international financial institutions were increasingly seeking private capital for large dam projects, as for municipal water supply. Civil society advocacy and campaigns shifted their focus to include explicit attention to private sector contractors and financiers. For example, campaigners bought shares in and introduced share-holder resolutions to place pressure on dam builders such as British firm Balfour Beatty, and challenged a bond issue and exerted public pressure on major European and American banks such as Credit Suisse First Boston and BankAmerica Corporation, in an effort to halt construction of large dam projects (Hilyard and Mansley 2001). An updated civil society statement on large dams formulated in Curitiba, Brazil just three years after Manibeli, while reiterating the earlier message, also reflects the changed priorities:

> The process of privatization which is being imposed by multilateral institutions . . . is increasing social, economic and political exclusion . . . we do not accept the claims that this process is a solution to corruption, inefficiency and other problems in the . . . our priority is democratic and

effective public control and regulation of entities that provide electricity
and water ...

(Declaration of Curitiba 1997)

In the re-drawn battle lines of global water politics, a movement drawn from civil
society sought both to contain market control over water, and also to draw on
state authority to re-embed the water domain in social context.

By the later 1990s, the global political campaign against large dams had
inflicted sufficient reputational damage and financial damage – in terms of time
and cost overruns – on the World Bank and private financiers to bring them to the
bargaining table (Dubash *et al.* 2001; McCully 2001). Under the auspices of the
World Bank and the World Conservation Union, a World Commission on Dams
was established to conduct an independent review of the experience with large
dams, and to develop criteria and guidelines to advise future decision-making on
dams (World Commission on Dams 2000). The WCD's twelve commissioners
spanned the range from a leading anti-dam activist from India to the CEO of one
of the world's largest engineering firms, to the Minister of Water for South Africa.
The extraordinary breadth of representation on the Commission, hard won
through intense negotiation and unprecedented in a body of this sort, lent it
considerable credibility as an arbiter of the dams debate. In addition, drawing on
a decade of organizing, activists mobilized very effectively to engage, track and
shape the Commission in every step of its work, with the result that "civil society
was better able to exploit this space than industry" (McCully 2001).

In its final report, the Commissioners located decisions over dams, and
implicitly all large development projects, firmly within a human rights
framework. In an eloquent preface, the chair of the WCD, South Africa's Water
Minister Kader Asmal asserted that: "We are much more than a 'Dams
Commission.' We are a Commission to heal the deep and self-inflicted wounds
torn open wherever and whenever far too few determine for far too many how best
to develop or use water and energy resources." To do so, the WCD articulated a
"rights and risks" framework whereby no rights should automatically be
considered superior to other rights. Where rights were in conflict, they called for
a negotiated process. Through this framework, the WCD sought to create the
political space for articulation of all interests, in counter-position to a perceived
history of systematic privileging of state and corporate interests. This framework
has the effect of obscuring the hitherto convenient distinction between technical
and political decisions that has allowed political exclusion in the name of
technical soundness (Bradlow 2001). Unsurprisingly, the WCD report was met
with criticism from many governments and industry groups as unbalanced and
unrealistic, and with a vague statement from the World Bank that the report
would serve as a "valuable reference" (Dubash *et al.* 2001).

Despite the failure of these institutions to embrace and put into practice the
WCD's ideas, the Commission has undoubtedly impacted the politics of the
water arena in multiple ways. To the extent the WCD signals the potential for

global policy-making based on the explicit construction of alternate frames, it points to the future shrinking of decision-making space for the state relative to civil society and private sector voices. With the credibility of its constituent Commissioners behind its bold formulation, the WCD has injected into the global water arena a set of specific ideas that create political space for marginalized populations and hold the promise of re-embedding economic decisions within a more explicitly social framework.

This brief review suggests that the idea of water as a commodity – so effectively challenged by re-framing water within a rights discourse – underlies the expansion of the market into the water arena. Treating water as a commodity alone, however, ignores the existing management systems and institutions within which current patterns of water control and use are embedded. It also ignores the ecological context within which water is harvested, managed and used. Instead, the deepening of commodification is a deeply political process through which markets are imposed on social and ecological context, or, more likely, markets are forced to accommodate a re-assertion of social practices and ecological realities. Critical to understanding this process of market expansion is an awareness of the local politics, institutions, and ecological circumstance within which water is used. In the remainder of this chapter, I examine and explain the functioning of local groundwater "markets" in western India, as a means of illustrating the embedding of water use in social and ecological context.

EMBEDDED SYSTEMS OF EXCHANGE FOR GROUNDWATER IN WESTERN INDIA

In recent years, patterns of groundwater use in the western Indian state of Gujarat have come under considerable scrutiny. Groundwater has long been thought of as a resource best developed through public intervention, or if through private means then only on a small scale. Consequently, the documentation of extensive systems of groundwater exchange in Gujarat, based on large and quite sophisticated privately owned deep "tubewells," has aroused considerable attention. Much existing work analyzes groundwater use in Gujarat as if it operates as a price-clearing market, with a normative goal of finding ways of making groundwater markets more competitive, and hence more efficient (Shah 1991; Shah 1993; Shah and Ballabh 1997; Shah and Raju 1988). This analysis of the Gujarat case forms an important building block of an emergent international literature on groundwater markets, which seeks to demonstrate the potential efficiency and welfare benefits of markets for groundwater (Easter, et al. 1998). By contrast, the discussion below shows how groundwater use in Gujarat has been shaped by the historical trajectory of groundwater use, by its natural context, and is regulated through sophisticated institutions forged through the rough and tumble of local politics.

This study is set in North Gujarat, a semi-arid zone in western India almost exclusively dependent on groundwater (Figure 8.1). The region lies above an

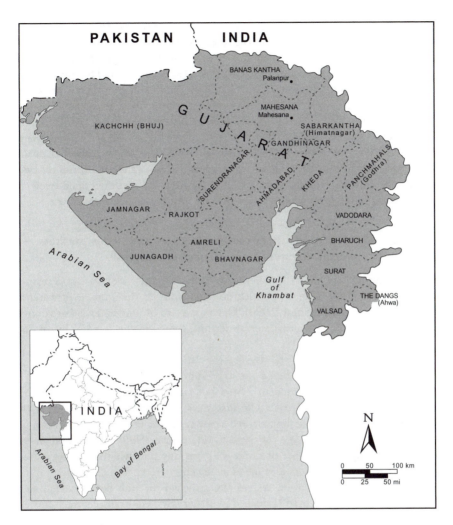

Figure 8.1 Map of Gujarat.

extensive alluvial aquifer that cuts from north to south, bounded by mountains in the east and by hard basalt rock formations in the west (Phadtare 1988). The villages of Ratanpura in Mehsana district, and Paldi in Banaskantha District (the village names are fictional), lie squarely within this alluvial belt. Composed of irregular layers of permeable sands and impermeable clays down to depths of 250 meters, this is an extraordinarily rich and prolific aquifer.

Over time farmers in this region have invested in deep tubewells occasionally exceeding 200 meters in depth, laid miles of underground pipelines in a sophisticated water infrastructure, and created complex institutional arrangements for groundwater transactions. Over time, groundwater transactions have come to assume enormous importance in the regional economy. As a result of this dependence on groundwater, there has been a dramatic decline in water levels, estimated to be 3 meters a year in some regions (Kavalanekar and Sharma 1992).

The narrative of groundwater use in Ratanpura and Paldi rests on two puzzling observations. First, there is wide divergence in the density of exchange across villages; Ratanpura illustrates a "thick" or dense system of exchange, while that in Paldi is relatively thin; what accounts for this difference? Second, and related, there is paradoxically far greater competition in Paldi, the thin market, than in Ratanpura, the thick market. These two observations guide the narrative of the two villages that follows. First, however, I explore the roots of the current open access to groundwater in Gujarat, which has set the stage for current patterns of use.

Legal and institutional pre-conditions for groundwater markets: groundwater law in the colonial period

Groundwater use in Gujarat is governed by a legal framework that ties access to groundwater to land ownership, with no legal limit to the amount of water a land-owner can extract (Singh 1990). Open access to groundwater is based on a strong link between land and water ownership, rooted in the colonial history of the region.

During the 1880s and the 1890s, the British administration sought to introduce a tax on wells, based on a presumption of state ownership over groundwater. This attempt led to a brief but significant skirmish between the cultivator elites and the British administrators in central Gujarat. Faced with cultivator protests, the British instituted a new system of taxing the water bearing quality or "sub-soil advantage" of the land (Hardiman 1998). This tax on water capacity was applied whether or not a well was dug and whether or not water used, which gave cultivators a tremendous incentive to construct wells.

The sub-soil water taxation system strengthened the link between land and groundwater, which had been enshrined in earlier laws. For example, the Transfer of Property Act of 1882 stipulated that rights to water could only be transferred along with land, and the Land Acquisition Act, 1894, tied interest in an easement affecting land, here water, to an interest in that land. At stake was not only a cultivator's right to invest in and reap the full benefits of irrigation development, but, implicitly, also ownership over groundwater.

During the colonial period, then, a precedent of private control and untrammeled access to groundwater was established, and the growth in wells was further accelerated through state support. In independent India, state irrigation acts have sought to undermine this precedent to allow for a state role in determining the use of groundwater. For example, the Bombay Irrigation Act of 1879 (which applies to the state of Gujarat as well) was amended in 1950 to extend the procedure of "notification" – which allows a measure of state control – to privately constructed wells and tubewells as "second-class irrigation works."

Notification allows an existing "record of rights" based on current use patterns, to be subsequently commuted by the state based on payment of money, land or alternative compensation as the officiating officer "may think fit" (Government of Gujarat 1963). More recently, the Indian government has urged states to adopt a bill that would introduce a system of licenses and other forms of limited regulation (Bhatia 1992; Moench 1998; Singh 1990). However all these steps have had little real impact on patterns of use, and on private rights of unchecked access to groundwater that were established in the colonial period. Thus, the precedent set by historical patterns establishes a *de facto* right to groundwater use tied to land ownership, with no corresponding obligations to maintain the quality or the quantity of water available to others.

The uneven "architecture" of groundwater markets

That farmers across the region have common access to a prolific aquifer, and under similar legal conditions, has by no means spawned homogenous water markets across the region. Instead, the forms, degree of importance, and transactions arrangements across the region vary widely from village to village. By examining in some detail the structure and institutions of water markets in two villages, Ratanpura, in Mehsana district, and Paldi, in Banaskantha district, I show how patterns of groundwater use and sale are embedded in social practice, and shaped by both natural characteristics and historical circumstance.

The pattern of well investment in Gujarat, as evidenced in Ratanpura and Paldi is, in many ways, quite exceptional. Widespread private ownership of technologically advanced "deep tubewells" in India has, until recently, been almost entirely dismissed as a possibility. Vohra (1982) cites land fragmentation as an insurmountable problem to deployment of deep tubewell technology. Dhawan (1982) starts with the same premise, and suggests there is a threshold farm size for private well ownership under a given set of geohydrological conditions. Beyond this threshold, the problems of coordinating shared use of water are sufficiently large to preclude individual investment. In a detailed study, Boyce (1987) confirms that this conclusion holds true for Bengal, in the context of a highly unequal agrarian structure. Based on these studies, a massive program of land consolidation, or an emphasis on public investment rather than private, appears to be the only way of accessing deep groundwater.

In the discussion below, I show that private investment in Ratanpura, and to a lesser extent in Paldi, demonstrates how a combination of caste-based cooperative action and the emergence of water sales have enabled farmers to invest in water beyond the limit dictated by plot size. The discussion below illustrates the scope of water sales in both villages, and examines the natural and social factors that have led to private investment in wells and a dependence on groundwater transactions.

To briefly introduce the agrarian structure in both villages using land as one important, if imperfect indicator of power, both Ratanpura and Paldi are societies with deep inequalities in power relations (Table 8.1).[2] In Ratanpura over a quarter

Table 8.1 Well ownership by land ownership (% in each category).

Land owned (bighas)	Well-owner	No well	Total
Ratanpura			
0	0	2	2
0.1–5.0	17	30	47
5.1–10.0	19	15	34
10.1–20.0	8	4	11
20.1–30.0	6	0	6
>30.1	0	0	0
Total	50	51	100
Paldi			
0	3	8	11
0.1–5.0	11	19	31
5.1–10.0	17	3	19
10.1–20.0	17	0	17
20.1–30.0	11	0	11
>30.1	11	0	11
Total	70	30	100

Source: Sample survey of Ratanpura (fifty-three households surveyed) and Paldi (thirty-six households surveyed), 1995–96. The sample was structured around water buyers and sellers and hence is under-representative of the landless.

Note: 1 bigha = 3/5 acre.

(29 percent) and in Paldi, almost half (47 percent) the households are landless. In addition, well ownership is skewed toward land-owners in both villages. This distribution of land and well ownership makes possible vibrant markets for groundwater.

It would not be overstating the case to say that groundwater transactions are central to economic life and agricultural productivity in Ratanpura. To give one indication of dependence on groundwater purchases, 90 percent of landed households depend in whole or part on purchased groundwater and 44 percent rely entirely on purchased groundwater for irrigation. Water is sold from all but one well in Ratanpura, and, depending on alternative forms of calculating this number, between 61 percent and 71 percent of the water pumped is sold. Curiously, even well-owners rely heavily on purchases of groundwater; 13 out of 15 well owners supplement their own water with purchased groundwater. Moreover, the scale at which wells operate is considerable. For example, in the village region with the densest exchange, each well irrigates on average forty-four different plots, with the largest well irrigating ninety-seven plots and the smallest twenty-one. This "thick" market suggests a highly complex system of exchange, which poses considerable problems of timing and coordination.

Turning to the second village, while a relatively high 61 percent of sampled households in Paldi are reliant in some way on purchased groundwater for irrigation, this number conceals a number of differences with the organization of groundwater transactions in Ratanpura. The crucial difference is that production relations in Paldi demonstrate what Wood (1995) calls a "mutiplex" nature, that defy easy categories of water buyer and seller. Specifically, many water transactions are bundled into existing landlord–tenant relationships, and in some cases water is transacted as part of a three-way relationship including tenant, landlord, and water provider. Thus, of the 61 percent of households who rely on purchased water, 39 percent do so through a tenancy arrangement. From a well owner's point of view, water sales are not as important in Paldi as they are in Ratanpura. Only five out of twenty wells sampled in Paldi sell water; if water provided to tenants is included, this number rises to eleven. By contrast with Ratanpura, where much of the water pumped is sold, only 7 percent of water is sold, and another 17 percent goes to tenants. While I was unable to collect data on the number of plots irrigated per well for Paldi, a comparable measure of transaction complexity is the number of buyers per well: 0.6 for water buyers, and 1.2 if tenants are included.[3]

Thus, by a variety of measures, Ratanpura demonstrates a very "thick" system of exchange, with considerable reliance on the smooth functioning of groundwater sales (Table 8.2). These sales are also significant in Paldi, if not as widespread as in Ratanpura, but additionally many of them are wrapped into ongoing tenancy relations. Moreover, the overall complexity of the system is far lower, in terms of the number of potential and actual buyers per well. Consequently, the need for institutional mechanisms for coordination is far lower in Ratanpura than in Paldi.

What explains the difference in the density of exchange across Ratanpura and Paldi? One important driver is the "architecture" of the groundwater exchange system – the spatial dimension of land use patterns and irrigation infrastructure that together determine the need and capacity to move water. As I discuss below, the architecture of exchange is determined by path dependent resolution of social and natural circumstance.

The divergence in market architecture is best illustrated by maps of each village which show the area cultivated in each village, and functioning wells and pipelines (Figures 8.2 and 8.3). Ratanpura demonstrates considerable scope for

Table 8.2 "Thickness" compared in Ratanpura and Paldi.

	Ratanpura	Paldi	
	Water buyers	*Water buyers*	*Tenants*
% of water pumped that is sold	61–71%	7%	17%
% h'holds who rely on purchased g'water	90%	22%	39%
Number of owners who buy	13 out of 15	3 of 25	
Complexity of water management	high	low	

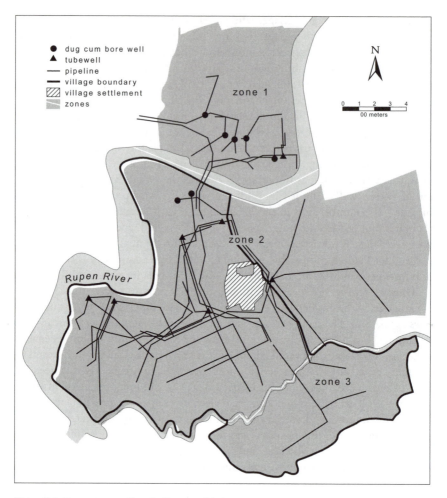

Figure 8.2 Ratanpura: wells, pipelines, and irrigation zones.

Source: Village outline: District Land Records Office, Mehsana, Gujurat. Well, pipeline, and irrigation data are from surveys carried out in Ratanpura in 1995–96.

competition among wells. While there are only a relatively small number of wells clumped along the river that cuts through the village, these wells have access to a large command area through an extraordinarily complex and dense network of pipelines. For example, the oldest tubewell in the village has a pipeline network 4.5 km in length. Due to this network of pipelines, the command areas of the various wells overlap, leaving considerable scope for competition. I have divided the irrigated area into three distinct zones. Of these, Zone II exhibits the most dense pipeline network, flagging it as an area of possible intense competition. By contrast, buyers in Zone III have minimal choice.

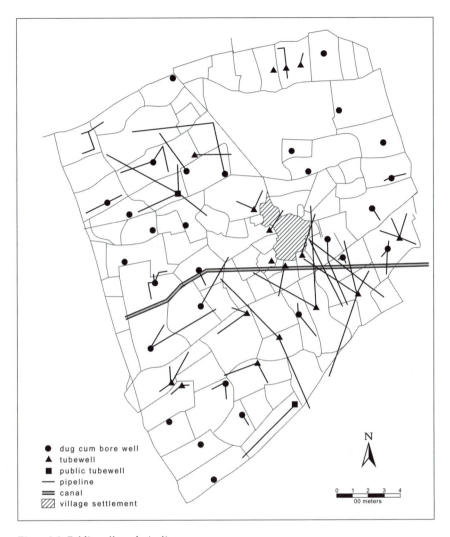

Figure 8.3 Paldi: wells and pipelines.

Source: Village outline: District Land Records Office, Palanpur, Gujarat. Well, pipeline, and irrigation data are from surveys carried out in Paldi 1995–96.

In Paldi, there are many more wells evenly distributed through the village lands, but each of these has limited pipeline networks. Command areas scarcely overlap, leaving little possibility for competition. I summarize these differences in Table 8.3.

The market architecture in the two villages has been shaped by the different trajectories of groundwater development in each. The location of wells along the

Table 8.3 Architecture compared in Ratanpura and Paldi.

	Ratanpura	Paldi
Plot size	Small	Large
Plot fragmentation	High	Low
Number of wells	Few (15)	Many (50)
Spatial distribution of wells	Clumped	Even
Density of pipelines	Dense in areas	Sparse
Command area of wells	Overlapping (esp. Zone II)	Little overlap

river in Ratanpura has been strongly shaped by hydrological considerations. In the early days of hand-dug wells, this location provided quick and efficient well recharge. Although with the introduction of diesel engines and deep tubewells this reason is no longer valid, pipelines and other infrastructure were laid decades ago based on hydrological advantage, locking in the current pattern of well location.

Pipeline networks were particularly important in Ratanpura because of a pattern of substantial land fragmentation. As Table 8.4 shows, by 1995–96, it was common for Ratanpura well owners to own between four and five separate fragments of land. Having invested in pipeline networks, it was then in the interest of well-owners to maximize use of their surplus water by selling it to other farmers whose land lay adjacent to their pipeline, providing considerable stimulus to water sales.

Over time, over-pumping of groundwater soon dropped groundwater levels beyond the range of diesel engines, requiring investment in more expensive and powerful electric tubewells. However, this hydrologic turning point occurred in Ratanpura in the late 1960s when the village did not have access to electricity. By the time an electricity connection was provided in 1973, there was a severe water shortage, considerable excess demand for water, and a perception that wells were highly risky driven by a rash of costly well failures. These perceptions stimulated

Table 8.4 Land ownership characteristics by well ownership.

	Ratanpura		Paldi	
	Well-owner	*Non-owner*	*Well-owner*	*Non-owner*
Average land owned (bighas)	8.2	4.0	17.3	2.2
Average no. of plots	4.5	3.0	2.0	1.0
Average plot size (bighas)	1.7	1.3	9.0	2.0

Source: Sample survey of Ratanpura and Paldi, 1995–96.

Note: These data do not distinguish between individual, family and joint ownership of wells.

an important collective response and institutional innovation. Drawing on kin networks, a group of fifteen members of the dominant Patel caste came together, sought access to credit from merchants in a nearby town, and invested in a deep tubewell, which at 200 meters was seven times deeper than the deepest existing dug well. A partnership arrangement allowed the partners to spread the risk of well failure, pool credit, and take advantages of economies of scale.

Partners were held together by the social bonds of *samaj* or marital groups, which, more than caste or kin, delineate social boundaries in Gujarat (Pocock 1972). It is significant that group-owned wells were organized around *samaj* ties rather than around contiguity of land-holdings. Here, affinal ties provided the glue to exploit economies of scale. Not all social groups, however, had equal access to capital. Since land ownership was the primary source of collateral for credit, well ownership closely followed contours of control over land. Only relatively wealthy groups were able to mobilize the capital to drill tubewells. In Ratanpura, for example, lower castes with marginal land-holdings were not only excluded from the group formed by various *samaj* of Patels, but were also unable to mobilize on their own. Falling water levels, then, placed a greater adjustment burden on small farmers (Table 8.5).

Those able to forge partnerships also had to consider the need for an expanded pipeline network. Instead of providing water to four or five dispersed plots of a single farmer, a single well now had to reach all fifteen partners' plots. This extended network enabled partnership wells to sell surplus water to other farmers, whose fields were located near the pipelines. Buying water was, for most, more attractive than investing in a well, given the shortage of water at the time, the

Table 8.5 Land and well ownership by caste.

Ratanpura	Patel	Harijan	Vaghri	Other	
% of village households	66	20	12	2	
% of village land owned	97	3	0	0	
Ratio of land % to pop %	1.5	0.2	–	–	
% of village wells owned	100	0	0	0	
Ratio of wells % to pop %	1.5	–	–	–	
Paldi	*Kunbi*	*Rabari*	*Koli*	*Harijan*	*Other*
% of village households	38	10	16	23	13
% of village land owned	59	14	7	7	14
Ratio of land % to pop %	1.6	1.4	0.5	0.3	1.1
% of village wells owned	64	10	8	4	14
Ratio of wells % to pop %	1.7	1.0	0.5	0.2	1.1

Source: Household numbers are based on field work estimates for Ratanpura, and from census information for Paldi. Land-ownership is based on unpublished 1991 census data. Well ownership is compiled from the records of the Gujarat Electricity Board.

large up-front financial cost, and the requirements of a suitably strong network and access to credit. By 1973, the first partnership tubewell was irrigating over a quarter of the total village land. This successful example was followed by others, sometimes based on large groups of partners, sometimes assembled by small groups of relatives, and occasionally by ambitious individuals. For many, however, purchased water remained a better option than investing in well ownership. Over time, the combination of deep groundwater, land fragmentation, the need for financial and social capital and the particular historical trajectory of groundwater deepening led to the division of farmers into a few well owners, and a large group of water purchasers.

By contrast, in Paldi, farmers were free to place wells in the location most convenient to their fields, since they were free of any strong hydrological constraints. Larger plot sizes and less land fragmentation also obviated the need for pipeline networks beyond the edges of their plots, limiting the scope for water sales. While in Paldi, as in Ratanpura, water levels fell beyond the point that hand-dug wells could reach; this occurred only in the mid-1990s, long after electricity was available. Without a similar perception of risk or return, farmers responded incrementally by investing in a slightly larger tubewell, sufficient to meet their own needs. Consequently, there was little impetus to sell surplus water, or to treat water sales as an entrepreneurial activity. The result is two very different sorts of "architecture" in the two villages.

As in Ratanpura, however, groundwater depletion was refracted unevenly through the agrarian structure. In the early 1990s, there was one well jointly owned by a group of lower caste Harijan smallholders, the only such arrangement in the village. Several partners were unable to raise their share of the additional funds, and had to withdraw from the arrangement. Another maintained a nominal stake in the well only by leasing his rights to water to a higher caste Kunbi. The result was a further concentration of ownership among existing landowners, but to a far smaller extent than occurred in Ratanpura. As with Ratanpura, and indeed throughout Gujarat, those without access to land are also without access to groundwater (Table 8.5).

Thick markets, then, are made more or less likely by how the path dependent interaction of hydrology and changes in land fragmentation patterns are resolved. The particular form of the resolution is dependent on social considerations such as access to credit markets, and kin and caste ties that facilitate joint action. Market thickness is in large part contingent on the path dependent outcomes of historical, social and natural contingencies.

Fluid contracts: groundwater regulation through contractual norms

Thick markets with many possible buyers and sellers, according to the neoclassical model of price-clearing markets, are a precondition for competition. If a seller were to seek a higher price, or a buyer were to demand a lower price, there would be other buyers, or sellers, ready to step in. The relatively thick water markets in Ratanpura, however, display little of this behaviour, while in the relatively thin markets of Paldi, we see well-owners compete for the custom of buyers. This section explains how exchange is institutionalized in Ratanpura, through a process whereby norms around exchange are negotiated and legitimated. This process of institutionalization has conferred structure and predictability on a complex exchange system. Indeed, limiting the scope of exchange has made possible a thick market.

Several observations illustrate the lack of competition in Ratanpura. The stated price of groundwater is uniform across all buyers and sellers in the village. Indeed, the equality of terms of exchange for water is much emphasized by both buyers and sellers. Neither caste nor class played a role in determining the stated price. The shared conception of the prevailing price as the cognitive basis for transactions is significant, as is the moral content with which the single price was imbued. This result is striking because, as we have seen, there is spatial variability in the density of pipelines across the village, and hence considerable variation in the scope for competition across the three zones. Glancing again at Figure 8.2, we would expect Zone II – a region criss-crossed by pipelines and hence for the potential of water sales – to be the site of heavy competition and hence lower prices, but find that this is not so.

Curiously, the terms of payment differ by crop and by season, but these differences are applied uniformly across all buyers. The most significant form of payment is *ucchak*, a fixed payment in kind for irrigation of a unit plot of land. A second form is a cash payment based on hourly provision of water. A third form is a one-third share of the crop payment, which was once the dominant form of exchange, but is now only used for a few minor summer crops.[4]

An exploration of past transaction arrangements helps account for this diversity of contractual forms. It also leads us to broaden the discussion beyond price to explore how stability and reliability in water supply is assured. In the early days of mechanized irrigation, widespread use of share payments for water brought well-owners into the production relationship as a *bagirdar* or partner, providing sellers with an incentive to provide timely water supply. *Bagirdari* or partnership arrangements would have been common as a basis for land tenancy, and faced with a new important input, water, it is likely that farmers adapted an existing institutional arrangement that served well.

Why, then, the shift to fixed payment in kind, or *ucchak*? Over time, the complexity of the water system in Ratanpura – recall that a single well often irrigated tens of fields – was such that there was considerable scope for buyer malfeasance. A well owner was hard pressed to sufficiently monitor crop yields to assure that he received his full share. Consequently, and sparked by particular incidents of malfeasance that live on in local memory, well-owners banded together to change the dominant contractual form from share to *ucchak*. Moreover, while advantageous for the seller, this shift undermined any incentives for sellers to supply timely water. This assured sellers a return, but placed all the risk of crop failure on the buyer, placing the latter at considerable disadvantage. In this case, a shift in contractual terms was brought about by the exercise of economic power.

The importance of local conditions in shaping exchange is reinforced by the observation that the prevailing price structure in Ratanpura – share, fixed in kind, and hourly cash payments, each for a particular crop and season – stops at the village border. Neighboring villages have their own and very different terms of exchange for water, with some dominated by share transactions, and others reliant on a different mix of cash and *ucchak* transactions. Thus, despite common cropping patterns, soil type and hydrology – the factors one might call upon to independently explain contract choice – contract terms differ from village to village.

Price is only one dimension of the terms on which water is exchanged. Equally important is the agreement on quality of service. As discussed above, one way to interpret share payments for water is that they provide contractual incentives for good quality service. In the absence of these contractual incentives under an *ucchak* arrangement, a framework of institutionalized norms that have emerged over time governs transactions in Ratanpura. These norms are by no means a formal and explicit framework of rules, but rather shared understandings undergirded by a moral basis (Elster 1989). Empirically, norms become visible in the rare cases when they are violated, and by their repeated invocation as a decision rule for settling disputes.

There are three central elements to these norms. First, water provision is contracted for an entire cropping season, even in the case of hourly cash contracts.[5] Second, buyers are assured of timely delivery through institutionalization of the *vara* or water turn. According to this system, water delivery follows a set rotation among all users, including the well-owners. This system ensures that water is delivered on a timely basis, or at the least, that the burden of unreliable supply is distributed evenly over all the users. Nonetheless, there remain possibilities for a seller to discriminate between buyers. One important factor is the place allocated in the *vara*, which is set at the time of the first irrigation of the season. If too early, the field may not be prepared; if too late, the early days of summer heat could damage the late-developing crop. The third norm provides guidelines for the exercise of such discrimination: long-term users are to be rewarded with preferential treatment and, in the event of a severe shortage, are to be given preferential access. These norms are particularly crucial to the effective functioning of a water market in Ratanpura, because irrigation capacity and demand are quite closely matched. With many buyers who seek water from the same seller for spatially disparate fields, establishing an assured supply of water is extremely important. Under the old share payment regime, once the seller had agreed to provide water, his stake in a successful crop provided all the necessary incentive to supply timely water and for the full season. Under alternative contractual forms, these incentives no longer hold, but are replaced by the institutionalized norms described here.

In Paldi, terms of exchange have little of the complexity that characterizes water exchange in Ratanpura. Instead, it is the production relations around water in Paldi that are complex. As discussed above, well-owners do three things with their water: they use it on their own land, they provide some of it to their tenants, and they sell some of it to other farmers who are not their own tenants (but who may be tenants of other landowners).

Looking first at direct sales of water to other farmers, water is sold for a share of the crop. By contrast with Ratanpura, there is considerable bilateral haggling over the price, as evidenced by a range of share ratios in use. In 1996, I observed water sales for one-third, two-fifths and one-half shares of the crop all in use at the same time, with most clustered around a two-fifths share. That a two-fifths share had become the most common was a change, I was informed, from a few years earlier when most water sales were based on a one half share. Both buyers and sellers explained this shift as a result of a shift to deep tubewells from older dug-cum-bore wells, with a resultant increase in water supply over the village as a whole. Consequently, in numerous cases buyers have negotiated down the price of water, or sellers have bid down the price to gain additional buyers in the immediate environs of their well. By contrast to Ratanpura, Paldi illustrates the potential for price variation across both space and time.

Turning to water transactions when a tenant is involved, across the village the tenant receives a one-quarter share of the crop, while the landlord and water provider receives the remaining three-quarters. Of this three-quarters, a one-half share is considered to be the water portion with the remaining one-quarter accruing for land. Tenants, therefore, have continued to pay one-half share for water even as buyers, who are typically better off, have negotiated a price reduction from one-half to two-fifths share.[6] In Paldi, then, the market for water is fragmented by social class, and since class and caste are closely tied in this village, also by caste.

Finally, under conditions of excess supply of water at the village level, and a mosaic of relatively distinct command areas, problems of coordination, timing, and assured access are simply not as relevant in Paldi as in Ratanpura. Moreover, share payments are the dominant contractual form, which carries built in incentives for timely and adequate supply of water. Hence, one does not observe the sort of institutionalized norms in Paldi that are readily apparent in Ratanpura.

How do we understand these differences in outcomes in the two villages (Table 8.6). In Paldi, we see well-owners competing, even in the context of little scope for competition, for the custom of buyers. In many ways, this bilateral bargaining and negotiation is what we would expect. In Ratanpura, however, where the scope for competition is far greater, there was a process of institutionalization of exchange as a way of imposing predictability on a complex exchange system. These institutionalized norms were structured around assuring reliable and timely access to groundwater. A lack of price competition among buyers is a crucial component of assuring this stability. Norms on price and quality of service proscribe competition on both these fronts.

Table 8.6 Terms of exchange compared in Ratanpura and Paldi.

	Ratanpura	*Paldi*
Contractual forms	Diverse (flat, hourly, share)	Share
Variation in terms of exchange	Negligible	Variation across class
Degree of institutionalization	High: institutionalized rules	Low: built in incentives

On what basis were these norms forged in Ratanpura? At one level, uniform prices can simply be interpreted as collusive behavior among sellers. However, buyers are by no means powerless, and are able to place considerable bounds on seller behavior. I suggest that price coordination in Ratanpura works within a village wide compromise, a shared understanding of how the exchange system works. There are at least three elements to this shared understanding.[7]

First, there is a moral economy underpinning price in Ratanpura. Price rises have thus to be explained and legitimized in moral terms, and sellers subject themselves to the moral calculus of fair and unfair prices. The weapons at the disposal of buyers are social censure and reputation loss, and in the extreme, the threat of damage to irrigation equipment. Significantly, the two partnership wells play an important role in mediating buyer and seller relations. The wells constitute a form of "public" provision of water that plays a considerable role in setting the bounds of acceptable pricing. Through their broad partnership base and structured management organized around annual meetings, these wells provide a form of representation of broader interests, which boosts their legitimacy over that of "private" wells. The partnership wells therefore serve as a barometer of buyer consent, a moderating force on prices, and a basis for reinforcing and undermining norms on service quality.

Second, uniform terms and conditions for all is the glue of legitimacy for this arrangement, and carries a strong moral weight. With water sellers and buyers both almost exclusively drawn from the ranks of Patels, this unity is often articulated in caste terms. Both buyers and sellers when asked the cost of or returns from irrigating an acre, will compute payments in terms of the common village price, demonstrating the "grip on the mind" quality that distinguishes such moral norms from rules that are maintained purely by a structure of sanctions (Elster 1989). Third, in keeping with the overtones of a moral economy right to subsistence, at times of scarcity access to minimal levels of water is widely assured.

Both buyers and sellers benefit from this compromise. Sellers benefit from assured demand in a context where they must cover high fixed costs of electricity supply, avoid potentially damaging price wars, and manage to keep prices at relatively high levels. Although unorganized, buyers manage to maintain price increases within a moral cost-plus calculus, and ensure stability of supply.

In sum, while water exchange is increasingly depersonalized in Ratanpura, the normative content of the exchange relationship has been retained through the creation of village level institutions such as price norms and timing and delivery norms. These institutions are forged through the political maneuverings of social groups of buyers and sellers who bring to the negotiation strengths and weaknesses that arise from their structural location in village society. In Paldi, the negotiating position of buyers and sellers is also determined by their structural location, but the negotiation operates bilaterally, free of any village regulatory structure. It is the very lack of price competition and the institutionalization of exchange around goals of stability that have enabled thick markets to flourish in Ratanpura. Instead of progressively floating free of social ties, it is by re-embedding in social context that thick markets have developed in Ratanpura.

CONCLUSION

Markets, as the case studies in this chapter suggest, are ultimately social institutions, that are shaped by local circumstance and ecological context. This embeddedness of local institutions accounts for the heterogeneity of groundwater markets in Western India, and carries broader lessons for the spread of markets as a mechanism of control and coordination in the water arena.

In Gujarat, the natural resource characteristics of water proved to be an important determinant of institutional form. Ecological variables such as aquifer conditions, the depth of groundwater levels, topography, and surface hydrology exert a strong influence on the spatial dimensions or "architecture" of groundwater markets, and the density of exchange within that market. These attributes, in turn, define the need for mechanisms of irrigation coordination and the scope for competition.

However, that environmental characteristics are important by no means suggest an ecological determinism. Instead, the interaction of nature and society in historically contingent ways has shaped groundwater markets in Ratanpura and Paldi. In both cases, water exchange has been depersonalized and, to some measure, freed from precapitalist social relations. In Ratanpura, however, these relations have by no means been replaced by the market processes and the price mechanism. Instead, groundwater exchange in Ratanpura has been re-embedded in social rules articulated at the village level. These institutions are not the aggregation of bilateral ties, but are forged through the political maneuvers of social groups of buyers and sellers. Notably caste continues to play an important role, as a force binding village unity around these shared understandings that govern groundwater exchange. The result is an instituted set of norms around price and timing, held together by a moral economy of water use.

By contrast, particularistic ties of caste and kin have been cast aside in Paldi, but only selectively. For some water buyers, growing access to water has led to intense competition over water, driving down water prices as a deepening cycle of commodification requires that precapitalist production relations are replaced by exchange relations organized around price. In other cases, high caste farmers have carefully preserved precapitalist ties in the form of sharecropping relations, which has enabled them to exploit their simultaneous control over water and land unimpeded. The development of commodification, then, has proceeded unevenly in Paldi.

Both cases together suggest that local institutions do much to shape how markets are governed. These institutions are themselves the outcomes of complex local politics, through which prior convention, economic function, and the exercise of power are mediated. A robust analytics of local institutions along these lines is more empirically relevant to an understanding of the spread of the market than is a market abstraction that presumes the unhampered functioning of the price mechanism.

The cases explored in this chapter also show us how markets can be built on exclusion. In both Ratanpura and Paldi, lower castes and classes progressively lost control over water in a competitive race to the bottom of the aquifer. Without economic or social capital, marginalized groups were excluded from processes of deepening wells or the formation of caste-based partnership wells, forcing them into participation in the water market.

Ultimately, the cases of Ratanpura and Paldi demonstrate that actually functioning systems of exchange operate quite differently from the price-clearing markets that occupy the imaginations of policy-makers who craft the global water "consensus." As the market is increasingly privileged in the ongoing reconfiguration of state, market and civil society roles, the detailed empirical record in particular places reminds us that these terrains being made ready for the market are already hotly contested. In the water arena, as elsewhere, the market is best understood as one amongst many mechanisms of social coordination, and one subject to the complex pressures of local ecology, local institutions and local politics.

NOTES

* Parts of this chapter previously appeared in N. Dubash (2002) *Tubewell Capitalism: Ground Water Development and Agrarian Change in Gujurat*, Bombay: Oxford University Press, and are reprinted by permission from the copyright holder, Oxford University Press, India.

1 Williamson (1994) provides a defining articulation of the "Washington Consensus."

2 The use of land as a measure of agrarian class structure is deeply contested. This measure does not account for differences in quality of land or whether it is irrigated, it assumes that production technologies are everywhere the same, and it relies on a fairly arbitrary translation from land-size categories to class categories. Finally, land ownership does not account for "secondary" relations of exploitation such as credit arrangements. The main merit of the measure is its ease of use compared to alternatives. I provide land ownership here as an indicative rather than as a conclusive measure. For further discussion of these issues see Patnaik (1987) and Athreya *et al.* (1990).

3 In addition to private sellers, there are two public tubewells in Paldi. The public wells have a reputation for inadequate and unreliable supply, but water from these wells is priced at far less than the private wells. It is important and significant that small farmers, most often also low caste farmers, depend heavily on the public tubewells for irrigation.

4 There are some important differences across the three contractual forms in terms of the incentives afforded to buyers and sellers. For example, share contracts place some of the production risk on water providers, giving the latter a stake in timely irrigation provision. Similarly, the two in kind payment forms insulate the water buyer from output price fluctuations as compared to the hourly cash rate. In addition, the hourly payments place the cost of leakage from the pipeline delivery system on the buyer, while in the other two systems, these costs rest with the seller. These sorts of arguments are central to the New Institutional Economics (NIE) literature, which focuses exclusively on choice of contractual form. While the functionalist arguments of the NIE shed some light on the choice of contractual form, the empirical data point to several cases where contractual choice runs counter to the suppositions of the theories and is additionally shaped by the exercise of power to benefit some groups over others, and by the effect of institutional norms and the legitimacy of particular contractual arrangements over others.

5 There are occasional cases where spot sales of water take place, but these are typically to substitute for the contractual arrangement in the event of equipment failure, for example.

6 There are also a few cases of three way transactions between a tenant, a landowner, and a well owner. The evidence suggests that in these cases the water arrangements are typically made directly with the landowner. In keeping with this observation, the water output share in these arrangements more closely mirrored the first transaction form described above, water sales to farmers, than the second, sales to tenants.

7 The literature on water markets does contain references to "social constraints" (Saleth 1998) operating on exchange, but there is little effort to explore what these might be, what their effect is and how they are created and reinforced.

REFERENCES

Athreya, Venkatesh B., Djurfeldt, Goran, and Lindberg, Steffan (1990) *Barriers Broken: Production Relations and Agrarian Change in Tamil Nadu*, New Delhi: Sage Publications.

Basu, Kaushik, and Bell, Clive (1991) "Fragmented duopoly: theory and applications to backward agriculture," *Journal of Development Economics*, 36: 145–65.

Bhatia, Bela (1992) "Lush fields and parched throats: political economy of groundwater in Gujarat," *Economic and Political Weekly* (December 19–26): A142–A170.

Blaikie, Piers, and Brookfield, Harold (1987) *Land Degradation and Society*, London: Metheun.

Boyce, James K. (1987) *Agrarian Impasse in Bengal: Institutional Constraints to Technological Change*, Delhi: Oxford University Press.

Bradlow, Daniel D. (2001) "The World Commission on Dams' contribution to the broader debate on development decision-making," *American University International Law Review*, 16 (1531).

Brubaker, Elizabeth (2001) *The Promise of Privatization*, Toronto, Ontario: Energy Probe Research Foundation.

Declaration of Curitiba (2003) *Declaration of Curitiba: Affirming the Right to Life and Livelihood of People Affected by Dams* (14 March 1997). Online. Available HTTP: <http://www.irn.org/ programs/curitiba.html> (accessed 19 February 2003).

Dhawan, B.D. (1982) *The Development of Tubewell Irrigation in India*, New Delhi: Agricole Publishing Academy.

Dubash, Navroz K., Dupar, Mairi, Kothari, Smitu, and Lissu, Tundu (2001) *A Watershed in Global Governance? An Independent Assessment of the World Commission on Dams*, Washington D.C.: World Resources Institute.

Easter, K. William, Dinar, Ariel, and Rosegrant, Mark W. (1998) "Water markets: transaction costs and institutional options," in K.W. Easter, M.W. Rosegrant, and A. Dinar (eds) *Markets for Water: Potential and Performance*, Boston: Kluwer Academic Publishers.

Elster, Jon (1989) "Social norms and economic theory," *Journal of Economic Perspectives*, 3 (4): 99–117.

Finnegan, William (2002) "Leasing the rain." *New Yorker* (8 April): 43–53.

Gleick, Peter, Wolff, Gary, Chalecki, Elizabeth L., and Reyes, Rachel (2002) "The new economy of water: the risks and benefits of globalization and privatization of fresh water," Oakland, CA: Pacific Institute for Studies in Development, Environment, and Security.

Global Water Partnership (GWP) (2000) *Toward Water Security: A Framework for Action to Achieve the Vision for Water in the Twenty-first Century*, Stockholm, Sweden: Global Water Partnership.

Government of Gujarat (1963) *Handbook of Basic Statistics, Gujarat State, 1963, Handbook of Basic Statistics*, Ahmedabad: Bureau of Economics and Statistics, Government of Gujarat.

Guha, Ramachandra (1989) *The Unquiet Woods: Ecological Change and Peasant Resistance in the Himalaya*, New Delhi: Oxford University Press.

Hall, David (2002) "The water multinationals 2002 – financial and other problems," London: Public Services International Research Unit.

Hardiman, David (1998) "Well irrigation in Gujarat: systems of use, Hierarchies of control," *Economic and Political Weekly*, 33 (25): 1533–44.

Hilyard, Nicholas, and Mansley, Mark (2001) "The campaigners' guide to financial markets," Sturminster Newton, Dorset, UK: Corner House.

International Conference on Water and the Environment (ICWE) 1992. Online. Available HTTP: International Conference on Water and the Environment (ICWE) (2003) *The Dublin Principles*, <http://www.wmo.ch/homs/icwedece.html> (accessed 19 February 2003).

International Consortium of Investigative Journalists (2003) *The Aquas Tango: Cashing in on Buenos Aries' Water Privatization* (7 February 2003). International Consortium of Investigative Journalists 2003. Online. Available HTTP: <http://www.icij.com> (accessed 9 February 2003).

—— (2003) *Metered to Death: How a Water Experiment Caused Riots and a Cholera Epidemic* (5 February 2003). International Consortium of Investigative Journalists 2003. Online. Available HTTP: <http://www.icij.com> (19 February 2003).

Jehl, Douglas (2003) "As cities move to privatize water, Atlanta steps back," *New York Times*, 10 February.

Kavalanekar, N.B., and Sharma, S.C. (1992) "Over-exploitation of an alluvial aquifer in Gujarat, India," *Hydrological Sciences Journal*, 37 (4): 329–46.

Keck, Margaret, and Sikkink, Kathryn (1997) *Activists Beyond Borders*, Ithaca: Cornell University Press.

Khagram, Sanjeev (2000) "Toward Democratic governance for sustainable development: transnational civil society organizing around big dams," in A. Florini (ed.) *The Third Force: The Rise of Transnational Civil Society*, Washington D.C.: Carnegie Endowment for International Peace.

Kolavalli, Shashi, and Chicoine, David L. (1989) "Groundwater Markets in Gujarat, India," *Water Resources Development*, 5 (1): 38–44.

McCully, Patrick (2001) "How to use a trilateral network: An activist's perspective on the World Commission on Dams," Paper read at Agrarian Studies Program Colloquium, January 19, Yale University.

—— (2001) *Silenced Rivers: The Ecology and Politics of Large Dams*, London: Zed Books.

Manibeli Declaration (2003) *Manibeli Delaration, Calling for a Moratorium on World bank Funding of Large Dams* (September 1994). Online. Available HTTP: <www.irn.org/programs/finance/manibeli.shtml> (accessed 19 February 2003).

Moench, Marcus (1998) "Allocating the common heritage: debates over water rights and governance structures in India," *Economic and Political Weekly*, 33 (26): A46–A53.

Patnaik, Utsa (1987) *Peasant Class Differentiation: A Study in Method with Reference to Harayana*. Delhi: Oxford University Press.

Phadtare, P.N. (1988) "Geohydrology of Gujarat state," Ahmedabad: Central Groundwater Board.

Pocock, David F. (1972) *Kanbi and Patidar*, Oxford: Clarendon Press.

Polanyi, Karl (1944) *The Great Transformation: The Political and Economic Origins of our Time*, Boston: Beacon Press.

—— (1957) "The economy as instituted process," in K. Polanyi, C.M. Arensberg, and H.W. Pearson (eds) *Trade and Market in the Early Empires: Economies in History and Theory*, Pearson. Glencoe, Illinois: The Free Press.

Saleth, R. Maria (1998) "Water markets in India: economic and institutional aspects," in K.W. Easter, M. W. Rosegrant, and A. Dinar (eds) *Markets for Water: Potential and Performance*, Boston: Kluwer Academic Publishers.

Shah, Tushaar (1991) "Water markets and irrigation development in India," *Indian Journal of Agricultural Economics*, 46 (3): 335–48.

—— (1993) *Groundwater Markets and Irrigation Development: Political Economy and Practical Policy*, Bombay: Oxford University Press.

Shah, Tushaar, and Ballabh, Vishwa (1997) "Water markets in North Bihar: six village studies in Muzaffarpur District," *Economic and Political Weekly*, XXXII (52): A138–A190.

Shah, Tushaar, and Raju, K.Vengama (1988) "Ground water markets and small farmer development," *Economic and Political Weekly*, VXXIII (13): A23–A28.

Sinclair, Scott (2000) *GATS: How the World Trade Organization's New "Services" Negotiations Threaten Democracy*, Ottawa, ON: Canadian Centre for Policy Alternatives.

Singh, Chhatrapati (1990) "Water rights in India," New Delhi: Indian Law Institute.

Vohra, B.B. (1982) "Land and water management problems in India" 8. New Delhi: Training Division, Department of Personnel and Administrative Reforms, Ministry of Home Affairs, Government of India.

Williamson, John (ed.) (1994) *The Political Economy of Policy Reform*, Washington D.C.: Institute for International Economics.

Wood, Geoff D. (1995) "From farms to services: agricultural reformation in Bangladesh." February 1995. Bath, UK: University of Bath.

—— (1995) "Private provision after public neglect: opting out with pumpsets in North Bihar." January 1995. Bath: Centre for Development Studies, University of Bath.

World Bank (1993) "Water resources management policy paper," Washington, D.C.: The World Bank.

World Commission on Dams (2000) *Dams and Development: A New Framework for Decision-Making*, London: Earthscan.

9

TRANSITION ENVIRONMENTS

Ecological and social challenges to post-socialist industrial development

Dara O'Rourke

The soldier comes to another front now, the environmental front ...
without environmental recovery, Vietnam cannot have economic
recovery.

(General Vo Nguyen Giap, cited in Beresford and Fraser 1992)

INTRODUCTION

Inching down the truck-clogged highways of Guangdong province in southern
China, or the even more polluted roadways of Dong Nai province in Vietnam,
past hillsides turned into gravel quarries, rice fields transformed into cities,
buildings converted to billboards for the world's leading brands, and factory after
factory producing goods destined for the world's markets, it is hard to avoid
feeling that these former bastions of socialism are now centers of capitalist
economic globalization. China and Vietnam have ironically become both symbols
of the triumph of capitalism and the "end of history" (Fukuyama 1992), and
central examples in debates about globalization, development, and sustainability
(Hertsgaard 1997; World Bank 2000).

A key turning point in these debates is whether the socialist "transition to the
market" is supportive or destructive of so-called "sustainable development," that
is, whether the marketization of these economies will help to root out
inefficiencies, irrationalities, and contradictions of state socialism (World Bank
1997a), or rather accelerate and exacerbate processes of resource extraction,
environmental exploitation, and pollution (Brown 1996). Starkly contrasting
analyses compete to explain post-socialist environmental futures. Market
triumphalists invoke the logic of the Kuznets curve (De Groot *et al.* 2001) to
argue that rising wealth, technological advances, and efficiency gains will support
gradual improvements in environmental quality and reductions in pollution

(Hettige *et al.* 1997). Market critics argue conversely that the rapidly transforming (and increasingly polluted) landscapes of China and Vietnam represent not the end of history, but rather the birth of a hybrid model of development that combines the worst of central planning and state control, with the worst of capitalist exploitation (Bello and Rosenfeld 1990; Muldavin 1996). From this perspective, the post-socialist transition, while clearly good for GDP growth and income levels, may actually be worse for the environment, communities, and workers (Vermeer 1998).

The goal of this chapter is to engage these debates and to move beyond simple dualisms regarding environmental trajectories in post-socialist development. Changes in the state, market, and civil society are so sweeping in post-socialist countries that we must simultaneously examine the different transitions underway, and their dynamic interactions, in order to understand development, environmental trends, and civic responses. Current politics are also being transformed in ways that are not totally predictable, and that may actually create openings for new debates and contestations around development and the environment.

This chapter draws on the frame of political ecology to analyze transitions in post-socialist development. This literature is particularly useful for analyzing struggles around environmental resources and the complex dynamics between states, firms, and civil society actors influencing environmental degradation and response (Blaikie and Brookfield 1987; Peet and Watts 1996). By shining the spotlight of political economy on the recesses of environmental disputes and decision-making, political ecology has helped to show that power, politics, and poverty matter as much as physical environments or populations for environmental trajectories. However, while this literature has done an excellent job of analyzing rural, natural resource-based environmental conflicts, it has all too rarely directed its light on industrial environmental issues, or on the most dynamic region of industrial development (and environmental degradation) in the world – the socialist transition economies of East Asia.

This chapter seeks to engage these two gaps in the literature, applying the lens of political ecology, and its focus on state, market, and civil society interactions around the environment, to assess the multiple transitions underway in industrial development in post-socialist economies. To narrow the analysis to a somewhat manageable scope, the chapter focuses on the case of Vietnam, with only occasional references to parallels in China. This analysis is nonetheless, extremely challenging and preliminary, as the research seeks to study a rapidly moving object.

The central argument of the paper is that while on-going state and market transitions in countries such as Vietnam and China are driving industrial expansions and resulting adverse environmental impacts, concurrent transitions in the public sphere and in the politics of development are creating space for community and worker resistance to the worst impacts of this industrialization. Local actors are challenging the environmental and social impacts of what is now seen by many as the most successful economic development model in the world.

Communities and workers are making surprising demands, often unsuccessfully, but occasionally effectively, on the state to constrain industrial activities and to protect workplace and community environments.

These new pressures for reform and regulation appear quite distinct from traditional state environmental reform strategies or western social movements. Environmental contestations are occurring at the local level, often away from traditional NGO campaigns, and are employing a distinctly socialist discourse to challenge state policies and priorities. Communities are challenging the claims (and very legitimacy) of the state and Communist Party to protect workers, cultures, and the environment from unbridled capitalism, and indirectly challenging justifications for continued single-party rule.

There are of course impediments and limitations to these emerging local demands for environmental and workplace protections. And the dynamic nature of the on-going transitions makes it difficult to assert a "successful model" exists for local responses to environmental degradation. Nonetheless, potential does seem to exist to learn from and potentially build on local struggles to transform the political economy of industrial development and environmental protection in transition economies, and to advance more democratic participation in governance of development and environmental issues.

The chapter begins with a brief discussion of existing theory on the role of the state in balancing development and the environment, and related transitions underway in state policies and programs in Vietnam. The chapter then reviews current market transitions in Vietnam, the general outlines of the resulting environmental transitions, and emerging civil society responses to these trends. The chapter then turns to a detailed case of one industry, its pollution, and community struggles to motivate the state to more effectively regulate this pollution in Vietnam. The chapter concludes with an assessment of the potentials and challenges for building on local community-based struggles in post-socialist economies, and the implications of these trajectories for longer-term debates about development, democracy, and sustainability.

STATE TRANSITIONS

States, as many have argued, are seldom monolithic or static, but rather, arenas in which competing interests struggle to advance their goals and objectives (Evans *et al*. 1985). This is particularly clear in controversies around development and the environment (Gould *et al*. 1996), as states seek to both facilitate economic development and protect the social and economic structures that support that development. In developing countries this means both promoting industrial expansion and trying to protect bases of local livelihood such as clean air and clean water.

The preponderance of empirical evidence on industrial regulation in developing countries points to an extremely poor track record of states in balancing development and environmental goals, or more simply even enforcing environmental laws (Desai 1998; Dwivedi and Vajpeyi 1995; Taylor 1995). A number of theorists posit that because the state is ultimately interested in the promotion of industry, the accumulation of capital, and taxing these activities, state power and even survival is tied to the promotion and protection of industrial production (Gould *et al.* 1996; Schnaiberg 1994). The state's interest in promoting industrial activities severely limits its ability to enforce environmental and other regulations that might decrease the profitability of these activities.

Within this skewed balancing act, policies to promote industry directly conflict with environmental policies. Subsidies to industry, for instance, drive patterns of inefficient resource use and increased pollution that overwhelm anti-pollution laws. Tax incentives to support resource-based industries can create new forms and levels of pollution that environmental agencies are not prepared to regulate. Weak environmental enforcement may actually attract industries fleeing stricter regulations elsewhere.

These conflicts are particularly stark in formerly socialist economies (Feshbach and Friendly 1992; Gille 2001; Jancar-Webster 1993; Ross 1998), where constraints on state action are exacerbated by the state's ownership of means of production (i.e., polluting industrial enterprises) and commitment to the promotion of industry as a primary strategy of growth and accumulation. Beresford and Fraser (1992: 10–11) assert further that

> the traditional socialist economy has inbuilt mechanisms leading to enormous waste of resources ... which clearly have important implications for the environment.... Moreover, "nature" (water, air, minerals and vegetation in its uncultivated state) is regarded in traditional socialist theory as a free resource ... there is no incentive to introduce innovations which would result in less wasteful use of raw materials and energy, or reduce pollution.

Given these contradictions between advancing development and environmental protection, one might wonder why states would ever regulate industrial externalities. O'Connor (1988, 1994) argues that states are motivated to regulate environmental problems essentially only when the material and social conditions necessary for further capital accumulation are threatened by pollution or resource depletion. In this analysis, the state's main function is the protection of the "conditions of production," and the state intercedes and regulates when these conditions are threatened. Lester (1995) proposes a similar, though more conventional, "severity" argument, in which only the most severe pollution crises motivate state regulation.

These internal state conflicts are quite pronounced in Vietnam. The state is not only interested in production and accumulation, but has historically advanced a broader definition of development than simply GDP growth, promoting a rhetoric that includes protection of workers, development of rural areas, social equity, and protection of the environment. Delivering on these promises has raised conflicts and contradictions. Political stability and state legitimacy – which are continuously being renegotiated with social and political reforms – also remain issues of major concern to the government and Communist Party.

Severe environmental problems are also rapidly emerging in Vietnam, motivating the government to institute changes in its legal and institutional frameworks for environmental management. As the former Minister for Science, Technology, and Environment of Vietnam recently asserted, "Industrial pollution is starting to impact greatly on the social and economic development of Vietnam" (quoted in Anonymous 2001). Since 1993, the Vietnamese government has thus promulgated an umbrella Law on Environmental Protection, standards on air and water quality, a decree on environmental fines and enforcement, a decree on the implementation of Environmental Impact Assessments (EIAs), and a long list of circulars and directives to advance environmental protection.

This combination of laws, decrees, and regulations forms the basis of a traditional "command-and-control" environmental regulatory system. The government has opted (and been supported by international donors and advisors) to develop a strategy that requires the state to implement comprehensive standards (both ambient and sectoral pollution standards), monitor compliance with these standards, and sanction non-compliance through fines and other punishments.

In order to advance this command-and-control environmental system, the government has established a new set of institutions responsible for environmental and natural resource issues. The main agency responsible for industrial pollution issues is the National Environment Agency (NEA) within the newly created Ministry of Natural Resources and Environment (MONRE). The NEA has ten divisions, but only roughly 80 staff working to implement environmental laws throughout the country. At the local level, the government has established Departments of Science, Technology, and Environment (DOSTEs) in each province. DOSTEs are responsible for implementing environmental laws for all but the largest firms, and generally have monitoring and inspection teams. DOSTE staff working on environmental issues have doubled in recent years from 120 nationwide in 1994, to over 260 in 2000. However, as Giang (2001) notes, this means that Vietnam still "has just four state employees in the field for every million people compared with Malaysia's 100, Cambodia's 55, Thailand's 30, and China's 20."

To date, the Vietnamese government has not made much progress on the implementation of its environmental laws. Institutional weaknesses in fledgling environmental agencies have led to slow progress on monitoring and enforcement. Inspection divisions in the NEA and the DOSTEs are under-staffed and poorly trained for their duties. And perhaps more importantly, the environment is still primarily treated by the government as a resource and sink to be exploited. This contradiction between environmental uses and protection is a major impediment (and underlying contradiction) to fully integrating environmental concerns into development planning and implementation.

The continued failure (or more optimistically, the slow implementation) of state environmental regulations highlights the limits of state-centric, command-and-control strategies for environmental protection in countries such as Vietnam. Standards are easy to promulgate, but costly and difficult to enforce. Punishments are easy to stipulate, but difficult to implement in the face of political opposition. And a lack of funds, trained personnel, and political influence severely constrain the effectiveness of state environmental agencies. More importantly, contradictions and conflicts within the state – between developmentalist and environmental concerns – create incentives against enforcement of environmental regulations.

Nonetheless, largely due to public pressures, the government has been forced to reassert its commitment to protecting Vietnam's ecosystems, workers, and urban environments. And the establishment of the DOSTEs has created at the minimum a target for local community pressures. However, the state alone does not control development or environmental trajectories. Market dynamics – and the decisions of private firms – increasingly play a critical role in driving patterns of industrial development, environmental degradation, and social responses.

MARKET TRANSITIONS

The post-socialist transition economies are, in a word, some of the fastest growing, most dynamic developing countries in the world. China and Vietnam are home to over a quarter of the world's population, and the fastest growing economies in the world during the 1990s. As recently as the early 1990s, Vietnam and China were two of the poorest countries in the world. This has changed rapidly however, and the two countries are now viewed as success stories of modern development.

In a number of regards, these countries have been literal battlegrounds for development theory over the last 50 years. From external colonizers (the French and British), to Communist alliances (the Russians and East Europeans), to both hot and cold capitalist wars (the Americans), and more recently to institutions of globalization (the World Bank, Asian Development Bank, and International Monetary Fund), Vietnam and China have been "models" for competing theories and programs of development.

The most recent round of development changes has involved the market-ization, privatization, and internationalization of these economies. In the case of Vietnam, a political and economic "renovation" process, known as *Doi Moi*, was initiated in 1986 and accelerated in 1988. (Similar reforms were implemented a decade earlier in China.) This political and economic process involved first creating conditions for private agricultural production, by allowing peasants to sell surplus production to the open market, and later conditions for private and foreign industrial production. Price controls for most goods were lifted in 1989. By the early 1990s the government was moving swiftly to support privatization of state and collective enterprises, and to open the economy to international investors (Fforde and de Vylder 1996).

In a very short time span, Vietnam moved from autarkic self-sufficiency with limited trading, to export-oriented production as its focus of economic development. The country has also turned outwards to foreign capital and western aid for financing and technology to develop its industrial base. Hundreds of state enterprises have been shut down, while others are being encouraged (or forced) to compete in international markets. Changes in the agricultural sector were so successful that Vietnam moved from being a rice importer to being the number three rice exporter in the world in a matter of years. Inflation fell from 300 percent to 35 percent in one year. Government revenues rose from new found taxes and expanding oil exports. A 1990 World Bank report called Vietnam's reforms "the most comprehensive and radical set of reform measures adopted by any socialist country at the time" (quoted in Ljunggren 1993: 49).

In addition to the *Doi Moi* reforms, the Vietnamese government has advanced a program of "Modernization and Industrialization" designed to respond to an employment crisis in the country (1 million new workers enter the job market every year). This philosophy of development has been supported by a wide range of policies, laws, and directives, and has led to major transformations in the country's economic systems. Since the lifting of the US trade embargo in 1994, and the subsequent normalization of World Bank and IMF lending, Vietnam's economy has begun to resemble the Tiger it aspires to be, albeit with fits and starts.

While Vietnam as a whole experienced GDP growth of over 8 percent per year throughout the 1990s, with industry growing by 13 percent per year, Vietnam's urban centers have grown at twice that rate (EIU 2000a). Even amidst the Asian economic crisis, when overall GDP growth slowed to approximately 4.8 percent in 1999, Vietnamese industry continued to expand by over 10 percent (EIU 2000b). In 2000, the economy regained momentum with overall GDP growing by 6.8 percent and industry growing by a whopping 15.5 percent. Table 9.1 shows overall GDP and industrial growth figures from 1993 through 2001.

A massive inflow of foreign direct investment (which has slowed since 1998) and foreign aid (which has stayed almost constant at $2 billion per year) has been driving a process of rapid urbanization and industrialization. The Vietnamese economy however, is not just growing rapidly, it is also being transformed. Economic reforms and foreign investment have led to a significant shift in the structure of the economy, and a troubling trend toward more toxic industrial

Table 9.1 GDP and industrial growth in Vietnam (%).

	1993	1994	1995	1996	1997	1998	1999	2000	2001
GDP Growth (%)	8.1	8.8	9.5	9.3	8.8	5.8	4.8	6.8	6.5
Industrial Growth (%)	13.2	14.0	13.9	14.4	13.2	11.0	10.4	15.5	10.5

Source: EIU (2001).

activities. Small and medium sized enterprises (SMEs) are multiplying in cities around Vietnam, while large foreign joint-ventures are concentrating in the country's more than 50 new industrial zones. Although the state has retained control over certain strategic industrial sectors, and has sought to create Korean-style conglomerates in key sectors, today the private domestic and foreign invested sectors are the fastest growing segments of the economy. In 2000, the emerging private sector recorded a growth rate of 18.6 percent, with foreign firms trailing slightly with 17.0 percent growth (EIU 2000b). The oil and gas, steel, chemicals, garments, footwear, and printing sectors in particular have all been growing by over 20 percent per year during the 1990s. Many of these industries are, not incidentally, serious polluters.

There has also been a shift in the management of industry (Andreff 1993). In the past, almost all major industries were controlled by either central or local state authorities, and central planning motivated factory managers to focus primarily on meeting production targets. Profits, in the capitalist sense, were not a critical factor in production decisions, although accumulation of surplus was a stated objective. In state-owned enterprises (SOEs), profits were separated from production decisions, so that political objectives could drive decisions on what to produce and in what quantities. Resource inputs were priced through planning decisions or considered "free" goods (such as water), resulting in a general atmosphere of inefficient use of resources and ineffective production methods. Environmental considerations were rarely included in production decisions.

Under Vietnam's economic reforms, subsidies to SOEs have been drastically reduced. Access to capital is now one of the most important issues facing industrial managers. While certain SOEs still receive some special privileges on access to capital, competition for scarce funds has become intense. The cost of capital has risen rapidly (particularly compared to a previous situation of grants for factories). Capital is now often only available for investments that increase productive output, sales, and profits in the short-term. Factories receive loans if they can demonstrate that investments will have strong short-term returns. A tacit constraint on capital thus exists to exclude investments in pollution control equipment or longer-term investments in pollution prevention technologies.

State enterprise incentives and goals are also changing. Ministerial and external pressures are driving a transition from focusing on output, to focusing more on profitability. Competition based on price and quality – the new bottom-lines for industry – is creating pressure to lower costs and improve product quality standards. A new focus on export-oriented production is also driving demand for new production technologies. This modernization of Vietnamese SOEs is making them more efficient, and driving them to expand their production and resource use. This transition, however, is also motivating firms to cut costs however possible, with significant implications for the environment and workplace conditions.

ENVIRONMENTAL TRANSITIONS

Vietnam's phenomenal economic development over the last ten years, while obviously positive in many regards, has had major impacts on the country's environment. Urban and industrial development activities have resulted in a wide range of impacts on land use patterns, resource depletion, biodiversity loss, and water, air, and soil pollution. Rapid growth in industry is creating new stresses on uses of natural resources and the environment. Growth of 10 to 14 percent per year in industrial activities (with some sectors more than doubling in the last few years) is requiring increased extraction of natural resources, increased production and use of energy, and more transportation and other infrastructure services, all of which result in more wastes and pollution. Changes in the scale and structure of the economy, the efficiency of industrial activities, and mechanisms of regulatory control, all affect rates of natural resource depletion and pollution levels. Increased demand for energy also leads to a wide range of impacts on land uses, natural resource extraction, and air and water pollution.

Pollutant emissions from industry in Vietnam are very high relative to state-of-the-art or even average international production technologies. Worker health and safety conditions are also quite severe throughout Vietnamese industry. A recent survey of working conditions showed that approximately 60 percent of workers were exposed to polluted air, excessive noise, and high temperature conditions in the workplace (MOH 1998). While very little capital was reinvested in the maintenance or upgrading of production equipment in factories, even less seems to have been invested in environmental or workplace protections.

Natural resources are clearly being impacted by these economic developments. Forest cover has declined from 44 percent of the country in 1943 to less than 20 percent today (Poffenberger 1998). Over 1.5 percent of forest land is deforested each year, a rate which will eliminate all remaining forests within 50 years (EIU 2000a). Marine ecosystems are similarly being damaged and depleted by over-fishing, while mangrove forests are being threatened by new shrimp farms.

Vietnam is also experiencing a shift in the structure of industry towards more polluting sectors, and from traditional pollutants (such as biochemical oxygen demand (BOD)) to complex toxic compounds (such as heavy metals and hazardous wastes). Extractive industries such as oil and gas production, and mining, which have been expanding rapidly, cause significant environmental impacts. Other resource-based industries such as food processing and aquaculture also result in increased pollution loads to rivers. In the future it is likely that highly polluting sectors such as petrochemical and electronics production will bring new hazards to Vietnam. The World Bank has estimated that "if Vietnam does not implement pollution prevention and control policies, its toxic intensity will increase by a factor of 3.8 over a ten-year period (2000–2010), equivalent to a 14.2 percent annual [pollution] growth rate" (World Bank 1997b).

Water pollution is a growing problem throughout Vietnam. Many rivers and canals near urban centers are burdened with municipal and industrial wastes.

Water emissions from sectors such as food processing, beverages, textiles, paper, and chemicals include: biochemical oxygen demand (BOD), chemical oxygen demand (COD), acids, chlorinated organics, and heavy metals. Groundwater in cities such as Hanoi are also contaminated. Virtually all of the domestic wastewater in Vietnam is discharged untreated into rivers. Industrial effluents are also often still discharged without proper treatment, affecting both ecosystems and human health. Water pollution from industry regularly leads to fish kills, crop damage, and a wide range of skin and stomach illnesses in communities living near industrial zones. A recent report by a Canadian consulting firm found concentrations of certain toxic chemicals (including pesticides) at 890,000 times the allowable levels for California drinking water standards (Hatfield 1998).

Air pollution in urban areas is worsening with increased manufacturing activities, energy production (particularly from coal), and vehicle emissions. Dust pollution (particulate matter) has reached alarming levels in many urban areas due to construction projects, increased cars, trucks and motorcycles, and industrial activities. Toxic air pollutants have been measured at unhealthful concentrations near industrial facilities located in residential areas. A Ministry of Science, Technology, and Environment report in 2000 reported over 4,000 industrial entities were releasing smoke, dust, and toxic gases above legal limits.

Solid and hazardous wastes are also on the rise with the growth in household and industrial consumption. Sectors such as steel, electronics, and chemical manufacturing are producing new sludges, acids, and toxic solvent wastes. The government estimates that industrial estates alone produce approximately 400 tons of solid waste per day. Vietnam as a whole generates approximately 7 million tons of industrial waste per year (Anonymous 2001). Unfortunately, there are no systems in place in Vietnam for handling, storage, or treatment of hazardous wastes. An alarming amount of hazardous wastes from industry and hospitals are thus released untreated (or illegally dumped) in urban areas (JICA 1998).

Volumes of residential waste are also increasing rapidly. Consumption habits are beginning to create the outlines of a Vietnamese throw-away society, while domestic solid waste collection and disposal remain grossly inadequate. Vietnam currently generates approximately 8 million tons of solid waste per year, with only approximately 50 percent of this waste collected and disposed of properly. Landfills that do exist do not meet international standards for safety and management.

Workplace environmental issues present severe problems in industries throughout Vietnam as well. Workers are commonly exposed to toxic air pollutants, noise, heat, and radiation, without proper protections (MoH 1998). An estimated 20,000 workers in Vietnam suffer from silicosis, and over 10 percent of all industrial workers are estimated to be exposed to harmful noise levels (MoH 1998). Four hundred and three workers were reported killed in industrial accidents in 2000, although the government admits the real figure may be ten times higher. Use of chemical fertilizers and pesticides has skyrocketed over the last decade, with chemical fertilizer use (per cropped hectare) increasing from 40 kg in 1982 to 223 kg in 1997 (EIU 2000a). Pesticide poisoning is growing among farm laborers as Vietnam still permits the spraying of toxic pesticides such as hexachlorobenzene, Dieldrin, and DDT. These pesticides, banned in most developing countries, also adversely affect groundwater and soil quality.

On virtually every metric, environmental problems appear to be worsening in Vietnam. China's environmental trends are even more staggering (Hertsgaard 1997; Murray and Cook 2002; Smil 1993), and of even more global import. The question these trends raise is whether the citizens of these countries can, will, or are, responding?

CIVIL SOCIETY TRANSITIONS

In countries such as China and Vietnam, there is seldom any official public participation in environmental regulatory processes, and little effort to achieve effective or inclusive forms of participation in environmental disputes. Vietnam currently does not allow even the most basic forms of participation, such as public review of environmental impacts statements. Historically, the Vietnamese state has also sought to penetrate civil society through the extension of Communist Party control to mass organizations such as the Women's Union, Peasant Union, Youth Union, Fatherland Front, and the Vietnam General Confederation of Labor. Nonetheless, in Vietnam and other post-socialist countries, spontaneous forms of public participation and protests do emerge, and local actions to oppose the siting of environmental hazards such as polluting factories, waste dumps, and unwanted land uses are increasingly common.

In many countries, nongovernmental organizations (NGOs) are now the central forces for advancing civic environmental concerns, as well as for influencing state and corporate decision making (Keck and Sikkink 1998; Princen and Finger 1994). In industrialized countries, environmental NGOs that employ scientists and lawyers have been particularly effective in leveraging state-authorized forms of participation and challenging technocratic policies and programs which previously served to exclude environmental and social concerns. In China and Vietnam, however, there are few independent NGOs working on pollution issues. In fact, in Government-Organized Nongovernmental Organizations (GONGOs) dominate these landscapes rather than independent NGOs (Mas 2000; Wu 2002). And to date, there have also been no mass movement campaigns or national protests regarding the environment. Strategies for addressing workplace environmental and health issues – such as through independent trade unions – have also been limited or blocked outright by the governments of China and Vietnam (Amnesty International 2002; Young 2002).

There has been, nonetheless, informal community participation around pollution issues in China and Vietnam, although these actions look quite different from western conceptions of environmental movements. An interesting body of work has been emerging regarding these types of actions within civil society in Vietnam. This research is challenging traditional notions of communist state–society relations. Kerkvliet (1995: 415) for instance, surveys the existing debate around Vietnamese state–village relations, grouping theorists into three camps: (1) those who believe the state is all dominating; (2) those who assert the state controls civil society through various means; and, (3) scholars who believe social resistance is central to state-village relations. He concludes that the third view is more and more accepted, as "a prominent theme in the recent history of rural Vietnam is debate, bargaining, and interaction between the state and villages." Thrift and Forbes (1986) arrive at similar conclusions for processes and controversies over urbanization, showing how communities play a major role in challenging state policies.

Collective action of course does not spring from the ether, but is based in existing institutions and networks – or "mobilizing structures."[1] In Vietnam these mobilizing structures include state controlled institutions such as People's Committees, non-state commune and village relations, churches, kinship relations, and even Party-controlled mass organizations. Although "civil society" appears poorly developed in Vietnam, long-standing social and kinship relations (which have weathered wars and major political changes), and de-centralized social networks (strengthened during war time), serve as the basis for thick formal and informal social institutions. These networks and organizations are critical to collective action in Vietnam.

The Communist Party has of course attempted to control these collective politics. However, it has at the same time been responsive to some local demands. The state has repressed certain individuals and organizations, but it has not been able to quell all community action. Even at the height of the "planned economy," vibrant markets existed beyond the reach of the state. As Kerkvliet (1995: 414) notes, "the State is sufficiently 'mass regarding' to be at least worried about the peasantry and responsive to persistent resistance." This concern for stability, combined with the Communist Party's own ideology and rhetoric, have been used as resources that allow local community actors to influence state action.

As mentioned above, under the current legal and institutional framework in Vietnam, there are numerous formal mechanisms, but few actual pressures for firms to comply with environmental laws. Fines are too low to motivate changes. The threat of closure is non-existent for any but the smallest firms. Monitoring and inspections are still all too rare. And even when inspectors do show up, they are often ill-prepared and underpaid so they can easily be bought off or blocked.

However, regulation and punishments do sometimes occur under essentially a complaint-based environmental protection system. In this system, the more a community complains, the higher priority a problem receives from government inspectors. Public pressures appear to be the primary means of motivating state agencies to pressure firms to reduce pollution. Because agency capacity is so weak, virtually the only time environmental agencies go out and actually enforce is when they are pressured by community members. Ironically, this public pressure also strengthens the position of these agencies, giving them cover to regulate, and supporting their own requests for more resources from the state.

Obviously, there are potential weaknesses and problems with a system of complaint-driven inspections and compensations. Weak capacity and poor coordination limit the scope and depth of regulation. Compensations never fully pay the costs of pollution. State officials still generally side with the polluters. And community members only focus on certain types of pollution problems. Nonetheless, these complaints can lead to real environmental improvements, as the case below describes.

INDUSTRIAL POLLUTION AND COMMUNITY RESISTANCE IN VIETNAM

The Dona Bochang Textiles factory and the industrial system it represents, appear not only to be on the cutting edge of global production, but also an almost completely un-regulatable element of that system. The firm is a profitable company with connections to large brand names, but without a reputation of its own that can be publicly damaged. No activist or consumer campaigner in the world has ever heard of Dona Bochang, or could even find this factory in the complex of the global production system. The company has set up operations in partnership with local government officials. The Communist Party of Dong Nai province (the province just northeast of Ho Chi Minh City) actually owns 12.5 percent of the factory. There are no local non-governmental organizations (NGOs) to put pressure on the firm. And to top it all off, the local regulatory agency – which is both young and weak – answers to the very same Party officials who own a portion of the factory. How could any government agency, let alone a weak one, regulate a multinational corporation like this as it plays off government agencies, and even countries against one another?

Dona Bochang Textiles appears to be one more example of globalization gone unchecked. The factory in Vietnam is one link in a long and mobile supply chain, a link that could be shut down and moved to China, or Bangladesh, or Guatemala literally tomorrow. This factory, which has been operating since the early 1990s, represents the new face of Vietnamese export-oriented industrialization. The factory serves as an export platform, spinning, weaving, knitting, dyeing, and printing imported cotton and polyester fibers into low-value products such as towels for export to developing countries around the region, part of a wave of Taiwanese, Korean, and Japanese factories at the forefront of the recent boom in foreign manufacturing in Vietnam.

And while Dona Bochang clearly produces benefits for some people, the factory also produces serious air and water pollution, and adversely impacts workers' and community health. The dyes used to make the company's towels blue (or green, or red, or whatever color dye the factory uses that day) also turn the local river that color. Local groundwater is now too polluted to drink. Air pollution from the factory's boilers regularly coat the surrounding community in a layer of soot.

Despite these problems, for more than six years, local government officials had done little to regulate the factory. From the outside, this case would seem to confirm the worst fears of critics of globalization and post-socialist economies: local governments have few incentives and even less power to enforce environmental or labor laws which might impinge upon a factory's profits. In interviews, several government officials made clear that they fear firms like Dona Bochang will move on to other, cheaper countries, if Vietnam regulates them too strictly, or if workers' wages rise too fast. Vietnam, they worry, is already pushing these limits with its $40 per month minimum wage and stricter regulatory environment than other governments in the region.

Community members, however, are less resigned to unregulated industrial activity and pollution. Through letters, meetings, and even protests, community members have repeatedly asked why the plant is allowed to operate in a residential area. The factory and the local residents are separated by no more than a three meter-high wall and a dirt road that runs along the perimeter of the factory. The plant is surrounded on two sides by small, tightly packed houses, on the third by the local Catholic church, and on the fourth by an important commercial road, with more houses across the street.

Community members on one side of the plant complain that the factory was built with no wastewater treatment system (something that would be illegal for a textile plant in Taiwan, the United States, and most other industrialized countries), and that their wells are now contaminated with toxic wastewater. On the other side of the factory, community members complain about air pollution from the plant, toxic releases in the middle of the night (when no one can see what is being emitted), and the health impacts on their children and older family members. The factory's air emissions, when not blowing into the residential area, blow directly into the church, which was until recently an open air building with little more than a roof, an altar, and rows of pews.

Few people would be surprised by a story of a developing country failing to regulate pollution during its drive to industrialize and connect to the global economy. It is almost taken for granted that governments will pass environmental laws and then fail to enforce them, choosing to postpone environmental protection in the interests of economic development. If this were the conclusion to this case, it would only provide further evidence of the failures of environmental regulation in developing countries and of the inability of local governments to regulate global firms. Fortunately, there is more to the story.

Even though local regulators say the air pollution from Dona Bochang is really not all that bad, the neighboring community has not been placated by the thought that other communities have it worse. In their view, pollution from the Taiwanese-run factory is a continuing assault on the neighborhood, affecting peoples' daily lives, disrupting social occasions, even defiling their center of worship. Pollution impacts such as respiratory problems, corroded roofs, and blackened plants have led to an escalation of community actions and pressures demanding the factory reduce its pollution or move out of the neighborhood and into an industrial estate.

The community around Dona Bochang is on the middle rung of Vietnamese urban communities, certainly not rich by western standards, but better off than rural communities. The community has prospered under the *Doi Moi* market reforms. Many families run their own businesses – either producing simple products like wood furniture or trading petty goods. A number of people in the community work in nearby industrial zones, adding an important source of income to their households. What is more surprising perhaps is how many people in the community choose not to work in these factories, as they can earn a higher income than the factory minimum wage.

As the Taiwanese managers found out, the people living around Dona Bochang are a tightly knit community. Approximately 90 percent of the community are Catholics who moved to the area in 1954, fleeing the Communist victory in the north. The church's location next to the factory has made it a critical base for discussions on the impacts of the factory, and for sharing grievances. While the priest went out of his way to make clear in interviews that he was not organizing the community around environmental issues, he did admit that the church had become a center of discussion for these concerns. But the church alone is not the leader of the community's actions. The real base of community mobilization around Dona Bochang is an informal social network which exists among community members. People live, socialize, worship, and sometimes work together. This community is also cohesive partly because of its past experiences of religious discrimination. So while Catholics may be more fearful of state oppression (and thus less likely to complain), they are also more unified and cohesive in their demands when they do mobilize.

There is clearly a very high level of community interaction in the neighborhood along the back wall of the factory. Contrary to stereotypes of Communist countries, communities in Vietnam can be surprisingly strong and confrontational in their dealings with government agencies and firms. The community around Dona Bochang Textiles is a case in point. After researching the environmental effects of the factory, I set out to interview impacted community members. In another country I would have very likely hopped on a motorcycle and driven out to the community. In Vietnam it is never so simple. The government works hard to control outside access to communities, requiring multiple permits, approvals, and often chaperones. After securing these approvals, I set off in a government Landcruiser (the vehicle of choice to attract the most attention), accompanied by a research assistant, two representatives from the Department of Science, Technology, and Environment (DOSTE), a police officer, two representatives of the local People's Committee, and a driver.

When we finally arrived at the first community member's home my entourage included five government officials (including one in uniform). Within minutes of sitting down, twenty or so curious neighbors had gathered in the room, spilling out through the doors and windows. And now, with a gallery of close to thirty people, I was to interview a local family on the sensitive issues of the factory's pollution, community complaints, and the government's response.

Despite the ridiculous conditions for the interview, the family was shockingly frank about local environmental problems and openly critical of both the factory management and government authorities. As the family expounded on their problems, it was impossible to miss the DOSTE chaperone scowling and taking careful notes of the criticisms leveled at his organization. But the family showed little fear. Mr. Nguyen (not his real name) carefully detailed the environmental and health impacts of the factory's pollution on the neighborhood. He complained that the government and the Taiwanese factory managers had ignored their complaints for years. A neighbor alluded to corruption blocking solutions. Other community members joined in on a free wheeling critique of the firm and the government.

To an outside observer it is surprising that community members in Vietnam would openly criticize local government officials, particularly in front of a foreigner. And it is even more surprising when these complaints turn into community actions against a factory (particularly one like Dona Bochang that is partially owned by the government). With the easing of state controls, substantial community action and bottom-up pressures for reform are emerging throughout Vietnam. Communities can be vibrant, aggressive, and sometimes extremely well-organized. High levels of community participation and bottom-up political pressures are playing important, though varied roles in state policies. While these processes are complex and often conflicted, they hold out significant potential for social, political, and environmental advances.

In my interviews, one family stood out as leaders of this community's actions. They seemed fairly well educated and quite well off for the community, running their own small household enterprise finishing wood furniture. Living and working just a few feet from the factory wall, the family had collected a thick file on the factory's pollution, including press clippings, letters they had sent to various government agencies, the responses they had received, and photographs of pollution impacts. They regularly drafted letters for others to sign. They had been on official delegations to the factory and to government meetings. They had even made a video of the pollution. In many ways, they seemed fearless in their quest to end the pollution, a quest they have yet to finish.

After several years of having their complaints ignored, the community was ignited by an incident in 1993. On the day of a local wedding, air pollution from the factory coated trays of food laid out for the reception in a layer of black soot. A group of community members considered this the last straw, marched to the front gate, and threatened to tear down the wall and shut down the factory if the manager did not come out to negotiate with them. A number of young people threw bricks at the factory to make clear their anger.

A factory representative did finally emerge and asserted that the company was doing all it could and promised the problems would be solved. The community forced the manager to sign a statement attesting to the level of pollution. Photographs were taken of the pollution impacts. Several months later, when nothing had changed, the community brought their complaints, the pictures, and the signed statement to the Dong Nai DOSTE and then to the media.

After newspaper reports questioned the failure of the government to regulate the pollution – one pithy headline read "Hope that Concerned Agencies Will Make Efforts to Solve the Environmental Pollution at the Dona Bochang Factory Area" – the DOSTE agreed to take action. It responded to the community complaints by organizing an inspection team and several meetings between community members and the factory. The community however, criticized this inspection process, charging that because it was a planned inspection, the factory would be able to turn off the polluting equipment before the inspectors arrived. Community members argued that their daily experiences were more accurate than the data collected from the inspection. Later, when pollution levels resumed, the community sent further written complaints to higher government authorities and the media. This renewed pressure motivated more meetings and finally resulted in the factory agreeing to install equipment to reduce its emissions.

As one community member explained (Ms. Lan, 6 May 1997):

> We have complained many, many times to the DOSTE, the newspapers, the TV. Many times. We have also complained to the factory many times. We had a meeting between the people, the factory managers, and the People's Committee in the *Phuong* [ward]. We have also had many meetings of our own. People around the factory have done many things. Young people have thrown rocks at the factory many times. Because of the rock throwing, the factory went to the *Phuong* People's Committee to try to solve the problem. But still the factory didn't change and the people still opposed the factory. The factory has even tried to hire the young people who threw the rocks and complain. Because everyone knows each other so well, we can have coordinated actions together. We have so many things in common.

By the fall of 1997, the neighbors of Dona Bochang had achieved a qualified victory over the factory. Since the wedding party incident, the factory had made three changes to reduce its air pollution. First, it built a taller smokestack – the classic solution to local environmental problems. When that did not reduce the local impacts, the factory changed its practice of "blowing the tubes" from its boiler (where built up soot is forced out of the smoke stacks) which was a major source of the black soot people. Finally, when the problems were still not resolved, the factory installed an air filtration system to capture the pollution. This process took several years, but resulted in a significant reduction in air emissions (according to the firm and the Dong Nai DOSTE). The community also agrees there have been improvements, but continues to pressure for further pollution reductions.

The community around Dona Bochang has only been partially successful in winning increased environmental protections. Complaints have succeeded in getting the factory to invest modest amounts of money in cleaning up its air pollution, but they have not been able to motivate the factory to invest the millions of dollars needed for wastewater treatment. The community has also never received any compensation for the pollution.

The state's role in this case is complicated as the Dong Nai People's Committee owns a percentage of the company. Community action is thus in conflict with the short-term economic interests of the provincial People's Committee, which controls the environmental regulatory agency. The community members' perception that they had to overcome this conflict of interest led them to extralocal actors such as the National Environment Agency and the media to help address their problems. It also strengthened the community's resolve to keep pressure on the factory and on the provincial authorities. Community members did not trust the state to take action without repeated pressure. At the same time, the fact that the factory is majority foreign-owned may have worked to the community's advantage: Vietnamese government agencies appear extremely sensitive about public perceptions that the state is privileging foreign capitalists over common people.

Some community members have framed the factory's pollution as a moral and political issue, decrying the fact that the search for profits is out-weighing all other considerations. One community member put it bluntly, "The factory doesn't care about pollution because they only think about profits." Another argued that the state had lost perspective on the purpose of socialist industrialization: "Factories are built to serve the people, not to harm people. If the factory is hurting people it is not meeting its purpose."

Within these debates, the head of the DOSTE inspection division reiterated how important public complaints are to the inspection process (Mr. Tien, 6 September 1997):

> If the complaints of a community are very strong, that factory will be inspected first. We have too many factories to inspect, so we prioritize based on complaints. The compensation process starts from complaints. If people lose crops but don't complain, then there is no compensation.

The director of the environmental management division explained further how delicate regulation can be (Mr. Het, 8 January 1998):

> Things are very sensitive with companies right now. Companies with operations in Indonesia can move their production there if they have a problem in Vietnam.... We understand that, they don't need to say it ... The province is very concerned now about the economic downturn in Asia. 10,000 workers have been laid off this year. Investment is down. Growth is slowing. We don't want to make conditions difficult for

foreign investors. We have to be even more careful about environmental regulation now. We don't want to hurt companies or scare them off.

The DOSTE is thus forced to try to balance pressures from below with pressures from above, and occasionally pressures from overseas. Essentially, they must prove they are serious about environmental protection without hurting industrial interests. This leads to a process of negotiated regulation, and responses to the most vocal community complaints.

The chairman of the local People's Committee claims that this has translated into the community having more influence over factories than he does (Mr. Viet, 6 July 1997):

> We cannot fine Dona Bochang. We can only contact them and make suggestions about issues to be resolved. The company is really afraid of bad publicity. The company does things to appease the peoples' complaints.

Another government official agreed, explaining (Mr. Het, 10 November 1997):

> Peoples' complaints have a great impact and they play an important role. The complaints force factories to take action, and they raise environmental issues for companies built before the Law on Environmental Protection. Whenever there are protests, the company [Dona Bochang] does something to appease the people. But then they stop. When there are more protests, the company makes one more change. They don't really want to solve the environmental problems.

Through a strategy of mobilizing community members, building linkages to the media, and monitoring and holding local government officials accountable, the community has achieved limited successes against a well-connected foreign firm. The community has strategically used recent state reforms, in particular the passage of the Law on Environmental Protection, and the creation of the DOSTE, to focus its demands. Increased wealth in the community, and resultant independence from state and foreign enterprise jobs, has strengthened the community's resolve to transform or move the factory altogether. Political openings now seem to permit, or at least not repress, protests and direct criticisms of the state on environmental issues. And continued use of socialist discourses of development and equity fuel community organizing and accountability campaigns.

Many of the improvements at Dona Bochang have been due to the cohesiveness, organization, and persistence of the community in using political openings. However, it should be noted that the community on its own has not been able to directly change the firm. Letters and meetings with the factory owners did not result in pollution reduction. Success came through the community exerting pressure on local and national government agencies both directly and through the media, and by employing the discourse of the state to demand more balanced, equitable, and environmentally sound development.

CONCLUSION

Broad political, economic, and social transitions underway in Vietnam are simultaneously driving industrial expansion, new environmental problems, and emerging civil society responses. Negotiating these controversies involves an on-going balancing act between government agencies, firms, and community actors. Contestations and competing interests are central to these interactions.

With the transition to "market-oriented socialism," the state is assuming new roles in the economy, exercising less control over direct production and more control over macroeconomic processes. One of the impacts of these changes is the state's decreased willingness to subsidize and protect state-owned enterprises. Some firms are being "equitized" – Vietnam's version of privatization – while others are being forced to compete on their own. These changes have opened up space for critical assessments of the costs and benefits of industrial activities and the specific impacts of state enterprises. Community members are increasingly able to question the trade-offs between state subsidies, jobs, and pollution.

The entrance of foreign multinational corporations into the Vietnamese economy has opened another arena for criticizing industry. Industrialization is no longer considered the panacea for all problems, and can be discussed separately from national development. The public is well aware – partly because Communist party rhetoric has hammered home this point for years – that some firms exploit workers and pollute the environment. The contradictions between socialist ideology and the lax regulation of foreign multinationals has opened a political debate in the country.

Since at least the 1940s, the Vietnamese Communist Party has understood that its main source of support and strength was rooted in the provision of economic and social benefits to the peasant class. The foot-soldiers of early liberation struggles were rural peasants. As Kolko (1995: 33) notes "the Party's success as a social movement was based largely on its responsiveness to peasant desires." The recent transition process, while clearly influenced by international economic and political processes, has also been influenced and constrained by populist concerns and demands.

It is said in Vietnam that "The writ of the king bows to the customs of the village" (*"phep vua thua le lang"*) or as Fforde and de Vylder (1996: 50) explain, "local autonomy was part of the accepted balance of power between the central Imperial Court and the local communes." Despite decades of central planning, a surprising amount of local autonomy has not only survived, but now flourishes in Vietnam.

This has led to a wide range of development decisions being contested in Vietnam. Peasants in Thai Binh province have recently opposed local taxation policies and corrupt government officials. Women in Dong Nai province have opposed government efforts to reallocate land. Farmers just outside Hanoi have violently struggled to protect their land from a government-supported golf course development. And most recently, villagers near Qui Nhon, in central Vietnam, actually stormed a factory, tore down its walls, and destroyed its equipment to protest local pollution (Anonymous 2002).

The Qui Nhon case is a particularly violent example of public concerns regarding the environmental impacts of industrialization. Since 1993, and the passage of the Law on Environmental Protection, complaints and protests around environmental issues have been growing exponentially.[2] With the formalization of the legal legitimacy of environmental complaints, the public has made clear that it is not willing to bear the environmental and social costs of industrial development while others profit from this development. As Beresford and Fraser (1992: 15) argued even before the passage of the LEP:

> What is increasingly being called into question is the ability of the traditional socialist model to deliver both economic growth ... and development broadly interpreted to include rising living standards, improved health, more leisure and more democracy for the majority of the population in whose name the socialist revolution was carried out. Containment and prevention of environmental damage is increasingly being seen as one of the necessary conditions for this goal.

As with other issues, the state has responded to public demands with a careful mixture of both rhetoric and action. The Vietnamese state, however, appears to be in a vulnerable position regarding environmental issues. By staking its claims to legitimacy on its ability to chart a development course that promotes economic growth while mitigating the adverse social and environmental impacts of unbridled capitalist industrialization, the state comes face to face with a number of internal contradictions. The state must both attract foreign investors (and the jobs they bring), and at the same time regulate these actors. The state must balance its support for the emergence of a capitalist class, with the need to continue protecting the interests of peasants and workers.

As Kolko (1995: 33) warns, "a Party that triumphed because of the social role and needs of the peasantry, and then alienates them, is risking suicide." The Party is thus charting a careful course that involves responding to certain community demands as a means to balance (or pacify those affected by) adverse impacts of industrial development. Demands for environmental quality are accepted as one trade-off in the transition to the market.

In this regard, the state's struggle for legitimacy, and local government sensitivity to accusations of corruption, have opened up new space for community action around environmental issues. Vietnam's socialist legacy has essentially provided a window of opportunity for community participation, and civil society more broadly, to play a constructive role in development and environment debates.

The Vietnamese government has also felt external pressures from foreign donors to establish environmental laws and institutions. The government signed all the major conventions of the Rio summit in 1992 and has been in line ever since for Capacity 21 funds from the United Nations to implement "sustainable development" programs and policies. These external agreements and pressures have helped push the government to pass laws and to accept certain forms of public participation. Some have argued that this external pressure is a new form of green colonialism (Goldman 2001). Nonetheless, in the Vietnamese case, it has at least supported the transformation of state environmental institutions, and created openings – and targets – for local environmental protest.

The passage of the Law on Environmental Protection was a critical step in creating these legal openings for community complaints and demands. Article 43 of the law states that "Organizations, individuals have the right to complain, denounce to the State management agency for environmental protection or other competent State agencies about activities in breach of environmental protection legislation" (SRV 1994). As Roodman (1999) points out, this right of "complaints from people living near factories, along with media pressure, have perhaps played at least as significant role as conventional regulatory measures in driving industrial pollution reductions in Vietnam."

Vietnamese citizens now literally file thousands of complaints against industrial polluters each year, and journalists produce hundreds of stories (Roodman 1999). Ho Chi Minh City's DOSTE reports receiving over 1,000 complaints per year. The Hanoi DOSTE similarly receives around 1,000 complaints per year, while Dong Nai receives approximately 200 complaints per year (O'Rourke 2004). The DOSTEs also appear particularly sensitive to media criticisms of their failure to enforce pollution laws. Accusations of corruption, incompetence, or capture by capitalist interests, appear to generate fairly strong responses.

The right of citizens to criticize the government and to protest development policies does of course have limits. Community members allude to unstated lines that must not be crossed. People generally avoid publicly accusing high officials of corruption. And the Law on Environmental Protection itself is quite restrictive in how the public can participate in environmental decisions, essentially only allowing complaints after pollution has occurred.

Nonetheless, public and media pressures are gradually raising the profile of environmental issues. Community pressures have helped to overcome agency resistance to implement laws that impact other state actors, and to motivate inspectors to simply do their jobs, a not insignificant feat as most inspectors are overwhelmed by their tasks, under-trained for their duties, and under-paid. Community pressures also help shine light on local-level corruption and increase the transparency of state decision-making.

A critical question however, is whether these local struggles – and occasional victories – for environmental protection can be scaled up or out? Peet and Watts

(1996: 36) are right to push analysts to consider the "conditions under which local movements transcend their locality, and hence contribute to the building of a robust civil society." While it is clearly difficult to win local struggles in Vietnam and China, it appears even more difficult to institutionalize emergent mechanisms of local environmental protection.

It is simply not clear at this point how cases of local activism and environmental struggles around specific development projects might be expanded or institutionalized in Vietnam. There is little civil society infrastructure upon which to build environmental movements or national campaigns. China and Vietnam continue to be hostile environments for the development of western-style social movements. (As one China critic recently quipped, "If you can't meditate in public in China without being beaten up and thrown in jail, what chance is there for real social movements.") China and Vietnam have also made clear that they have no intention of following Russia's lead on opening politics to multiple parties or pluralist interest groups. Western models of environmental politics, in which NGOs and local activists play off a "big state" against "big capital" also seem to hold little potential in post-socialist countries. The state's continued participation in production and joint-ventures with foreign firms blurs the lines between the regulators and regulated. China and Vietnam also appear largely impervious to international pressures or shaming on human rights, labor, or environmental issues.

So does this mean Vietnam and China are doomed to a low-road path of environmentally destructive development and severely constrained democratic participation? Perhaps. But perhaps not. As the Dona Bochang case shows, the seeds of resistance and reform exist within these transition environments. Dynamics on the ground in Vietnam and China are driving both environmental degradation and emergent responses to these environmental and health problems. Local struggles are at the heart of current resistance. But these struggles also challenge broader state policies and legitimacy. On-going state transitions – and continuing environmental reforms within the state – may open further points of leverage for local demands.

While post-socialist economies have been rightly criticized as rife with environmental problems, the challenge for Vietnam is not simply to shake off this past and trust in the miracle of the market, but rather to learn from, and build on, the strengths of its history. If Vietnam can strengthen local autonomy and government responsiveness to environmental and social problems, and at the same time support mechanisms for community participation, the country may yet be able to chart a less destructive path towards industrialization. In the longer-term, one could imagine the development of policies and programs that do more than just support pollution reductions at individual factories. Policies could promote broader efforts to balance economic development with environmental protection by developing mutually reinforcing programs that help build state agency capacity, public awareness and involvement, and incentives for firms to innovate and improve environmental performance.

The post-socialist economies will continue to be battlegrounds for development theory in the coming years. It is thus critical to carefully assess the multiple transitions underway in these countries in state, market, and civil society actors, and resulting changes in the political economy of development. For those interested in the environmental futures of the post-socialist world, it is also critical to understand (and potentially support) on-going contestations and struggles over better balancing development and the environment. The success of the post-socialist economies in balancing development, environmental livability, and equity may determine not only their environmental and social futures, but also the coming history of "sustainable development."

NOTES

1 McAdam *et al.* (1996) define mobilizing structures as "those collective vehicles, informal as well as formal, through which people mobilize and engage in collective action," and in particular "organizations and informal networks that comprise the collective building blocks of social movements."
2 While there is no mass environmental "movement" in Vietnam, Vietnamese press reports and my own interviews indicate that complaints and protests regarding environmental issues have become more widespread over the last five years.

REFERENCES

Amnesty International. (2002) *People's Republic of China – Labour Unrest and the Suppression of the Rights of Freedom of Association and Expression*, ASA 17/015/2002, 30 April, London: Amnesty International.

Andreff, Wladimir. (1993) "The Double Transition from Underdevelopment and from Socialism in Vietnam," *Journal of Contemporary Asia*, vol. 23, no. 4.

Anonymous. (2000) "Environmental Crisis Takes Over Vietnam's Economic Zones," *Asia Times*, 24 February 2000.

—— (2001) "Vietnam: Amid Industrialization, Country Choking on Development," *Interpress Service*, 22 November 2001.

—— (2002) "Angered by Pollution, Villagers Damage a Factory in Vietnam," *Associated Press*, 23 October 2002.

Bello, W. and Rosenfeld S. (1990) *Dragons in Distress – Asia's Miracle Economies in Crisis*, San Francisco: Food First.

Beresford, M. and Fraser, L. (1992) "Political Economy of the Environment in Vietnam," *Journal of Contemporary Asia*, vol. 22, no. 1, pp. 3–19.

Blaikien, P. and Brookfield, H. (1987) *Land Degradation and Society*, London: Methuen.

Brown, L. (1996) "China's Challenge to the United States and to the Earth," *World Watch Magazine*, vol. 9. no. 5. September/October.

Buffett, S. (1996) *Reconciling Disparate Paradigms: Environmental Sustainability and Economic Growth in Viet Nam*, Masters Thesis, American University International Development Program, Washington, D.C. June 1996.

De Groot, H., Withagen C., and Minliang Z. (2001) "Dynamics of China's Regional Development and Pollution," Rotterdam: Tinbergen Institute.

Desai, U. (ed.) (1998) *Ecological Policy and Politics in Developing Countries*, Albany: State University of New York Press.

Dwivedi, O.P., and Vajpeyi D.K. (1995) *Environmental Policies in the Third World: A Comparative Perspective*, Westport: Greenwood Press.

Economist Intelligence Unit (EIU) (2000a) *Vietnam Country Profile 2000*, London: EIU.

—— (2000b) *Vietnam Country Report*, October 2000, London: EIU.

—— (2001) *Vietnam Country Profile 2001*, London: EIU.

Evans, H., Rueschemeyer, D., and Skocpol T. (eds) (1985) *Bringing the State Back In: New Perspectives on the State as Institution and Social Actor*, New York: Cambridge University Press.

Feshbach, M. and Friendly, A. (1992) *Ecocide in the U.S.S.R*, New York: Basic Books.

Fforde, A. and Vylder S. (1996) *From Plan to Market – The Economic Transition in Vietnam*, Boulder: Westview Press.

Freeman, R. (1995) "Are Your Wages Set in Beijing?" *The Journal of Economic Perspectives*, vol. 9, Issue 3, Summer 1995, pp. 15–32.

Fukuyama, F. (1992) *The End of History and the Last Man*, New York: The Free Press.

Giang, T. (2001) "Nation Scores Low in 'Green Guardian' Stakes," *The Vietnam Investment Review*, 5 November 2001.

Gille, Z. (2001) "State and Society in Hungarian Waste Politics: A Local Historical Perspective," in P. Evans (ed.) *Livable Cities: The Politics of Urban Livelihood and Sustainability*, Berkeley: University of California Press.

Goldman, M. (2001) "Constructing an Environmental State: Eco-Governmentality and other Transnational Practices of a 'Green' World Bank," *Social Problems*, 48: 3.

Gould, K., Schnaiberg A., and Weinberg A. (1996) *Local Environmental Struggles: Citizen Activism in the Treadmill of Production*, New York: Cambridge University Press.

Hatfield Consultants (1998) *Preliminary Assessment of Environmental Impacts Related to Spraying of Agent Orange Herbicide During the Viet Nam War*, consultant report vol. 1 and 2, Vancouver, Canada: Hatfield Consultants.

Hertsgaard, M. (1997) "Our Real China Problem," *The Atlantic Monthly*, November.

Hettige, H., Mani, M., and Wheeler D. (1997) "Industrial Pollution in Economic Development: Kuznets Revisited," Development Research Group. Washington, D.C.: World Bank.

Jancar-Webster, B. (1993) *Environmental Action in Eastern Europe: Responses to Crises*, New York: M.E. Sharpe.

Japanese International Cooperation Agency (JICA) (1998) *Household Survey on Environmental Awareness of Hanoi Citizens*, prepared for the Study for Environmental Improvement for Hanoi City, Hanoi, November.

Keck, M. and Sikkink K. (1998) *Activists Beyond Borders – Advocacy Networks in International Politics*, Ithaca: Cornell University Press.

Kerkvliet, B., J., and T. (1995) "Village-State Relations in Vietnam: The Effect of Everyday Politics on Decollectivization," *The Journal of Asian Studies*, vol. 54, No 2.

Kolko, G. (1995) "Vietnam Since 1975: Winning a War and Losing the Peace," *Journal of Contemporary Asia*, vol. 25, no. 1, pp. 3–49.

Lester, J. (1995) *Environmental Politics and Policy – Theories and Evidence*, Durham: Duke University Press.

Ljunggren, B. (ed.) (1993) *The Challenge of Reform in Indochina*, Cambridge: Harvard Institute for International Development.

McAdam, D., McCarthy, J., and Zald M. (1996) *Comparative Perspectives on Social Movements*, New York: Cambridge University Press.

Mas, S. (2000) *The Legal Environment for Non Profit Organizations in China and Its Impact*, Kunming, China: Center for Biodiversity and Indigenous Knowledge.

Ministry of Health (MoH). (1998) *Proceedings of the Third National Scientific Conference on Occupational Health*, Hanoi, December 1998.

Muldavin, J. (1996) "The Political Ecology of Agrarian Reform in China," in R. Peet and M. Watts (eds) *Liberation Ecologies: Environment, Development, and Social Movements*, New York: Routledge.

Murray, G. and Cook, I.G. (2002) *Green China – Seeking Ecological Alternatives*, London: RoutledgeCurzon.

O'Connor, J. (1994) "Is Sustainable Capitalism Possible?" in M. O'Connor (ed.) *Is Capitalism Sustainable? Political Economy and the Politics of Ecology*, New York: Guilford Press.

—— (1988) "Capitalism, Nature, Socialism: A Theoretical Introduction," *Capitalism, Nature, Socialism*, vol.1, no.1, Fall.

O'Rourke, D. (2004) *Community-Driven Regulation: Balancing Development and the Environment in Vietnam*, Cambridge, MA: MIT Press.

Peet, R. and Watts M. (eds) (1996) *Liberation Ecologies: Environment, Development, and Social Movements*, New York: Routledge.

Poffenberger, M. (1998) *Stewards of Vietnam's Upland Forests*, Berkeley, California: Asia Forestry Network.

Princen, T. and Finger M. (1994) *Environmental NGOs in World Politics*, New York: Routledge.

Roodman, D.M. (1999) "Vietnam: The Paradox of Public Pressure," *Worldwatch Magazine*, vol. 12, no. 6, November/December.

Ross, L. (1998) "The Politics of Environmental Policy in the People's Republic of China," in U. Desai (ed.), *Ecological Policy and Politics in Developing Countries*, Albany: State University of New York Press.

Schnaiberg, A. (1994) "The Political Economy of Environmental Problems and Policies: Consciousness, Conflict, and Control Capacity," *Advances in Human Ecology*, vol. 3.

Smil, V. (1993) *China's Environment: An Inquiry into the Limits of National Development*, Armonk, NY: M.E. Sharpe.

Socialist Republic of Vietnam (SRV). (1994) *Law on Environmental Protection*, Hanoi: National Political Publishing House.

Taylor, B.R. (ed.) (1995) *Ecological Resistance Movements: The Global Emergence of Radical and Popular Environmentalism*, Albany: State University of New York Press.

Thrift, M. and Forbes, D. (1986) *The Price of War: Urbanization in Vietnam*, 1954–85, Boston: Allen and Unwin.

United Nations Industrial Development Organization (UNIDO). (1998) *Incentives for Pollution Prevention and Control in Dong Nai*, Report to the project VIE/95/053, 6 August 1998.

Vermeer, E. (1998) "Industrial Pollution in China and Remedial Policies," in R. Edmonds (ed.), *Managing the Chinese Environment*, Oxford: Oxford University Press.

World Bank. (1993) *Viet Nam – Transition to the Market*, Washington, D.C.: World Bank.

—— (1997a) *Clear Water, Blue Skies*, Washington, D.C.: World Bank.

—— (1997b) *Vietnam – Economic Sector Report on Industrial Pollution Prevention*, Agriculture and Environment Operations Division, Washington, D.C.: World Bank.

—— (2000) *Greening Industry – New Roles for Communities, Markets, and Governments*, A World Bank Policy Research Report, Washington, D.C.: World Bank.

Wu, F. (2002) "New Partners or Old Brothers? GONGOs in Transnational Environmental Advocacy in China," *China Environment Series*, Issue 5, Washington, D.C.: Woodrow Wilson International Center for Scholars.

Young, N. (2002) "Three C's: Civil Society, Corporate Social Responsibility, and China," *The China Business Review*, January–February.

Part IV

CONFLICT AND STRUGGLE

10

VIOLENT ENVIRONMENTS

Petroleum conflict and the political ecology of rule in the Niger Delta, Nigeria[1]

Michael Watts

Blood may be thicker than water, but oil is thicker than either.

(Anderson 2001: 30)

Oil, more than any other commodity, illustrates both the importance and the mystification of natural resources in the modern world.

(Coronil 1997: 49)

The annals of oil are an uninterrupted chronicle of naked aggression, genocide and the violent law of the corporate frontier. Iraq was born from this vile trinity. In their own way, the awful spectacle of oil-men parading through the corridors of the White House, the rise of militant Islamism across the Q'uran belt, and the carnage on the road to Baghdad, all bear out the dreadful dialectics of blood and oil. Paul Wolfowitz' confession, to the Asian Security Summit in Singapore in June, 2003 that the Iraq war was driven not by the fiction of weapons of mass destruction but by the "simple fact" that the "country swims on a sea of oil" (Wright 2003) is consistent at least with the last eighty years of US foreign policy (Painter 1986). But there is another oil story in train, bearing all the hallmarks of the long, ugly history of petrolic violence. In 2001, Vice President Dick Cheney predicted that Africa would become the fastest growing source of oil for the American market (as much as 25 percent of US imports by 2015) and it is hardly a surprise to learn there where oil reigns supreme the military is sure to follow. The *Wall Street Journal* reported that the Pentagon, in the most radical deployment of American forces since the end of the Cold War, will move troops from Germany to the Caucasus and West Africa to "protect key oil reserves" (Okonta 2003b). More oil, more blood.

Protection demands, of course, keeping the oil flowing by working hand-in-hand with a phalanx of African dictators and political psychopaths on the one side, and supermajors like ExxonMobil and ChevronTexaco on the other who, citing confidentiality agreements, refuse to disclose the fees, royalties and other services (paramilitaries and security forces among them) made to the phalanxes of well-placed African *nomenklatura*. Such is the scale of the decrepitude and venality – $300 million in Equatorial Guinea, billions in Nigeria over the last two decades – that Tony Blair is now in a trans-Atlantic tug of war to have oil majors disclose their payments for leases and concessions, a proposal fiercely resisted by American Big Oil and the Bush administration. In a separate proposal George Soros calls for mandatory reporting as a prerequisite for listing of oil companies on the world's stock exchanges (Luft 2003). In the last year a raft of new reports inventory the appalling record of oil-based economies in relation to corruption, economic growth and poverty alleviation.[2] Oil, as Anderson says, is thicker than blood or water.

ANOTHER OIL STORY

A year before the events of 11 September 2001, the US Department of State in its annual encyclopaedia of "global terrorism" identified the Niger Delta – the ground zero of oil production in Nigeria – as a breeding ground for increasingly militant "impoverished ethnic groups" for whom terrorist acts (abduction, hostage taking, kidnapping and extra-judicial killings) against foreigners were legion.[3] The CIA concurred (2000), laying emphasis on the catalytic effects of "environmental stresses" in the oil-rich southern Delta on "political tensions." At this time, Nigeria – the thirteenth largest producer of petroleum (which provides 80 percent of government revenues, 95 percent of export receipts, and 90 percent of foreign exchange earnings) – was providing almost 14 percent of US American petroleum consumption.[4] At about the same time, the Petroleum Finance Company (PFC) presented to the US Congressional International Relations Committee Sub-Committee on Africa a report of the strategic and growing security significance of West African oil whose high quality reserves and low cost output – coupled with massive new deepwater discoveries – required, in the view of PFC, serious attention, and substantial foreign investment. In the wake of the Al-Qaeda attacks, and on the larger canvas of the crisis in Venezuela and the Iraq war, West Africa has emerged as the site of "the new Gulf oil states" (Servant 2003). Indeed by January 2002 the Institute for Advanced Strategic and Political Studies was providing a forum for the Bush oil-administration to declare that African oil is "a priority for US national security."[5] In the last year, Africa's black gold – in Gabon, Sao Tome, Angola, Equatorial Guinea – and its ugly footprint are rarely off the front pages. Oil and blood, as Jon Anderson put it, are ubiquitous (Anderson 2000). With the additional *frisson* of terrorism: the "nightmare," as the *New York Times* noted, of "sympathisers of Osama Bin Laden sink[ing] three oil tankers in the Straits of Hormuz."[6]

The mythos of oil and oil-wealth has been central to the history of modern industrial capitalism. But in Nigeria the discovery of oil, and annual oil revenues of $40 billion currently, has ushered in a miserable, undisciplined, decrepit, and corrupt form of "petro-capitalism." After a half century of oil production from which almost $300 billion in oil revenues have flowed directly into the Federal exchequer (and perhaps fifty billion promptly flowed out only to "disappear" overseas), Nigerian per capita income stands at $290 per year. For the majority of Nigerians, living standards are no better now than at independence in 1960. A repugnant culture of excessive venality and profiteering among the political class – the Department of State has an entire website devoted to so-called 419 fraud cases – confers upon Nigeria the dubious honour of sitting atop Transparency International's ranking of most corrupt states. Paradoxically, oil-producing states in the federation – the Niger Delta – have benefited the least from oil-wealth. Devastated by the ecological costs of oil spillage and the highest gas flaring rates in the world, the Niger Delta is a political tinderbox. A generation of militant "restive" youth, deep political frustrations among oil-producing communities, and pre-electoral thuggery all combine to prosper in the rich soil of political marginalization. The massive rigging of elections across the Delta in the April 2003 elections simply confirmed the worst for the millions of Nigerians who have suffered from decades of neglect.

The Middle East scholar Robert Vitalis (2002) has recently suggested that the rapid, complete and irreversible rise of American dominance in Saudi Arabia can shed much light on why "the Niger Delta is currently in crisis." And indeed it is. Since 12 March 2003, mounting communal violence accounting for at least fifty deaths, and the levelling of eight communities in and around the Warri petroleum complex, has prompted all the major oil companies to withdraw staff, to close down operations and reduce output by over 750,000 barrels per day (almost half of national output). President Obasanjo has dispatched large troop deployments to the oil-producing creeks, prompting Ijaw militants, incensed over indiscriminate military action and illegal oil bunkering in which the security forces were implicated, to threaten the detonation of eleven captured oil installations. The strikes on the offshore oil platforms – a long-festering sore that rarely reaches the media – were quickly resolved but nobody seriously expects that the deeper problems within the oil sector will go away. Relatively new to delta politics, however, are a series of assassinations, the most shocking being the killing of Chief Marshall Harry, a senior member of the main opposition party and leading campaigner for greater resource allocation to the oil-producing Niger Delta. Fallout from the Harry assassination has already become a source of tension in his native oil-producing state of Rivers where supporters of the main opposition party, the ANPP and another opposition grouping of activists and politicians, the Rivers Democratic Movement, have linked the ruling party to the assassination. With good reason, the business-as-usual character of the gubernatorial election victories across the oil-producing states, has led some to believe that Nigeria is another Colombia in the making (Cesarz et al. 2003).

The strategic significance of Nigeria is incontestable. One of every five Africans is a Nigerian. Nigeria is the world's seventh largest exporter of petroleum and a key player in African regional security, most recently in Sierra Leone. Nigeria is an archetypal oil nation. Three quarters of government revenues, and almost all export earnings, flow from black gold. A long-time member of OPEC, and the fifth largest supplier of oil exports to the US, Nigeria pumps oil much coveted for its "lightness" and "sweetness," yielding more gasoline and diesel than the "sour" crude from the Middle East. It is also home to a vast Muslim community. Since the oil boom of the 1970s, political power has shifted from the conservative Sufi brotherhoods to well-organized modern Islamist groups like the Yan Izala founded in 1978. Islamic Law (Sharia'a) of a dogmatic and literalist sort, has been adopted and implemented in twelve of the populous northern states, amidst considerable political acrimony and international censure. At least 350 people were killed in four days of terrible rioting in northern Nigeria triggered by protests against US military action in Afghanistan, including particularly bloody clashes between Muslims and Christians in Kano, Kaduna and Jos.

Olesegun Obasanjo's presidential victory in 1999, in the wake of the darkest period of military rule in Nigeria's forty-year post-Independence history, held much promise. An internationally recognized statesman and diplomat imprisoned during the brutal Abacha years, he inherited the mantle of a massively corrupt state apparatus, an economy in shambles, and a federation crippled by longstanding ethnic enmity. Committed to reforming a corrupt and undisciplined military – the largest in Africa – and to deepening the process of democratization, Obasanjo was confronted within months of his inauguration by militant ethnic groups speaking the language of self-determination, local autonomy and resource control (meaning a greater share of the federally allocated oil revenues). In an incident widely condemned by the human rights community, some 2000 persons were slaughtered in Odi, Bayelsa State after federal troops were dispatched in response to clashes between local militants and the police. Obasanjo has consistently refused to apologize for the murders and there has been no full inquiry. Last year the military was involved in yet another massacre, this time in the Middle Belt, in Benue and Taraba States, the most serious communal conflict since the clashes preceding the outbreak of the Biafran civil war in 1967. On President Obasanjo's watch, over 10,000 have perished in ethnic violence. He has failed miserably to address the human rights violations by the notoriously corrupt Nigerian security forces. Early in his tenure, Obasanjo did retire a number of senior officers and embark upon an anti-corruption campaign, aided and abetted by the US's role in "retraining" the Nigerian armed forces. President Clinton committed foreign assistance to "reprofessionalize" the Nigerian army in 1999, including the equipping and training of seven battalions at a cost of over $1 billion. During the Bush imperium, the presence of 200 special forces in Nigeria, including on-site training grounds in some of the most sensitive areas of the Muslim north, has generated enormous suspicion and now vocal opposition. Not unexpectedly, a number of powerful Nigerian constituencies see a beleaguered and corrupt Obasanjo regime as simply another miserable US oil colony.

The *zeitgeist* of oil – its mythic and spectacular qualities in relation to the modern – is an essential expression of contemporary hydro-carbon capitalism. Oil's fetishistic appeal was not lost on the great Polish journalist Ryszard Kapucinksi who, in his reflections on oil-rich pre-revolutionary Iran, sardonically observed that,

> Oil creates the illusion of a completely changed life, life without work, life for free.... The concept of oil expresses perfectly the eternal human dream of wealth achieved through lucky accident.... In this sense oil is a fairy tale and like every fairy tale a bit of a lie.
>
> (1982: 35)

It is this lie, one might say, that currently confronts West African oil producers, and Nigeria in particular.

A RESOURCE CURSE?

In virtue of the geo-strategic significance of oil to contemporary capitalism – and to US hegemony in particular – relations between natural resources (and oil in particular) and economic growth, democracy, and civil war have emerged as an object of substantial scholarly attention (operating under the sign of "resource politics"), not least by economists and political scientists.[7] The IMF and its stenographers have posited a strong association between resource-dependency, corruption and economic performance. Sachs and Warner (1995) argue that one standard deviation increase in the ratio of natural resource exports to GNP is associated with a decrease of just over 1 percent in the growth rate (irrespective of the endogeneity of corruption, commodity price variability and trade liberalization). Leite and Weidemann (1999) believe that for fuels the figure is 0.6 percent and due "entirely to the indirect effect of corruption" (1999: 29). For Michael Klare (2001), writing from a very different vantage point, oil is a dwindling, key-strategic resource that will necessarily be generative of inter-state conflict (see also Homer-Dixon 1999).[8] This is a line of argumentation developed by Paul Collier, who, in his work with the World Bank, uses resource dependency as a way of thinking about rebellion (especially in Africa), with oil posited as central to the *economics of civil war*. It encourages extortion and looting through resource predation (at least up to the point where 26 percent of GDP is dependent on resource extraction). And the feasibility of predation by states or rebel groups determines the risk of conflict. For Collier, the risks are greater because of resource dependency than ethnic or religious diversity. Oil is a "resource curse." Ross elaborates on this claim seeing oil in terms of its *rentier effect* (low taxes and high patronage dampen pressures for democracy); its *repression effect* conferred by the direct state control over sufficient revenues to bankroll excessive military expenditures and expanded internal security apparatuses; and a *modernization effect*, namely the "move into industrial and service sector jobs render them less likely to push for democracy" (2001: 357).[9] This comes terribly close to a sort of commodity determinism confirming perhaps Coronil's point that oil, more than any other commodity, "illustrates both the importance and the mystification of natural resources in the modern world" (Coronil 1997: 49). But if oil hinders democracy (as though copper might liberate parliamentary democracy?), one needs to appreciate the centralizing effect of oil and the state in relation to the oil-based nation-building enterprises that are unleashed in the context of a politics that pre-dates oil.

Much of the resource curse analysis runs the risk of imputing enormous powers to oil (without grasping its specificity), conflating petroleum's purported Olympian powers with pre-existing political dynamics, and, as in the case of Collier, misidentifying a predation-proneness for what is in fact the dynamics of state and corporate enclave politics (Leonard and Strauss 2003). What is striking in so much of what passes as "resource politics" is the total invisibility of both transnational oil companies (which typically work in joint ventures with the state) and the specific forms of rule associated with petro-capitalism. My analysis charts the relations between oil and violence, but does so through examining how forms of governable (or non-governable spaces) are created through the analytics of "authoritarian governmentality" (Dean 1999) growing out of the soil of petro-capitalism. Rather than see oil-dependency as a source of predation or as a source of state military power, I explore how oil capitalism produces particular sorts of enclave economies and particular sorts of governable spaces characterized by violence and instability. To do so, the qualities of oil in relation to predation matter (oil and diamonds are after all very different sources of predation: see Le Billon 2001). But so do the powers of transnational oil companies, the character of what I call "the oil complex," and the ways in which oil as a territorially based and nationalized commodity can become the basis for making claims. Unlike the work of Collier and others, I seek to trace the varieties of violence engendered by oil, to elaborate the ways in which resources, territoriality and identity can constitute forms of rule (or unrule), and understand the genesis of economies of violence that emerge from differing sorts of governable (or ungovernable) spaces.

BLACK GOLD AND THE NIGER DELTA

The Niger delta is a vast sedimentary basin constructed through successive layers of sediments dating back 40–50 million years to the Eocene epoch. A classic arcuate delta covering almost 70,000 square kilometers, the Niger delta is also endowed with very substantial hydrocarbon deposits. Crude oil production currently runs at almost 2 million barrels per day, roughly 90 percent by value of Nigerian export revenues. Nigeria is not only the largest producer of petroleum in Africa and is among the world's top ten oil producers but, in the wake of

11th September and the current Middle East crisis, it is being pursued by the Bush Administration as a major supplier for the US market.

The contemporary geo-strategic significance of the Niger delta has emerged from an astonishing ethnic and linguistic complexity, and from a recent history of economic and political irrelevance. There are five major linguistic categories (Ijoid, Yoruboid, Edoid, Igboid and Delta Cross), but each embraces a profusion of ethno-linguistic heterogeneity. The establishment of the Nigerian colony and the imposition of Indirect Rule in the early 1900s marked an end to the brief period of commercial vitality associated with the commercialization of palm oil across the region in the nineteenth century. For most of the first half of the twentieth century, the delta was an economic and political backwater. In the gradual transition to Independence in the 1950s, the so-called ethnic minorities voiced their concerns to the departing British administrators that their interests in a Nigerian federation dominated by three ethnic majorities (the Hausa, the Yoruba and the Ibo) were to all intents and purposes, invisible. What was true at the moment of imperial departure only became more so as the postcolonial period got under way.

The onset of commercial petroleum production in the heart of the delta in 1956 – discovered in Oloibiri in current Bayelsa State – seemed to hold out the promise of rapid development for the ethnic minorities. But instead, the presence of transnational oil companies in joint-ventures with the Nigerian State (through the Nigerian National Petroleum Company [NNPC]) produced enormous environmental despoliation and a crisis in extant forms of livelihood. By the 1970s and 1980s, a number of ethnic communities had begun to mobilize against the so-called "slick alliance" of oil companies and the Nigerian military. Most famously, the Movement for the Survival of the Ogoni People (MOSOP) led by Ken Saro-Wiwa challenged Shell for its environmental despoliation and human rights violations and the Nigerian state for its unjust control of "their oil." Saro-Wiwa and the MOSOP leadership were executed by the Nigerian military in 1995, but since that time the Niger delta has become a zone of intense conflict as more oil-producing minorities (for instance, the Adoni, the Itsekiri, the Ijaw) clamor for compensation and for the recognition of their claims for resource control. As I write, substantial coverage in the world press has been devoted to a group of Delta women who have occupied Chevron oil refineries, demanding company investments and jobs for indigenes.[10]

I want to make three important points about oil in Nigeria. The first is that oil capitalism operates through what I call an "oil complex" (comparable in, say, Venezuela or Gabon or Indonesia) involving a statutory monopoly over mineral exploitation (the 1969 Petroleum Law in Nigeria, reinforced by a number of key laws including the 1978 Land Use Decree), a nationalized oil company (NNPC) that operates through joint ventures with oil majors who are granted territorial concessions (blocs), the security apparatuses of the state (and the companies) to ensure that costly investments are secured, the oil producing communities, and apolitical mechanism (in Nigeria called the "derivation principle") by which federal oil revenues are distributed to the states (Anugwom 2001, Ibeanu 2003). This oil complex – rather than a laundry list of oil-attributes *à la* Ross, is key to understanding the relations between oil and violence. The second point is that this oil complex matters profoundly to the character and dynamics of Nigerian development. Oil is, of course, a biophysical entity (a subterranean fluid capable of being pumped and transmitted); it is also a commodity that enters the market with its price tag, and as such is the bearer of particular relations of production. And, not least, oil harbors fetishistic qualities: it is the bearer of meanings, hopes, expectations of unimaginable powers. Not unexpectedly, oil is a constant in the popular Nigerian imagination (see Watts 2000), resplendent with all manner of brilliant and unctuous qualities. The third point is that Nigerian petro-capitalism contained a sort of double-movement, a contradictory unity of capitalism and modernity captured in the fact that oil production in Nigeria has always been a joint venture, currently with fourteen transnational companies, in which joint operating agreements determine the distribution of royalties and rents. On the one hand, oil has been a centralizing force that has rendered the state more visible and globalized, underwriting a process of secular nationalism and state building. On the other, a corrupt and undisciplined oil-led development, driven by an unremitting political logic of ethnic claims-making, has fragmented and discredited the state and its forms of governance. It has produced a set of conditions that have compromised and undermined the very tenets of the modern nation state. Coronil (1997) refers to this conundrum as "the Faustian trade of money for modernity," which in Venezuela brought "the illusion of development." In Nigeria, too, the double movement brought illusion and produced forms of governable spaces that questioned the very idea of Nigeria, spaces that generated forms of rule, conduct and imagining at cross purposes with one another, antithetical to the very idea of a coherent modern nation-state that oil, in the mythos of the West, represented.

ECONOMIES OF VIOLENCE AND GOVERNABLE SPACES

Government, for Michel Foucault (2000), referred famously to the "conduct of conduct," a more or less calculated and rational set of ways of shaping conduct and securing rule through a multiplicity of authorities and agencies in and outside the state and at a variety of spatial levels. In contrast to forms of pastoral power in the Middle Ages from which a sense of sovereignty was derived, Foucault charted an important historical shift, beginning in the sixteenth century, toward government as a right manner of disposing of things "so as to not lead to the common good, but to an end that is convenient for each of the things governed" (2000: 211). The new practices of the state, as Mitchell Dean (1999: 16) says, shape human conduct by "working through our desire, aspirations, interests and beliefs for definite but shifting ends." It was Foucault's task to reveal the genealogy of government, and the origins and modern power, the fabrication of a modern identity. The conduct of conduct – governmentality – could be expressed as pastoral, disciplinary or as bio-power. Modern governmentality was rendered distinctive by the specific forms in which the population and the economy were administered, and specifically by a deepening of the "governmentalization of the state" (how sovereignty comes to be articulated through the populations and the processes that constitute them). What was key for Foucault was not the displacement of one form of power by another, nor the historical substitution of feudal by modern governmentality, but the complex triangulation involved in sustaining many forms of power put to the purpose of security and regulation.[11]

On this theoretical canvas, I seek to explore the relations between two interrelated aspects of governmentality.[12] One is what Foucault explicitly refers to as relations between men and resources (in my case, people and oil in the Niger delta) as an expression of his complex notion of the governance of things. As he put it:

> On the contrary, in [the modern exercise of power], you will notice that the definition of government in no way refers to territory: one governs *things*. But what does this mean? I think this is not a matter of opposing things to men, but rather of showing that what government has to do with is not territory but, rather, a sort of complex composed on men and things. The things, in this sense, with which government is to be concerned are in fact men, but *men in their relations, their links, their imbrication with those things that are wealth, resources, means of subsistence, the territory with its specific qualities, climate, irrigation, fertility, and so on; men in their relation to those other things that are customs, habits, ways of acting and thinking and so on; and finally men in relation to those still other things that might be accidents and misfortunes such as famines, epidemics, death and* so on. What counts is essentially this complex of men and things; property and territory are merely one of its variables.
>
> (2000: 208–9, emphasis added)

The other aspect of governmentality that I use is taken from Rose's notion of "governable spaces" as they emerge from the analytics of government detailed above. For Rose, governable spaces and the spatialization of government, are "modalities in which a real and material governable world is composed, terraformed, and populated" (Rose 1999: 32). The scales upon which government is "territorialized" – territory is derived from *terra*, land, but also *terrere*, to frighten – are myriad: the factory, the neighborhood, the commune, the region, the nation. Each of these governable spaces has its own topology and is modeled, as Rose puts it – through systems of cognition and remodeled through government practice – in a way that frames how such topoi have emerged: the social thought and practice that has territorialized itself upon the nation, the city, the village or the factory. The map has been central to this process as a mode of objectification, marking and inscribing but also as "a little machine for producing conviction in others" (1999: 37). But in general it was geography that formed "the art whose science was political economy."[13] Modern space and modern governable spaces were produced by the biological (the laws of population which determine the qualities of the inhabitants) and the economic (the systems of the production of wealth). Governable spaces necessitate the territorializing of governmental thought and practice but are simultaneously produced as differing scales by the cold laws of political economy.

The Nigerian oil complex can be grasped in territorial terms, taking as my cue Nikolas Rose's point about enclosures:

> Governmental thought territorializes itself in different ways ... We can analyse the ways in which the idea of a territorially bounded, politically governed nation state under sovereign authority took shape..... One can trace anomalous governmental histories of smaller-scale territories ... and one can also think of these [as] spaces of enclosure that governmental thought has imagined and penetrated ... how [does it] happen that social thought territorializes itself on the problem of [for example] the slum in the nineteenth century?
>
> (Rose 1999: 34–6)

I want to think about the genesis of differing sorts of governable spaces in Nigeria as part of a larger landscape of what Dean calls "authoritarian governmentality," that is to say an articulation of generalized uses of the instruments of repression with bio-politics. As he says "it regards its subjects' capacity for action as subordinate to the expectation of obedience" (1999: 209). I want to root these spaces and forms of power in the logic of petro-capitalist development, that is to say a particular sort of extractive development which is generative of differing sorts of scale, or the "politics of scale" as Neil Smith (1992) calls it. My analysis conversely charts the relations between oil and violence, but does so through examining how forms of governable (or un-governable) spaces are generated by the baneful twins of authoritarian government and petro-capitalism. To do so, I will turn briefly to governable spaces, and to three in particular that I shall refer to as the space of chieftainship, the space of indigeneity, and the space of the nation-state.

The space of chieftainship

Nembe community[14] in Bayelsa State stands at the originary point of Nigerian oil production. In the 1950s, the Tennessee Oil Company (a US Company) began oil explorations there but oil was not found until much later when Shell D'Arcy unearthed the Oloibiri oil field in Ogbia. Subsequent explorations led to the opening of the large and rich Nembe oil fields near the coast in Okpoama and Twon-Brass axis. Currently, the four Nembe oil fields produce approximately

150,000 barrels of high quality petroleum through joint operating agreements between the Nigerian National Petroleum Company (NNPC), AGIP and Shell. If Nembe is the ground zero of oil production, it is also a theatre of extraordinary violence and intra-community conflict, the result of intense competition over political turf and the control of benefits from the oil industry. The violence can be traced back to the late 1980s when the Nembe Council of Chiefs acquired power from then King, Justice Alagoa Mingi IX, to negotiate royalties and other benefits with the oil companies. The combination of youth-driven violence and intense political competition has transformed Nembe's system of governance and set the stage for further challenges to the traditional authority of chieftainship (see HRW 2002; Kemedi 2000).[15]

Oil became commercially viable in the 1970s, but to grasp its transformative effects on Nembe politics and community – that is to its genesis as a distinctive governable space – requires an understanding of chieftainship in the Delta. Indirect rule in the colonial period certainly left much of the Niger Delta marginalized and isolated, but it also, in the name of tradition, built upon and frequently invented chiefly powers of local rule which in the Nembe case were grafted onto a deep and complex structure of kingship and gerontocratic rule. To understand the dynamics of Nembe as a governable space recall that land lay in hands of customary authorities (notwithstanding the fact that the 1969 Petroleum Law granted the state the power to nationalize all oil resources). Land rights and therefore claims on oil royalties were, from the outset, rooted in the *amayanabo* (king), and derivatively the subordinate powers, namely the Council of Chiefs and the Executive Council. Historically, the Nembe community possessed a rigid political hierarchy consisting of the *amayanabo* presiding over, in descending order, the Chiefs (or heads of the war canoe houses[16]) elected by the entire war canoe houses constituted by their prominent sons. Although the Chiefs were subservient to the *amayanabo*, they acted as his closest advisors, supporting him in the event of military threat and, in turn, were responsible for electing the *amayanabo* from the Mingi group of Houses, or the royal line. The current Nembe Council of Chiefs is the assemblage of the recognized Chiefs of Nembe "chalked" by the King.[17]

Accordingly, in 1991, the Nembe monarch's ineffectiveness in dealing with the oil companies led to a radical decentralization of his powers to the Council of Chiefs, headed by Chief Egi Adukpo Ikata. Insofar as the Council now dealt directly with Shell, and handled large quantities of money paid by the oil companies, competition for election to the Council intensified as various political factions struggled for office. By 2000, the Council had expanded from twenty-six to ninety persons. Coeval with the evisceration of kingly powers, the deepening of the Council mandate, and the expansion of the Council members, was a subtle process of "youth mobilization." In an age-graded society like the Nembe Ijaw, youth refers to persons typically between their teens and early forties who, whatever achievements they may have obtained (university degrees, fatherhood and so on), remain subservient to their elders. Central to any understanding of the emergence of a militant youth in Nembe town was the catalytic role played by a former company engineer with Elf Oil Company named Mr. Nimi B.P. Barigha-Amage. He deployed his knowledge of the oil industry to organize the youths of the Nembe community into a force capable of extracting concessions from the oil companies, in essence, by converting cultural organization into protection services. Chief Ikata was quick to exploit the awareness and restiveness of the youths to pressure Shell into granting community entitlements. A pact between Chief Ikata and the young engineer was in effect instituted; the engineer supplied the youths with information regarding community entitlements, and the Chief deployed his knowledge of military logistics to organize the shutting down of flow stations, the seizure of equipment and sabotage (Alagoa 2001; HRW 2002).

Armed with insider knowledge of the companies and an understanding of a loosely defined set of rules regarding company compensation for infringements on community property, Barigha-Amage pushed for the creation of youth "cultural groups" which gradually, with the support of some members of the Council of Chiefs, intermediated with oil companies and their liaison officers, and manipulated the system of compensation in the context of considerable juridical and legal ambiguity. Liaison officers, colluding with community representatives, invented ritual or cultural sites that had ostensibly been compromised or damaged by oil operations, for which monies exchanged hands. As the opportunities for appropriating company resources in the name of compensation became visible through the success of the cultural groups, other sections of the youth community began to organize in turn around clan and familial affiliations. In 1994, for example, a group called "House of Lords" (*Isongoforo*) was created by a former university lecturer, Lionel Jonathan, and a year later in 1995, Mrs Ituro-Garuba, wife of a well-placed military officer, established *Agbara-foro.* Inevitably, with much at stake financially, and control of the space between community and company in the balance, conflicts within and among youth groups proliferated and deepened. In turn, growing community militancy spilled over into often-violent altercations with the much-detested mobile police ("Mopos") and local government authorities. The regional state and Governor attempted to intervene as conditions deteriorated but a government report, on which such action was predicated, was never released for political reasons. A subsequent banning of youth groups had, as a result, no practical effect (HRW 2002).

Slowly, the subversion of royal authority, the strategic alliances between youth and Chiefs, and the growing (and armed) conflict between youth groups for access to Shell resulted in the ascendancy of a highly militant *Isongoforo*. In an environment of rampant insecurity and lawlessness, occupation and closure of flow stations, and tensions between the companies, the service companies and local security forces, *Isongoforo* was provided "stand by" payments by the companies (that is to say, it was hired for protection purposes), even as it colluded with the community liaison officers to invent compensations cases. *Isongoforo* occupied the center of a new governable space which it ruled through force rather than any sense of consent or customary authority. This quasi-mafia was funded by the large quantities of monies that it commanded from the companies, and by the arms which it controlled. This volatile state of affairs collapsed dramatically as local resentments and struggles proliferated. In February 2000, a "People's Revolution" overthrew *Isongoforo*, ostensibly precipitated by the humiliation of the Council of Chiefs at the hands of Shell (backed by the intimidating *Isongoforo* forces). The Chiefs now orchestrated the occupation of flow stations and undermined the powers of *Isongoforo* by recruiting and supporting other youth groups. By May 2000 *Isongoforo* had been sent into exile but it was promptly replaced in the wake of Barigha-Amage's return as High Chief of Nembe, by his own "cultural group" *Isenasawo/Teme*. *Teme* instituted a rule of terror and chaos far worse than its predecessors. It too proved unstable in the context of excessive youth mobilization and split into two factions, which subsequently produced two "counter coups" and much bloodshed. A government Peace Commission was established in January 2001 in a desperate effort to bring peace to one of the jewels in the oil-producing crown (Alagoa 2001; NDWC 2000).

Much of this later violence (after 1996) could not be regulated by the state authorities because of its concurrence with the 1999 elections in which some of the key youth leaders were expected to deliver votes for the incumbent gubernatorial race. In the creation of what, in effect, was a sort of vigilante rule, there were complex complicities between Chiefs, youth groups, local security forces, and the companies. Plans to occupy oil flow stations (for purposes of extortion) were often known in advance and involved collaboration with local company engineers; youths were *de facto* company employees providing protection services, and local compensation and community officers of Shell and AGIP produced fraudulent compensation cases and entitlements. Nembe, a town with its own long and illustrious history and politics, had become a sort of company-town in which authority had shifted from the king to warring factions of youth who were in varying ways in the pay of, and working in conjunction with, the companies. The Council of Chiefs stood in a contradictory position, seeking to maintain control over revenues from the companies and yet was intimidated and undermined by the militant youth groups on whom it depended. In the context of a weak and corrupt state, the genesis of this power nexus bears striking resemblances to the genesis of the Mafia in nineteenth-century Sicily (Blok 1974).

What I have described is the displacement of a specific form of power (chieftainship) by a governable space of civic vigilantism, a thickening of civil society that does not necessarily imply the basis of the kind of governance put forth by Granovetter, Putnam and others (see Putnam 2000) – that is, the self-organizing networks that arise out of the interactions between a variety of organizations and agencies. Civic powers have expanded by overthrowing a territorial system and a gerontocratic royal order. Youth mobilization – whose political affiliations and ambitions in any case were complex because they reflected an unstable amalgam of clan, family and local electoral loyalties – had thrown up an identity and subject that was indisputably revolutionary, representing an unholy alliance between civic organizations (presenting themselves as cultural organizations) and private companies. Rule in Nembe is a realm of privatized violence, a form of consent by a form of force. Government here turns on what Foucault (2000: 208–9) calls "men in their imbrication with wealth and resources" – the government of men and things, as opposed to territory. It is institutionalized through forms of calculability, *techne*, and visibility that emerge from the legal and company dispositions to regulate local populations backed by the forces of repression. The governable subject is *de facto* a sort of employee, and rule is a Gramscian "war of position." Culture serves as the form by which company rule is experienced – violent youth groups – but in a way that renders the space increasing *ungovernable*.

The space of indigeneity

The Niger delta is a region of considerable, even bewildering, ethno-linguistic complexity. The eastern region of which the delta is part is dominated statistically by the Ibo majority, but there is a long history of excluded ethnic minorities in the delta dating back at least to the 1950s when the Willinck Commission took note of the inter-ethnic complexity of the region. Throughout the colonial period prior to the arrival of commercial oil production, there had been efforts by various minorities, who saw themselves as dominated by the Ibo, to established Native Authorities of their own. In the 1960s, prior to the outbreak of civil war, two charismatic local figures, both Ijaw – Nottingham Dick and Isaac Boro – declared a Delta Republic, a desperate cry for some sort of political inclusion that lasted a mere twelve days. The ill-fated Delta People's Republic of 1966 was the forerunner of what is now a prairie fire of ethnic mobilization by the historically excluded minorities – now tagged as "indigenous" in order to capture the political and legal legitimacy conferred by the International Labour Organization of the United Nations (ILO 169) (see Brysk 2000; Nelson 1999). The paradigmatic case in the delta is the struggle by Ken Saro-Wiwa and the Movement for the Survival of the Ogoni people (MOSOP). Their case reveals a rather different sort of governable space, one marked by ethnic subjects and indigenous territory.

The Ogoni are typically seen as a distinct ethnic group, consisting of three sub-groups and six clans dotted over 404 sq miles of creeks, waterways, and tropical forest in the northeast fringes of the Niger delta. Located administratively in Rivers State, a Louisiana-like territory of some 50,000 sq. kilometers, Ogoniland is one of the most heavily populated zones in all of Africa. The most densely settled areas of Ogoniland – over 1500 persons per sq. km. – are also the sites of the largest oil wells. Ogoniland's customary productive base was provided by fishing and agricultural pursuits until the discovery of petroleum, including the huge Bomu field, immediately prior to Independence. Part of an enormously complex regional ethnic mosaic, the Ogoni were drawn into internecine conflicts within the delta region, largely as a consequence of the slave trade and its aftermath, in the period prior to arrival of colonial forces at Kono in 1901. The

Ogoni resisted the British until 1908 (Naanen 1995) but thereafter were left to stagnate as part of the Opopo Division within Calabar Province. As Ogoniland was gradually incorporated during the 1930s, the clamor for a separate political division grew at the hands of the first pan-Ogoni organization, the Ogoni Central Union, which bore fruit with the establishment of the Ogoni Native Authority in 1947. In 1951, however, the authority was forcibly integrated into the Eastern Region. Experiencing tremendous neglect and discrimination, integration raised longstanding fears among the Ogoni of Ibo domination.[18] Politically marginalized and economically neglected, the delta minorities feared the growing secessionist rhetoric of the Ibo and consequently led an ill-fated secession of their own in February 1966. Ogoni antipathy to what they saw as a sort of internal colonialism at the hands of the Ibo led to their support of the federal forces during the civil war. While a Rivers State was established in 1967 – which compensated in some measure for enormous Ogoni losses during the war – the new state recapitulated in microcosm the larger "national question." The new Rivers State was multi-ethnic but presided over by the locally dominant Ijaw, for whom the minorities felt little but contempt.[19]

During the first oil boom of the 1970s, Ogoniland's fifty-six wells accounted for almost 15 percent of Nigerian oil production[20] and in the past three decades an estimated $30 billion in petroleum revenues have flowed from this Lilliputian territory. It was, as local opinion had it, "Nigeria's Kuwait." Yet according to a government commission, Oloibiri, where the first oil was pumped in 1958, has no single kilometer of all-season road and remains "one of the most backward areas in the country" (cited in Furro 1992: 282; see also Okonta and Douglas 2001). Rivers State saw its federal allocation fall dramatically in absolute and relative terms. At the height of the oil boom, 60 percent of oil production came from Rivers State but it received only 5 percent of the statutory allocation (roughly half of that received by Kano, Northeastern States and the Ibo heartland, East Central State). Between 1970 and 1980 it received in revenues one fiftieth of the value of the oil it produced. Few Ogoni households have electricity, there is one doctor per 100,000 people, child mortality rates are the highest in the nation, unemployment is 85 percent, 80 percent of the population is illiterate and close to half of Ogoni youth have left the region in search of work. Life expectancy is barely 50 years, substantially below the national average. If Ogoniland failed to see the material benefits from oil, what it *did* experience was an ecological disaster – what the European Parliament has called "an environmental nightmare." The heart of the ecological harms stems from oil spills – either from the pipelines which criss-cross Ogoniland (often passing directly through villages) or from blowouts at the wellheads – and gas flaring. As regards to the latter, a staggering 76 percent of natural gas in the oil producing areas is flared (compared to 0.6 percent in the US). As a visiting environmentalist noted in 1993 in the delta, "some children have never known a dark night even though they have no electricity."[21] Burning twenty-four hours per day at temperatures of 13–14,000 degrees Celsius, Nigerian natural gas produces 35 million tons of CO_2 and

12 million tons of methane, more than the rest of the world (making Nigeria probably the biggest single cause of global warming). The oil spillage record is even worse. There are roughly 300 spills per year in the delta and in the 1970s alone the spillage was four times that than the much-publicized Exxon Valdez spill in Alaska. In one year alone, almost 700,000 barrels were soiled according to a government commission. Ogoniland itself suffered 111 spills between 1985 and 1994 (Hammer 1996). Figures provided by the NNPC document 2676 spills between 1976 and 1990, 59 percent of which occurred in Rivers State (Ikein 1990: 171), 38 percent of which were due to equipment malfunction.[22] Between 1982 and 1992 Shell alone accounted for 1.6 million gallons of spilled oil, 37 percent of the company's spills worldwide. The consequences of flaring, spillage and waste for Ogoni fisheries and farming have been devastating. Two independent studies completed in 1997 reveal total petroleum hydrocarbons in Ogoni streams at 360 and 680 times the European Community permissible levels (HRW 1999, Rainforest Action Network 1997).

The hanging of Ken Saro-Wiwa and the "Ogoni Nine" in November 1995 – accused of murdering four prominent Ogoni leaders – and the subsequent arrest of nineteen others on treason charges, represented the summit of a process of mass mobilization and radical militancy which had commenced in 1989. MOSOP necessarily built upon previous cultural and political organizations like the Ogoni Klub and Kagote (both elite organizations) and most especially the Ogoni politician Naaku Paul Birabi who established in 1950 the Ogoni State Representatives Association to promote Ogoni interests in the new eastern Region Government. The civil war hardened the sense of external dominance among Ogonis. A "supreme cultural organization" called Kagote, which consisted largely of traditional rulers and high-ranking functionaries, was established at the war's end and in turn gave birth in 1990 to MOSOP. A new strategic phase began in 1989 with a program of mass action and passive resistance on the one hand and a renewed effort to focus on the environmental consequences of oil (and Shell's role in particular) and on group rights within the federal structure. Animating the entire struggle was, in Leton's words, the "genocide being committed in the dying years of the twentieth century by multinational companies under the supervision of the Government" (cited in Naanen 1995: 66). A watershed moment in MOSOP's history was the drafting in 1990 of an Ogoni Bill of Rights (Saro-Wiwa 1992). Documenting a history of neglect and local misery, the Ogoni Bill took head on the question of Nigerian federalism and minority rights. Calling for participation in the affairs of the republic as "a distinct and separate entity," the Bill outlined a plan for autonomy and self determination in which there would be guaranteed "political control of Ogoni affairs by Ogoni people ... the right to control and use a fair proportion of Ogoni economic resources ... [and] adequate representation as of right in all Nigerian national institutions" (Saro-Wiwa 1989: 11). In short, the Bill of Rights addressed the question of the *unit* to which revenues should be allocated, and derivatively the rights of minorities (HRW 1999; Okonta 2002). At the heart of Saro-Wiwa's vision was an Ogoni state.

In spite of the remarkable history of MOSOP between 1990 and 1996, its ability to represent itself as a unified pan-Ogoni organization remained an open question. There is no pan-Ogoni myth of origin (characteristic of some delta minorities), and a number of the Ogoni subgroups (clans) engender stronger local loyalties than any affiliation to Ogoni "nationalism." Gokana clan for example was the most populous and well-educated and its elites wielded disproportionate influence in Ogoni. Conversely, the Eleme clan-head did not even sign the Ogoni Bill of Rights and Eleme's leading historian has argued that they are not in fact Ogoni (Ngofa 1988). In 1994, Eleme leaders proposed the creation of Nchia state which comprised non-Ogonis from Bonny, Andoni, Opobo and Etche, thereby turning their backs on Saro-Wiwa's goal (Okonta 2003a). Furthermore, the MOSOP leaders were actively opposed by elements of the traditional clan leadership, by prominent leaders and civil servants in state government, and by some critics who felt Saro-Wiwa was out to gain "cheap popularity" (Osaghae 1995: 334). Some Ogoni notables (Edward Kobani, Dr. Leton) aspired to participate in conventional politics by running for the two major parties rather than assisting in the birth of a nation. MOSOP, moreover, was a political movement of the elite led by the elite; it was not a mass movement and youth and women were not represented on its first steering Committee. Gradually the youth wing of MOSOP, which Saro-Wiwa had used, emerged as militants but the leadership was often incapable of controlling it. MOSOP, in short, was wracked by tensions. There were as Okonta says, "cracks in the pot" (2003a: 12).

Saro-Wiwa built upon over fifty years of Ogoni organizing and upon three decades of resentment against the oil companies, to provide a mass base and a youth-driven radicalism – and an international visibility – capable of challenging state power. Yet at its core the indigenous subject – and the indigenous space – was contentious and problematic. Ike Okonta (2002) has brilliantly shown, how in the Ogoni case, indigeneity unraveled into fragments of class, clan, generation and gender. These tensions came to the fore after Saro-Wiwa's death and MOSOP declined as rapidly as it had ascended.

What sort of articulation of indigenous identity and political subjectivity did Saro-Wiwa pose? What sort of governable space did Ogoniland represent? It was clearly one in which territory and oil were the building blocks upon which ethnic difference and indigenous rights were constructed. And yet it was an unstable and contradictory sort of articulation. First, there was no simple sense of "Ogoniness," no unproblematic unity, and no singular form of political subject (despite Saro-Wiwa's claim that 98 percent of Ogonis supported him). MOSOP itself had at least five somewhat independent internal strands embracing youth, women, traditional rulers, teachers and Churches. It represented a fractious and increasing divided "we," as the splits and conflicts between Saro-Wiwa and other elite Ogoni confirms (Ogoni Crisis 1996).[23] Second, Saro-Wiwa constantly invoked Ogoni culture and tradition yet he also argued that war and internecine conflict had virtually destroyed the fabric of Ogoni society by 1900 (1992: 14). His own utopia then rested on the re-creation of Ogoni culture and suffered like all *ur*-histories from a quasi-mythic invocation of the past. Third, ethnicity was the central problem of post-colonial Nigeria – the corruption of ethnic majorities – and for Saro-Wiwa its panacea (the multiplication of ethnic minority power). To invoke the history of exclusion and the need for ethnic minority inclusion as the basis for federalism, led Saro-Wiwa to ignore the histories and geographies of conflict and struggle among and between ethnic minorities. Saro-Wiwa's brilliance, then, was to make MOSOP a green, indigenous movement (with international backing and visibility) and to take the movement to the poor and the young to secure a powerful identity, in the face of elite opposition (and his own marginal position in the 1980s). Saro-Wiwa's crowning glory was Ogoni national day on 4 January 1993 when he presided over the birth of the Ogoni flag, the Ogoni anthem and the National Youth Council of the Ogoni People.

Paradoxically MOSOP surfaced as a foundational indigenous movement even though Ogoni's significance as an oil-producing region was diminishing. By the late 1990s moreover as a movement it had fallen apart and inter-group struggles deprived it of much of its previous momentum and visibility. But it gave birth to what one might call indigenous movements among oil producing communities. The same forces have spawned a raft of self-determination indigenous movements among Ijaw (INC, IYV), Isoko (IDU), Urhobo (UPU), Itsekiri (INP), Ogbia (MORETO) among others (Obi 2001). MOSOP itself fell apart precisely as these other movements gained power. Since the return to civilian rule in 1999, there has been a rash of such minority movements across the Delta calling for "resource control," autonomy and a national sovereign conference to rewrite the Nigerian constitution. At the same time the Delta has become ever more engulfed in civil strife: militant occupations of oil flow stations, pipeline sabotage, intra-urban ethnic violence, and the near-anarchy of state security operating in tandem with company security forces. The shock troops of many of these indigenous movements are youth and women, and the multiplication of ethnic youth movements is one of the most important political developments in contemporary Nigeria. And it is here that the politics of oil-producing communities meet up with the politics of oil-producing indigenous groups.

What does the Ogoni case reveal, then, as a governable space? Particular "populations" have been constructed as indigenous. As I explain below, this construction emerged from the nationalist struggle as customary rights were added to a discourse of citizenship. But the process received enormous energy as indigeneity as a political category garnered international support in the last part of the twentieth century, a resource that Saro-Wiwa deployed brilliantly (Bob 2002). The emergence of a national debate in Nigeria over resource control in the late 1990s is precisely a product of indigenous claims-making on the state, a process by which ethnic identifications must be discursively and politically produced. The Ogoni case shows that there is no pre-given ethnic identity, but complex and unstable genealogies of identification that have emerged in the last century (see Li 1996). The indigene has to be made – interpollated – around a strong sense of territory and tradition but in the context of cultural, economic and political heterogeneity. This was achieved through an imbrication of things and people – oil and ethnicity – and it has been generative of a profusion of indigenous movements. Indigeneity has in this sense unleashed the huge political energies of ethnic minorities who recapitulate in some respects the postcolonial history of spoils politics in Nigeria. The effect of this multi-ethnic mobilization was the production of political and civic organizations and new forms of governable space, a veritable jigsaw of militant particularisms. The Kaiama Declaration in 1999 indicates that there is a pan-ethnic solidarity movement in the works, but its contours are at present limited (see Okonta and Douglas 2001; ERA 2000). As the Ogoni case shows, much of this visibility and identification turned on the invention and reinvention of tradition and local knowledge, with an eye to the Nigerian constitution and international politics (Nelson 1999). This is a case of the multiplication of governable spaces which stand in some tension or even contradiction with each other – they account for the explosion of inter-ethnic tensions in the delta – and within the national space of Nigeria, to which I now turn.

The space of nationalism

One of the striking aspects of the governable spaces of indigeneity as they emerge in the delta is that they become vehicles for political claims, typically articulated as the need for a local government or in some cases a state. Indigeneity necessarily raises the question of a third governable space, that of the nation state, an entity that pre-existed oil and came to fruition in 1960 at Independence. Oil in this sense became part of the nation-building process – the creation of an "oil nation." Nature and nationalism become inextricably linked. But how did petro-capitalism – understood as a state-led, and thoroughly globalized, development strategy – stand in relation to the creation of the governable space called modern Nigeria?

As Mamood Mamdani (1996, 2000) has observed, colonial rule and decentralized despotism were synonymous. The Native Authorities consolidated local class power in the name of tradition (ethnicity) and sustained a racialized view of civic rights. The Nationalist movement had two wings, a radical and a mainstream. Both wished to de-racialize civic rights but the latter won out and reproduced the dual legacy of colonialism. They provided civic rights for all Nigerians, but a bonus of "customary rights" for indigenous people. The country had to decide which ethnic groups were indigenous and which were not a basis for political representation, a process that became constitutionally mandated in Nigeria. Federal institutions are quota-driven for each state but only those indigenous to the state may apply for a quota. As Mamdani puts it:

> The effective elements of the federation are neither territorial units called states nor ethnic groups but ethnic groups with their own states.... Given this federal character every ethnic group compelled to

seek its own home its NA, its own state. With each new political entity the non-indigenes continues to grow.

(1998: 7)

Once law enshrines cultural identity as the basis for political identity, it necessarily converts ethnicity into a political force. As a consequence, in Nigeria, clashes in the postcolonial period were ethnic, and such ethnic clashes, which dominated the political landscape in the last three decades, are always at root about customary rights to land and, derivatively, to a local government or to a state that can empower those on the ground as ethnically indigenous.

Into this mix enters oil, that is to say a valuable, centralized (state-owned) resource. It is a *national* resource on which citizenship claims can be constructed. As much as the state uses oil to build a nation and to develop, so communities use oil-wealth to activate community claims on what is seen popularly as unimaginable wealth – black gold. The governable space of Nigeria is as a consequence re-territorialized through ethnic claims-making (Adebayo 1993, Suberu 2001). The result is that access to oil revenues amplifies what I call sub-national political institution-making; politics becomes then a massive state-making machine. This partly explains how, between 1966 and the present, the number of local governments has grown from fifty to almost 1000, and the number of states from three to thirty-six! Nigeria as a modern nation state has become a machine for the production of ever more local political institutions. The logic is ineluctable and terrifying.

What sort of national governable space emerges from such multiplication in which, incidentally, the political entities called states or LGAs (local government areas) become vehicles for massive corruption and fraud – or the disposal of oil revenues? The answer is that it works against precisely the creation of an imagined community of the sort that Benedict Anderson (1998) saw as synonymous with nationalism. Nation building, whatever its style of imaging, rests in its modern form on a sort of calculation, integration, and state and bureaucratic rationality which the logic of rent-seeking, petro-corruption, ethnic spoils politics, and state multiplication work to systematically undermine. Lauren Berlant has said that every nation – and hence every governable national space – requires a "National Symbolic"; a national fantasy which "designates how national culture becomes local through images, narratives and movements which circulate in the personal and collective unconsciousness" (1991: 61). My point is that the Nigerian National Symbolic grew weaker and more attenuated as a result of the political economy of oil. There was no sense of the national fantasy at the local level; it was simply a big pocket of oil monies to be raided in the name of indigeneity. At Independence, Obafemi Awolowo, the great western Nigerian politician, said that Nigeria was not a Nation but a "mere geographical expression." Forty years later this is still true.

What we have in other words is not nation building – understood in the sense of a governable space – but perhaps its reverse: the "unimagining" or deconstruction of a particular sense of national community. Nicos Poulantzas (1978) said that national or modern unity requires a historicity of a territory and a territorialization of a history. Oil capitalism (and its attendant governmentality) in Nigeria has achieved neither of these requirements. The "fictional" governable space called Nigeria was always something of a public secret. Forty years of postcolonial rule has made this secret both more public as ethnic segregation has continued unabated and undermined the very idea of the production of governable subjects. The double-movement of petro-capitalism within the frame of a modern nation state has eviscerated the governable space of the nation; it has compromised it and worked against a sense of governable subject. The same, incidentally, might be said of the impact of oil on the Muslim communities of Nigeria (Watts 1998; 2000). Oil and identity – people and things – have produced an unimagined community on which the question of Nigeria's future hangs.

BLOOD AND OIL

The entire history of petroleum is, as Daniel Yergin (1991) details in his encyclopaedic Whig account of the industry, *The Prize*, replete with criminality, violence and the worst of frontier capitalism. Graft, autocratic thuggery, and the most grotesque exercise of imperial power are its hallmarks. As the US armored divisions roll up the Iraqi oil corridor around Basra, this point hardly needs further empirical confirmation. And it is to be expected in an age of unprecedented denationalization and market liberalization – to say nothing of the horrific rise of the gas-guzzling Sports Utility vehicle in the United States – the mad scramble to locate the next petrolic El Dorado continues unabated. Eastern Russia looks ever more like a slice of Mafiosi sovereignty. Petro-violence is rarely off the front pages of the press. The Caspian basin reaching from the borders of Afghanistan to the Russian Caucasus is a repository of enormous petro-wealth; Turkmenistan, Kazakhstan, Azerbaijan, Georgia and the southern Russian provinces (Ossetia, Dagestan, Chechnya) have however become, in the wake of the collapse of the Soviet Union, a "zone of civil conflict and war."[24] The oil companies jockey for position in an atmosphere of frontier vigilantism and what the Azerbaijani President calls "armed conflict, aggressive separatism and nationalism."

Based on the Nigerian case, I have suggested that petro-capitalism operates through a particular sort of oil complex (a configuration of firm, state and community). The complex is strongly territorial, operating through local oil concessions. The presence and activities of the oil companies as part of the oil complex constitute a challenge to forms of community authority, inter-ethnic relations, and local state institutions principally through the property and land disputes that are engendered, and via forms of popular mobilization and agitation to gain access to (i) company rents and compensation revenues, and (ii) the petro-revenues of the Nigerian state largely through the creation of regional and/or local state institutions. The oil complex generates differing sorts of governable spaces in which contrasting identities and forms of rule come into play; in some cases youth and generational forces are key, in some cases gender, in others the clan or the kingdom or the ethnic minority (or indigenous peoples), in some cases local chiefly or governmental authorities, and in others the forces of the local state. A striking aspect of contemporary development in Nigeria is the *simultaneous* production of differing forms of rule and governable space – different politics of scale (Smith 1992) – all products of similar forces and yet which work against, and often stand in direct contradiction to one another. I have focused on youth, the indigene or ethnic minority, and the nation. These idioms are inseparable from oil, but their forms of identification and the robustness of their spaces are often incompatible. Standing at the center of each governable space is a contradiction: at the oil community, the overthrowing of gerontocratic authority but its substitution by a sort of violent Mafia youthful rule. At the level of the ethnic community is the tension between civic nationalism and a sort of exclusivist militant particularism. And at the level of the nation, one sees the contradiction between oil-based state centralization and state fragmentation, or multiplication, as oil becomes a sort of generalized equivalent put to the service of massive corruption. I have tried to root these contradictions in the double-movement of petro-capitalism which is generative of an authoritarian governmentality constituted by the three forms of governable space that I have described. Such is the heart of the so-called crisis of the postcolonial state in Africa. It is in this sense that I invoke the idea of "economies of violence" to characterize governmentality in contemporary Nigeria.

NOTES

1 I am grateful to Oronto Douglas, Ike Okonta, Von Kemedi, Amita Baviskar, and Donald Moore for their suggestions and assistance.
2 See Gary and Karl (2003); Global Witness Resources (2001) Pendleton *et al.* (2001) Renner (2002), OXFAM (2001).
3 See http://www.state.gov/s/ct/rls/pgtrpt/2000 (accessed April 2003).
4 See http://www.eia.gov/emeu/cabs/nigeria.html (accessed April 2003).
5 See http://www.iasps.org (accessed April 2003).
6 *New York Times*, 14 October 2001: III, 1
7 Economists typically distinguish direct (so-called Dutch Disease) effects in which resource booms lead to recession, and indirect effects through rent seeking and institution building.
8 For an elaborate critique of this position, see Peluso and Watts (2001).
9 Ross provides a shopping list of the consequences of oil: limited backward linkages, inelastic demand, labor extensive, subject to boom and bust cycles, subject to rent-seeking and so on. At the very least one needs to be clear how oil differs from other commodities (in order to be able to distinguish what is peculiar to oil as opposed to extraction in general), and to be able to distinguish what is it about the resource (and not the political context into which it is inserted) that can explain the so-called "paradox of plenty" (see Gary and Karl (2003) as a case in point).
10 *New York Times*, 13 August 2002.

11   Accordingly, we need to see things not in terms of the replacement of a society of sovereignty by a disciplinary society and the subsequent replacement of a disciplinary society by a society of government; in reality one has a triangle, sovereignty-discipline-government, which has as its primary target the population, and as its essential mechanism the apparatuses of security ... *I want to demonstrate the deep historical link between movement that overturns the constants of sovereignty on consequence of the problem of choices of government; the movement that brings about the emergence of population as a datum, a field of intervention ... the process that isolates the economy as a specific sector of reality; and political economy as the science and the technique of intervention of the government in that reality.* Three movements – government, population, political economy – that constitute from the eighteenth century onward a solid series, one that even today has assuredly not been dissolved.

(Foucault 2000: 219, emphasis added)

12 Some of these Foucauldian ideas have already been productively deployed in the understanding of nature and resource management – what one might call "green governmentality" – and the relations between nature and nationalism. See Braun (2000).

13 Rhein quoted in Rabinow (1989: 142).

14 Nembe in its macro-usage refers to six towns (Bassimbiri, Ogbolomabiri, Okpoama, Odioma and Akassa) that are among the sixteen towns that comprise Nembe Kingdom. For the purposes of this paper, however, Nembe town refers to Ogbolomabiri only.

15 The data for the case study was collected during a visit to the Niger delta in January and February 2001. I also rely on the assistance on Von Kemedi and his work (Kemedi 2002) and the Nembe Peace Commission (Alagoa 2001).

16 The war canoe houses were the units of the kingdom's defense forces. A war canoe house consisted of the head of the house and a formidable number of able-bodied men who were responsible for defending the house and the King.

17 There is a long running dispute over kingly authority that has spilled over into the establishment of local government areas (LGAs). In this paper I do not address the conflicts between Bassambiri and Ogbolamabiri (two contiguous towns) over the authority of their respective paramount chiefs, and disputes over LGA territory (and hence access to oil rents).

18 As constitutional preparations were made for the transition to home rule, non-Igbo minorities throughout the Eastern Region appealed to the colonial government for a separate Rivers state. Ogoni representatives lobbied the Willink Commission in 1958 to avert the threat of exclusion within an Ibo-dominated regional government which had assumed self-governing status in 1957 but minority claims were ignored (Okilo 1980, Okpu 1977).

19 The Ogoni and other minorities petitioned in 1974 for the creation of a new Port Harcourt State within the Rivers State boundary (Naanen 1995: 63).

20 According to the Nigerian government estimates, Ogoniland currently (1995) produces about 2 percent of Nigerian oil output and is the fifth largest oil-producing community in Rivers State. Shell maintains that total Ogoni oil output is valued at $5.2 billion before costs.

21 *Village Voice*, 21 November 1995, pp. 21.

22 The oil companies claim that sabotage accounts for a large proportion (60 percent) of the spills, since communities gain from corporate compensation. Shell claims that 77 of 111 spills in Ogoniland between 1985 and 1994 were due to sabotage (Hammer 1996). According to the government commission, however, sabotage accounts for 30 percent of the incidents but only 3 percent of the quantity spilled. Furthermore, all oil producing communities claim that compensation from the companies for spills has been almost non-existent.

23 Saro-Wiwa was often chastised by Gokana (he himself was Bane) since most of the Ogoni oil was in fact located below Gokana soil. In other words, on occasion, the key territorial unit became the clan or clan territory rather than a sense of pan-Ogoni territory.
24 *San Francisco Chronicle*, 11 August 1998: A8.

REFERENCES

Adebayo, A.(1993) *Embattled Federalism*, New York: Peter Lang.

Alagoa, M. (2001) *The Report of the Nembe Peace and Reconciliation Committee*, Port Harcourt.

Anderson, B. (1998) *Imagined Communities*, New York: Verso.

Anderson, J. (2000) "Blood and Oil," *The New Yorker*, 14 August, 46–59.

Anderson, P. (2001) "Scurrying Towards Bethlehem," *New Left Review*, 10: 5–30.

Anugwom, E. (2001) "Federalism, Fiscal Centralism and the Realities of Democratization in Nigeria," paper presented at the Conference on Africa at the Crossroads, UNESCO. Available http: www.ethnonet-africa.org/pubs/crossroadsed1.htm (accessed April 2001).

Berlant, L. (1991) *The Anatomy of National Fantasy*, Chicago: University of Chicago Press.

Blok, A. (1974) *The Mafia in a Sicilian Village*, Waveland: Prospect Heights.

Bob, C. (2002) "Merchants of Morality," *Foreign Policy*, 131, Feb./March 36–45.

Braun, B. (2000) "Producing Vertical Territory," *Ecumene*, July: 7–46.

Brysk, A. (2000) *From Tribal Village to Global Village*, Palo Alto: Stanford University Press.

Cesarz, E., Morrison, S., and Cooke, J. (2003) "Alienation and Militancy in the Nigeria Delta," Center for Strategic and International Studies, Washington DC, Africa Notes, 16, May.

C.I.A. (Central Intelligence Agency) (2000) "Nigeria: Environmental Stresses and their Impacts over the Next Decade," CIA DCI Environmental Center.

Collier, P. (2000) *The Economic Causes of Civil Conflict and Their Implications for Policy*, Washington DC: The World Bank.

Coronil, F. (1997) *The Magic of the State*, Chicago: University of Chicago Press.

Dean, M. (1999) Governmentality, London: Sage.

ERA (Environmental Rights Action) (2000) *The Emperor Has No Clothes*, Benin: Environmental Rights Action.

Foucault, M. (2000) *Power*, New York: The New Press.

Furro, T. (1992) "Federalism and the Politics of Revenue Allocation in Nigeria," Ph.D. Dissertation, Clark Atlanta University.

Gary, I. and Karl, T. (2003) *Bottom of the Barrell*, Baltimore, MD: Catholic Relief Services.

Global Witness Resources (2001) *Conflict and Corruption*, London: Global Witness.

Hammer, J. (1996) "Nigerian Crude," *Harper's Magazine*, June, 58–68.

Homer-Dixon, T. (1999) *Environment, Scarcity, and Violence*, Princeton: Princeton University Press.

Human Rights Watch (1995) *The Ogoni Crisis*, Report, 7/5, New York: Human Rights Watch.

—— (1999) *The Price of Oil*, Washington DC: Human Rights Watch.

—— (2002) *The Niger Delta: No Democratic Dividend*, Washington DC: Human Rights Watch.

—— (2003) *Testing Democracy*, Washington DC: Human Rights Watch.

Ibeanu, O. (2003) "(Sp)oils of Politics," paper presented at the Conference on Oil and Human Rights, University of California, Berkeley, 24–26 January.

Ikein. I. (1990) *The Impact of Oil on a Developing Country*, New York: Praeger.

Ikporukpo, C. (1993) "Oil Companies and Village Development in Nigeria," *OPEC Review*, 83–97.

—— (1996) "Federalism, Political Power and the Economic Power Game: Control over Access to Petroleum Resources in Nigeria," *Environment and Planning*, 14: 159–77.

Kapucinksi, R. (1982) *Shah of Shahs*, New York: Vintage.

Kemedi, V. (2002) "Oil on Troubled Waters," Berkeley: Environmental Politics Working Papers.

Klare, M. (2001) *Resource Wars*, Boston: Beacon Press.

Le Billon, P. (2001) "Angola's Political Economy of War," *African Affairs*, 100: 55–80.

Leite, C. and Weidemann, J. (1999) "Does Mother Nature Corrupt?" IMF Working Paper. Washington DC: IMF.

Leonard, D. and Strauss, S. (2003) *Africa's Stalled Development*, Boulder: Westview.

Li, T.M. (1996) "Images of Community," *Development and Change*, 27: 501–27.

Luft, Gal (2003) "Africa drowns in a pool of oil," *Los Angeles Times*, 1 July.

Mamdani, M. (1996) *Citizen and Subject*, Princeton: Princeton University Press.

—— (1998) "When Does a Settler become a Native?" Inaugural Lecture, University of Cape Town, manuscript.

—— (2000) *When Victims becomes Killers*, Princeton: Princeton University Press.

Naanen, B. (1995) "Oil Producing Minorities and the Restructuring of Nigerian Federalism," *Journal of Commonwealth and Comparative Politics*, 33 (1): 46–58.

NDWC (Niger Delta Wetlands Commission) (2000) *Mediation and Conflict Resolution of the Crisis in Nembe*, Port Harcourt: Niger Delta Wetlands Commission.

Nelson, D. (1999) *Finger in the Wound*, Berkeley: University of California Press.

Ngofa, O. (1998) *Eleme Traditions*, Eleme: Rescue Publications.

Obi, C. (2001) *The Changing Forms of Identity Politics in Nigeria*, Uppsala: Africa Institute.

Ogoni Crisis (1996) "Lagos, Ministry of Information, Nigerian Federal Government," Lagos.

Okilo, M. (1980) *Derivation: A Criterion of Revenue Allocation*, Port Harcourt: Rivers State Newspaper Corporation.

Okonta, I. (2002) "The Struggle of the Ogoni for Self-Determination," D.Phil. thesis, Oxford University.

—— (2003a) "When Citizens Revolt," unpublished manuscript, Berkeley: University of California.

—— (2003b) "Nigeria and the World," *This Day*, 22 June, Sunday editorial.

—— and Douglas, O. (2001) *Where Vultures Feast*, San Francisco: Sierra Club.

Okpu, U. (1977) *Ethnic Minority Problems in Nigerian Politics*, Stockholm: Wiksell.

Osaghae, E. (1995) "The Ogoni Uprising," *African Affairs*, 94: 325–44.

Oxfam (2001) *Extractive Sectors and the Poor*, Boston, Mass: Oxfam.

Painter, D. (1986) *Oil and the American Century*, Baltimore: Johns Hopkins University Press.

Peluso, N. and Watts, M. (2001) *Violent Environments*, Ithaca: Cornell University Press.

Pendleton, A., Stuart, L., Davison, J., and Bishop, S. *Fuelling Poverty*, London: Christian Aid.

Poulantzas, N. (1978) *State, Power, Socialism*, London: New Left Books.

Putnam, R. (2000) *Bowling Alone*, New York: Harpers.

Rabinow, P. (1989) *French Modern: Norms and Forms of the Social Environment*, Cambridge: MIT Press.

Rainforest Action Network (1997) *Human Rights and Environmental Operations Information on the Royal Dutch/Shell Group of Companies*, London.

Renner, M. (2002) *The Anatomy of Resource Wars*, Washington DC: Worldwatch Institute.

Rose, N. (1999) *Powers of Freedom*, London: Cambridge University Press.

Ross, M. (1999) "The Political Economy of the Resource Curse," *World Politics*, 51: 297–322.

—— (2001) "Does Oil Hinder Democracy?" *World Politics*, 53: 325–61.

Sachs, J. and Warner, A. (1995) "Natural Resource Abundance and Economic Growth," NBER Working Paper 5398. Cambridge, MA.: National Bureau of Economic Research.

Saro-Wiwa, K. (1989) *On A Darkling Plain*, Port Harcourt: Saros Publishers.

—— (1992) *Genocide in Nigeria*, Port Harcourt: Saros International Publishers.

—— (1995) *A Month and A Day*, London: Penguin.

Servant (2003) *Le Monde Diplomatique*, 13 January.

Smith, N. (1992) "Geography, Difference and the Politics of Scale," in J. Doherty, E. Graham, and M. Malek (eds) *Postmodernism and the Social Sciences*, London: Macmillan.

Suberu, R. (2001) *Federalism and Ethnic Conflict in Nigeria*, Washington DC: USIP.

Vitalis, R. (2002) "Black Gold, White Crude," *Diplomatic History*, 26 (2): 185–213.

Watts, M. (1998) "Islamic Modernities?" in J. Holston (ed.) *Cities and Citizenship*, Durham, Duke University Press.

—— (2000) *Struggles over Geography*, Heidelberg: University of Heidelberg, Hettner Lectures.

Whelch, C. (1995) "The Ogoni and Self Determination," *Journal of Modern African Studies*, 33 (4): 635–50.

Wright, George (2003) "Wolfowitz: Iraq war was about oil," *Manchester Guardian*, 4 June.

Yergin, D. (1991) *The Prize*, New York: Random House.

11

GENDER AND CLASS POWER IN AGROFORESTRY SYSTEMS

Case studies from Indonesia and West Africa

Richard A. Schroeder and Krisnawati Suryanata

> [A]groforestry initiatives ... have been sheltered in the discursive shade
> of trees as symbols of green goodness.
>
> (Rocheleau and Ross 1995: 408)

Agroforestry systems are widely touted for their prodigious capacities. From a production standpoint, intercropping trees with underlying crops can fix nitrogen and improve nutrient cycling, enhance chemical and physical soil properties, add green manure, conserve moisture, and make generally efficient use of a range of limited yield factors. Similarly, from the standpoint of environmental stabilization, agroforestry systems may reduce erosion, provide alternate habitat for wildlife, and shelter a diverse range of plants; they are also sites where the critical knowledge systems of indigenous peoples are reproduced. In the context of contemporary environmentalism, an agroforestry approach that simultaneously boosts commodity production and contributes to stabilizing the underlying resource base is constructed as an unambiguous and unalloyed "good" (Rocheleau and Ross 1995; Schroeder 1995, 1999). Institutional actors in forestry and environmental agencies, as well as the major multilateral donor agencies such as the World Bank, have accordingly joined forces to promote and preserve agroforestry in many parts of the world.

This chapter challenges the assumption that environmentalist policies and development practices related to agroforestry are universally beneficial to local interests. In doing so, we follow the lead of an increasing number of scholars who have demonstrated the usefulness of combining fine-grained political analysis with a detailed understanding of socialized ecological processes (Castree and Macmillan 2001; Cronon 1983; Fitzsimmons and Goodman 1998; Gadgil and Guha 1992; Neumann 1998; Peluso and Watts 2001; Pulido 1996; Whatmore 1999). Specifically, we seek to redirect attention to agroforestry as a site of contentious political struggle, a domain in which "regional discursive formations" (Peet and

Watts 1996) focused on environmental stabilization are heavily contested. Proponents of agroforestry stress that trees are assets that enhance the value and quality of land resources and vary the scope and seasonality of income streams. The problem with such an idealized view is that it assumes stable ecological equilibria and minimizes the internal workings of property and labor claims, despite ample evidence that these are pivotal to successful management (Fortmann and Bruce 1988; Raintree 1987). In agroforestry, the longevity of trees and the different mechanisms that grant the rights to trees and land often combine to create multiple and conflicting claims over resources (Berry 1993; Rocheleau 1997). As these successional systems mature, one species, and hence one set of property claims, supersedes all others (Bryant 1994; Goswami 1988; King 1988; Peluso 1992). Moreover, where agroforestry approaches are commercialized, they tend to reinforce the tenure rights of tree growers *vis-à-vis* competing claimants, such as cultivators of underlying crops, forest product collectors, and pastoralists (Millon 1957; Raintree 1987). With such social dynamics embedded in combinations of tree and understorey crops, the design and implementation of agroforestry systems, and especially the actions of tree holders, must be carefully analyzed.

We seek to draw attention to the fact that, in addition to favorable production and environmental capacities, agroforestry approaches sometimes open up critical options for otherwise disenfranchised groups. Rocheleau (1987), for example, demonstrates how women mobilize agroforestry strategies to make the best use of the minimal landholdings allotted to them (cf. Leach 1994). Proponents of cultural survival have argued that indigenous peoples use agroforestry systems to perpetuate livelihood practices and safeguard key components of cultural identity (Clay 1988). And analysts of agrarian transitions have argued that the diversity and complexity of so-called "home garden" agroforestry systems, which incorporate a wide range of cultivars with high use-value but low exchange-value crops, provide peasant groups with the means to effectively resist the extractive propensities of the state (Dove 1990). Thus, a focus on gender, cultural and class relations require alternate readings of agroforestry. At a minimum, there is a need to move beyond technocratic and managerial classification systems (Farrell 1987; Nair 1989, 1990) and toward a greater appreciation of the social nature of agroforestry systems (cf. Castree and Braun 2001). For systems such as those described by Dove, Clay, Rocheleau, and others are fundamentally different in scope and purpose than narrowly drawn economic, forest management, and environmental projects bent simply on merging environmental and commodity production objectives.

This chapter applies these insights to two contemporary agroforestry initiatives in The Gambia and upland Java, which illustrate the importance of understanding both the *political* and the *ecological* dimensions of environmental interventions. Both systems involve the production of tree commodities, and have been hailed as bold steps toward environmental stabilization. In The Gambia, agroforestry projects have been implemented in attempts to reverse the cumulative effects of drought and deforestation; in Indonesia, they have been employed to reduce erosion on steeply sloped hillsides in order to reduce the silting up of reservoirs. In both cases, an environmental discourse has served to mask the exclusionary objectives of fruit tree holders – male mango growers in The Gambia, and a new class of "apple lords" in Java – which are ultimately directed at entrepreneurial gain and control over key production resources.

Our first argument is that while these agroforestries often contribute in some measure to ecological goals, they nonetheless can also be seen as deliberate strategies of dispossession and private accumulation. Second, we demonstrate that both polarization and environmental change are mediated by property struggles within the political arenas of the household and the village society. Our studies reiterate the ambiguity of property relations (Schlager and Ostrom 1992; Vandergeest 1997), highlighting the fact that such relations are sometimes expressed through the specific local dynamics of intercropping. In the two case studies, the commoditization of tree cropping has driven a wedge between holders of tree and land/crop rights, which in turn produced a range of agro-ecological and social contradictions. Such dynamics erode moral economies and replace them with a morally indifferent (not to say bankrupt) stance, which elevates profit taking above all other objectives, *including* ecological stability (Schroeder 1995).

GENDERED AGROFORESTRY IN GAMBIAN GARDEN/ORCHARDS

Rights over resources such as land or crops are inseparable from, indeed are isomorphic with, rights over people.

(Watts 1992: 161)

Since the mid-1980s, agroforestry efforts in The Gambia have primarily been focused on adding trees to hundreds of low-lying women's gardens originally established under the guise of "women in development" initiatives. A veritable boom in market gardening by women's groups grew out of a conjuncture of poor climatic conditions, foreign investment in women's programs, and numerous unconscionable national budget reductions mandated by a World Bank structural adjustment program. Average annual rainfall along the river basin has declined approximately 25–30 percent over a twenty-year period. During that time, the respective fortunes of the male and female agricultural sectors have reversed: hundreds of thousands of dollars have been invested in the women's garden sector by donors interested in promoting better nutrition and an increase in female incomes. while prices for male peanut producers (gardeners' husbands) have stagnated on the world market (Carney, Chapter 12 in this volume; Schroeder 1993). Despite the fact that women's gardens have become the basis for household reproduction in many areas, they have since come under threat from male landholders interested in planting fruit orchards in the same locations.

Customary land law among the Mandinka residents of The Gambia's North Bank Division, where research for this chapter was conducted in 1989, 1991 and 1995, preserves a basic distinction between matrilineal and patrilineal land. Women's landholding rights are almost exclusively limited to swampland, where plots originally cleared by women are heritable property passing from mother to daughter. Patrilineal land, by contrast, consists both of upland areas, where men control virtually all arable land and grow groundnuts, millet, and maize, and some swampland, where rice is grown by female family members for joint household consumption. Such land is nominally controlled by men who are relatively senior in the lineage structure, although practical day-to-day production decisions are often taken by junior kin who are either delegated responsibility for cultivation or are granted use rights to plots prior to acceding to full landholding status as they grow older. Women's gardens, ranging in size from a fraction of a hectare to nearly 5 hectares, are almost all constructed on lineage land. Rights of access are granted on a usufruct basis to groups, although individual women operate separate plots within the communally fenced perimeters. The gardens are thus vulnerable to being reclaimed by landholders interested in planting tree crops. According to Mandinka custom, trees belong to those who plant them. Under circumstances such as the gardens in question, where the tree planter is also the landholder, the tree crop takes precedence over other forms of cultivation. (This is clearly an empirical question, however. Tree crops may take precedence even in systems where the tree planter is *not* the landholder, as in the Javanese case outlined below.)

On the face of it, this situation appears clear cut: two groups of commodity producers vie for control of the same land and labor resources, as well as the development largesse generated through their respective production systems. Neither group has total power over the garden/orchard spaces (Schroeder 1999): gardeners are dependent upon usufruct rights to land controlled by senior male members of landholding lineages, and would-be orchard owners are dependent upon the labor of women's groups, not just for irrigation, but for maintenance of fences and wells, clearing brush from garden/orchard plots, and protection from livestock incursions. The potential for conflict between gardeners and landholders is thus manifest in every production decision taken within the fence perimeters which bound the system. Each relocation of the fence line, each tree planted, each year's planting sequence and plot layout can be read as a strategic and spatial embodiment of power (Schroeder 1999).

Conjugal conflict and intensified land use

Work in the horticulture sector has generated incomes for women gardeners that are roughly equivalent to the rural per capita income in The Gambia, and female household members have consequently taken on major new financial responsibilities. Of the women in the sample, 57 percent had purchased at least one bag of rice in 1991 to supplement home-grown food supplies; 95 percent buy all their own clothes, 84 percent buy all their children's clothes, and 80 percent had purchased Islamic feast day clothes for at least one member of their family – all responsibilities borne either solely or primarily by men prior to the garden boom. While all cash earned from vegetable sales is nominally controlled by women, growers' husbands have, nonetheless, devised a complex system of tactics for alienating female earnings, or otherwise directing them toward ends of their own choosing. These include a range of loan-seeking strategies, each carrying its own measure of commitment for repayment, and its own underlying threat of reprisal if the loan is not forth-coming. Gardeners' husbands also increasingly default on customary financial obligations they fed their wives can assume due to improved financial circumstances. The key point here is that the social pressure for women to share garden incomes with other family members mounted steadily throughout the early stages of the garden boom, and vegetable growers responded by both expanding and intensifying production (Schroeder 1996, 1999).

Attempts to resolve *intra-household* tensions often displace the conflict to the spatial arena of the garden perimeters. The technical innovations accompanying the garden boom included replacement of poor quality stick and thorn fences and hand-dug, unlined wells serving individual plots with communal wire and concrete structures that do not have to be replaced on an annual basis. These enhancements reduced prohibitive recurrent expenses, removed some of the threat of encroachment by grazing livestock, and improved access to ground-water. While such improvements stabilized the vegetable production system in several key respects, the narrow selection of crops cultivated and relatively poor market returns meant that gardeners were unable to adequately meet their husbands' demands for greater financial support. Moreover, even as marginal increases were achieved, a strongly "pulsed" income stream left women vulnerable to their husbands' loan requests. Growers consequently reverted to more complicated intercropping strategies that prolonged the market season and spread income over several months. Planting fruit trees and production of new crops such as cabbage, bitter tomatoes, and sweet peppers opened up sizable new markets and improved the seasonality of the income returns from gardens. The potential of these intercropping strategies could only be met with an expansion of garden territory, however. Requests to enclose new areas for gardening purposes and the *de facto* conversion of garden space into a more complex agroforestry system caused male landholders to re-evaluate the garden boom and its long-term effects. From the landholders' perspective, fruit production in the gardens threatened to confer a sense of permanence and legitimacy upon women's usufruct rights. Like the Javanese case below, the interests of tree holders and landholders began to diverge, with tree holders – in this instance, women gardeners – apparently holding the upper hand.

Shady practice

When an expatriate volunteer was posted in the area in 1983, local gardeners seized upon the opportunity to lobby for material support to expand two existing garden sites. Ensuing efforts to implement plans to rebuild and enlarge the community's two primary fenced perimeters were thwarted, however, when the landholder on one of the sites objected to the fact that his landholding prerogatives were being violated by the provisions of the proposed project.

Increasing tensions eventually resulted in the detention of three garden leaders and a spontaneous protest demonstration on the part of several hundred gardeners, which resulted in the issuing of a temporary injunction against gardening on the site. In the court's ruling, nearly all substantive claims by the vegetable growers were upheld. The sole exception involved allegations made by the landholder that the women had planted dozens of fruit trees within the perimeter without authorization. His insistence that they be removed won the court's backing, and women were ordered to remove all trees at his request. Within a day or two of the decision, the land-holder visited the garden and ordered several dozen trees removed. Then, in an action that foreshadowed much of what was to come in the North Bank's garden districts, he immediately replanted several dozen of his own trees within the perimeter. By locating seedlings directly on top of garden beds already allocated to vegetable growers, his expectation was that water delivered by women farmers to the vegetable crop would support his trees until the ensuing rainy season (a sort of indirect subsidy).

This controversy marked a watershed in the political ecology of gardening on the North Bank. Not only were several hundred women involved in the demonstration at the police station, but the case also received attention from politicians at the highest levels of government. Every step taken by the landlord and every aspect of the women's claims to use rights were carefully scrutinized and debated throughout the area. This led other landholders to reappraise their own stance with respect to their management of low-lying land resources. Most telling, it set a precedent for landholders in the attempted use of female labor to establish private fruit tree orchards.

Within a few years of this incident, both gardeners' and landholders' attitudes toward agroforestry practices had changed. From the gardeners' perspective, the relative economic benefits of tree planting and vegetable growing shifted decisively in favor of gardens. As the leader of one of the oldest garden groups in the area put it: "We are afraid of trees now.... You can have one [vegetables or *fruit*] or you can have the other, but you can't have both." Thus, in order to minimize shade effects, growers began cutting back or chopping down trees – in many cases, trees which they themselves had planted – in order to open up the shade canopy and expose their vegetable crops to sunlight. At the same time, landholders saw a new opportunity developing for themselves. Whereas they had initially resisted tree planting on the grounds that it reduced their future land-use options, the "capturing" of a female labor force to water trees, manure plots, and guard against livestock incursions within the fenced perimeters led landholders to wholeheartedly embrace fruit growing.

In 1983, a new garden site was established immediately adjacent to an older site where gardeners had already begun to feel the effects of shade canopy closure.

Given the land pressure at the time, many women from the older site took second plots in the new site. Under what was then still a somewhat novel arrangement, the garden was converted into a garden/orchard, with a dense stand of trees in a grid pattern over the entire area. The understanding was that ownership of the trees would be divided between the landholder and gardeners on an alternating basis; every other tree, in effect, belonged to the landholder.

Within five or six years, however, the prospect of shade canopy closure appeared in the new garden. Gardeners had already determined that vegetables brought them a greater return than any harvest they could expect from their trees. Consequently, many of the maturing trees were either drastically trimmed or simply removed, including, apparently, many of the trees belonging to the land-holder. In response, the landholder banned tree trimming in his garden, only to find his young trees still being destroyed as women burned crop residues to clear plots for each new planting season. While some of this destruction was doubtless accidental, the landholder claimed that growers deliberately hung dry grass in tree branches so that fires set to clear plots would fatally damage trees. A survey of tree density on the site revealed that fully half of the original orchard no longer exists, so it is clear that vegetable growers were at least partially successful in defending their use rights.

By 1991, the situation regarding garden/orchard tenure was somewhat uncertain. Survey data from a dozen gardens show clear trends toward tighter control of garden spaces by orchard entrepreneurs, and a major emphasis within orchards on mango trees – the species most likely to cause shade problems for gardeners sharing the space. Landholders opening new gardens in the late 1980s tended to do so only under the strict conditions that women agree in advance to water the landholder's tree seedlings and vacate their temporary use rights when the trees matured. Of the twelve sites surveyed, only three remained solely under gardeners' control. All others had either already been, or were about to be, planted over with tree crops. Some 60 percent of the prime low-lying land in the vicinity of the communities surveyed was thus at risk of being lost to shade within the decade.

In sum, this brief comparison of the North Bank's garden/orchards establishes that trees can be used as a means for claiming both material and symbolic control over garden lands. Tree planting on garden beds, moreover, is a mechanism for landholders to alienate surplus female labor and subsidies embodied in concrete-lined wells and permanent wire fences. In this respect, the Gambian case differs from the apple-based agroforestry system in Java described below, where landholders often lack the capital to build the infrastructure necessary to convert their lands to orchards. At the same time, shade effects from tree planting threaten to undermine the productivity of gardeners, who now play key roles in providing for the subsistence needs of their families.

On balance, the agroforestry system practiced by women gardeners seems of greater value than the successional systems landholders have imposed. Viewed from a production standpoint, garden-based agroforestry practiced by women appears to generate a greater absolute income than a monocrop mango system, as well as a more seasonally varied income stream, one better suited to meeting the myriad financial challenges rural families face throughout the year. From an environmental standpoint, since the orchards in the successional schemes are small, they have little impact on climate change and deforestation problems they were ostensibly intended to address. On the micro-scale, the women's systems are clearly more diverse than the men's. Soil quality is typically better, by dint of the incorporation of countless head pan-loads of compound sweepings and manure. Moreover, the evidence shows that, given the chance, gardeners routinely incorporate fruit trees into their crop mix, and that they effectively manage the ecological competition between vegetables and trees implied by intercropping, *if* they actually control decisions over the selection of species, the location of trees, and rights of trimming or removal, which is to say, the substance of the labor process and property rights. Such social relations are precisely what is overlooked in theories of agroforestry that construct all forms of tree planting in the same terms, namely as beneficial interventions with unambiguous stabilizing effects on local environments.

AGROFORESTRY AND CLASS RELATIONS IN A JAVANESE VILLAGE

Conventional wisdom suggests that upland Java faces an imminent ecological crisis under increasing population pressure. Poor, subsistence households seek to increase their immediate income by using cropping patterns that accelerate soil erosion from their rain-fed farms (USAID and Government of Indonesia 1983). Rainfall intensities are extremely high in Java, contributing to severe soil erosion (Carson 1989). One survey in the mid-1980s estimated that 2 million hectares, or one-third of Java's cultivated uplands were severely degraded, and that the problem was increasing at a rate of 75,000 hectares annually (Tarrant *et al*. 1987).

Since the early 1980s, however, dramatic economic and land-use changes have occurred in many upland villages in Java. As urban incomes have risen, improving the market for fresh fruit, upland farmers have expanded cultivation of commercial fruit trees. A Jakarta-based newspaper reported that throughout the 1980s, domestic demand for fruit increased at the rate of 6.5 percent per annum (*Pelita*, 1 September 1991). Development planners concerned with stabilizing the environment of upland Java viewed this with optimism, as tree planting and agroforestry have always been associated with lower soil erosion rates.

Agroforestry has indeed been an essential component for upland development programs in Indonesia (Mackie 1988). Nonetheless, adoption of tree cropping in response to these programs was modest at best. Conversion of upland farming systems depended heavily on government subsidies (Huszar and Cochrane 1990; McCauley 1988), and farmers often reverted to old practices soon after a project ended. Soil erosion rates from Java's uplands remained high, much to the confusion of planners who failed to understand how peasant-based agroforestry programs could meet with so little success in a country famed in environmental circles for its home gardens.

Where the more narrowly constructed environmental initiatives failed to arrest erosion, however, a commercial "fruit boom" had dramatic stabilizing effects. The following case study examines the development of apple-based agroforestry in a high mountain village in the upper watershed of the Brantas River in East Java. In much of this region, economic depression during the 1930s, followed by war in the 1940s, and subsequent disease outbreaks and soil fertility exhaustion (Hefner 1990), have caused widespread poverty and land degradation. Since the introduction of apples in the late 1970s, however, many farmers have adopted sophisticated soil conservation measures to support fruit production.

Unprotected sloping soils in this region erode at the rate of 2 cm per year, exposing and destroying roots within the lifetime of apple trees (Carson 1989). Construction of bench terraces is thus a prerequisite to apple farming, and smallholders and large growers alike have built terraces in the anticipation of growing apples. By the time apple seedlings are planted, the completion of backsloping terraces and closed ditches between terraces has accounted for roughly 1,000 person days of labor investment per hectare. During heavy rainfall, virtually all mud carried by water runoff collects in the ditches of each terrace bench. After the rain, farmers return the mud to the terraces, thus minimizing the loss of topsoil and fertilizers.

Apple-based farming has markedly changed the agronomic and conservation scene. Approximately three-quarters of the land in the village has been converted into terraced apple orchards or apple-based agroforestry. Of the remaining lands, about half have already been terraced. Overall, close to 90 percent of lands in the village have been "stabilized" in this manner within the last two decades. Government officers both at district and provincial levels, struggling in their efforts to reduce soil erosion from Java's upper watersheds, have applauded this development, and this village has often been cited as a model of successful upland management practices (Carson 1989; KEPAS 1988).

Changing social relations of apple-based production

> There is no landlord in this village, but we have plenty of apple-lords. This is a good arrangement because nobody loses all means to make a living. A small farmer can still grow vegetables even when the trees on his land are leased-out.
>
> (former village head, 1991)

Temperate fruit fills a particular, albeit small, niche in the urban market of Indonesia, and apples are the most important temperate fruit crop in Indonesia. In 1980, the Indonesian government banned the imports of many categories of food, including most fresh fruits. As a result, domestically produced temperate fruit such as apples enjoyed a buoyant market. In the few areas suitable for growing apples, such as in this study village, an economic boom followed. One of the challenges in growing temperate fruit in the tropics is finding ways to prevent bud dormancy in the absence of variation in temperature and day length. Workers must defoliate and modify plant architecture to stimulate buds to flush. Cultivation of apple trees in the tropics also relies on the frequent application of heavy doses of pesticides and fungicides. With heavy inputs of labor and fertilizers, apple trees in Java can be harvested twice each year.

In 1991, 94 percent of the lands in the village were owner operated, with an average holding of 0.53 hectare. A few large farms of more than 2 hectares belonged to the richest 6 percent, and they covered only about a quarter of lands in the village, which is fairly typical of the region (cf. Hefner 1990). While the seemingly egalitarian distribution pattern indicates that the most recent economic boom has not resulted in land accumulation by richer peasants, this finding belies the ongoing struggle, not over land, but over the utilization of space beneath the apple trees. Just as in The Gambia, boom conditions produced tensions and competition between fruit growers and vegetable gardeners.

Apples are intercropped with underlying vegetable crops, including leeks, scallions, garlic, cabbages, and potatoes. By 1991, close to 80 percent of all landowning households have planted apple trees in their vegetable gardens. Customary law in Java does not distinguish land tenure rights along gender lines, hence tree planting does not cause intra-household tensions as in The Gambia. Instead, conflicts develop along class lines, as capital-rich farmers favor apple trees. The high commercial value of apples has reinforced the separation of tree tenure from land tenure. Apple trees constituted a valuable asset with higher marketability than land itself, and were often exchanged independently of land. In times of emergency, rights over trees, especially mature trees at fruit-bearing stage, can quickly be liquidated to raise cash. In 1991, only 69 percent of trees four years and older were owner operated, as compared to 91 percent of the younger ones.

Tree transfers under such circumstances contribute to a process of rapid economic differentiation without apparent land accumulation (Suryanata 1994). Despite the fact that the pattern of land distribution remains relatively undisturbed, a new class of "apple lords" emerged as the village's dominant power. The richest 15 percent controlled only 50 percent of the land in the village, but 80 percent of the apple harvest. Similarly, although only 21 percent of the village's households were landless, 68 percent did not have any access to apple harvest. Despite the fact that the largest landholding was only 5 hectares, the largest apple farmer operated close to 15,000 trees growing on 20 hectares of land.

Mechanisms for the transfer of tree assets varied. Tree seedlings themselves were sometimes sold and transplanted, but the transfer of rights to trees *and* the space they occupy was more common. Although the land tenancy rate in this village was only 6 percent of all individual landholdings, close to 20 percent were operated under some form of *tree* tenancy, and that figure appeared to be growing.

By transferring only the tree tenure, a landowner retained the rights *to* other uses of the land. A structural tension was nonetheless created between the two land management systems.

Two specific forms of tree transfer emerged. The institution of tree sharecropping *(maro apel)* began about a decade before in this village, and is a modified form of a credit arrangement, once common among vegetable growers. Sharecroppers provide the capital, and in most cases, the labor and skills necessary for the cultivation of apple trees. Landowners provide the land but retain the rights to grow annual crops underneath the trees until it is prohibitively difficult to do so. The terms of tree sharecropping specify how profit from apple production is to be divided, and rules on other access *to* the land where the trees are standing. In contrast to vegetable sharecropping, the longevity of apple trees and their permanent tenure preclude terminating the contract at a season's notice, unless landowners compensate their tenants for the trees, a practical impossibility in most cases given their high value.

Tree leasing *(sewa apel)* is a post-boom phenomenon. As capital-rich apple growers begin to acquire management skills and reduce production risks, they increasingly favor fixed-rent leasing. Persistent credit needs of smaller-scale owner-operators have created a market for apple trees separately from the lands. The typical arrangement involves capital-rich growers leasing apple trees from landowning, capital-poor peasants. Invariably, the reason for leasing out trees is a pressing need for cash, which may arise from crises or basic demands of household reproduction, such as the illness or death of a family member, children's education, and house construction expenses. It may also arise from the desire to possess luxury goods such as motor vehicles that have become more common as the new prosperity contributed toward changes in consumption patterns (cf. Lewis 1992). Most often, the need to lease out apple trees arises from the inability to maintain young trees that have absorbed investment capital, but not yet produced any return. Renting out the trees is the *only* option if a farmer does not want to lose the investment made thus far. If a farmer owns several fields, tree leasing of one plot may be a way to raise capital to finance the operation costs for another field. The rent is typically negotiated and payable in advance, albeit within the context of a renter's market. In most cases, the liquidity crisis puts the lessor in a disadvantaged position, resulting in a very low rent relative to the potential yield.

The duration of the lease contract ranges from one harvest to as long as fifteen years (thirty harvests under a double crop regime). If a lessor needs extra cash before the contract expires, the lessor can choose to extend the contract in return for an agreed sum of money, or a share of the net profit of an agreed number of harvests. The lessor's bargaining position then, however, is far weaker than when the contract was first established. The lessee is in a position to negotiate a lower rent, impose more restrictions on growing field crops, or advance a permanent tenure claim to the trees. With the reduced amount of resources available to a lessor household after it enters into the contract, the likelihood of needing further credit extensions before the lease term expires is fairly high. Of the twenty-nine cases of tree leasing in the study, more than half have renegotiated their contracts before the original terms expired, resulting in increased benefits for tree lessees. As one lessee put it in 1991:

> In 1984 I rented 900 apple trees from my neighbor for twenty harvests. Five harvests into the lease, he wanted to borrow more money. In return, he would stop growing vegetables on this land. I agreed to suspend the lease for one season, and share the net profit of the sixth harvest. Because of this adjustment, when the lease expires I gain the right to sharecrop the trees even though I did not plant them.

Agroforestry and labor control

After a long string of failures in stabilizing the environment in Java's sloping uplands, improved market incentives for tree products have enhanced the adoption rate of tree planting. At the outset, apple-based agroforestry seemed to offer a sustainable and equitable solution to the problems of poverty and soil erosion that characterized the village twenty years before. Indeed, the case appeared to counter arguments that link agricultural commoditization with environmental degradation (Blaikie 1985; Grossman 1981), insofar as apple cultivation provided incentives for land improvement and rehabilitation, while simultaneously bringing economic prosperity.

In sharp contrast to this vision, however, the new land-use system was neither environmentally sound nor equitable. Instead of developing into a system with a high biological diversity that requires low inputs, apple-based farming systems were increasingly simplified, requiring extensive use of chemicals. While this system did play a role in reducing soil erosion, the reduction did not come from the vertically intermingled plant cover as in traditional home gardens, but from the heavy labor input for constructing and maintaining terraces.

As apple trees matured, spatial conflict and competition between apple trees and vegetables increased. Village surveys showed that in owner-operated fields, apple trees continued to be intercropped with vegetable crops, in spite of expanding canopies and intensive maintenance of the trees. In fields under contracts, however, the ecological competition between vegetables and trees is pivotal in the struggle over resource control. Tree lessees seek to limit access by landowners in order to protect apple crops from accidental bruising. Landowners, on the other hand, wish to defend their residual rights of growing vegetable crops even after giving up the rights to apple trees. In such struggles, tree lessees invariably come out as winners. Their advantages are exercised either through formal term in the contract extensions or through the reckless practices of apple workers, most important of which is the frequent trampling of vegetable crops. As a result, just as in The Gambia, many fields effectively turned into monoculture apple orchards, which deprived landowners of access to their own land.

The system's equity soon deteriorates as input costs are driven up by the increasing demand for a controlled environment. The spatial conflict peculiar to the configuration of apple-based agroforestry also serves as a means of labor control for the "apple lords." Labor need is highest during the first ten weeks of each season when the trees are defoliated, fertilized, and pruned. Competition for hiring wage laborers escalates during peak operations. Apple lords growing more than 1,000 trees secure laborers in dependent wage-labor relations akin to patron–client relationships. Patrons offer benefits that included loan provisions with low or no interest, access to fodder from patrons' fields, or year-round guarantees of employment. Under such terms in 1991, about 24 percent of the lessor/landowners also worked as paid laborers for their tree lessees. These arrangements provide landowners with the opportunity to personally ensure that apple maintenance does not cause trampling damage to the vegetable crops. The landowner's residual rights are thus appropriated by the apple patron and *returned* to the landowner as part of a labor contract. Thus, while the labor relation may partially mitigate the effect of lost control over trees, it does so only under terms that increase the dependency of landowners on their creditors/tree lessees, deepening the imbalance of power between them.

A combination of tenure multiplicity and intercrop dynamics unique to agroforestry has actually facilitated economic polarization in this village. Tree leasing in particular slowly dispossessed capital-poor landowners from any land-based production, as access to growing field crops is increasingly suppressed by the lessees. In addition, apple cultivation often pushes vegetable growers into dependent wage–labor relationships. Despite their formal landowning status, they have formed a new class of "propertied labor" (cf. Watts 1994) as the original multi-purpose agroforestry system has given way to monoculture apple orchards, controlled by the richest few.

CONCLUSION

It is easy to invoke the environmental crisis and the poor people's energy crisis to open up new avenues for reductionism and commodity production. (see Shiva 1988.)

We have argued in this chapter that agroforestry approaches are not always the unalloyed good they are sometimes made out to be. In practice, "stabilization" efforts involving tree crops are often highly ambiguous. Our two case studies examined agroforestry practices premised on the commoditization of tree crops and the assumption that market incentives enhance the rate of tree planting (Murray 1984). Both cases, however, show the contradictions of efforts to stabilize the environment through the market as commoditization leads to shifting patterns of resource access and control. In each place, this process takes on different characteristics, producing different forms of social friction and resistance depending on local social structure and institutions. In The Gambia, gender conflict between husbands and wives has grown out of multiple tenure claims to patrilineal land, which intensified with the commoditization of fruit trees. Similarly, the tree boom in upland Java was the cause of *inter-class* tenure conflict as commercialization polarized the village's peasantry. Both case studies illuminate the need to recognize basic political ecological considerations, such as identifying clearly on whose behalf stabilization efforts are undertaken, specifying who is in the position to define stability and determine when in fact it is achieved. In addition, these cases show the centrality of agro-ecological dynamics in shaping how various interests define and negotiate access and control over agroforestry resources.

In the case of The Gambia's garden boom, in each of the hundreds of garden perimeters springing up over two decades from roughly 1975–1995, the ecological and economic significance of wells, fences, soil improvements, and tree stands must be assessed in light of competing local, national, and international interests. Wells, fences, and soil improvements provide the necessary conditions for vegetable production and thus serve the needs of both vegetable growers and their families heavily dependent on vegetable incomes. But such improvements also tie female labor to a specific spatial domain, thereby stabilizing conditions that allow landholders to establish orchards. The addition of the tree crop, in turn, negates the value of the infrastructure for gardeners, effectively destabilizing their productive base, and actually compounding problems within a broader political economic context by attracting the intervention of outside donors interested in claiming the land improvements as their own (Schroeder 1993). Similarly, Javanese upland farmers have built elaborate terrace systems to stabilize their land resources and accommodate commercial apple-based farming. The presence of high-value apple trees, however, is conducive for the development of tree-leasing contracts and a gradual dispossession of land resources, and thereby helps capital-rich apple lords to establish and accumulate apple orchards. As a result, while the threat of soil erosion to downstream interests may have been reduced, the value of this "stable" environment to the landowners themselves has been shrinking.

Viewed from a slightly broader perspective, the loan-seeking behavior of men in The Gambia's North Bank District has forced their vegetable-growing wives to intensify horticultural production through expansion of fence enclosures and tree planting. Landholders – a select group of men who hold senior positions in family lineages – have finessed the issue of enclosure in a way that allows them to control women's labor and capture subsidies intended for the construction of garden infrastructure. Non-governmental donor agencies use landholders' leverage over vegetable growers to meet their own objectives of land stabilization via tree planting (Lawry 1988; Mann 1989; Norton-Staal 1991; Thoma 1989; Worldview International Foundation 1990). And the state and multilateral donors build on NGO successes to meet national goals in environmental stabilization, agricultural diversification, and full-scale economic readjustment (Agroprogress International 1990; Government of The Gambia n.d., 1990; Thiesen *et al.* 1989; Thoma 1989; USAID 1991). This implies, quite simply, that developers at all levels pin their hopes, indeed stake their very legitimacy in some cases, on the continued mobilization of unpaid female labor. Once again, Java offers a striking parallel. After decades of failure in promoting tree cropping by upland smallholders, district and provincial governments point to the recent growth of fruit-based agroforestry as an indicator of success in meeting the goals of environmental stabilization and economic development. Commercial agroforestry has become a model for upland development, and donor-assisted programs have funded new research and development efforts directed at fruit trees suitable for upland farming. At the national level, the government is interested in exploiting the growing international markets for tropical fruit and thereby increasing its non-traditional exports. As in The Gambia, these various interests are premised on the development of a new class of "fruit lords" who can mobilize the labor of capital-poor landowners to their own ends.

We contend on the basis of this evidence that there is a contradiction at the heart of commercial agroforestries undermining their effectiveness as strategies of resource stabilization. The strengths of agroforestry systems do not lie exclusively in the ways they enhance productivity or reverse degradation; they also rest in the opportunities afforded for sheltering a multiplicity of claims and uses. From a political ecological point of view, agroforestry systems are strongest when people can manage their resources independently, beyond the scope of powerful interests that often converge when commercial incentives increase the rigidity and exclusivity of claims.

REFERENCES

Agroprogress International (1990) *Project Preparation Consultancy for an Integrated Development Programme {European Community} for the North Bank Division*. Bonn: Agroprogress International.

Blaikie, P. (1985) *The Political Economy of Soil Erosion in Developing Countries*. London: Longman.

Bryant, R. (1994) "The rise and fall of *taungya* forestry: social forestry in defense of the Empire," *The Ecologist* 24 (1): 21–6.

Carson, B. (1989) "Soil conservation strategies for upland areas in Indonesia." East–West Center Environment and Policy Institute, Occasional Paper 9. Honolulu.

Castree, N. and B. Braun (eds) (2001) *Social Nature: Theory Practice, and Politics*. Malden, Mass. and Oxford: Blackwell Publishers.

Castree, N. and T. Macmillan (2001) "Dissolving dualisms: Actor-networks and the reimagination of nature," in N. Castree and B. Braun (eds) *Social Nature: Theory, Practice, and Politics*. Malden, Mass. and Oxford: Blackwell Publishers pp. 208–24.

Clay, J. (1988) *Indigenous Peoples and Tropical Forests: Models of Land Use and Management from Latin America*. Cambridge, MA: Cultural Survival Inc.

Cronon, W. (1983) *Changes in The Land: Indians, Colonists, and The Ecology of New England*. New York: Hill and Wang.

Dove, M. (1990) "Socio-political aspects of home gardens in Java," *Journal of Southeast Asian Studies* 21 (1): 155–63.

Farrell, J. (1987) "Agroforestry systems," in M. Altieri (ed.) *Agroecology: The Scientific Basis of Alternative Agriculture*, Boulder: Westview Press.

Fortmann, L. and J. Bruce (eds) (1988) *Whose Trees? Proprietary Dimensions of Forestry.* Boulder: Westview Press.

FitzSimmons, M. and D. Goodman (1998) "Incorporating nature: environmental narratives and the reproduction of food," in B. Braun and N. Castree (eds) *Remaking Reality: Nature at the Millenium.* London and New York: Routledge pp. 194–220.

Gadgil, M. and R. Guha (1992) *This Fissured Land: an Ecological History of India.* Berkeley: University of California Press.

Goswami, P. (1988) "Agro-forestry: practices and prospects as a combined land-use system," in L Fortmann and J. Bruce (eds) *Whose Trees? Proprietary Dimensions of Forestry.* Boulder: Westview Press.

Government of The Gambia (1990) *National Natural Resource Policy.* Banjul: Government of The Gambia.

—— (n.d.) *Executive Summary: Programme for Sustained Development. Sectoral Strategies.* Banjul: Gambia Round Table Conference.

Grossman, L. (1981) "The cultural ecology of economic development," *Annals of The Association of American Geographers* 71(2): 220–36.

Hefner, R. (1990) *The Political Economy of Mountain Java.* Berkeley: University of California Press.

Huszar, P. and H. Cochrane (1990) "Subsidization of upland conservation in West Java: the Citanduy II Project," *Bulletin of Indonesian Economic Studies* 26(2): 121–32.

KEPAS (1988) *Penelitian agroekosistem lahan kering Jawa Timur.* Bogor, Indonesia: KEPAS.

King. K. (1988) "Agri-silviculture (*taungya* system): the law and the system," in L. Fortmann and J. Bruce (eds) *Whose Trees? Proprietary Dimensions of Forestry.* Boulder, CO; Westview Press.

Lawry, S. (1988) *Report on Land Tenure Center Mission to The Gambia.* Madison: University of Wisconsin Land Tenure Center.

Leach. M. (1994) *Rainforest Relations: Gender and Resource use Among the Mende of Gola, Sierra Leone.* Washington, DC: Smithsonian Institution Press.

Lewis, M. (1992) *Wagering the Land: Ritual, Capital and Environmental Degradation in the Cordillera of Northern Luzon, 1900–1986.* Berkeley: University of California Press.

McCauley, D. (1988) *Citanduy Project Completion Report, Annex v: Policy Analysis.* USAID: Jakarta.

Mackie, C. (1988) *Tree Cropping in Upland Farming Systems: An Agroecological Approach.* USAID/ Indonesia, Upland Agriculture and Conservation Project.

Mann. R. (1989) "Africa: the urgent need for tree-planting," Methodist Church Overseas Division, unpublished manuscript.

Millon. R. (1957) "Trade, tree cultivation and the development of private property in land," *American Ethnologist* 57: 698–712.

Murray. G. (1984) "The wood tree as a peasant cash crop: an anthropological strategy for the domestication of energy," in R. Charles and A. Foster (eds) *Haiti Today and Tomorrow.* New York: University Press of America.

Nair, P. (1989) *Agroforestry Systems in the Tropics.* Boston: Kluwer Academic Publications and ICRAF.

—— (1990) "The prospects for agroforestry in the tropics," World Bank Technical Paper No. 131. Washington, DC: World Bank.

Neumann, R.P. (1998) *Imposing Wilderness: Struggles over Livelihood and Nature Preservation in Africa.* Berkeley: University of California Press.

Norton-Staal, S. (1991) *Women and their Role in the Agriculture and Natural Resource Sector in The Gambia.* Banjul: US Agency for International Development.

Peet, R. and M. Watts (eds) *Liberation Ecologies.* London: Routledge.

Peluso, N. (1992) *Rich forests, Poor People: Resource Control and Resistance in Java*. Berkeley: University of California Press.

Peluso, N.L. and M. Watts (eds) (2001) *Violent Environments*. Ithaca, N.Y. and London: Cornell University Press.

Pulido, L. (1996) *Environmentalism and Economic Justice: Two Chicano Struggles in the Southwest*. Tucson: University of Arizona Press.

Raintree, J. (1987) *Land, Trees and Tenure*. Nairobi, Kenya and Madison, Wisconsin: ICRAF and University of Wisconsin Land Tenure Center.

Rocheleau, D. (1987) "Women, trees and tenure: Implications for agroforestry research and development," in J. Raintree (ed.) *Land, Trees and Tenure*. Nairobi, Kenya and Madison, Wisconsin: ICRAF and University of Wisconsin Land Tenure Center.

Rocheleau, D. and L. Ross (1995) "Trees as tools, trees as text: struggles over resources in Zambrana Chacuey, Dominican Republic," *Antipode* 27(4): 407–28.

Schroeder, R. (1993) "Shady practice: gender and the political ecology of resource stabilization in Gambian garden/orchards," *Economic Geography* 69(4): 349–65.

—— (1995) "Contradictions along the commodity road to environmental stabilization: foresting Gambian gardens," *Antipode* 27(4): 325–42.

—— (1996) "'Gone to their second husbands': marital metaphors and conjugal contracts in The Gambia's female garden sector," *Canadian Journal of African Studies* 30(1): 69–87.

—— (1999) *Shady Practices: Agroforestry and Gender Politics in The Gambia*. Berkeley: University of California Press.

Schlager, E. and E. Ostrom (1992) "Property rights regimes and natural resources: A conceptual analysis," *Land Economics* 68: 249–62.

Shiva. V. (1988) *Staying Alive: Women, Ecology and Development*. London: Zed Books.

Suryanata, K. (1994) "Fruit trees under contract: tenure and land use change in upland Java," *World Development* 22(10): 1567–78.

Tarrant, J., E. Barbier, R. Greenberg, M. Higgins, S. Lintner. C. Mackie. L. Murphy and H. van Veldhuizen (1987) *Natural Resources and Environmental Management in Indonesia: An Overview*. USAID: Jakarta.

Thiesen, A., S. Jallow, J. Nitder and D. Philippon (1989) "African food systems initiative," project document. The Gambia: US Peace Corps.

Thoma, W. (1989) *Possibilities of Introducing Community Forestry in The Gambia. Pt. 1*. Gambia-German Forestry Project, Deutsche Gesellschaft for Technische Zusammenarbeit (GTZ). Feldkirchen, Germany: Deutsche Forstservice.

USAID (1991) *Agricultural and Natural Resource Program. Program Assistance Initial Proposal*. Banjul: Government of The Gambia.

USAID and Government of Indonesia (1983) *Composite Report of the Watershed Assessment Team*. Baryul: Government of The Gambia.

Vandergeest, P. (1997) "Rethinking property," *The Common Property Digest* 41: 4–6

Watts, M. (1992) "Idioms of land and labor: producing politics and rice in Senegambia," in T. Bassett and D. Crummey (eds) *Land in African Agrarian Systems*. Madison: University of Wisconsin Press, pp. 157–93.

—— (1994) "Life under contract: contract farming, agrarian restructuring, and flexible accumulation," in P. Lime and M. Watts (eds) *Living under Contract: Contract Farming and Agrarian Transformation in Sub-Saharan Africa*. Madison: University of Wisconsin Press.

Whatmore, S. (1999) "Hybrid geographies," in D.B. Massey, J. Allen and P. Sarre (eds) *Human Geography Today*. Cambridge, UK and Malden MA: Polity Press; Blackwell, pp. 22–40.

Worldview International Foundation (1990) *WIF Newsletter* 3,1: 4.

12

GENDER CONFLICT IN GAMBIAN WETLANDS

Judith Carney

The startling pace of environmental change in the Third World during recent decades has directed increasing attention in regional studies to political ecology. This research framework combines the broad concerns of ecology with political economy, especially the ways that institutions like the state, market, and property rights regulate land use practices (Blaikie and Brookfield 1987; Peet and Watts 1996). The growing association of environmental change with female-based social movements and gender conflict within rural households, however, suggests the need for improved understanding of gender relations and the domestic sphere since class as well as non-class struggles over resources are frequently mediated in the idiom of gender (Carney and Watts 1991; Hart 2002; Pearson and Jackson 1998; Shiva 1989). This poststructuralist emphasis on gender and household relations offers a political ecological conceptualization of the complex and historically changing relations that shape rural land-use decisions (Peet and Watts 1996).

A central insight of poststructuralist research in "Third World" agrarian studies over the past fifteen years is the need to extend the definition of politics from the electoral politics of the state and/or between classes to one that includes the political arenas of the household and workplace (Guyer and Peters 1987; Hart 1991: 95). This emphasis brings attention to the crucial role of family authority relations and property relations in structuring the gender division of labor and access to rural resources (Carney and Watts 1991; Moore 1988). However, as development interventions, environmental transformations, and markets place new labor demands and value on rural resources, these socially constructed relations of household labor and property rights often explode with gender conflict. Struggles over labor and resources reveal deeper conflicts over meanings in the ways that property rights are defined, negotiated, and contested within the political arenas of household, workplace, and state (Hart 1991: 113). As an active structuring force, gender informs the non-capitalist class relations that frequently attend projects of modernization in Third World societies (Gibson-Graham and Ruccio 2001; Hart 2002). By linking property rights and

gender conflict to environmental change, this chapter extends the poststructuralist concern with power relations and discourse to political ecology (Peet and
Watts 1996).

The environmental transformation of the wetlands of Gambia, a small country
in West Africa, provides the setting for this chapter, an examination of the
gender-based resource struggles that have accompanied the development of
wetlands to irrigation schemes. I use multiple case studies of two forms of
wetland conversion – irrigated rice schemes and horticultural projects – to trace
the disputes that have surfaced over the past fifty years in Mandinka households
over women's land rights. The analysis reveals repeated gender conflicts over rural
resources as male household heads concentrate landholdings in order to capture
female labor for surplus production. Mandinka gender conflicts of the wetlands
involve disputes over women's traditional land rights within the common
property system, thereby illustrating the significance of struggles over meaning
for contemporary struggles over labor and resources.

Building upon previous research in several Gambian wetland communities,
the chapter is divided into five sections. The first presents the environmental
context of the Gambian wetlands, the extent and significance of wetland farming,
as well as women's labor in ensuring its productive use. The next section provides
an historical overview of environmental and economic changes modifying
women's access to Gambian wetlands. This follows an account of recent policy
shifts addressing the country's environmental and economic crisis. Two case
studies then detail the relationship between economic change and the forms of
women's resistance to the process of land concentration. The chapter concludes by
analyzing how wetland commodification has made women's access to resources
increasingly tenuous despite income gains.

THE ENVIRONMENTAL CONTEXT
OF THE GAMBIAN WETLANDS

Gambia, a narrow land strip 24–50 kilometers (14–30 miles) wide and nearly
500 kilometers (300 miles) long, encloses a low-lying river basin that grades
gradually into a plateau where the altitude seldom exceeds 100 meters (325 feet)
(Figure 12.1). The plateau forms about one third of the country's land base and
depends upon rainfall for farming. Precipitation during the months of June to
October averages 800–1100 milliliters (31–43 inches) and favors the cultivation
of millet, sorghum, maize, and peanuts. As with neighboring Sahelian countries,
the Gambian rainfall regime fluctuates considerably between year and within a
season. Between the 1960s and 1980s, for example, annual rainfall declined by
15–20 percent and became increasingly distributed in a bi-modal seasonal
pattern (Hutchinson 1983: 7). The recurrence of a two week, mid-season dry spell
during the month of August increased cropping vulnerability on the uplands and
dependence on lowland farming (Carney 1986: 25–30).

317

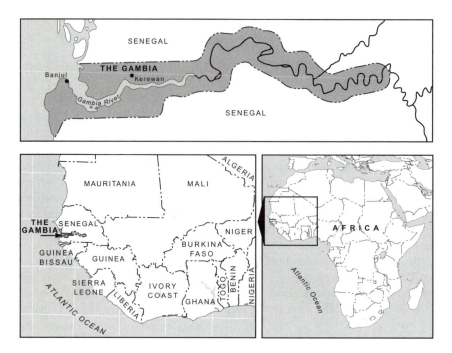

Figure 12.1 Gambia.

These wetlands are critical for understanding human livelihood and survival in the unstable rainfall setting of the West African Sudano-Sahelian zone. Lowland environments permit a multiple land-use cropping strategy which utilizes other forms of water availability, thereby freeing agricultural production from strict dependence of rainfall. Constituting nearly 70 percent of the country's landmass the Gambian wetlands make available two additional environments for agriculture: (1) the alluvial plain flooded by the river and its tributaries; and (2) a variety of inland swamps which receive water from high water tables, artesian springs, or occasional tidal flooding (Figure 12.2). Farming these critical environmental resources enables en extension of crop production into the dry season or even year-round. Gambian wetlands are traditionally planted to rice, although vegetables are frequently grown with residual moisture following the rice harvest.

While Gambia abounds in lowland swamps, not all are suitable for farming. Riverine swamps coming under marine tidal influence are permanently saline within 70 kilometers (42 miles) of the coast, seasonally saline up to 2,150 kilometers (150 miles) and fresh year round only in the last 150 kilometers (90 miles) of the Gambia River's course (Carney 1986: 33). The suitability of inland swamps for crop production, moreover, depends on the influence of differing moisture regimes for groundwater reserves. Consequently, although

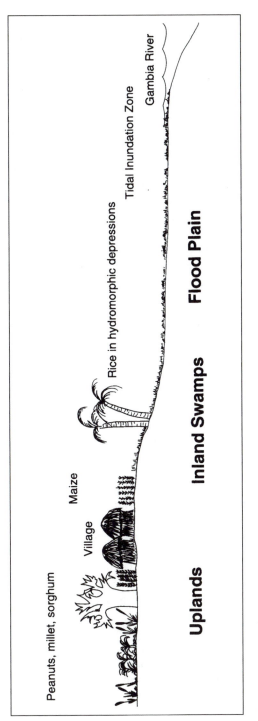

Figure 12.2 Agro-ecological zones, Central Gambia.

Source: Carney (1986).

Gambia contains over 100,000 hectares (247,000 acres) of lowland swamps, only about a third can be reliably planted (CRED 1985: 127). Until the mid-1980s most of the available swampland was farmed to rice, with about 20,000–25,000 hectares (49,400–61,750 acres) planted along the river flood plain and another 6,000–8,000 hectares (14,800–19,760 acres) cultivated in inland swamps (FAO 1983: 17).

Wetland cultivation is thus pivotal to the Gambian farming system, enabling crop diversification over a variety of micro-environments and reduction of subsistence risk during dry climatic cycles. Wetland agriculture, however, requires considerable attention to forms of water availability as well as edaphic and topographic conditions. In Gambia this knowledge is embodied in women who have specialized in wetland cultivation since at least the early seventeenth century and have adapted hundreds of rice varieties to specific micro-environmental conditions (Carney 1991: 40; Jobson 1904 [1623]: 9). This cumulative *in situ* knowledge of lowland farming underlies Gambia's regional importance as a secondary center of domestication of the indigenous West African rice, *Oryza glaberrima*, cultivated in the area for at least 3,000 years (Portères 1970: 47).

GENDER, ENVIRONMENT, AND ECONOMY

Although lowland swamps and rice production are traditionally women's domain, prior to the mid nineteenth century both men and women were involved in upland and lowland cropping systems. Men assisted in the fields clearing for rice cultivation while women weeded upland cereal plots (Carney and Watts 1991: 657; Weil 1982: 45–6). The abolition of the Atlantic slave trade and the turn to "legitimate commerce" in the nineteenth century led to Gambia's incorporation into the world economy thorough commodity production. By the 1830s peanut cultivation proliferated on the uplands. The imposition of British colonial rule by the end of the century brought taxation and fiscal policies, thereby accelerating reliance of peanuts as a cash crop. These political–economic changes resulted in an increasingly specialized use of agricultural space and a more gendered division of labor. The effects of colonial rule became most evident among the rice growing Mandinka, Gambia's dominant ethnic group and principal wetland farmers.

By the end of the nineteenth century, Mandinka men's growing emphasis in peanut cultivation resulted in a reduction in millet and sorghum production for household subsistence (Weil 1973: 23). Women compensated for upland cereal shortfalls by augmenting rice production in lowland swamps. The gender division of labor became increasingly spatially segregated with the cash crop concentrated on the uplands under male control and women's farm work largely oriented to lowland rice, which emerged as the dietary staple (Carney 1986: 89–91; Weil 1973) (see Figure 12.2). The specialized use of agricultural land and concomitant disruptions in the gender division of labor accompanying nineteenth-century commodity production provide the setting for understanding twentieth-century gender conflicts among the Mandinka over commodification of the Gambian wetlands.

Policy interests in wetland environments began in the early decades of the twentieth century when local officials began documenting farming practices in diverse lowland settings. The objective was to improve household subsistence security and generate rice surpluses which would feed an expanding pool of migrant male laborers whose seasonal influx (c. 20,000) accounted for the pace of peanut expansion in the uplands (Carney 1986; Swindell 1977). Initial colonial efforts aimed at improving swamp accessibility, opening up new areas for cultivation, and increasing rice yields through improved seeds. By the 1960s swamp development projects had culminated in an expansion of rice planting to some 26,000 hectares (65,000 acres) (Carney 1986: 178). But limits had been reached on the degree to which women could carry the subsistence burden. Further gains in food availability rested on altering the gender division of labor by drawing men into rice growing. The colonial government's inability to persuade Mandinka men to take an active part in rice cultivation prompted Governor Blood to lament: "[Rice] is still almost entirely a woman's crop and I doubt whether more woman hours can be devoted to this form of cultivation" (NAG 1943). With absolute limits reached on available female labor for rice cultivation, swamp rice projects came to a halt (Carney and Watts 1991: 660).

In 1949 the colonial government initiated another approach to surplus rice generation by implementing a large-scale irrigation scheme on the site of the present-day Jahaly Pacharr project. The Colonial Development Corporation (CDC) scheme departed from the earlier swamp rice improvement project in one important way: land was removed from female growers through a 30-year lease program in an effort to force men into rice cultivation (Carney 1986: 126; Carney and Watts 1991: 666). The project failed due to a poorly designed irrigation system and a lack of male and female interest in wage work; yet the CDC scheme is notable for foreshadowing the post-independence emphasis on irrigation as well as the gender-based conflicts that would surface in subsequent wetland development projects.

These conflicts center on resistance by male households heads to repeated attempts by the colonial and post-colonial state to force them into rice cultivation, a change in household labor patterns crucial for producing a marketable surplus of the dietary staple. Gambian rice projects have accordingly exploded with gender conflicts as males attempt to deflect the labor burden in rice growing onto their wives and daughters while simultaneously making new claims to the surpluses produced by female labor. This objective has been facilitated by manipulating customary tenure "laws" to reduce women's individual land rights in developed wetlands and thus, female remuneration from the value produced by their labor. The steady erosion of women's crop rights over the past fifty years has erupted into mounting female militancy in technologically improved wetlands.

The stage for the ensuing conflict was set in the 1940s when colonial development policies improved swampland access and rice productivity. Male household heads and village elites responded to swamp development by calling into question women's long-standing rights on the wetlands. In one case that reached the colonial authorities, Mandinka men articulated emerging norms, arguing that "if women mark the land and divide it, it would become 'women's property' so that when a husband dies or divorces his wife, the wife will still retain the land, which is wrong. Women must not own land" (Rahman 1949: 1). Women's land rights were being contested on the grounds that wetland development would enable females to alienate developed swampland from the domestic land holding.

Such claims, however, ignored women's pre-existing rights to individual plots within the customary tenure system. Presenting female land rights as a threat to domestic property relations obfuscated the real issue at stake with colonial rice development: control over the surplus produced by female labor. Under the guise of safeguarding the domestic land holding from a female land grab, male village leaders got colonial officials to acquiesce by severing individual female land rights (*kamanyango*) in developed swamps. New land development was placed under another customary tenure category, *maruo*, or land whose product contributes to household subsistence. A similar conflict arose following the CDC project failure: women claiming the benefits of the plots they farmed (*kamanyango*) with household heads declaring the improved project land *maruo* (Dey 1980: 252–3). The significance of the *maruo* designation for resource struggles is that females no longer receive the income benefits from their farm work. The surplus, instead, comes under the control of the male household head who appropriates the value of its sale.

A brief review of the meaning of the two Mandinka terms for property rights and control over surplus elucidates the issue in dispute. Land in rural Gambia is held in communal tenure but controlled at the village level by male lineages who trace their decent to the village's first settlers. As unmarried daughters, women receive land rights from their father's landholding lineage. When they marry and move to their husband's household, land rights are exercised through their husband's kinship group.

The struggles over land use on improved wetlands derives from the multiple meanings assigned to the Mandinka term *maruo*. At the most general level, *maruo* refers simultaneously to the household landholding as well as to the labor obligations of family members towards collective food production. Thus, as men argued in the swampland development projects, the family landholding cannot be alienated from the household. But this claim conflates discussion of family property control with the intricacies of the communal tenure system which, by custom, accords all able family members a few individual plots (*kamanyango*) in return for laboring on the greater number of fields dedicated to household subsistence (*maruo*). The salient issue is that such individual plots grant junior male and female family members the right to keep the benefits from their labor and to sell the product if they so desire. Men traditionally meet their maruo work obligations on the uplands through cultivation of millet, sorghum, and maize as well as peanuts which are often sold for cereal purchases. Male *kamanyango* production is usually devoted to the cultivation of peanuts for sale. Mandinka women, whose work chiefly occurs on the wetlands, fulfill both their *maruo* and *kamanyango* production through rice cultivation.

Kamanyango plots are therefore a critical issue in Gambia, where rural society is largely polygynous. Young men compete with their fathers for wives, male and female budgets are frequently separate, and each co-wife is customarily responsible for the purchases of clothing and supplemental foods crucial for the well-being of her children. While the steady erosion in female *kamanyango* rights with subsequent rice development projects has resulted in males assuming some of the responsibilities formerly met by their wives, women often experience acutely the loss of their economic independence. The shift in wetland resource control frequently exacerbates intra-household conflict between co-wives, their husbands and fathers-in-law as females are forced into negotiating for the allocation of household resources received from their labor in rice swamps.

THE ENVIRONMENTAL AND ECONOMIC CRISIS:
POLICY SHIFTS

Since independence in 1965, Gambia has experienced rainfall declines and accelerated environmental degradation of its uplands, a massive influx of foreign aid for development assistance (1968–1988), policy shifts favoring the cultivation of commercial crops on the wetlands, and International Monetary Fund (IMF) structural adjustment programs (1985–1995). These changes have shaped post-independence accumulation strategies and gender conflicts among rural households.

Gambia entered independence with a degraded upland resource base and a vulnerable economy. The results of the longstanding monocrop export economy were evident throughout the traditional peanut basin, once mantled with forest cover but substantially deforested during the colonial period (Mann 1987: 85). Reliance on peanuts to finance mounting rice imports grew more precarious in the years after independence: peanut export values fluctuated considerably but through the 1980s, grew less rapidly than the value of food imports (FAO 1983: 4). Farmers responded to declining peanut revenues through an intensification of land use – namely, by reducing or eliminating fallow periods in peanut cultivation. Land degradation accelerated, especially along the river's north bank, nearer Senegalese peanut markets, which paid higher producer prices (see Figure 12.1).

These combined factors brought renewed attention to the wetlands. The 1968–1973 Sahelian drought coincided with an escalation of capital flows from multilateral banks and financial institutions to the Third World (Thrift 1986: 16). The changing pattern of global capital accumulation impacted the Gambian wetlands in the form of river basin development and irrigated farming.

International development assistance brought far-reaching changes to the critical wetland food production zone. Nearly 4,500 hectares (11,115 acres) of riverine swamps were converted to irrigation schemes and another 1,000 hectares (2,470 acres) of inland swamp to horticultural projects (Carney 1992: 77–8). Although affecting about one-tenth of the country's swamps, these conversions in land use have had profound consequences for food production, female labor patterns, and access to environmental resources. Irrigated rice schemes and the introduction of technology to implement year-round cultivation, however, have not reversed the country's reliance on imported rice. By the 1990s, only 40 percent of the land under this new development strategy remained in production and just 10 percent were double-cropped (Carney 1988). As domestic production lags, milled rice imports steadily climb, now accounting for more than half the country's needs (Government of Gambia 1973–2000). Population growth rates exceeding 3 percent per annum suggest demographic pressure on agricultural land; yet the failure to achieve food security is not the result of a Malthusian squeeze. Rather, it is the outcome of the changing use and access to rural resources which concentrates land within the communal tenure system while diminishing female benefits from improved rice production.

By the 1980s, women's economic marginalization in irrigated rice schemes resulted in non-governmental organizations (NGOs) targeting them for horticultural projects developed on inland swamps. The policy emphasis on horticulture intensified with the debt crisis of the 1980s and the implementation of an IMF-mandated structural adjustment program in 1985 to improve foreign exchange earnings and debt repayment. Economic restructuring has reaffirmed Gambia's "comparative advantage" in peanuts while favoring inland swamp conversion to horticulture (McPherson and Posner 1991).

The respective policy emphases of the past twenty years have commodified Gambian wetlands and accelerated changes under way in customary use and access to environmental resources. As the irrigation schemes provide new avenues for income generation within rural communities, women's access to improved land for income benefits is increasingly contested. The next two sections present an overview of the two post-drought wetland policy shifts, illustrating how customary laws are reinterpreted to reduce female access to productive resources, and the forms of women's resistance to a deteriorating situation.

"DROUGHT-PROOFING" THE ECONOMY: IRRIGATED RICE DEVELOPMENT

In 1966 the Gambian government, with bilateral assistance from Taiwan, initiated a wetland development strategy aimed at converting tidal floodplains into irrigated rice projects. The rationale for this development was to promote import substitution by encouraging rice production. Rice imports had reached 9,000 tons per annum, and foreign exchange reserves had seriously eroded with declining world commodity prices for peanuts. The 1968–1973 Sahelian drought revived late colonial interest in irrigation and mobilized foreign aid for investment in river-basin development and irrigated agriculture (CILSS 1979; CRED 1985: 17). Hailed as a way of buffering the agricultural system from recurrences of a similar disaster, irrigation projects also created demand for imported technical assistance, machinery, spare parts, and other foreign inputs. This "drought-proofing" strategy prioritized irrigated rice schemes in The Gambia. From the 1970s to the mid-1980s, the World Bank, the mainland Chinese government and the International Fund for Agricultural Development (IFAD) followed the Taiwanese irrigated rice model by developing more than 4,000 hectares of tidal floodplain swamps traditionally cultivated by women (Figure 12.3).

Despite differences in the donors' ideological perspectives, the irrigated rice projects adhered to a remarkably similar course by introducing a Green Revolution package for increased production to male household heads (Dey 1981: 109). Developed at a cost of US $10,000–$25,000 per hectare, and imposed on a gendered division of labor and land use, rice development consistently failed to deliver its technological promise. The rural economy became even more vulnerable as dependence deepened on imported fuel oil, fertilizers, and spare parts whose prices were rising (Carney 1988; CRED 1985: 273). Donor production targets required double cropping and thus a shift in agricultural production to year-round farming. While male household heads were taught how to grow irrigated rice, the cropping calendar could only be followed if male and female family members provided field labor. By placing men in charge of technologically improved rice production, the donors hoped to encourage male participation; instead, they unwittingly legitimized male control over the surpluses gained from growing two crops of rice annually.

As in the era of colonial swamp development schemes, gender-based conflicts erupted in project households over which family members were to assume the labor burden in irrigated rice cultivation. Male household heads claimed female labor under the customary category, *maruo*, but irrigation projects now meant that the claim was invoked for year-round labor. *Maruo* labor claims for household subsistence had historically evolved within the confines of a single agricultural season. There was no precedent for women to perform *maruo* labor obligations during *two* cropping periods when production would yield men a marketable surplus (Carney 1988: 306). Irrigation projects were commercializing rice production, but income gains depended on female labor availability.

Women contested the changing lexicon of plot tenure and the enclosure of traditional *kamanyango* and *maruo* swamp into irrigation schemes. For them "development" meant the delivery of female labor for intensified rice farming without concomitant income gains. The reinterpretation of customary tenure by male household heads and village elites aimed at ensuring continued female access to rice land, but only as workers on plots whose benefits would flow to men as disposable surpluses. The donors' uninformed view of the Gambian household-based production system was to prove the projects' nemesis.

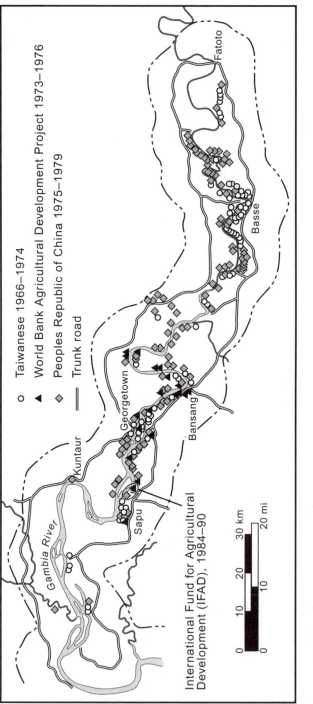

Figure 12.3 Irrigated rice projects in Gambia.

Source: Carney (1986).

Female rice farmers responded in three principal ways to loss of *kamanyango* plots and efforts to augment their labor burden: (1) by relocating *kamanyango* production to unimproved swamplands where they could generate small surpluses for sale; (2) when alternative swampland for rice farming was not available, by agreeing to perform *maruo* labor obligations on irrigated rice plots during the dry season cropping cycle in exchange for using the same plot without irrigation during the rainy season for *kamanyango* production; or (3) by working year round in the schemes but demanding remuneration in rice for labor performed for one of the cropping seasons (Carney 1994). The first two responses involved an attempt by women to re-establish *kamanyango* rights while the third focused on substituting the value of those rights with claims to part of the surplus.

In the earliest phase of Gambian irrigation schemes, donor agencies developed village swampland on a small scale (c. 30 hectares). Decisions over land use remained with community households. This system facilitated female efforts to reassert *kamanyango* land rights, even though anticipated productivity rates suffered during the rainy season when women either planted individual fields outside the scheme or within the project, without irrigation. A third response characterized the 1,500-hectare Jahaly-Pacharr irrigation scheme, implemented by IFAD in 1984, which removed land from village control. Deploying a legal mechanism reminiscent of the earlier colonial CDC project, the IFAD project negotiated a thirty-year lease. As most of the available swampland of forty contiguous villages became absorbed within the large-scale project, thereby proletarianizing numerous women from the floodplains where they customarily grew rice. The lease served as an instrument to discipline household labor for year-round irrigated rice production. Land-use decisions were placed under project management jurisdiction. While participation in the scheme was made more attractive by raising the producer price of rice, access to project plots depended upon prompt repayment of inputs and mechanization charges advanced on credit. These rates were based on anticipated productivities which could only be achieved by carefully adhering to the irrigation labor calendar (IFAD 1988). Households unable to comply with these terms faced eviction.

The project's mandate to grow two rice crops per annum as a condition for plot access placed intense pressure on household labor, which the subsequent designation of project land as *maruo* could not easily resolve. Previous irrigation schemes had frequently accommodated women's *kamanyango* claims at the cost of year-round pumped production. Confronted by a legal mechanism that threatened eviction for households falling short of production targets, female household members now faced enormous pressure to augment their labor burden. As the IFAD scheme had incorporated most of the region's available swampland, pre-existing *kamanyango* land access came to an end. Gender-based conflicts exploded throughout the project area as women resisted the erosion of their right to derive benefits from a greatly augmented work burden. While ethnicity, class, and differences in types of irrigated land within the IFAD project shaped the patterns of conflict resolution, Mandinka women responded by demanding either *kamanyango* plots within the project or 10 percent of the rice harvest as payment for their work (Carney 1994). Nevertheless, about one quarter of Mandinka households failed to honor female demand for individual plots or remuneration in kind for year-round *maruo* labor. The result was women's outright refusal to work on the family's irrigated plots.

These dispossessed women then pursued two complementary economic strategies for income generation: the formation of work groups to carry out the project's labor-demanding tasks of transplanting, weeding, and harvesting and a shift in *kamanyango* production to upland farming. By organizing work groups for hire, women managed to bid up their daily wage rate within the project and take advantage of peanut land made available as men intensified their work in the now more remunerative rice scheme (Webb 1989: 66). But their efforts to obtain upland *kamanyango* plots were not always successful as they came into direct competition with the claims of junior males for individual land rights. Women consequently placed considerable effort in capturing the support of NGOs to help them develop village vegetable gardens for income generation (Carney 1992).

In summary, Gambian rice development unfolded initially on riverine floodplains. As these areas became technologically improved, male household heads reinterpreted women's pre-existing crop rights and benefits in order to access female labor to carry out the intensified work burden. Irrigated rice development undermined women's customary access to rice land for income generation while enabling male household heads to capture surplus value. In the effort to discipline the Mandinka peasantry for domestic rice import-substitution, the Gambian state had ruptured the relationship between women's knowledge systems and agronomic expertise that had regulated wetland cultivation for centuries. The gender conflicts underlying project labor demands contributed to repeated delays in cropping schedules and lower yields.

Project management efforts to evict farmers for non-payment of advanced credits suffered a setback from 1985 when the Gambian government underwent successive IMF structural adjustments. The producer rice price dropped in favor of peanuts while fertilizer prices quadrupled. Reminiscent of prior irrigated rice schemes, the IFAD project operated principally during the dry season. IMF reforms lured males back to peanut cultivation during the wet season and grow rice chiefly as a dry season crop (Carney 1994, fieldwork). Project yields continue to decline with rising fertilizer costs and the meager application of a critical input for green revolution rice varieties. Despite the underutilization of project land, few households have faced eviction as the costs of rice production fail to attract new recruits to the scheme. Efforts to remove unproductive households from project land also proved difficult as a consequence of the project's negative reception in rural Gambia. The Jahaly-Pacharr scheme is viewed as an example of an uncaring central government attempting to "take" land from peasants.

COMPARATIVE ADVANTAGE AND
HORTICULTURAL DEVELOPMENT

Shortly after the 1968–73 Sahelian drought the Gambian government promoted economic ventures in inland swamps that proliferated over the years into a major focus of development assistance. During the 1970s, the government encouraged onion-growing schemes among village women's groups as a means to increase household incomes in the peri-urban corridor and north bank district, geographically proximate to the nation's capital (Ceesay *et al*. 1982). Over the following decade women's vegetable gardens emerged as a major focus of donor support within the country (see Schroeder and Suryanata, this volume Chapter 11). By the 1990s over 340 small (0.5–2 ha) and medium-scale (5–15 ha) vegetable gardens were developed by NGOs and multilateral donors (DeCosse and Camara 1990; Nath 1985; Sumberg and Okali 1987). The entry of private growers into the burgeoning horticultural sector accounts for the expansion of market gardening in the country.

The boom in market gardening on Gambian wetlands results from the confluence of several policy directions over the past fifteen years. Following independence, Gambia began developing its pristine beaches for international tourism; by the 1990s over 100,000 Europeans were taking the six-hour flight to vacation along the Gambian coast between November and April each year (N'Jang 1990). The initial onion projects successfully linked local production to the tourist sector and awakened donor agencies to the possibilities of expanding vegetable production to meet the dry season tourist demand. These developments meanwhile were unfolding against a growing clamor within the international donor community for women in development (WID) projects. The emergent WID focus in Gambia was pioneered by NGOs who countered male control over irrigation schemes by implementing horticultural projects on unimproved inland swamps previously sown to rice. The donors aimed to bolster female income earning opportunities by improving seasonally wet swamps with wells for dry season vegetable cultivation.

Policy support for diversifying wetland agriculture into horticulture received additional impetus with IMF structural adjustment programs. Geographic proximity to Europe encouraged policy makers to exploit Gambia's comparative advantage in winter fruit and vegetable production, as did favorable tariffs and the removal of export taxes on fresh produce (Jack 1990). Seeds of non-traditional horticultural crops such as lettuce, tomatoes, green peppers, carrots, eggplants, beans, cabbages, and tropical fruits were distributed, and marketing strategies focused on hotels, the expatriate community and export opportunities in neighboring Senegal and Europe.

By the 1990s horticultural production had expanded to rain-fed areas in the peri-urban corridor located near the international airport, with boreholes dug to reach underground aquifers. With few exceptions, these projects are operated by the state, senior government officials, and resident Lebanese and Indian landowners, and are oriented to European export markets. In the five years following IMF economic reforms, annual fruit and vegetable exports to Europe alone grew to 3,000 tons, a value exceeding US $1 million (Jack 1990). The same period witnessed growing involvement by multilateral donors (EEC, Islamic Development Bank, UNDP, and the World Bank) in women's horticultural production and marketing along the coastal corridor (Barrett and Browne 1991; Carney 1992). Despite this most recent form of donor support, Gambian women's horticultural projects remain concentrated in rural areas, on inland swamps of small areal extent (0.5–2 ha), and oriented to local and regional markets.

Although the policy emphasis on converting inland swamps to horticulture dates from the 1970s, Gambian women have long been involved in vegetable production (Schroeder 1999). By the sixteenth century, they were observed marketing vegetables during the dry season, while eighteenth-century travelers noted the cultivation of vegetables in inland swamps right after the rice harvest (Adanson 1759; Crone 1937: 48: Park 1983). Dry season horticultural production received encouragement during the colonial period, but its expansion was limited by the elementary technology employed for irrigation, namely *shadufs*, in which river water is lifted by hand with a pole and bucket (Carney 1986: 144). Although colonial horticultural programs targeted men, their failure kept vegetable growing in female hands. Women remained the country's principal producers, using residual moisture from inland rice swamps early in the dry season to cultivate traditional crops such as bitter tomatoes, okra, cowpeas, sorrel, and hibiscus for subsistence.

Donor support for well construction from the 1980s had facilitated the extension of the vegetable-growing period in inland swamps. Deep dug, concrete-lined wells on the uplands revolutionized Gambian horticultural production by tapping water tables for dry season cultivation. Vegetable gardening no longer remains a seasonal activity, as it was prior to donor involvement. Women's village gardens receiving NGO assistance grow vegetables during the entire dry season and in some areas, year round.

The provision of reliable water supplies through well-digging is central to NGO efforts to implement a rural development strategy aimed at improving women's incomes. By promoting village gardens among women's groups interested in commercialized vegetable cultivation, NGOs have launched a development strategy that targets females who were ignored in the previous wetland policy approach. NGO support for well construction remains crucial to women's negotiations with male landowners and village elites for access to land for a communal vegetable garden. At a cost of US $3,000–4,000 per hectare, these wells provide communities with a valuable infrastructure to ensure a permanent water source for dry season agricultural production (Nath 1985: 6; Schroeder 1999; Sumberg and Okali 1987).

Arrangements to secure female access to improved village gardens, however, vary between communities and depend on the availability of land locally, as well as the swamp's land-use history. Consequently, in rural communities with

NGO-supported gardens, women are granted either year-round usufruct or *kamanyango* dry season rights. Once female land access is assured, NGOs provide funding to build the concrete-lined wells and barbed wire fences (for protection from livestock damage). Once the infrastructure is in place, female growers are credited the seeds and tools for vegetable farming.

An examination of three areas of market garden development in the country underscores the significance of water availability for rural income opportunities and its relationship to contemporary gender conflicts. These include: (1) the region centered around Kerewan on the river's north bank, the site of the pilot onion schemes; (2) the peri-urban corridor close to the capital, Banjul, where tourist hotels, government offices, and expatriate community are concentrated; and (3) the Jahaly-Pacharr project, 260 kilometers up-river from the capital on the river's south bank (see Figure 12.1). The first two areas, the original foci for the country's horticultural development, enjoy numerous marketing opportunities not available to female growers in the Sapu area, who rely principally on weekly markets *(lumos)* for vegetable sales. Their incomes averaged only US $15–35 for dry season production in the mid-1990s (Carney fieldwork).

Each of these areas offers men different income opportunities. Jobs in government, the tourist sector, petty commerce, and transportation are concentrated in the capital. In the Jahaly-Pacharr project and north-bank areas, men derive incomes chiefly from agriculture. As noted above, male control over irrigation schemes and peanut farmland in the project has severely limited women's access to village land for vegetable gardens. The explosion over the past decade of banana groves planted along riverbanks and in inland swamps has provided men yet another income source. However, there is little land available for women to garden.

Female growers in the capital face an altogether different situation. The concentration of tourist hotels, an expatriate community, and the airport provide numerous marketing opportunities. But the proliferation of vegetable projects and excessive number of female participants in each scheme limit their income benefits. Local markets are saturated with women selling vegetables, which results in modest returns and a continuous search for new outlets. An important outcome of the explosion of vegetable gardens within the peri-urban corridor is the growing involvement of women's groups in contract farming production relations with large growers. While the latter arrangement provides a marketing outlet, large-scale growers and traders, who set conditions, drive down prices and control distribution networks (Carney 1992: 80–2).

Vegetable gardening nonetheless remains attractive to women whose alternative income-earning prospects are limited. While structural adjustment programs have led to a 10 percent reduction in employment within the government sector catapulting men into increasing involvement in horticultural production, women have generally maintained usufruct to village land for gardening because donor representatives, located in the capital, are poised to defend them. The case of the north bank, detailed by Schroeder and Suryanata

(Chapter 9 in this volume), differs, however, and provides yet another illustration of the complex intersection of gender with environmental transformation and economic change.

CONCLUSION

The structural dislocation of a mono-crop export economy and attendant food shortages brought government attention to the Gambian wetlands during the late colonial period. The pattern of swamp development implemented during colonial rule foreshadowed a large-scale emphasis on the wetlands that materialized with the influx of foreign capital coincident with the Sahelian drought. During the past twenty-five years, wetland development through irrigation projects has transformed Gambian agriculture from a seasonal to year-round activity, enabling agricultural diversification, surplus cereal production, and new avenues of income generation among rural households.

The promise of irrigated agriculture, however, depends upon the ability of peasant households to restructure family labor to the dictates of irrigated farming – a labor regime that requires a greater work burden during the entire calendar year. As claims to family labor evolved in the context of a limited wet season, institutional mechanisms within the household-based production system were deformed to mobilize family labor for year-round agriculture. Use of the term *maruo* for technologically improved swamps proved central for obtaining a female labor reserve for the intensified work burden in irrigated farming. The term strengthens the prevailing patriarchal power relations while facilitating the concentration of benefits produced by female labor within the household.

Women contest the semantics of *maruo* precisely because it provides a mechanism for the loss of their customary rights, They are acutely aware that the rules of access to and control over environmental resources are not a codification of immemorial tradition, but rather the outcome of struggle and negotiation with husbands, male community leaders, state and donor officials (Berry 1989; Okoth-Ogendo 1989: 14). The case studies of gender conflict in Gambian wetland projects reveals the co-existence of non-capitalist relations of production with capitalist class relations, while emphasizing the uneven process of capitalist transformation in third world projects of modernization. Gambian women are not engaging men in mere semantic discussions as they fight for *kamanyango* rights irrigated rice and vegetable projects. Rather, they mobilize non-capitalist idioms to contest the increasing labor burdens involved with wetland development schemes.

A process of land concentration is occurring with Gambian irrigation projects. Concentration is not the result of absolute land scarcity and overpopulation but rather a response to increased demands on household labor and new income opportunities with irrigated agriculture. The process of claiming irrigated plots as *maruo* reveals how land is enclosed to create an artificial scarcity for accessing female labor. This unusual type of land enclosure grants women access to irrigated plots while denying them full benefits from their work. Land concentration consequently involves the conversion of wetlands from land with multiple female rights to the surplus product to land with a single claim over the surplus produced by multiple female laborers.

In contrast to the pattern described by Humphries (1990: 38–9) for eighteenth-century England, land enclosure in Gambia has resulted in very different outcomes for women. While land enclosures in England pushed rural men into waged work and left their wives and children to defend traditional rights to rural resources, it is women in contemporary rural Gambia who are increasingly proletarianized, as men gain control of both customary rights over female labor and the income from irrigated lands. The forms of economic change promoted by national and international organizations are intensifying household labor demands in wetland cultivation, thereby spearheading a form of enclosure that uses the *maruo* designation to weaken women's customary rights to rural resources so that male heads of households can capture their labor for individualized accumulation.

In outlining the social and historical processes of changing land-use strategies on the Gambian wetlands, this chapter reveals that more than the environment is being transformed. So too are the social relations that mediate access to, and use of, land within rural households. Gender provides a window for seeing how non-capitalist social relations mediate projects of modernization in remote areas of the global economy. The contemporary pattern of accumulation unfolding in the Gambian wetlands depends on limiting access to technologically improved swamps by dissolving women's customary land rights and by imposing new work routines that undervalue and intensify female labor on irrigated land. Women, however, are resisting their assigned role as cheap labor reserves by invoking a non-capitalist discourse while negotiating labor and crop rights with household and village heads and appealing to donor and government representatives to defend them. The outcome of this phase of capitalist expansion on the Gambian wetlands is by no means certain.

REFERENCES

Adanson, M. (1759) *A Voyage to Senegal, the Isle of Gorée and the River Gambia.* London: Nourse.

Barrett, H. and A. Brown (1991) "Environment and economic sustainability: women's horticulture production in The Gambia," *Geography* 776: 241–8.

Berry, S. (1989) "Social institutions and access to resources," *Africa* 59, 1: 41–55.

Blaikie, P. and H, Brookfield (1987) *Land Degradation and Society.* New York: Methuen.

Bryant, R. (1992) "Political ecology: an emerging research agenda in Third World studies," *Political Geography* 11, 1: 12–36.

Carney, J. (1986) "The social history of Gambian rice production: an analysis of food security strategies," Ph.D. dissertation, University of California, Berkeley.

—— 1988. "Struggles over crop rights within contract farming households in the Gambian irrigated rice project," *Journal of Peasant Studies* 15: 334–49.

—— (1991) "Indigenous soil and water management in Senegambian rice farming systems," *Agriculture' and Human Values* 8: 37–58.

—— (1992) Peasant women and economic transformation in The Gambia," *Development and Change* 23: 67–90.

—— (1994) "Contracting a food staple in The Gambia," in P. Little and M. Watts (eds) *Peasants under Contract: Contract Farming and Agrarian Transformation in Sub-Saharan Africa*. Madison: University of Wisconsin Press, pp. 167–87.

Carney, J. and M. Watts (1991) "Disciplining women? Rice. mechanization, and the evolution of Mandinka gender relations in Senegambia," *Signs* 16: 651–81.

Ceesay, M., O. Jammeh and I, Mitchell (1982) *Study of vegetable and Fruit Marketing in The Gambia*. Banjul, Gambia: Ministry of Economic Planning and Industrial Development and the World Bank.

Center for Research on Economic Development (CRED) (1985) *Rural Development in the Gambian River Basin*. Ann Arbor: CRED.

CILSS (Permanent Interstate Committee for Drought Control in the Sahel) (1979) *Development of Irrigated Agriculture in Gambia: General Overview and Prospects. Proposals for a Second Program 1980–1985*, Paris: Club du Sahel.

Crone, G.R. (1937) *The Voyage of Cadamosto*. London: The Hakluyt Society.

DeCosse, P. and E. Camara (1990) *A Profile of the Horticultural Production Sector in The Gambia*. Banjul, Gambia: Department of Planning and Ministry of Agriculture.

Dey, J. (1980) "Women and rice in The Gambia: the impact of irrigated rice development projects in the farming system," Ph.D. dissertation, University of Reading, UK.

—— (1981) "Gambian women: unequal partners in rice development projects?" *Journal of Development Studies* 17: 109–22.

Food and Agriculture Organization (FAO) (1983) *Rice Mission Report to The Gambia*. Rome: FAO.

Gibson-Graham, J.K. and D. Ruccio (2001) "'After' development: re-imagining economy and class," in J.K. Gibson-Graham and S.A. Resnick and R.D. Wolff (eds), *Representing class: essays in postmodern Marxism*. Duham, NC: Duke University Press, pp. 158–81.

Government of The Gambia (GOG) (1973–2000) Sample surveys of agricultural production. Banjul: Public Planning and monitoring Unit.

Guyer, J. and P. Peters (1987) "Introduction" to Special Issue: Conceptualizing the Household: Issues of Theory and Policy in Africa, *Development and Change* 18, 2: 197–214.

Hart, G. (1991) "Engendering everyday resistance: gender, patronage and production politics in rural Malaysia," *Journal of Peasant Studies* 19, 1: 93–121.

—— (2002) "Geography and Development: development/s beyond neoliberalism? Power, culture, political economy," *Progress in Human Geography*, 26, 6: 812–22.

Humphries, J. (1990) "Enclosures, common rights, and women: the proletarianization of families in the late eighteenth and early nineteenth centuries," *Journal of Economic History* 2, I: 17–42.

Hutchinson, P. (1983) *The Climate of The Gambia*. Banjul: Ministry of Water Resources and the Environment.

International Fund for Agricultural Development (IFAD) (1988) *Small-scale Water Control Program*. Rome: IFAD.

Jack, I. (1990) *Export Constraints and Potentialities for Gambian Horticultural Produce*. Report prepared for the National Horticultural Policy Workshop. Banjul: Ministry of Agriculture.

Jackson, C. (1993) "Women/nature or gender/history? A critique of ecofeminist 'development'," *Journal of Peasant Studies* 20, 3: 389–419

Jobson, R. (1904) [1623] *The Golden Trade.* Devonshire: Speight & Walpole.

McPherson, M. and J. Posner (1991) "Structural adjustment in sub-Saharan Africa: lessons from The Gambia," paper presented at the 11th annual symposium of the Association for Farming Systems Research-Extension, Michigan State University.

Mann, R. (1987) "Development and the Sahel disaster: the case of The Gambia," *The Ecologist* 17: 84–90.

Moore, H. (1988) *Feminism and Anthropology.* London: Polity.

Nath, K. (1985) *Women and Vegetable Gardens in The Gambia: Action Aid and Rural Development.* Working Paper No. 109, African Studies Center. Boston: Boston University.

N'Jang. A. (1990) *Characteristics of Tourism in The Gambia.* Banjul: Ministry of Information and Tourism.

National Archives of The Gambia (NAG) (1943) Department of Agriculture files, 52: 47/50.

Okoth-Ogendo. H. (1989) "Some issues of theory in the study of tenure relations in African agriculture." *Africa* 59. 1: 56–72.

Park. M. (1983) [1799] *Travels into the Interior of Africa.* London: Eland.

Pearson, Ruth and Cecile Jackson (1998) "Introduction: Interrogating development: feminism, gender and policy," in Jackson and Pearson (eds) *Feminist Visions of Development.* London: Routledge, pp. 1–16.

Peet, R. and M. Watts (eds) (1996) *Liberation Ecologies* London: Routledge.

Portères. R. (1970) "Primary cradles of agriculture in the African continent." in J. Fage and R. Oliver (eds) *Papers in African Prehistory.* Cambridge: Cambridge University Press, pp. 43–58.

Rahman. A.K. (1949) Unpublished notes on land tenure in Genieri. courtesy of David Gamble.

Schroeder. R. (1999) *Shady Practices: Agroforestry and Gendered Politics in The Gambia.* Berkeley: University of California Press.

Shiva, V. (1989) *Staying Alive.* London: Zed Books.

Sumberg. J. and C. Okali (1987) *Workshop on NGO-sponsored Vegetable Gardening Projects in The Gambia.* Yundum: Department of Agriculture Horticultural Unit and Oxfam America.

Swindell K. (1977) "Migrant groundnut farmers in The Gambia: the persistence of a nineteenth-century labor system," *International Migration Review* 11(4): 452–72.

Thrift, M. (1986) "The geography of international economic disorder," in R.J. Johnston and P.J. Taylor (eds) *A World in Crisis?* New York: Basil Blackwell, pp. 12–67.

Webb. P. (1989) *Intrahousehold Decisionmaking and Resource Control. The Effects of Rice Commercialization in West Africa.* Washington. DC: International Food Policy Research Institute (IFPRI).

Weil, P. (1973) "Wet rice, women and adaptation in The Gambia," *Rural Africana* 19: 20–9.

—— (1982) "Agrarian production. intensification and underdevelopment: Mandinka women of The Gambia in time perspective," in *Proceedings of the Title XII Conference on Women in Development.* Newark: University of Delaware.

Part V

MOVEMENT

13

ENVIRONMENT, INDIGENEITY AND TRANSNATIONALISM[1]

Tania Murray Li

How does a transnational category such as "indigenous people" come to mobilize people in a particular place? How does it enter into their struggles and reconfigure alliances? How do the "environmental imaginaries" associated with indigeneity translate on the ground? Who is privileged by the indigenous frame, and who is left out? These are the questions explored in this chapter. My focus is on Indonesia, where answers to the question "who is indigenous" are far from obvious, yet a social movement supporting indigenous people is taking hold. I seek to explore how that has happened, and some of the consequences.

My argument is that a group's self-identification as tribal or indigenous is not natural or inevitable, but neither is it simply invented, adopted, or imposed. It is, rather, a *positioning* which draws upon historically sedimented practices, landscapes, and repertoires of meaning, and emerges through particular patterns of engagement and struggle. The conjunctures at which (some) people come to identify themselves as indigenous, realigning the ways they connect to the nation, the government and their own, unique tribal place, are the contingent products of agency and the cultural and political work of *articulation*.

Articulation is a key word of political ecology because it recognizes the structured character of distinct entities (means of production, social groups, ideologies) but highlights the contingency of the ways in which they are brought together – articulated – at particular conjunctures. *Conjuncture* is another key word: it challenges us to examine unique histories, geographies, and the micro-politics of situated lives and struggles without losing track of their intersection with constructions of landscape, livelihood and identity generated across different spatial and temporal scales, and the associated fields of force: the terrain explored so productively by Donald Moore (1998a, 1998b, 2000). Comparison is a useful tool for unpacking conjunctures and I use it here, drawing a contrast between one site in which the concept of indigeneity has indeed taken hold, and another, where it has not.

THE POLITICS OF INDIGENEITY IN INDONESIA

It was the official line of Suharto's regime that Indonesia has no indigenous people, or that all Indonesians are equally indigenous.[2] The concept of indigeneity that took shape in the white settler colonies and is reflected in the International Labor Organization's Convention 169 simply had no equivalent in this sprawling archipelagic nation. The national motto "unity in diversity" and the displays of Jakarta's theme park, Taman Mini, presented the acceptable limits of Indonesia's cultural difference. Development efforts were directed at improving the lot of "vulnerable population groups," including those deemed remote or especially backwards. Rural citizens were expected to express their desire for development through the approved channels of bottom-up planning and supplications to visiting officials. National activists and international donors who argued for the rights of indigenous people were dismissed as romantics imposing their primitivist fantasies upon poor folk who wanted, or should have wanted, to progress like "ordinary" Indonesians.

Since Suharto fell in 1998, the concept of indigeneity has begun to receive some official recognition in matters such as land and forest law and provisions for increased regional autonomy. But many government departments, policies and individual officials continue to give the whole concept short shrift. They argue that if anything distinguishes the people advocates would call indigenous, it is their backwardness, still a problem to be overcome. Moreover government programs and advocates who do recognize that category "indigenous people" in principle still face the practical problem of identifying who fills the slot. The Alliance of Indigenous Societies of the Archipelago (Aliansi Masyarakat Adat Nusantara or AMAN), formed in 1999, defines its constituency as "communities which have ancestral lands in certain geographic locations and their own value systems, ideologies, economies, politics, cultures and societies in their respective homelands." Broadly interpreted, this definition could refer to tens of millions of rural Indonesians, almost anyone who still lives in the region of their own ethno-linguistic group. Narrowly interpreted, it could refer to the two or three million who live in especially isolated areas, and follow the lifeways of their ancestors largely unrevised. In between these two poles, there is a range of possibilities for claiming, assigning or denying indigenous identities by highlighting particular criteria.

The two case studies through which I explore who comes to occupy Indonesia's indigenous slot are both drawn from the hilly interior of Central Sulawesi.[3] In earlier centuries, these two locations were inhabited by rather similar people: scattered swidden farmers loosely organized into family groups, threatened by slave raiders and by sometimes hostile neighbors, and involved in important but tense and unstable trade and tribute relationships with coastal powers. Today, one of these regions is peopled by prosperous, literate, Christian farmers growing irrigated rice and coffee, whose children aspire to government jobs, while in the other, very few people can read or speak the national language, swidden cultivation is the norm, housing and nutrition are poor, and livelihoods and health precarious. Yet it is in the former location – Lake Lindu – that a collective, indigenous identity has been persuasively articulated. The immediate context of this articulation was a national and international campaign to oppose the construction of a hydro-power plant at the lake, but the preconditions which enabled it have deep historical roots. In the Lauje area, by contrast, while no one would question that the hill farmers are the original inhabitants of their land, the specificity of their identity has not been made explicit, nor does it serve to conjoin local projects to national or global ones.

The contrast between these two locations offers some powerful insights, but highlighting it does present some risks. In view of the perspective still prominent in many official circles that indigenous people are figments of an NGO imagination unduly influenced by imported ideas, the contrast between the two sites could be taken to imply that the indigenous identity articulated at Lindu has been "invented" or adopted strategically – that it is opportunistic and inauthentic. So too might discussions of ethnic identity framed in individualist terms, which seem to suggest that goal-oriented "actors" switch or cross boundaries in pursuit of their ends, approaching questions of identity in consumer terms, as a matter of optimal selection. Equally problematic from another perspective are theoretical positions which might suggest that one or other of the groups is suffering from false consciousness: the Lindu perhaps for articulating a tribal position rather than one defined in class terms, or the Lauje for their apparent failure to mobilize at all.

The tense politics around indigeneity and its serious consequences for rights, resources, and the formation of alliances highlight the need for new ways to conceptualize identity that are theoretically more adequate to the diversity of conditions and struggles in the Indonesian countryside, and alert to the risks and opportunities posed by particular framings. For reasons of history and social structure which I discuss later, anthropologists have not tended to use the term "tribe" in reference to Indonesia, and legal scholars (e.g. Kingsbury 1998) are uncertain about whether the term "indigenous people" fits the Asian scenario.[4] But these are mobile terms which have been reworked and inflected as they have traveled, and as they have been used to engage with, and envision alternatives to, the models of development promoted by Indonesia's New Order regime. They have taken on new meanings in relation to quite specific fields of power.

The concepts of articulation and positioning, which I draw from Stuart Hall, are central to my analysis, and I discuss them in the next section. I then go on to describe the fields of power within which the discourse on indigenous people is taking shape in Indonesia, focusing upon the ways in which government departments and NGOs characterize, and seek to transform, the rural populace of frontier spaces potentially envisaged as indigenous or tribal. Following this I explore the historical and contemporary processes at work in the formation of collective identities in the two study areas, seeking the reasons why the discourse on indigenous people has taken hold in one place but not another. Finally,

I discuss issues of risk and opportunity, indicating what is at stake for those who might occupy Indonesia's tribal slot, as well as for those who seek to support their struggles and devise alternatives to the New Order development regime.

ARTICULATION AND POSITIONING

Stuart Hall alerts us to the dual meaning of the term "articulation." It is the process of rendering a collective identity, position, or set of interests explicit (articulate, comprehensible, distinct, accessible to an audience), and of conjoining (articulating) that position to definite political subjects. For Hall,

> An articulation is . . . the form of the connection that *can* make a unity of two different elements, under certain conditions. It is a linkage which is not necessary, determined, absolute and essential for all time. You have to ask under what circumstances *can* a connection be forged or made? So the so-called "unity" of a discourse is really the articulation of different, distinct elements which can be rearticulated in different ways because they have no necessary "belongingness." The "unity" which matters is a linkage between that articulated discourse and the social forces with which it can, under certain historical conditions, but need not necessarily, be connected. Thus, a theory of articulation is both a way of understanding how ideological elements come, under certain conditions, to cohere together within a discourse, and a way of asking how they do or do not become articulated, at specific conjunctures, to certain political subjects . . . [It] asks how an ideology discovers its subject rather than how the subject thinks the necessary and inevitable thoughts which belong to it; it enables us to think how an ideology empowers people, enabling them to begin to make some sense or intelligibility of their historical situation, without reducing those forms of intelligibility to their socio-economic or class location or social position.
>
> (Hall 1996: 141–42)

Hall's formulation offers a framework for addressing both the empirical and the political dimensions of my problem. In relation to the empirical question of how the tribal slot is defined and occupied, the concept of articulation usefully captures the duality of positioning which posits boundaries separating within from without, while simultaneously selecting the constellation of elements that characterize what lies within. Further, it suggests that the articulation (expression, enunciation) of collective identities, common positions, or shared interests must always be seen as provisional. Cultural identities, as Hall argues elsewhere, "come from somewhere, have histories. But far from being eternally fixed in some essentialized past, they are subject to the continuous 'play' of history, culture and power" (Hall 1990: 225). They are "unstable points of identification or suture. . . .

Not an essence but a *positioning*" (1990: 226). While the "cut" of positioning is what makes meaning possible, its closure is arbitrary and contingent, rather than natural and permanent. This feature renders any articulation complex, contestable, and subject to rearticulation. Positively asserted on the one hand, articulations are also limited and pre-figured by the fields of power or "places of recognition" which others provide (c.f. Hall 1995: 8, 14).

In relation to the political dimensions of my problem, Hall's argument that identities are *always* about becoming, as well as being, but are never simply invented, offers a way out of the impasse in which those who historicize the identities or traditions of "others" are accused of undermining subaltern political projects founded upon originary, perhaps essential truths.[5] In rejecting the idea of a necessary correspondence between social or class position and the discourses through which people make sense of their lives, Hall moves beyond the concept of false consciousness. At the same time, his attention to history and structure suggests a notion of agency quite different from that found in transactionalist accounts (e.g. Barth 1981). While there is a tactical element in the cut of positioning which may become explicit at times of heightened politicization and mobilization, the flow of meaning from which an articulation is derived and the fields of power with which it is engaged transcend that temporary fixity. The concept of articulation is thus alert to the unevenness of conjunctures and conditions of possibility, but offers no simple recipe for assessing degrees of determination or the points at which everyday understandings and practices shade into consciously selected tactics. It points rather to the necessity of teasing out, historically and ethnographically, the various ways in which room for maneuver[6] is present but never unconstrained. Finally, rather than focus on the identity dilemmas of the individual subject, Hall draws attention to those articulations which have the potential to define broad constellations of shared or compatible interests, and mobilize social forces across a broad spectrum.[7]

LOCATING THE TRIBAL SLOT IN SHIFTING FIELDS OF POWER

Simplification and stereotyping are characteristic modes of apprehending the symbolic and material space of a nation's frontiers, the space at the cutting edge of capitalist expansion and state territorial control (Shields 1991; Watts 1992: 116–17). Indonesia is an archipelagic state, whose frontiers are the hilly and forested interiors of the larger islands and the smaller islands of Eastern Indonesia. The populations that occupy these spaces are classified by the state according to two rather distinct frames of meaning and action, and classified by social and environmental activists according to a third, competing frame. Each of these frames narrows or simplifies the field of vision in its own particular way, highlighting some aspects of the landscape and its inhabitants and overlooking others. The tribal slot, like the savage slot described by Trouillot (1991), is a simplified frame of this sort. As my comparative study will later demonstrate, the predominance of a particular frame at a particular time and place depends not upon essential differences between the populations themselves, but upon the regimes of representation or "places of recognition" that preconfigure what can be found there, together with the processes of dialogue and contestation through which identifications are made on the ground.

State programs for interior and upland frontiers

The New Order government unilaterally classified about one million rural people as "estranged and isolated" – *masyarakat terasing, masyarakat terpencil* (Department of Social Affairs 1994). After Suharto's fall, these people were renamed with the slightly less insulting title "isolated native groups" (*kelompok asli terpencil*, KAT), but the program logic did not substantially change. The official program designed to civilize such people views them as generic primitives, occupants of a tribal slot which is negatively construed. Their ethnic or tribal identities, cultural distinctiveness, livelihood practices, and ancient ties to the places they inhabit are presented in program documents as problems, evidence of closed minds and a developmental deficit that a well-meaning government must help them to overcome. This is to be accomplished by means of a resettlement program, a successor to Dutch efforts, which attempts to narrow the distance (in time, space and social mores) between *masyarakat terasing* and the "normal average Indonesian citizen" (Koentjaraningrat 1993). The cultural distinctiveness they are encouraged to retain is of the song and dance variety.

Resettlement program guidelines specify that *masyarakat terasing* can be recognized by their tendency to move from place to place, as well as by their lack of a world religion, strong commitment to local customs and beliefs, and deficient housing, clothing, education, diet, health, and transportation facilities (Department of Social Affairs 1994). But there is, as I have argued elsewhere (Li 1999b), a problem with this list. Elements of the description could apply to almost all the rural population outside Java, especially to the tens of millions engaged in swidden agriculture or living in or near forests.[8] Identifying suitable subjects to be classified as *masyarakat terasing* is, therefore, a matter of interpretation and negotiation. Considerations include the need for the department responsible for resettlement to meet its quota; the distribution of construction contracts and associated forms of state largesse; pressures to reallocate land to more lucrative ventures; and the interest of the subjects themselves in access to the short or long-term benefits promised to them.

In contrast to the few classified as *masyarakat terasing* whose ethnic distinctiveness is acknowledged, and whose unique cultural characteristics are officially marked (albeit negatively), the majority of people occupying forested, mountainous, or other types of frontier land are classified simply as village folk, *orang kampong*. Under Suharto, the development programs designed for such people ignored ethnic differences and assumed, at the same time as they sought to create, homogenous forms of family and village life and a common administrative structure throughout the archipelago.[9] New regional autonomy legislation lifts the emphasis on cultural homogeneity, but the normalizing logic of "development" is still intact. Development programs often encourage or enforce mobility across the rural landscape. In the past few decades Indonesians have moved from one place to another as migrants, transmigrants, or workers attracted to, or ejected from, boom/bust industries (Brookfield *et al.* 1995). They have been forced to move when the state, which claims control over most of Indonesia's land (approximately 75 percent of it under the Ministry of Forestry), allocates their lands to other uses and users (Evers 1995; Moniaga 1993; Zerner 1990). Few rural people outside Java have formal title to their lands. Regardless of the depth of their attachments to a particular place, most of the people who are rural and poor are deemed to be illegal squatters, subject to expulsion and other sanctions (Departmen Kehutunan 1994). To be an "ordinary villager" is, therefore, to belong to a homogenized or simplified category of people whose localized commitments are officially unrecognized and often seen as contrary to national laws, policies, and objectives. In keeping with this official view of the countryside, data on matters such as the numerical size of ethnic or linguistic groups, their regional concentrations, or the relative proportion of migrants and original inhabitants is "important but scarce" (Peluso 1995: 399).

NGO visions and agendas

Counterposed to these two official frames for defining and managing rural space and populations is the category of indigenous people whose presence in the Indonesian countryside has been highlighted by social and environmental activists since the 1990s. Activists draw upon the arguments, idioms, and images supplied by the international indigenous rights movement, especially the claim that indigenous people derive ecologically sound livelihoods from their ancestral lands and possess forms of knowledge and wisdom which are unique and valuable. But the discourse on indigenous people has not simply been imported. It has, rather, been inflected and reworked as it has traveled. While it is significant that some Indonesian activists writing in their own language continue to leave the English term "indigenous people" untranslated, others use a range of terms such as *masyarkat adat*, *masyarakat tradisional*, *masyarakat asli* and *penduduk asli*, each of which is contextualized in particular struggles, some of them decades old.

Support for indigenous or tribal people is widespread in the Jakarta activist community, where their plight is seen as one among many ways in which the promises of Indonesian democracy and nationhood remain unfulfilled. The population that is envisaged to fit the indigenous or tribal slot differs according to the agenda and activities of the NGO in question.[10] For urban activists concerned to critique and redirect Indonesian modernity, indigenous people are the embodiment of pure forms of Indonesian cultural heritage unsullied by encounters with colonialism, westernization, and modernity.[11] Some activists focus their concern upon especially isolated or exotic groups, who conform to the slot imagined by international promoters of tribal environmental wisdom. These are the same people who would readily be classified by the government as *masyarakat terasing* or KAT: some NGOs refer to the number published by the Department of Social Affairs (i.e. about one million) to identify the subjects of their concern. Their goal is to reverse the negative valorization that the government has placed upon the traditions of those in the tribal slot, and defend their right to maintain their distinctive ways of life, rejecting state-defined environment and development imperatives that involve displacement or forced and rapid change.[12]

For other activists, the term "indigenous people" can be applied not only to especially isolated or exotic groups, but to the majority of Indonesia's rural citizens outside Java. At their most radical, these broader definitions amount to an attempt to roll back the state's territorial, social, and political control over the countryside, and empower tens of thousands of rural communities to manage their own affairs.[13] A key objective for many activists is implementation of the provisions in the Basic Agrarian Law of 1960, which recognizes rights to land based upon *adat* or custom. They do not restrict their attention to those groups formally recognized by the Dutch as "*adat* law communities," but rather argue that any rural community can qualify under the provisions of the Basic Agrarian Law if their rights to land derive from and are recognized under local custom. Distinctive cultural styles which substantiate the idea of "a customary law community," and local sites and signs which provide proof of ancient ties to a place strengthen a claim but, according to some activists at least, are not essential to it. As one activist explained, "*Adat* is dynamic. So long as local people manage their land and resources in an orderly way, they can be said to have a customary tenure system."

Within this array of state and activist positions, there are many criteria for specifying which groups fill the tribal or indigenous slot, just as there are many agendas for their future. Rural people in Indonesia have some room to maneuver as they situate themselves in relation to the images, discourses, and agendas that others produce for or about them. On the one hand, if they are to fit the preconfigured slot of indigenous people they must be ready and able to articulate their identity in terms of a set of characteristics recognized by their allies and by the media that presents their case to the public. But the contours of the tribal slot are themselves subject to debate, as I have shown. Agency is involved in the selection and combination of elements that form a recognizably indigenous identity, and also in the process of making connections. Under some conditions, the room for maneuver may be quite limited. Struggles over resources, which are simultaneously struggles over meaning, tend to invoke simplified symbols fashioned through processes of opposition and dialogue, which narrow the gaze to certain well-established signifiers and traits. In contests that pit marginalized populations against the state it may be the case that only one story can be presented. Whichever story this is, its audibility increases to the extent that it fits a familiar, pre-established pattern.[14] But power is seldom so singular, and articulations are correspondingly complex. They are contingent but not random; provisional and indeterminate, but not without form. It is not possible, just by surveying the rural scene, to predict which articulations will in fact be made. Nevertheless, it is possible to gain some understanding of the processes involved. To this end, I focus upon particular conjunctures – in this case, the two contrasting sites in the Sulawesi hills.

ARTICULATING INDIGENOUS IDENTITY: WHERE AN IDEOLOGY FINDS ITS SUBJECT

Power and the production of cultural difference

In the western popular imagination fed by *National Geographic*, and also in the minds of some activists, tribes are naturally bounded, culturally distinct groups occupying spatially continuous and usually remote terrain. Tribes so imagined are hard to find in Indonesia, where analysis of history and social structure points, rather, to the political nature of group formation processes. The bilateral kinship system found in much of the archipelago lends itself more easily to the inclusion of others than to their exclusion. While there are some unilineal and hierarchical groups at the western and eastern extremes of Indonesia, in most areas kin loyalties are diffuse and residence patterns flexible. More common than sharp ethnic boundaries are patterns of continuous variation on familiar themes (Kahn 1999; Kipp and Rodgers 1987: 8). Therefore, when tribal or ethnic boundaries *are* clearly marked, they can usually be traced to specific histories of confrontation and engagement.[15] Kipp and Rodgers (1987: 1) argue that the distinctive ancestral customs claimed by Indonesia's more ethnicized groups are often "less ancestral than exquisitely contemporary . . . a system of symbols created through the interaction of small minority societies, their ethnic neighbours, colonial administrations, the national governments, and the world religions, Islam and Christianity."

Precolonial coastal kingdoms were not much interested in the details of cultural variation and ethnic affiliation in the uplands and interiors of their domains. Their principal goal was to monopolize trade and, in some cases, to control labor through direct enslavement or debt bondage. Costal powers were often thwarted in both these endeavors by the capacity of interior peoples to subsist on their swidden fields, avoid trade engagements, and retreat to inaccessible areas when faced with violence or unreasonable demands. Muslim coastal powers therefore relegated most of the inhabitants of the interior to a tribal slot which they characterized by animism, backwardness, and savagery. Interior peoples, meanwhile, developed positive identities stressing independence, autonomy, and their capacity to carve a livelihood out of their hilly, forested terrain.[16] Domination and difference therefore emerged within a single political and cultural system, as distinctive identities began to be attributed to, imposed upon, and forged by interior populations through a complex and resistance-permeated process, which Gerald Sider (1987: 17) terms "create and incorporate." Where definite, tribe-like social units were found in the interior, their emergence could often be traced to conditions of warfare and conflict.[17] In the absence of such encounters and confrontations, loosely structured, decentered, often scattered populations did not view themselves as distinct ethnic groups or tribes, and their identities remained only vaguely specified.[18]

The Dutch colonial authorities played an important role in ethnicizing or traditionalizing the Indonesian interior. In frontier areas where the indigenous political structures were amorphous, they set about consolidating people into tribe-like groups under centralized, hierarchical leadership.[19] They used the notion of tradition quite deliberately to legitimate colonial policies of indirect rule, and to help consolidate the authority of the Dutch-appointed "traditional" leaders through whom this rule would be exercised.[20] To this end, local practices or customs (*adat*) were codified by scholars and officials.[21] The Dutch concept of the *adat* law community (*masyarakat hukum adat*) assumed, as it simultaneously attempted to engineer named, bounded, and organized groups. It was a concept that resonated differently with the local social formations that existed across the archipelago.[22] Ironically, but not surprisingly, it corresponded better to the formations that arose as a *result* of colonial interventions (including the *adat* codification process itself) than it did to those that existed prior to Dutch control. In regions of little interest to the Dutch, the process of traditionalization did not occur or was incomplete, and identities, practices, and authority in matters of custom remained – and in some cases still remain – flexible and diffuse.[23]

Dutch efforts to systematize *adat* preconfigured the contemporary "indigenous peoples" slot, and their uneven reach continues to be reflected in the differential capacity of frontier peoples to articulate collective identities and positions. In the precolonial period, both of the highlands I am describing were peripheral to the concerns of the coastal chiefdoms which claimed nominal control over them. It was in the colonial period that a marked divergence occurred in their historical trajectories, laying the basis for the distinct spatial, political and social configurations that characterize them today.

The mountain Lauje: development supplicants, cynics, or tribe manquée?[24]

The Lauje, currently numbering about thirty thousand, occupy the hilly interior and the narrow coastal strip of the peninsula to the north of Tomini Bay. They are concentrated in the present day subdistricts of Tomini and Tinombo. Their language (Lauje) shades gradually into Tiaolo and Tajio, the languages of their neighbors, and no ethnicizing signs mark the borders of the Lauje domain. The Lauje hills are fairly densely settled and cultivated but not especially fertile, so they have not attracted outsiders. The Lauje have therefore not been provoked into articulating collective identities and associated boundaries in order to claim or defend their territory (Li 1996).

According to Nourse's (1999) account of local history, in precolonial times most Lauje kept to the hills for fear of slave raiders and pirates, although they traded jungle produce. Those occupying the drier lower slopes produced tobacco for regional markets. Lauje who moved down to the coast during the nineteenth century constituted themselves as a class of aristocrats, and intermarried with traders who moved in from other parts of Sulawesi: mainly Bugis, Mandar, and Gorontalo. The Lauje area was of only peripheral interest to the Dutch. It contained little natural wealth, and the coastal aristocrats were quiescent and easily co-opted, posing no threat to Dutch authority. A halfhearted attempt was made early this century to move the interior population to the coast, but it was clear that the land base was insufficient and they were soon allowed to return to their scattered mountain homes. Some undertook forced labor service, working on the construction of the coastal road and bridges, while others moved further inland to evade such obligations. Dutch revenues from the area, such as they were, came from taxing the owners of coconut groves that had been planted along the coast at Dutch insistence.

The coastal chiefs' minimal obligation to their Dutch overlords was to keep peace in the interior, and prevent feuding and bloodshed. Their model for governance was to select highlanders of renown and make them responsible for maintaining order. Since the expectations associated with rule over the interior were relatively light, the Dutch had no need to discover, constitute, or record Lauje customary practices or traditional law (*adat*). The mechanisms for accomplishing rule in the postcolonial period became somewhat more systematic but did not fundamentally change. Since the borders of the lowest level administrative units (*desa*) were defined to crosscut the terrain from the coast to the hills, the coastally-based *desa* heads continued the practice of appointing hillside leaders to be responsible for the maintenance of order in their vicinity. These leaders occupy the official positions of hamlet chief (*kepala RT*) and chief of customary affairs (*kepala adat*). The task of the latter is to adjudicate marriage arrangements and local disputes in the hillside hamlets to which they belong.

According to several of those holding responsibility for "customary affairs," the procedures, rules, and fines they administer in their hillside hamlets were not handed down by the ancestors but, rather, established by the coastal authorities earlier this century in order to overcome the anarchy and feuding that previously prevailed in the hills. They consider their own authority to settle disputes to be a power granted by the *desa* administration, ultimately backed by the civil, police, and military authorities of the district. They do not articulate a sense of *adat* as something distinctive, autochtonous, locally derived, or essential to Lauje identity. There are, of course, many beliefs and practices of a spiritual nature relating to ancestors as well as to features of the landscape, but these are described as matters of personal, family, or at most hamlet-wide conviction, rather than pan-Lauje tradition.

Desa officials regard the hill people and their farming practices as backward, and generally show little interest in them. *Desa* maps portray the hills in spatially compressed form, while depicting the houses and public facilities on the narrow coastal strip in minute detail. Some *desa* officials describe the hilly interior as "empty," even when more than half the *desa* population lives up above (Li 1996). When pressed to discuss the mountain population, they emphasize their primitive, unruly nature and their status – not as noble savages, but as awkward and annoying ones. They sometimes refer to the mountain dwellers as *orang dayak*, a term they have picked up through media exposure to the apparently wild and primitive people of the Kalimantan interior, and now use to label and characterize their own backwoods. Many *desa* officials are themselves Lauje, but they, like the rest of the coastal Lauje elite, regard their shared ancestry with the heathen and backwards interior as a source of embarrassment. Some coastal Lauje have tried to highlight distinctions between Lauje and "foreigners" (Bugis, Mandar, and others), but their goal has been to bolster their own claims to aristocratic status, rather than to foster an overarching Lauje identity uniting coast and hills (Nourse 1999). Meanwhile, officials from the Ministry of Education and Culture bemoan their assignment to an area of Indonesia so patently lacking in the kinds of songs, dances, and handicrafts that they are expected to identify and turn into emblems of the local, for display in provincial or national fora. No sympathetic outsiders have yet come looking for indigenous people.

Engagements between the state and the Lauje people have been framed within, rather than outside or in opposition to, the state's discourse of development. This does not mean that there is consensus on who or what needs to be developed, or how development should be accomplished. For their part, *desa* officials readily classify the mountain Lauje as *masyarakat terasing* when planners from the provincial capital visit to ask about local development needs. In so doing, they seek to absolve themselves of responsibility for the onerous tasks of trying to count, monitor, or control, let alone provide services to, a mountain population which, they stress, is continuously on the move. They also hope to attract resettlement projects to their *desa*, massive deployments of state attention and expenditure which would help to resolve their administrative difficulties and potentially their financial ones. To this end they have helped to generate long lists of names of people who should be resettled; the Department of Social Affairs (1994: 89–92) has it on record that there are 912 households of Lauje *masyarakat terasing* in need of government attention, in addition to the eighty that have already been resettled under the Department's program. But the Department receives many more requests for resettlement programs than it can handle. Numbers alone do not make a compelling case. The Department was already exposed to embarrassment when all the Lauje abandoned a resettlement site and returned to the hills within a year. Moreover, the Lauje are considered rather dull folk, lacking in the paint and feathers expected of true primitives. As one senior official observed in an interview, "sometimes we look at them and say these are not indigenous people, they are village people."[25] There are other groups in

Central Sulawesi, such as the Wana, who better fit the bill.[26] The mountain Lauje, who are not especially exotic and have no serious competitors for their hilly terrain, have therefore been left pretty much to their own devices.[27]

Generally, the mountain Lauje agree that their part of the province, and the hills in particular, suffer from a development deficit. This is a deficit they mostly attribute not to their own primitiveness or recalcitrance, but to the indifference, corruption, and greed of local elites, who direct state facilities, programs, and benefits away from them. Those who have heard about the official resettlement program oppose it on practical grounds. While they would be happy to receive new houses and rice rations as gifts from the government, they are rightly skeptical about livelihood prospects on the coast, and insist that they would have to remain where they are.[28] Not having been exposed to the overtly coercive dimensions of state power, nor to the threat of having land and livelihoods removed from them in the name of development, they have not articulated collective positions on these matters. Their engagements with state authority and development occur mostly through unremarked, "everyday" patterns of action and inaction. Some participate in mandatory public works days (*kerja bakti*), while others do not. Some hike down to the *desa* office when called to pick up free cocoa seedlings, while others surmise that any handouts offered to them will probably be of poor quality or purloined by coastal elites, and make their own arrangements. Some pay land taxes, while others claim they are too poor to pay, and count on officials to be lenient. Like the coastal elites, they bring a well-honed cynicism to these everyday encounters. They have learned the parameters of what can be requested from the government, the list of things (schools, seedlings, roads, or footpaths) that fit within the official purview of development. These are indeed things that many Lauje feel they want and need, although they do not define their lives as chronically deficient in the absence of such things, nor do they sit passively waiting for the government to secure their futures. They are, however, willing to adopt the position of supplicants in the hope that some of the desired goods and facilities will come their way.

So far, there has been no conjuncture, no context, site, event, or encounter, in which the mountain Lauje have articulated a collective position as indigenous people. No hillside leaders have been interested in, or capable of, articulating territorial claims beyond the level of their own hamlet, still less a generic Lauje identity. There are respected shamans living both in the hills and on the coast, but their agendas do not appear to be political. The pretensions of coastal Lauje "aristocrats" discussed by Nourse (1999) are largely unheard or ignored. The main authority hill folk acknowledge is that of *desa* officials, but, as noted above, the mountain Lauje are rightly suspicious of the motives of this group's motives and resent the unfair treatment they receive at their hands. The mountain Lauje are not anti-development. Indeed, they are taking their own initiatives to improve their chances of being *included* in state development agendas which have hitherto passed them by. They engage with the state in a discourse consistent with their knowledge of themselves, their needs and aspirations, and their understanding of what it is possible to demand and expect in that relationship. The ideology of indigenous people has not found its subject in the Lauje hills because, under current conditions, it would not help people to make sense of their situation, nor would it help them to improve it.

Sulawesi tribe opposes Lake Lindu dam project

So stated a headline in the *Jakarta Post* (an English-language daily) on 11 September 1994. The article quoted Gesadombu, "Tribal Chief of the Lindu plains," on the centrality of the lake to the Lindu tribe's livelihood; the "strong traditional and practical ties the Lindu people had with the land they live on"; and the certain loss of traditional values should the people be forced to move out. Accompanying the Chief were "twenty-three other fellow Lindu indigenous people, non-governmental activists, students and nature-lovers from Central Sulawesi." They were visiting Jakarta to meet with state officials and present their case against the construction of a hydropower plant at the lake. The article also quoted activists on the ecological soundness of the Lindu people's traditional resource management practices, on the need for the government to learn about land and water management from the people, and on the right of the Lindu people to express their culture.

Every component of this news story is familiar: the presence of tribes, tribal leaders, tribal ecological wisdom, and a specific tribal place central to the group's identity and culture, plus the presence of allies and sympathizers, and of a massive external force poised for destruction. It is a story for which the conceptual frame or "place of recognition" already exists, and for which the intended readership has been prepared. Nevertheless, the telling of this story in relation to Lindu or any other place in Indonesia has to be regarded as an accomplishment, a contingent outcome of the cultural and political work of articulation through which indigenous knowledge and identity were made explicit, alliances formed, and media attention appropriately focused.

The historical preconditions for this situation were established at Lindu at the turn of the century, when, according to Acciaioli (1989) the area was subjugated by the Dutch, and the scattered hill farmers (numbering about six hundred) were forced to form three concentrated settlements beside the lake. There they were converted to Christianity by the Salvation Army mission, educated in mission schools, and encouraged to view custom as matter for display at celebrations overseen by an officially recognized "customary" leadership, the *adat* council. The subsequent arrival of migrants from neighboring districts and Bugis from the south gave the Lindu people some (often bitter) experience in articulating claims to their "ancestral, customary or village land" (Acciaioli 1989: 151). Resource struggles thus provided the stimulus to articulate (select, formulate, and convey) a set of Lindu *adat* rules which *ought* to be acknowledged by outsiders – a process which in turn reworked the significance, and substance, of Lindu knowledge and identity. Even before the discourse of indigenous people became available to them, the preconditions that would suggest its relevance were firmly in place. Moreover, unlike the mountain Lauje whose aristocratic elite and *desa* administrators are located far away on the coast, in a distinct class position and ecological niche, the leaders of the consolidated lakeside villages at Lindu experienced the threat posed by newcomers in the same way as their covillagers. Thus mission-educated, literate *desa* officials, schoolteachers and prosperous farmers played a central role in the articulation of Lindu identity, rights, and claims.[29]

The identity of the Lindu as indigenous people with valuable knowledge and ancestral rights to their land was firmly established in the context of opposition to the hydro plan and the threat of forced resettlement. According to Sangadji's account (1996), the campaign involved confrontational encounters with the authorities, media attention, collaboration with national and international NGOs, and activities organized by Lindu leaders to heighten awareness within the community. NGO campaigning and support began in 1988. In 1992, at a dialogue with NGOs in Palu, a Lindu leader stated that he and his people would rather die than be removed from their ancestral lands. A youth group was formed at Lindu to research Lindu tradition and work for its preservation. Many journalists and officials visited the site, and *adat* leaders reiterated the preference to die rather than lose their culture. Security forces warned the people that activists, whose values were western and contrary to the official national ideology (*pancasila*), were misleading them. An environmental assessment was carried out by consultants in 1993, but invited no public input. The delegation mentioned above then visited Jakarta to meet with top officials, and was told that an amended design would avoid the necessity for resettlement. Currently, the hydro plan is on hold, though the Lindu and their supporters remain vigilant.[30]

The scale of the threat to local lives and livelihoods, the dramatic nature of a dam as a stage for NGO action, the location of the dam within a national park, and the massive economic implications of the project explain why Lindu attracted so much attention. But it remains to be explained how and why the Lindu have come to articulate their identity, present themselves, and be represented by their supporters in terms consistent with both national and international expectations concerning indigenous people or tribes.

The news coverage and documents prepared in the course of the campaign shed some light on the "how" question. Members of the NGO coalition worked with Lindu leaders to produce documents informing the public and policy makers about the Lindu people and the negative impacts of the dam. These documents present Lindu as a unique, tribal place, its integrity basically intact. They note that the Lindu are the only speakers of the Tado language (related to Kaili), and that they are an autonomous group who have managed their own affairs (*hidup mandiri*) for hundreds of years (Laudjeng 1994: 150–52; Sangadji 1996: 19). There is little mention of the impact of Dutch rule, or of the presence of Bugis and other non-Lindu at the lake. The documents focus upon cultural features which confirm the uniqueness of the Lindu people, their environmental wisdom, and their spiritual attachment to the landscape. Culture is substantiated through a focus upon "traditional" costumes, major annual feasts, and marriage arrangements.

Lindu capacities for environmental management are demonstrated through the existence of the *adat* council, which is said to have jurisdiction over the Lindu people's collective territory – an area extending to the peaks of all the mountains surrounding the lake. Management rules include the exclusion of outsiders from the use of Lindu resources except with permission from the *adat* chiefs, and the zoning of land according to specified uses. The documents pay considerable attention to the existence of named zones for farming, hunting, and grazing, and of sacred sites in which all forms of activity (tree-cutting, gathering, etc.) are strictly forbidden (Laudjeng 1994: 155–60). They also state that each clan – and within the clan, each household – has fishing rights over specific portions of the lake. Filtered and interpreted through a "green lens" (Zerner 1994), these land use categories are presented as similar to, but more efficient than, the land use zones imposed by the state through its forest and national park regulations (Sangadji 1996: 26–28).

Finally, the documents emphasize Lindu people's attachment to their place by naming features of the landscape: hills, sacred spots, grazing areas, and the sacred island in the lake, which is associated with the magical culture hero Maradindo. Although these place names mean nothing to a reader without a map, they assert and confirm that the Lindu are thoroughly familiar with their territory. Between the named zones and the specific named places, the point is made that there is no undifferentiated or unclaimed space, but rather an orderly system of land use designed and managed by the indigenous people of Lindu.

A finer reading reveals many subtleties in these accounts. They present a selective picture, but one which is complex rather than simple, positioning the Lindu in relation to multiple fields of power. They emphasize that the Lindu are "traditional" people, but in no sense are they primitive. The mention of Christianity confirms their nationally acceptable religious standing, yet little is made of the influence of ninety years of missionary work upon their "traditional" rituals and practices. They are shown to be in touch with nature and bearers of tribal wisdom, but by emphasizing the orderliness of the Lindu land use system it is made clear that there is nothing wild about this scene. The accounts emphasize subsistence uses of the forest, such as the collection of building materials and medicinal plants (Sangadji 1996: 44). They make less of the presence within these forests of the hillside coffee groves that provide the Lindu people with a significant source of cash. It is noted that the Lindu people are not poor. They have an adequate standard of living, though not luxurious, and are satisfied with their lot. Thus they are sufficiently similar to "ordinary villagers" not to be in need of drastic changes or improvements (framed as development), still less the civilizing projects directed at *masyarakat terasing*. Yet they are unlike "ordinary villagers" in their uniqueness, their special knowledge, and their attachment to their place.

When these documents are read through the prism of the Lindu history presented in Aciaioli's thesis (researched prior to the dam conjuncture), and in relation to the fields of power and opportunity presented by the Lindu people's NGO and government interlocutors, they reveal how group boundaries were defined, and how elements from the local repertoire of cultural ideas and livelihood practices were selected and combined to characterize the group. They reveal, that is, the "cut" of positioning, its arbitrary closure at a highly politicized moment. They point to the uniqueness and contingency of articulation, and its necessary occlusion of the larger flows of meaning and power, the practices of everyday life and work, the differences according to gender or class position, and the structures of feeling which form the larger canvas within which positioning occurs.

The efficacy of framing Lindu people's position in terms of the arguments and images associated with indigenous people was not guaranteed. It was effective in the NGO campaign, as activists were able to use the environmental soundness of the Lindu's livelihood practices to argue against the dam and also to support their arguments on behalf of other indigenous people in Indonesia. In activist circles, Lindu became an exemplary case, which was both framed within – and helped to frame – broader struggles.[31] But not all non-government organizations recognized the tribal uniqueness of Lindu. In 1992, while the Lindu campaign was underway, a parks-focused international conservation NGO described the population in the many villages bordering the national park as ethnically diverse, with a mix of "local" or "traditional" people and newcomers. It observed that the area's inhabitants were subsistence farmers, only weakly integrated into markets, and often exploited and displaced by aggressive immigrants. It also noted that they were rather lacking in handicrafts with a tourist potential (Schweithelm *et al.* 1992: 39–47). So described, they fit the state category of "ordinary villagers." But the NGO's report contains no suggestion that the border villages in general, or Lindu in particular, are populated by tribal people who have ancient ties to the forest, or who possess unique environmental wisdom.

Media receptiveness to the idea of Lindu as indigenous people was also mixed. The English-language news coverage cited earlier picked up the tribal angle, as the headline clearly shows. The coverage of opposition to the dam in a major Indonesian language newspaper (*Kompas*) was more equivocal. An article ("Masyarakat Lindu," 11 September 1993) described the Lindu people not as a tribe but as a subgroup of Kaili. It acknowledged their environmental wisdom, but observed that – the satisfaction expressed by residents notwithstanding – the area does suffer from a development deficit, signaled by the 17 kilometer hike from the nearest road, the muddy village paths, and the incomplete electrical service. Most of the media coverage skillfully analyzed by Sangadji (1996) supported the hydro plan on the grounds of development, and did not address the issue of indigenous people.

Throughout the campaign, the government agencies promoting the power plant neither accepted nor rejected the notion that the Lindu are indigenous people: they simply did not engage with it. Refusing, or not recognizing, the discursive terrain developed by the Lindu people and their allies, officials maintained their focus upon the need for electricity to promote modernization and industrial development in the Palu valley (Sangadji 1996: 54). They also made the argument that the resettlement of the Lindu would make them more developed, but this was difficult to justify. Livelihoods at Lindu are, in provincial terms, rather good, as the government itself previously acknowledged when it brought new settlers into the area to share in its prosperity and help develop the potential for irrigated rice production (Sangadji 1996: 44). For these reasons, the development argument was consistently rejected by Lindu spokesmen. Indeed, it was their overt rejection of the idea that they were in need of any form of state-directed development, as much as their emphasis upon the unique character of their tribal place, that was notable in their campaign.

In view of the weak case made by the state, various approaches could have been used to frame opposition to the project. A materialist case, focusing upon the loss of good livelihoods, and a political case, focusing upon the rights of the Lindu people to fair treatment as citizens were indeed argued. But the most prominent form of articulation – that which clarified positions and made connections – was focused upon the loss of a unique tribal identity and way of life. The reasons for this had to do with the fields of power and opportunity surrounding the concept of indigenous people at that juncture. The possibility of articulating local concerns with national and international agendas was clearly present. Situations which set indigenous people up against big projects and the state are guaranteed attention, and they set up predictable alliances (Sangadji 1996: 13, 16). Also significant is the way in which an indigenous or tribal identity asserts the unity of people and place, addressing an issue at the heart of state-society relations in the Indonesian countryside. According to the state model, which sees rural people as "ordinary villagers," those that must be moved to facilitate national development can be compensated in cash, or given new land to replace the old. If the Lindu people were simply villagers, their livelihoods could, in theory, be recreated elsewhere. Indeed, the future planned for them was to join the (technically troubled) transmigration scheme at Lalundu (Sangadji 1996: 20) as homogenized quota-fillers, names on a list. Only indigenous or tribal people can claim that their very culture, identity, and existence are tied up in the unique space that they occupy (Cohen 1993). There can be no compensation. This was the point argued repeatedly by the Lindu and their supporters (Sangadji 1996: 16).

Finally, the tribal slot opens up some maneuvering room unavailable to ordinary villagers. Obstinate peasants can be labeled communists, as they often are in Java (Sangadji 1996: 15) but communist tribesmen are somehow less plausible. Their concerns seem to be somewhat different from those of the mass of rural people reacting to the contempt and arrogance with which they are treated by their government. Indigenous people and their nature-and-culture loving supporters are differently positioned in relation to the field of power. The sacred shrine of the Lindu's heroic and supernatural ancestor Maradindo is located on an island within the lake. When Maradindo is angered he causes accidents, bizarre events of which the Lindu can cite recent examples (Sangadji 1996: 32, 41–2). The Lindu tell a powerful story: ignore Maradindo at your peril.[32]

ARTICULATING INDIGENOUS IDENTITY: CONDITIONS, RISKS AND OPPORTUNITIES

Conditions for articulation

The contrast between my two examples highlights some of the conditions and conjunctures that have enabled the articulation of "indigenous" identity in contemporary Indonesia. A summary of the factors present at Lindu, but not in the Lauje case, includes the following: competition for resources, in the context of which group boundaries were rendered explicit and cultural differences entrenched; the existence of a local political structure that included individuals (elders, leaders) and an *adat* council mandated to speak on behalf of the group; a capacity to present cultural identity and local knowledge in forms intelligible to outsiders – an activity undertaken in this case by a literate elite of teachers, local officials, prosperous farmers, and entrepreneurs; an interest on the part of urban activists in discovering and supporting exemplary indigenous subjects, and documenting indigenous knowledge which fit the niche preconstituted in national and international environmental debates; and, finally, heightened interest in a particular place, arising from a conflict which pit locals against the state or state-sponsored corporations.[33]

My comparative study also illustrates the contingent aspects of articulation and the significance of human agency. It was not predetermined which articulations would be made at the conjunctures described: by some of the obvious criteria, the Lauje were more qualified for the tribal slot. Every articulation is a creative act, yet it is never creation *ex nihilo*, but rather a selection and rearticulation of elements structured through previous engagements. It is also, as Hall points out, subject to contestation, uncertainty, risk, and the possibility of future rearticulation.

Contestation and risk

The potential for contestation is easy enough to identify, since the different interests at play in any articulation could always lead to its unraveling. At Lindu, for example, the Bugis and other settlers who currently go along with the indigenous position could object to, or find themselves threatened by, the potential exclusivism of "the Lindu Tribe," and identify alternative positions and alliances from which to oppose the dam. Lindu people themselves have different stakes in *adat* and its contemporary articulations, and are situated unevenly in relation to the power of *adat* chiefs.[34] NGOs do not always agree on visions, priorities, or the forms in which connections should be made and actions taken. Many activists are aware of the differential benefits that would accrue from a strengthening of customary land rights. Losers would include those who fail to fit a clear cut ethnic and territorial niche, whose family background or patterns of geographic and class mobility have removed them from any material connection to a specific tribal place. Several observers have noted that it is displaced, landless people – mainly Javanese, not indigenous people – who are Indonesia's most vulnerable group (Brookfield *et al.* 1995; Evers 1995: 11). The whole concept of indigenous people, and the idea that they have particular rights, can be – and is, in some quarters – contested on these grounds. Others see the possibility of broadening and redefining interests and visions to create even stronger alliances.

Risk is apparent at many levels. Under Suharto, risk was endemic to any form of political organizing. The government commonly saw activists as fomenting trouble, or, in the standard language used to refer to subversive activities, acting as an (unspecified) "third party," misleading and manipulating simple rural folk and creating "politics" where there is none. But, despite the risk, support for indigenous people provided activists with an opportunity, a space where they could act. The grounding of association and mobilization in culture and tradition, and its affinity with conservation agendas, became crucial to the (precarious) political acceptability of community organizing in the Indonesian countryside (Zerner 1994). It also provided a space in which some rural people could affirm positive identities, and articulate, substantiate, and defend their claims.

Conjunctures at which rural people have identified themselves, and become identified, as indigenous people are moments at which global and local agendas have been conjoined in a common purpose, and presented within a common discursive frame. But the tribal slot fits ambiguously with the lives and livelihoods of people living in frontier areas. It is not an identity space that every local group is able or willing to occupy. They may present themselves as indigenous people, or they may emphasize their standing as ordinary villagers. Too much like primitives and they risk to be classified as *masyarakat terasing*, delinquents in need of re-education and resettlement by the Department of Social Affairs. On the other hand, as "ordinary villagers," they are vulnerable to arbitrary removal under another set of government programs. Candidates for the tribal slot who are found deficient according to the environmental standards expected of them must also beware.[35] The majority of Indonesia's swidden farmers have long been committed to producing for the market, and many are more interested in expanding commercially-oriented agriculture than in conserving forests. Some are interested in profits from the sale of timber, and not just the non-timber forest products usually deemed appropriate to them (Dove 1993). Neither good tribes nor good peasants, they are in an ambiguous position which, rather than allowing them room for maneuver, may instead restrict their scope, and make it difficult to isolate opponents and identify allies and arenas for action.

Uncertainty and contingency

One of the most significant uncertainties in the articulation of indigenous identities concerns whether or not connections can actually be made. At Lindu, government officials refused to engage with the issue of indigenousness. They simply repeated the development argument regardless of evidence that it was inappropriate. Environmentalists, journalists, or other social and political activists searching for indigenous knowledge find it more easily in some places than others, as the contrast between Lauje and Lindu clearly reveals. For people in a hurry, it is easier to seek out conjunctures at which the articulations they seek are readily forthcoming and connections easily made. Such places then become exemplars, visited by many people, and are increasingly reified as they are written about, quoted, and cited in ever-broadening circuits of knowledge and action (c.f. Keck 1995; Rangan 1993). The process is similar to that which Robert Chambers has dubbed "rural development tourism" (1983) although in the case of tribes the main issue drawing outsiders is not development success but conflict, especially when it pits locals against the state. Struggles over access to mundane resources like schools and roads, and the strategies of those who seek to position themselves closer to the state, go relatively unremarked.

The circumstances of my research at Lake Lindu and in the Lauje hills can usefully illustrate the uneven channels through which outsiders connect to "the local." I point this out not in confessional mode, but because reflexivity, in this instance, brings to light issues of a general nature (Herzfeld 1997). NGO friends in Jakarta who were active in the campaign against the hydro project suggested I should visit Lindu, and put me in touch with their partner NGO in Palu. Contacts easily made, I was able to make a two-day visit to Lindu at the end of a five-week stint in the Lauje hills. When I arrived at Lindu, a group of community leaders gathered to talk to me. The contrasts with the Lauje area I had just left were palpable: a much higher standard of living, an educated, Indonesian-speaking population, and a leadership with a clearly articulated collective position. Moreover, the clarity of their discourse, together with the set of documents and press clippings given to me by the NGO, made it possible for me to write about them even without conducting field research.

Connecting with the hillside Lauje is much more difficult. Very few people speak Indonesian, illiteracy is almost total, and there are precious few documentary sources. The hillside population has no obvious spatial or social center, no hierarchy of leadership that would suggest to a visitor (especially one in a hurry) where they should go, or whom they should talk to. The historical reasons for this are deep but contingent, as I have shown. The possibilities for research, writing, and connecting are also preconfigured, and have real political effects. While I can protest that more attention should be paid to the Lauje and people like them, as well as to the historical contexts of meaning and action and the more subtle workings of power, it was usually the dramas at Lindu that captured the imagination of readers of this paper in its earlier drafts. My accounts of the Lauje are more nuanced, but also fuzzier, more equivocal, less easily picked up and read by outsiders in search of a tribal place.

Articulation versus imposition

Many locally produced images, counterimages, inversions, and inventions receive little attention on the global stage as a result of the unequal power relations within which processes of representation occur. One could mention here the shaman/ leader described by Tsing (1993), whose project for defining Meratus identity and reordering community life enthralled local audiences, although it would surely be dismissed by outsiders as the ravings of a mad woman. Her articulations fail to forge connections to wider circuits of meaning. Thomas (1994: 89) also draws attention to the problem of uneven privileging:

> constructions of indigenous identities almost inevitably privilege particular fractions of the indigenous population who correspond best with whatever is idealized: the chiefly elites of certain regions, bush Aborigines rather than those living in cities, even those who appear to live on ancestral lands as opposed to groups who migrated during or before the colonial period.[36]

As my studies in Central Sulawesi suggest, "correspondence" is itself a product of articulation. Few places could be more "bush" than the Lauje hills, and yet, as I have shown, the people and their concerns do not easily connect.

There has been much written about how subaltern struggles are distorted by representations created and imposed by outsiders. DuPuis and Vandergeest (1996) decry the simplified spatial images (wilderness, countryside) imposed upon rural people through policy processes (and their green counterpoints) pursued in ignorance of the complexity of local histories, livelihoods, and aspirations. Similarly, Fisher (1996) and Hecht and Cockburn (1990) are troubled by the way political space for Amazonians has been circumscribed by contemporary antidevelopment in the shape of environmentalism. Lohmann (1993: 203) argues that "green orientalism" compels locals to act out assigned roles which they can, at best, only "twist and subvert" to their own advantage. Similar effects result from *indigenismo* and images of the "hyperreal Indian" (Ramos 1994). Rangan (1993) has recounted the damage done by an externally generated image of the Garhwal Himalayas, home of Chipko, as an ecological utopia.

This is an important critique. However, it treats representation as a one-sided imposition. By paying attention to the process of articulation it is possible to appreciate opportunities as well as constraints, and the exercise of agency in these encounters. Simplified images may be the result of collaborations in which "natives" have participated for their own good reasons. According to Eder (1994), Batak highlanders in the Philippines see themselves simultaneously as a deprived underclass lacking the resources (but not the desire) to pursue lowland Filipino lifeways, and also as proud bearers of a tribal identity. The latter has become emphasized through their collaboration with NGO allies, as they have discovered the value of ethnic claims for obtaining desired outside resources. Neumann (1995) describes the way Tanzanian pastoralists have made productive political use of an environmentalist rhetoric even as it was deployed to displace them. Jackson (1995) describes Tukanoans in the Vaupes "orientalizing themselves" to acquire more Indianness. Complexity, collaboration, and creative cultural engagement in both local and global arenas, rather than simple deceit, imposition, or reactive opportunism, best describe these processes and relationships.

Connecting social forces

As Hall observed, the most important articulations go beyond the "cut" through which localized groups position themselves, to connect with broader social forces. Like a localized group, a social movement also needs to select some issues from a broader canvas if it is to position itself and build alliances. From this point of view, images of environmentally friendly tribes in exemplary places may be necessary, at least as a starting point. But there are limitations to a social movement built around such images. To the extent that they highlight primordial otherness, separating us from them, traditional from modern, and victim from aggressor or protector, they reinforce differences and channel alliances along binary pathways. Moreover, ideal candidates for the tribal slot are difficult to find in Indonesia, and their identification is, as I have indicated, a contingent matter. Taking advantage of such ambiguities, the government could set out new rules to identify and accommodate a few "primitives" or traditional/indigenous people, and even acknowledge their rights to special treatment, without fundamentally shifting its ground on the issue that affects tens of millions: recognition of their rights to the land and forest on which they depend. Some people would gain from official recognition of their "indigenous peoples" status, but the result might be heightened tensions as neighboring or intermingled populations find themselves differently affected.[37]

On the other hand, too much fuzziness, or too broad an agenda, makes it difficult to forge connections. It is not obvious to me, for example, that substituting a discourse of class for one about indigenous identities and practices, as proposed by Rouse (1995), would necessarily have formed a broader coalition or more effectively "found its subject" in the Indonesian countryside over the past decade. Rouse exposes the politics of identification in the US as the effect of routinized micropower and attempts by the ruling regime to deflect opposition potentially formulated in broader, class terms. In Indonesia under the New Order, in contrast to the US and also in contrast to the *adat*-making endeavors of the Dutch colonial period, ethnic identity was most decidedly *not* the chosen ideological terrain of the state. Although colorful cultural signs have always been acceptable, localized identities, histories, and commitments were consistently unmarked and derecognized in favor of a homogenizing discourse of development.

Positioned in relation to this particular field of power, an articulation that focused attention on the tribal slot was able to make important connections. But articulations are, as Hall argues, not given or fixed for all time.

The broader visions framed by the discourse on indigenous people have been attempts to rework the meanings of democracy, citizenship, and development. These are visions which could incorporate Lauje, Lindu, and millions of other rural Indonesians.[38] Often they note, but then proceed to blur the distinctions between indigenous people, local people, and other rural folk, including migrants, stressing the common concerns that arise from the grounding of livelihoods in particular places, and the need to contest arbitrary state power to displace and impoverish. These visions do not reject the idea of development, but hold the state accountable. They engage with the state at its most vulnerable point: when its promises are tested by routine or spectacular development failures, and its *raison d'etre* called into question. The Lindu rejected the idea that the state could or would bring them development, and mobilized accordingly. The cynical reflections of the Lauje are the product of decades of experience with official greed, incapacity, and indifference. They know full well that their future does not lie in state handouts – a knowledge which renders the exaggerated claims of state programs vulnerable to exposure and critique.

CONCLUSION

The discourse on indigenous people in Indonesia has emerged from new visions and connections that have created moments of opportunity, but there are no guarantees. There is the potential for the development of a broad social movement, in which urban activists and rural people can begin to articulate shared interests. There are also risks. Articulation, in Hall's formulation, is a process of simplification and boundary-making, as well as connection. The forms it takes are not predetermined by objective structures and positions, but emerge through processes of action and imagination shaped by the "continuous play of history, culture and power."

Seeking to negotiate the political dangers of attributing either too much or too little agency to those who would claim the tribal slot as their own, I explored contrasting conjunctures to expose the conditions and processes which made particular articulations possible. The Lindu came to position themselves in the tribal slot at a moment of crisis, but their articulations drew upon experiences of boundary-making and selection, sedimented over more than a century. The Lauje have engaged with more diffuse forms of power, and their positions have not been collectively defined. They do not easily fit into the tribal slot defined for them in some activist agendas. In their work on behalf of tribal and indigenous people, NGOs have also articulated their positions to engage quite specific fields of power. As agendas and positions are recalibrated in the post-Suharto era, no doubt the risks and opportunities associated with the tribal slot will be reassessed by those it potentially engages as subjects, and by those who seek to place the resource struggles and aspirations of Indonesia's frontier peoples at the center of a broad social movement.

NOTES

1 This is a slightly revised and updated version of an article that first appeared in *Comparative Studies in Society and History* 42(1): 149–79 (2000) under the title "Articulating Indigenous Identity in Indonesia: Resource Politics and the Tribal Slot." It is reprinted by permission from the copyright holder, Cambridge University Press.

    I thank Tim Babcock, Victor Li, Donald Moore, Anna Tsing, Bruce Willems-Braun and participants of the Environmental Politics Seminar, Institute of International Studies, University of California, Berkeley for incisive readings of an earlier draft. My research in Indonesia has been generously funded by Canada's Social Science and Humanities Research Council (1989–92, 1995–2000), and by the Canadian International Development Agency through a Dalhousie University linkage with Indonesia's Ministry of State for Population and Environment (1990–94). The usual disclaimers apply.

2 Sarwono, K. (1993), Minister of State for the Environment, addressing an NGO forum.

3 In one of these locations, the Lauje area, I have carried out fieldwork for a total of about seven months, spread over a period of seven years. For the other, Lake Lindu, I rely mainly on secondary sources.

4 I use the terms "indigenous" and "tribal" interchangeably in my general discussions, while drawing attention to nuances in the deployment of these terms and the meanings they invoke in particular contexts. Kingsbury (1998: 450) takes a "constructivist" position on indigenousness, arguing that this identification will emerge and shift in relation to international discourses, national policies and local dynamics. Gray (1995) argues that the term "indigenous" lacks descriptive coherence in relation to Asia, but signals a process and phenomenon which occurs in struggles that pit localized groups against encompassing states. Therefore, millions of people in Asia who actually or potentially experience this scenario fall within its compass.

5 See the polemics over this matter in the journal *Identities* (1996, volume 3, 1–2). See also Friedman (1992).

6 For an elucidation of the phrase "room for maneuver" and an insightful ethnographic account, see Tsing (1999).

7 The formulation of articulation in the "modes of production" literature of the 1970s focused upon the process of conjoining, but not on that of "giving expression to" (Foster-Carter 1975: 53). For an account of how Hall positions his concept of articulation in relation to the work of Althusser, Foucault, Lacan, and others, see Hall (1985).

8 Lynch and Talbott (1995: 22) estimate that Indonesia has eighty to ninety-five million people directly dependent upon forest resources, of whom forty to sixty-five million live on land classified as public forest.

9 See Colchester (1986a and b) for a discussion of transmigration and other programs which are explicitly designed to homogenize the rural population and eliminate ethnic distinctions. Much criticism has focused upon the *Desa* Administrative Law No. 5 (1979) which sought to standardize villages and weaken *adat* institutions concerned with social organization and leadership. See Moniaga (1993a: 33–5).

10 I draw here upon a set of interviews I carried out with the staff of Jakarta NGOs in 1996 and 2000, as well as upon their published documents and internet discussion groups. Where the subject matter might be sensitive, I do not identify the organizations to which I am referring in my discussion. I provide a more detailed examination of the origins and contours of movement, the role of the indigenous peoples' alliance AMAN, and the dilemmas of recognition in Li (2001a).

11 See, for example, Moniaga (1993b) and "Ekistensi Hukum" (*Kompas*, 27 March 1996).

12 Simply reversing the images is also problematic, as NGOs increasingly recognize. An NGO campaign against transmigration and large-scale plantations on the island of Siberut argued that the island's residents were so traditional they could not mix with newcomers, or adapt to rapid and major change. But the very same image of an extreme gulf between an isolated and primitive "them" and a modern Indonesian "us" was used by Transmigration Minister Siswono to argue that development must proceed, because the Siberut people cannot be left in a stone-age state. See "Siberut Island" (*Jakarta Post*, 14 February 1996) and "Skephi opposes" (*Jakarta Post*, 17 February 1996).

13 See critiques of the government for its refusal to recognize customary land rights in Moniaga (1993a, b); Skephi and Kiddell-Monroe (1993); "Semoga" (*Kompas*, 29 March 1993), "Indigenous Peoples" (*Kompas*, 29 April 1993), and "Eksistensi Hukum Adat" (*Kompas* 27 March 1996). See Evers (1995) for an overview of the legal status of customary land rights, the difficulties of specifying who should be included in the category of indigenous people in Indonesia, and an attempt to reconcile these questions with World Bank policies. For a discussion of the difference between the Dutch colonial concept of a traditional-law society (*masyarakat hukum adat*) and the internationally recognized concept of indigenous people, and the (lack of) resonance of these concepts with forestry law, see Safitri (1995).

14 See O'Brien and Roseberry (1991: 13); Cohen (1993: 203).

15 See Thomas (1992: 65), Scott (1992: 376) and Gupta and Ferguson (1992: 16) for general arguments along these lines; see Kahn (1993: 23) and Tsing (1993) for Indonesian examples.

16 For a summary of the large literature on upland-lowland relations in the precolonial era see Li (1999a), and references cited therein.

17 In Northern Sulawesi, for example, Henley characterizes the indigenous political geography in terms of "aterritoriality, fluidity and fragmentation" (1996: 143). He notes that local kin-based groups, or *walak*, became more strongly bounded and endogamous under warlike conditions, although they could still fragment and realign (1996: 26, 35).

18 See, for example, Tsing's (1993) description of the mountain dwellers of Southeast Kalimantan, for whom she had to coin a singular name, the Meratus.

19 For Sulawesi examples, see Acciaioli (1989: 66, 73); Henley (1996).

20 See Kahn (1993), von Benda-Beckmann and von Benda-Beckmann (1994), Ruiter (1999); for a more general discussion of colonial practices of discipline and rule, see Cooper and Stoler (1997).

21 See Kahn (1993: 78–110) for an extended discussion of the intellectual, economic, and political rationales for the Leiden School of *adat* law associated with van Vollenhoven, influential in the codification of *adat* in the period 1911–55. See also Ellen (1976).

22 For example Kahn (1993: 180; 1999) observes that in the nineteenth century the term "Minangkabau" did not have the sense of a discrete, bounded, distinctive cultural unit; this developed in the colonial period and subsequently.

23 See, for example, the discussion of Meratus identity, leadership and *ad hoc adat*-making processes in Tsing 1993.

24 I offer a more detailed account of the relations between Lauje highlanders and coastal elites historically and in relation to "development" in Li (2001b).

25 Thanks to Dan Paradis for access to transcripts of interviews with provincial officials in 1994. Because the transcripts had been translated, I do not know which Indonesian expression was translated here as "indigenous people."

26 To illustrate his point, the official showed photos of a Wana medicine man conducting a ritual. Prominently displayed in the Palu office are "before and after" pictures of near-naked Wana who are subsequently clothed, revealing the contradictory impulses of nostalgia and development.

27 This situation has begun to change in the past five years, as coastal elites see the economic potential for hillside cocoa and clove gardens. For a discussion of the local and regional class dimensions of this process, see Li (1996, 2002b).

28 Many people were reluctant to talk to me when I first started field work in the Lauje hills because they feared my research would lead to their resettlement. They were especially nervous about anything that looked like a list of names.

29 This did not mean they always spoke with one voice: disputes arose over the issue of who among "the Lindu" had the right to confer upon outsiders permission to use Lindu resources.

30 The redesign would still require a green belt around the lake, restricting access to both fisheries and farmland. Sangadji's (1996) research continues to highlight the ways in which the Lindu are, and must remain, anchored to very specific spots on the landscape, including fishing spots that are the preserve of particular families. During my visit to Lindu an NGO was facilitating a community mapping process in which the Lindu leaders who had traveled to Jakarta were key participants. They had been informed by the Minister of State for Environment that their case would be strengthened by representing their customary zones and places on maps which outsiders could read. On the politics of mapping and countermapping in Indonesia, see Peluso (1995).

31 See Moniaga (1993a: 33) and "Kearifan Masyarakat" (*Kompas*, 13 September 1993). The Institute of Dayakology also presents generic Dayak as environmentalists (Bamba 1993). For critiques of the claim that natives are naturally nurturant of nature, see Ellen (1986) and Stearman (1994).

32 Opposition to the hydro project at Lindu was widespread in the community, so there was a common interest in the success of the campaign. On other matters, including the relevance of indigenous environmental knowledge to everyday lives and practices, and the role of the *adat* council in controlling resources, there are bound to be differences of opinion among people differently situated by class, gender, and ethnic origins. Since I have not carried out field research at Lindu I am not in a position to discuss these.

33 For other Indonesian conjunctures in which some or all of these factors were also relevant see Tsing (1999) and Zerner (1994).

34 See Acciaioli (2001) for an update on how the politics of indigeneity are playing out at Lindu. On the non-egalitarian aspects of *adat* see von Benda-Beckman and von Benda-Beckman (1994); on "lairdism" or the risks associated with concentrating power in the hands of *adat* chiefs, Colchester (1994: 87); on the ways in which concentrated *adat* power becomes more easily enmeshed in or subverted by the projects of the colonial and post-colonial states, Zerner (1994).

35 For a good discussion of this point in the Philippine context see Brown (1994). Note, however, that ecological soundness is a relative matter: smallholders expanding into old-growth forests threaten biodiversity, but the resulting mosaic of land uses is vastly more biodiverse than the industrial-scale oil palm or timber plantations programmed to displace small holdings under state-sponsored schemes.

36 See also Carrier (1992), Friedman (1987), and Scott (1992: 387).

37 I explore the limits of recognition in Li (2001a). I explore some of the risks posed by the unilaterial reclaiming of indigenous territories in the absence of due process and the rule of law in Li (2002a). See also Peluso and Harwell (2001).

38 Outstanding here is the work of the Consortium for Agrarian Reform (KPA 1998).

REFERENCES

Acciaioli, G. (1989) "Searching for Good Fortune: The Making of a Bugis Shore Community at Lake Lindu, Central Sulawesi," Ph.D. Thesis, Australian National University.

—— (2001) "Grounds of Conflict, Idioms of Harmony: Custom, Religion, and Nationalism in Violence Avoidence at the Lindu plain, Cetral Sulawesi," *Indonesia* 72: 81–114.

Anonymous (1993) "Kearifan Masyarakat Adat dalam Konservasi Alam Sangat Tinggi," *Kompas*, 13 September 1993.

—— (1993) "Masyarakat Lindu Menolak Rencana Pembangunan PLTA," *Kompas* 11 September 1993.

—— (1994) "Sulawesi Tribe Opposes Lake Lindu Dam Project," *Jakarta Post* 11 September 1994.

—— (1996) "Siberut Island Likely to have New Settlement Areas," *Jakarta Post* 14 February 1996.

Babcock, T. and Ruwiastuti, M. (1993) "'Indigenous Peoples' dan Penguasaan atas Tanah," *Kompas* 29 April 1993.

Bamba, J. (1993) "The Concepts of Land Uses among the Dayaks and Their Contribution to the Sustainable Management of the Environment," in H.P. Arimbi (ed.) Proceedings, *Seminar on the Human Dimensions of Environmentally Sound Development*, Jakarta: WALHI and Friends of the Earth, 37–49.

Barth, F. (1981) *Process and Form in Social Life*, London: Routledge and Kegan Paul.

Brookfield, H., Lesley P., and Yvonne, B. (1995) *In Place of the Forest: Environmental and Socio-economic Transformation in Borneo and the Eastern Malay Peninsula*, Tokyo: United Nations University Press.

Brown, E. (1994) "Grounds at Stake in Ancestral Domains." in J. Eder and R.Youngblood (eds) *Patterns of Power and Politics in the Philippines*, Tempe: Arizona State University, 43–76.

Carrier, J. (1992) "Occidentalism: the World Turned Upside-Down," *American Ethnologist*, 19(2): 195–212.

Chambers, R. (1983) *Rural Development: Putting the Last First*, New York: Longman.

Cohen, A. (1993) "Culture as Identity: An Anthropologist's View," *New Literary History*, 24: 195–209.

Colchester, M. (1986a) "Unity and Diversity: Indonesian Policy towards Tribal Peoples," *The Ecologist*, 16(2–3): 89–97.

—— (1986b) "The Struggle for Land: Tribal Peoples in the Face of the Transmigration Programme," *The Ecologist*, 16(2–3): 99–110.

—— (1994) "Sustaining the Forests: The Community-based Approach in South and South-East Asia," *Development and Change*, 25(1): 69–100.

Cooper, F. and Ann S. (eds) (1997) *Tensions of Empire: Colonial Cultures in a Bourgeois World*, Berkeley: University of California.

Department of Social Affairs (1994) *Isolated Community Development: Data and Information*, Jakarta: Directorate of Isolated Community Development, Directorate- General of Social Welfare Development, Department of Social Affairs.

Department Kehutanan (1994) *Pentujuk Teknis Inventarisasi dan Indentifikasi Peladang Berpindah dan Perambah Hutan*, Jakarta: Direktorat Reboisasi.

Dove, M. (1993) "A Revisionist View of Tropical Deforestation and Development," *Environmental Conservation*, 20(1): 17–24, 56.

Dupuis, E.M. and Vandergeest P. (1996) "Introduction," in E.M. Dupuis and P. Vandergeest (eds) *Creating the Countryside: The Politics of Rural and Environmental Discourse*, Philadelphia: Temple University Press, 1–25.

Eder, J. (1994) "State-Sponsored 'Participatory Development' and Tribal Filipino Ethnic Identity," *Social Analysis*, 35: 28–38.

Ellen, R.F. (1986) "What Black Elk Left Unsaid: On the Illusory Images of Green Primitivism," *Anthropology Today*, 2(6): 8–12.

—— (1976) "The Development of Anthropology and Colonial Policy in the Netherlands: 1800–1960," *Journal of the History of the Behavioural Sciences*, 12: 303–24.

Emilia, S. (1996) "Skhepi Opposes Siberut Resettlement," *Jakarta Post* 17 February 1996.

Evers, P. (1995) "Preliminary Policy and Legal Questions about Recognizing Traditional Land in Indonesia," *Ekonesia*, 3: 1–24.

Fisher, W. (1996) "Native Amazonians and the Making of the Amazonian Wilderness: From Discourse of Riches to Sloth and Underdevelopment," in E. Melanie Dupuis and P. Vandergeest (eds) *Creating the Countryside: The Politics of Rural and Environmental Discourse*, Philadelphia: Temple University Press, 166–203.

Foster-Carter, A. (1975) "The Modes of Production Controversy," *New Left Review*, 107: 47–77.

Friedman, J. (1987) "Beyond Otherness of: The Spectacularization of Anthropology," *Telos*, 71: 161–70.

—— (1992) "The Past in the Future: History and the Politics of Identity," *American Anthropologist*, 94(4): 837–59.

Gray, A. (1995) "The Indigenous Movement in Asia," in R.H. Barnes, A. Gray and B. Kingsbury (eds) *Indigenous Peoples of Asia*, Michigan: Association for Asian Studies, 35–58.

Gupta, A. and Ferguson J. (1992) "Beyond 'Culture': Space, Identity, and the Politics of Difference," *Cultural Anthropology*, 7(1): 6–23.

Hall, S. (1985) "Signification, Representation, Ideology: Althusser and the Post-Structuralist Debates," *Critical Studies in Mass Communication*, 2(2): 91–114.

—— (1990) "Cultural Identity and Diaspora," in J. Rutherford (ed.) *Identity: Community, Culture, Difference*, London: Lawrence and Wishart, 222–37.

—— (1995) "Negotiating Caribbean Identities," *New Left Review*, 209: 3–14.

—— (1996) "On Postmodernism and Articulation: An Interview with Stuart Hall," edited by L. Grossberg, in D. Morley and K. Chen (eds) *Stuart Hall: Critical Dialogues in Cultural Studies*, London: Routledge, 131–50 (reprinted from *Journal of Communication Inquiry* (1986) 10(2): 45–60).

Hecht, S. and Cockburn A. (1990) *The Fate of the Forest*, London: Penguin.

Henley, D. (1996) *Nationalism and Regionalism in a Colonial Context: Minahasa in the Dutch East Indies*, Leiden: KITLV Press.

Herzfeld, M. (1997) "Anthropology: a practice of theory," *International Social Science Journal*, 153: 301–18.

Jackson, J. (1995) "Culture, genuine and spurious: the politics of Indianness in the Vaupes, Colombia," *American Ethnologist*, 22(1): 3–27.

Kahn, J. (1993) *Constituting the Minangkabau: Peasants, Culture and Modernity in Colonial Indonesia*, Providence: Berg.

—— (1999) "Culturalising the Indonesian Uplands," in T.M. Li (ed.) *Transforming the Indonesian Uplands: Marginality, Power and Production*, London: Routledge, 79–101.

Keck, M. (1995) "Social Equity and Environmental Politics in Brazil: Lessons from the Rubber Tappers of Acre," *Comparative Politics*, 27(4): 409–24.

Kingsbury, B. (1998) "'Indigenous Peoples' in International Law: A Constructivist Approach to the Asian Controversy," *The American Journal of International Law*, 92(3): 414–57.

Kipp, R.S. and Rodgers S. (1987) "Introduction: Indonesian Religions in Society," in R.S. Kipp and S. Rodgers (eds) *Indonesia Religions in Transition*, Tucson: University of Arizona Press, 1–31.

Koentjaraningrat. (1993) "Pendahuluan" and "Membangun Masyarakat Terasing," in Koentjaraningrat, (ed.) *Masyarakat Terasing di Indonesia*, Jakarta: Gramedia with Departement Sosial, 1–18, 344–50.

KPA (1998) *Usulan Revisi Undang-Undang Pokok Agraria*, Jakarta: Konsorsium Reformasi Hukum Nasional (KRHN) dan Konsorsium Pembaruan Agraria (KPA).

Kusumaatmadja, S. (1993) "The Human Dimensions of Sustainable Development," in H.P. Arimbi (ed.) *Proceeding, Seminar on the Human Dimensions of Environmentally Sound Development*, Jakarta: WALHI and Friends of the Earth, 12–15.

Laudjeng, H. (1994) "Kearifan Tradisional Masyarakat Adat Lindu," in A. Sangadji (ed.) *Bendungan Rakyat dan Lingkungan: Catatan Kritis Rencana Pembangunan PLTA Lore Lindu*, Jakarta: WALHI, 150–63.

Li, T.M. (1996) "Images of Community: Discourse and Strategy in Property Relations," *Development and Change*, 27(3): 501–27.

—— (1999a) "Marginality, Power and Production: Analysing Upland Transformations," in T.M. Li (ed.) *Transforming the Indonesian Uplands: Marginality, Power and Production*, London: Routledge, 1–44.

—— (1999b) "Compromising Power: Development, Culture, and Rule in Indonesia," *Cultural Anthropology*, 14(3): 1–28.

—— (2001a) "Masyarakat Adat, Difference, and the Limits of Recognition in Indonesia's Forest Zone," *Modern Asian Studies*, 35: 645–76.

—— (2001b) "Relational Histories and the Production of Difference on Sulawesi's Upland Frontier," *Journal of Asian Studies*, 60: 41–66.

—— (2002a) "Ethnic Cleansing, Recursive Knowledge, and the Dilemmas of Sedentarism," *International Journal of Social Science*, 173.

—— (2002b) "Local Histories, Global Markets: Cocoa and Class in Upland Sulawesi," *Development and Change*, 33: 415–37.

Lohmann, L. (1993) "Green Orientalism," *The Ecologist*, 23(6): 202–4.

Lynch, O.J. and Talbott K. (1995) *Balancing Acts: Community-Based Forest Management and National Law in Asia and the Pacific*, Washington: World Resources Institute.

Moniaga, S. (1993a) "The Systematic Destruction of the Indigenous System of Various Adat Communities throughout Indonesia," in H.P. Arimbi (ed.) *Proceeding, Seminar on the Human Dimensions of Environmentally Sound Development*, Jakarta: WALHI and Friends of the Earth, 31–6.

—— (1993b) "Toward Community-Based Forestry and Recognition of Adat Property Rights in the Outer Islands of Indonesia," in J. Fox (ed.) *Legal Frameworks for Forest Management in Asia: Case Studies of Community–State Relations*, Honolulu: Environment and Policy Institute, East–West Center.

Moore, D. (1998a) "Clear Waters and Muddied Histories: Environmental History and the Politics of Community in Zimbabwe's Eastern Highlands," *Journal of Southern African Studies*, 24: 377–403.

—— (1998b) "Subaltern Struggles and the Politics of Place: Remapping Resistance in Zimbabwe's Eastern Highlands," *Cultural Anthropology*, 13: 1–38.

—— (2000) "The Crucible of Cultural Politics: Reworking 'Development' in Zimbabwe's Eastern Highlands," *American Ethnologist*, 26: 654–89.

Neumann, R. (1995) "Local Challenges to Global Agendas: Conservation, Economic Liberalization and the Pastoralists' Rights Movement in Tanzania," *Antipode*, 27(4): 363–82.

Nourse, J. (1999) *Conceiving Spirits: Birth Rituals and Contested Identities among Lauje of Indonesisa*, Washington: Smithsonian Institution Press.

O'Brien, J. and Roseberry W. (1991) "Introduction," in J. O'Brien and W. Roseberry (eds) *Golden Ages, Dark Ages*. Berkeley: University of California Press, 1–18.

Peluso, N. (1995) "Whose Woods Are These? Counter-Mapping Forest Territories in Kalimantan, Indonesia," *Antipode*, 27(4): 383–406.

Peluso, N. and Harwell E. (2001) "Territory, Custom, and the Cultural Politics of Ethnic War in West Kalimantan, Indonesia," in N. Peluso and M. Watts (eds) *Violent Environments*, Ithaca: Cornell University Press, pp. 83–116.

Ramos, A.R. (1994) "The Hyperreal Indian," *Critique of Anthropology*, 14(2): 153–71.

Rangan, H. (1993) "Romancing the Environment: Popular Environmental Action in the Garhwal Himalayas," in J. Friedmann and H. Rangan (eds) *In Defense of Livelihood: Comparative Studies on Environmental Action*, West Hartford: Kumarian Press, pp. 155–81.

Rouse, R. (1995) "Personhood and Collectivity in Transnational Migration to the United States," *Critique of Anthropology*, 15(4): 351–80.

Ruiter, T. (1999) "Agrarian Transformations in the Uplands of Langkat: Survival of Independent Karo Batak Smallholders," in T.M. Li (ed.) *Transforming the Indonesian Uplands: Marginality, Power and Production*, Amsterdam: Harwood Academic Publishers, 279–310.

Safitri, M. (1995) "Hak dan Akses Masyarakat Lokal pada Sumberdaya Hutan: Kajian Peraturan Perundang-undangan Indonesia," *Ekonesia*, 3: 43–60.

Sangadji, A. (1996) *Menyorot PLTA Lore Lindu*, Palu: Yayasan Tanah Merdeka.

Schweithelm, J. *et al.* (1992) *Land Use and Socio-Economic Survey Lore Lindu National Park and Morowali Nature Reserve*, Sulawesi Parks Program, Directorate General of Forest Protection and Nature Conservation, Ministry of Forestry, and The Nature Conservancy.

Scott, D. (1992) "Criticism and Culture: Theory and Post-colonial Claims on Anthropological Disciplinarity," *Critique of Anthropology*, 12(4): 371–94.

Shields, R. (1991) *Places on the Margin: Alternative Geographies of Modernity*, London: Routledge.

Sider, G. (1987) "When Parrots Learn to Talk, and Why They Can't: Domination, Deception, and Self-deception in Indian–White Relations," *Comparative Studies in Society and History*, 29: 3–23.

Singarimbun, M. (1993) "Semoga Hak Ulayat Dihargai," *Kompas* 29 April 1993.

Skephi and Kiddell-Monroe, R. (1993) "Indonesia: Land Rights and Development," in M.Colchester and L. Lohmann (eds) *The Struggle for Land and the Fate of the Forests*, Penang: World Rainforest Movement, 228–63.

Sodiki, A. (1996) "Eksistensi Hukum Adat Dewasa Ini," *Kompas* 27 March 1996.

Stearman, A.M. (1994) "Revisiting the Myth of the Ecologically Noble Savage in Amazonia: Implications for Indigenous Land Rights," *Culture and Agriculture*, 49: 2–6.

Thomas, N. (1992) "Substantivization and Anthropological Discourse," in J. Carrier (ed.) *History and Tradition in Melanesian Ethnography*, Berkeley: University of California Press, 64–85.

—— (1994) *Colonialism's Culture*, Cambridge: Polity Press.

Trouillot, M.R. (1991) "Anthropology and the Savage Slot: The Poetics and Politics of Otherness," in R. Fox (ed.) *Recapturing Anthropology*, Santa Fe: School of American Research, 17–44.

Tsing, A.L. (1993) *In the Realm of the Diamond Queen*, Princeton: Princeton University Press.

—— (1999) "Becoming a Tribal Elder and other Green Development Fantasies," in T.M. Li (ed.) *Transforming the Indonesian Uplands: Marginality, Power and Production*, London: Routledge, 159–202.

von Benda-Beckmann, F. and K. (1994) "Property, Politics, and Conflict: Ambon and Minangkabau Compared," *Law and Society Review*, 28(3): 589–607.

Watts, M. (1992) "Space for Everything (A Commentary)," *Cultural Anthropology*, 7(1): 115–29.

Zerner, C. (1990) *Community Rights, Customary Law, and the Law of Timber Concessions in Indonesia's Forests: Legal Options and Alternative in Designing the Commons*, Forestry Studies UTF/INS/065.

—— (1994) "Through a Green Lens: The Construction of Customary Environmental Law and Community in Indonesia's Maluku Islands," *Law and Society Review*, 28(5): 1079–122.

14

FROM CHIPKO TO UTTARANCHAL

The environment of protest and development in the Indian Himalaya

Haripriya Rangan

In November 2000, ten districts in the Himalayan tracts of northwestern Uttar Pradesh (UP) were carved out to form a new state of Uttaranchal within the Indian Union. Its creation culminated after nearly a decade of popular protest that demanded statehood for the Himalayan regions of Garhwal and Kumaon. For the men and women involved in the struggle, statehood was the necessary condition for extricating their region from its backwardness. They claimed that the hill regions were afflicted by poverty and high unemployment because the politicians and bureaucrats of Uttar Pradesh were mainly concerned with the interests of the plains (*maidan*) and cared little for the distinctive culture and needs of the mountains (*pahar*). Garhwal and Kumaon lacked development because a callous and inefficient state administration merely saw them as sites of resource exploitation rather than of economic investment. Separation from Uttar Pradesh, the advocates for statehood argued, was the only way in which development would occur in this mountainous region (Dhoundiyal *et al.* 1993; *Himachal Times* 1990a; Jayal 2000; Kumar 2000).

Popular protests centered on issues of economic development or regional identity are neither new nor rare in post-independence India (Hauser 1993; Mitra 1992). Nor is it extraordinary that the Indian government acceded to the demands for statehood for the Uttarakhand region. As Atul Kohli (2001) points out, Indian democracy, working within the framework of a centralized yet accommodating state, has enabled regional forces to successfully press their demands for maintaining their territorial or linguistic identity, and bargain for their share of economic resources. After all, Uttaranchal was only one of three new states created in November 2002 (the others being Jharkhand and Chattisgarh), and similar arguments regarding development and regional identity had been advanced in their cause (cf. Corbridge 2002).

What is curious about the struggle for Uttaranchal, though, is its apparent disjuncture from the rhetoric commonly associated with Chipko, the movement that had emerged in this region during the mid-1970s. The Chipko movement had been hailed by academics and environmental activists throughout the world as a powerful critique of development, and described as a grassroots struggle against a modernizing Indian state whose pursuit of development destroyed local ecologies and traditional ways of peasant life (Bahuguna 1982, 1987; Bhatt 1988, 1991; Berreman 1989; Dogra 1980, 1983; Gadgil and Guha 1992; Guha and Gadgil 1992; Shiva 1989; Shiva and Bandhyopadhyay 1986, 1987, 1989; Weber 1988). Chipko, they claimed, symbolized a new ecological consciousness that challenged the conventional, reductionist approaches to development (Shiva 1991), and shone an alternative path towards a green earth and a true civilization (Weber 1988). Yet within the following two decades, the region that gave rise to the Chipko movement was consumed by another struggle that, rather than calling for environmental protection or alternatives to development, positively embraced conventional notions (i.e. modernization-as-development, cf. Simon 2003) of development and demanded the creation of a separate state for promoting economic growth.

This chapter illustrates how Chipko and subsequent struggles, including the Uttarakhand movement for statehood, have been linked by a central preoccupation with the need for economic development in this Himalayan region. The Chipko movement emerged in the Garhwal Himalayas during the 1970s in response to national economic policies and regulations that introduced stricter controls over access to resource extraction and constrained opportunities for local economic development. Chipko's initial attempts to alter regulations and demand economic concessions were overwhelmed by its subsequent celebration as a grassroots environmental movement. Its new-found fame in ecological guise resulted in the introduction of new environmental regulations that further restricted development activities in the region. I go on to show how subsequent protests and the movement for statehood, rather than being supposedly different from, or more multi-dimensional than, Chipko (Hannam 1998), continued to struggle for the same demands that were initially articulated by the Chipko movement: the need for policies and conditions for generating local employment opportunities and fostering regional economic development. With the goal of statehood achieved, the issue of development in Uttaranchal is not likely to recede into the background, but continue to be the focus of struggle in the years to come.

In this chapter I also take a critical look at the development debates that have, over the past decade or more, taken the somewhat baggy shape of the *anti-*, *post-*, and *alternatives to-* development literature. More specifically, I examine the claims of some scholars who view that new social movements in poor countries and regions as grassroots struggles seeking alternatives *to* development (see, for example, Dwivedi 1998; Escobar 1995a; Routledge 1993). I argue that, contrary to such views, social movements in most poor regions of the world are not necessarily new, but rather emerge in *different* forms and historical circumstances from the familiar tensions between territory and function (Friedmann and Weaver

1979). Or, in Harvey's words, from the tensions between territorial and capitalistic logics of power (1982, 2003) that produce uneven geographical development. The idea of development, rather than being dismissed or subverted by these regionalized protests, becomes an integral part of an expanded moral economy (Thompson 1993; Wells 1994) of communities struggling to overcome political and economic marginality. Contemporary social movements in poor countries of the world, I argue, are not *against* the idea of development, they are *part* of it. Hence even the most radical ideas celebrating the flowering of new social movements and calling for sustainable use of resources need to recognize that the broad notion of development, as a means of gaining access to social equality, economic well-being, and political recognition, remains crucial to any project of a liberation ecology.

DEVELOPMENT AND ITS CRITICS

It is no great surprise that development is a contested word. The term, from the very beginning of its usage, has carried within it a tension between *process* and *concerted action*. The "great transformation" (Polanyi 1944) that wrought dramatic changes to lives and landscapes appeared to observers in the eighteenth and nineteenth centuries as an inexorable force which found expression in words such as development and progress (Cowen and Shenton 1995). Yet many of them were also provoked enough by its ravages to claim that social and urban reforms by the state were integral to the process of development and the idea of progress. As Hart observes, far from being external to each other, the two opposing tendencies are contained *within* the development of capitalism as a geographically uneven and profoundly contradictory set of historical processes (2001: 650).

Much of the anti, post- and alternatives to development literature reflects little awareness of this fundamental tension. Critics in this genre have drawn on postmodern and post-structuralist theories to argue that development is a totalizing and hegemonic discourse which perpetuates social and economic inequalities between North and South (Escobar 1995b; Mies and Shiva 1993; Shiva 1989; Trainer 1989). Emerging in the post-World War II era, as Escobar has claimed, it was an invention and strategy produced by the First World about the underdevelopment of the Third World. Development has been the primary mechanism through which these parts of the world have been produced and have produced themselves, thus marginalizing or precluding other ways of seeing or doing (Escobar 1995a: 212). Development, they said, brought destruction and misery to the Third World (Esteva 1987; Trainer 1989); it either needed to be abandoned altogether, or replaced by building on practices of new social movements in the Third World for creating alternative visions of democracy, economy, and society (Escobar 1995a).

When seen in historical perspective, these criticisms seem entirely consistent in reflecting the opposing tendencies contained within the concept of development. But there is a cruel tautology (cf. Said 1994:xviii) in the anti-, post-alternatives to development critiques. By representing development as the hegemonic discourse of the West, every opposition to it only serves to confirm its overwhelming power; all alternatives to it are doomed from the start. What these critiques do not acknowledge is that the tensions contained within development have, in fact, produced a complex genealogy encompassing a diversity of meanings (Williams 1983). It is a term that is as dynamic as life's processes, taking shape as an idea, changing over time, diversifying in meaning, becoming a contested terrain, diffusing through translation and reemerging in different forms in different regions. Imposing a reductionist definition of instrumental or hegemonic control on the term not only denies its historicity but also ignores the diverse ways in which ideas of development, despite their origins in the West, have been translated, appropriated, refashioned, and reconfigured by local circumstances (Watts 1993). Surely it is in these geographical movements and translations that the liberating potentials of development (rather than wholesale abandonment, or alternatives to it) can be reclaimed in different ways.

There are two additional points to make about new social movements in the Third World. Although these have been celebrated as external challenges and alternatives to development, they are in fact very much internal to, and produced by, the processes of uneven geographical development, and articulated within the context of state action. It is through collective identification with the experience of *spatialized differences* in social and economic opportunities that gives rise to regional movements within particular political or administrative boundaries. While these struggles may not be intent on capturing the state, they are, nevertheless, centrally concerned with the distribution of economic and political power and governance. Their mobilization depends on using the legitimizing discourse of development often employed by the states within which they are based, to press for better access to resources or social rights of belonging, or demand that problems of uneven geographical development are addressed. The regionalized identities mobilized through such social movements are not free from the contradictions inherent in development. They may claim to oppose the state, or even partially succeed in forcing reform of the structures or modes of governance, but these actions only reconfigure the shifting geographies of state power and uneven development.

DEVELOPMENT, DEMOCRACY AND PLANNING IN INDIA

It has never been possible to ignore the realities of difference: regional, cultural, linguistic, religious, ethnic, class, and caste in India. Here, the idea of development (referred to in the vernacular as *vikas*, which means dawn, and alludes to both a new social era and the process of moving towards the dawn of a new social era) has been used as a secular, democratic means for opening the political arena to the

374

claims of various groups in civil society. Development is charged with the promise of change towards greater social equality and prosperity for all citizens. It has been taken up by disadvantaged groups as a means of gaining political recognition and access to economic empowerment. Social transformation is a glacial process in India, but the vocabulary of development combined with universal franchise has allowed poorer classes and socially disadvantaged groups to question, with growing assertiveness, structures of social and economic inequality such as the caste system (Bardhan 1985; Corbridge and Harris 2000; Hauser 1993; Naipaul 1990). The slow process of social change has its historical roots in the way ideas of development and democracy were refigured and reworked in India as part of the post-independence struggles.

Development emerged as the *leitmotif* of political discourse in post-independence India as groups on the Left, Right, and Center confronted each other with their interpretations of the "developmental imperative" for the country (Chakravarty 1987; Mitra 1992). There was little debate or argument over the precise definition of the term. Nationalist elites agreed that development, broadly meaning the reorganization of state, market, and social institutions, was necessary for overcoming persistent economic stagnation, high levels of poverty, and structural weaknesses in the Indian economy. Influenced in part by the Soviet experience and Keynesian economics, the nation's leaders saw the state as the central actor in achieving this purpose. State intervention in the economy was to follow a mixed socialist–capitalist approach, with central planning as a means of avoiding the unnecessary rigors of an industrial transition in so far as it affected the masses resident in India's villages (Chakravarty 1987: 3). The Indian Planning Commission was established for the purpose of systematically studying and monitoring the national economy, identifying sectoral priorities and working out a coherent set of investment policies for economic development in its Five-Year Plans.

Rational state-led planning was, however, framed by parliamentary democracy in postcolonial India. This presented a different set of challenges that were unlike anything experienced by Western liberal democracies (Corbridge and Harriss 2000). On the one hand, the Indian government sought to engage in national development, a process that inevitably produced or intensified social and economic differentiation within society (Herring 1989; Weiner 1989). On the other hand, the guarantees of universal franchise and social equality promised by the national Constitution meant that the government was also required to correct these imbalances and maintain its democratic credentials for ensuring continued political support of the populace. The nation's political leaders recognized that the dual commitment to development and democracy not only produced tensions but also allowed room for the state to exercise some degree of flexibility. Development, conceived as a process of nation-building, ensured the Indian government a central role in the economy and, as a consequence, a key role in political accommodation between dominant classes and other groups within civil society. The Indian state emerged as the mediator, appealing to democratic traditions and invoking the collective goal of national development, whenever conflicts in civil society grew violent or gave rise to secessionist demands. In the process, it also created, partly by design and part unintentionally, conditions of social fluidity that allowed less-privileged groups to maneuver for greater political representation and access to the benefits of economic development (Bardhan 1985; Corbridge and Harriss 2000; Kohli 2001; Mitra 1992).

Development thus came to be accepted by most of India's population as a legitimate activity promoted by the state, but its meanings, values, and benefits have been, and continue to be, constantly contested and renegotiated in the public realm. Since social and economic change is seen as part of the state's agenda, political parties seeking power have found it necessary to employ the language of development for mobilizing support among India's vast rural constituencies. Thus development has become the means by which political allegiance of rural elites is gained. Rural elites function as intermediaries in the development process, their status depending in large part on their ability to traverse the interstices between state, market, and civil society, and also on their ability to direct the flow of developmental resources from the state to their localities (Bayly 1976; Hauser 1993; Mitra 1992). In most parts of the country, development has become a generic term that typically refers to government-sponsored activities such as the construction of roads and bridges, provision of public utilities, social welfare programs, and investment subsidies for promoting economic growth. If local elites are seen as failing to bring development to their communities, their status and authority may be challenged by other contending social/caste groups who wish to increase their influence in the local or regional political arena. Because the discourse of development in India carries a broader symbolism of social justice and economic well-being, it muddles (or perhaps condenses within it) the conceptually rigid boundaries between state, markets, and civil society. It simultaneously creates the space for institutional participation and provides the language for radical critique. Social protests and movements in postcolonial India have typically not argued against development but have always been part of its process.

Development planning as nation-building reflected the economic and political debates and exigencies of its times and geopolitical circumstances. Even though critics of India's development policies argue that the Gandhian approach was a viable alternative to the path of modernization chosen by Nehru (Gadgil and Guha 1992), it is important to recognize that Nehru's inclinations towards Fabian socialism and enthusiasm for catching up with the West were tempered by the political lessons he learned from Mahatma Gandhi and from active participation in the nationalist movement. And while Gandhi's ideas were indeed drawn from close experience and understanding of the problems of rural India, there were few scholars or policy makers at the time who could have confidently demonstrated that his economic thought was based on a substantive theoretical foundation. Even fewer would have argued, soon after Independence, for low levels of production and consumption implicitly assumed in Gandhian economic thought. In contrast, India's development planners concentrated on assessing the structural weaknesses in the nation's economy, on land and social reforms that would break from the fetters of the past, and on developing a strong industrial sector that would establish the foundations for a healthy and self-reliant economy (Ambedkar 1945; Chakravarty 1987).

The state-led, rational-planning approach to economic growth and social transformation was pursued through the national five-year plans and encountered a predictable array of problems: contingent events such as wars, droughts, the oil crisis, as well as reflexive consequences of its policies. But what further complicated the rational approach to development was the problem of interregional disparity (uneven geographical development), and the shifting bases of populist power. As was to be expected, the five-year plans when translated into policies, were substantially reshaped by political pressures exercised by different classes and electoral constituencies. Sectoral policies aimed at promoting industrial and agricultural growth led to concentration of investment in a few states or sub-regions, to the exclusion of others. Spatially oriented policies for developing economically backward regions intensified social differentiation within them. Populist programs devised by ruling political parties at the central or state levels aimed at appeasing rural or regional elites undermined other policies and inevitably increased fiscal expenditure.

Development as nation-building combined with democracy ensured that development planning process constantly grappled with the contradictions of allocating resources between function (sectors), territory (regions), and populist power (electoral constituencies, or vote-banks). Complicated and messy, the development process in India has nevertheless been a dynamic force, reconfiguring landscapes, social relations, and forging new trajectories as regions, markets, and institutions within civil society contest and renegotiate its meanings.

UNEVEN DEVELOPMENT AND ITS OUTCOMES IN THE GARHWAL HIMALAYAS

At the time of independence in 1947, the major economic activities in the Garhwal Himalayas centered around subsistence agriculture, forestry, seasonal employment in pilgrimage services, and, to a smaller extent, in transHimalayan trade (Pant 1935; Rawat 1983; Walton 1910; Williams 1874). Forests, comprising nearly 60 percent of the region (Stebbing 1932), were largely under the control of the Uttar Pradesh (UP) State. The State Forest Department controlled approximately two-thirds of this forested area, with the rest distributed under the authority of the State Revenue Department, village institutions, and a few private holdings (UP Forest Department 1989; Rangan 1997). During the first and second five-year plans (1951–1955, 1956–1960) there was little economic investment in the Himalayan districts. The eastern districts in the Garhwal region received a few Community Development projects aimed at providing basic infrastructure and encouraging the formation of village-level cooperatives for artisanal production (Khan and Tripathy 1976). Half-hearted attempts at land reform by the UP State government led to the allocation of small plots to lower-caste and landless households on degraded state-owned lands. Statutory ceilings imposed on land ownership resulted in private forest owners stripping their forests of all valuable commodities before their lands were acquired by the State government (Bora 1987; Dhoundiyal *et al*. 1993; Joshi *et al*. 1983; Khanka 1988; Swarup 1991). (See Figure 14.1.)

The border wars between India, China, and Pakistan (1962–1965) had a more direct effect on the Himalayan region within Uttar Pradesh (Woodman 1969). Following the war, the border with Tibet became a national security concern. Army bases and depots were rapidly established and connected by new roads along India's Himalayan frontiers. The region's economy was profoundly affected; nearly 10 percent of UP State lands and forests in the region were transferred to the Indian government for defense purposes (UP Forest Department 1989), thus withdrawing access for communities that depended on wage earnings in forestry or petty-commodity extraction of forest resources. TransHimalayan trade, an activity that provided many border communities in the region with a substantial proportion of their income (Atkinson 1882; Pant 1935; Rawat 1983, 1989; von Fürer Haimendorf 1981; Walton 1910; Webber 1902), came to an abrupt halt with the closing of the Indo–Tibetan border. Few economic alternatives remained. Subsistence agriculture was neither adequate nor remunerative for the majority of households in the Garhwal Himalayas. More than 75 percent of landholdings were less than 1 hectare in size (Government of Uttar Pradesh 1984), mostly rainfed plots on steep, terraced mountain slopes where productivity largely depended on applications of cattle manure (Pangtey and Joshi 1987; Singh and Berrry 1990). The scope for expanding cultivation of commodity crops was also limited because it largely depended on the extent to which profits from transHimalayan trade and other activities such as small-scale timber extraction were reinvested by households in regionally specialized cash crops such as ginger, turmeric, chillies, and opium (Atkinson 1882; Rangan 2000; Walton 1910). Most households without access to capital or alternative employment within their localities were reduced to depend primarily on subsistence cultivation, augmenting their incomes with remittances by males who migrated in search of work (Bora 1987; Dobhal 1987; Joshi 1983; Khanka 1988).

Green Revolution policies introduced during the 1970s were mainly geared towards increasing agricultural productivity in the Indo-Gangetic plains and other grain-producing regions in India, and had little impact on the Garhwal Himalayas. Mountainous terrain, lack of infrastructure, and fragmented landholdings distributed across different ecological and altitudinal zones were hindrances to the introduction of Green Revolution techniques. Although touted as scale-neutral by scientists and planners, the successful adoption of Green Revolution techniques crucially depended on access to both capital and irrigation (Raj 1973). Enormous risks were involved in using capital-intensive inputs if steady and well-timed supply of irrigation was not assured. It was difficult for

Figure 14.1 The new Uttaranchal state.

households in Garhwal, already reduced to dependence on subsistence cultivation and remittances for survival, to raise the necessary capital or collateral for credit, to obtain chemical fertilisers, invest in pumped irrigation (extremely expensive in mountainous terrain), and purchase high-yielding seeds (Rangan 2000).

Economic marginalization due to the combined effect of the border wars, closure of transHimalayan trade, and lack of investable surplus in the region was accelerated by the implementation of forestry policies proposed in the Fourth Plan (1969–1973). The Plan directed State Forest Departments across the country to assume *de facto* authority over all forested and open lands owned by states (including forests and wastelands controlled by Revenue Departments) and expand plantations of fast-growing tree species that would provide the raw-material needs for industry (Government of India 1976). Under the new system, the Forest Department limited rights of access and concessions for extraction from forests to fuelwood and fodder for household consumption, thereby precluding all forms of small-scale commodity extraction of forest resources. Costs of competing in Forest Department auctions paralleled the growing market demand for timber and other forest products escalating at an average rate of 8 percent per annum between 1950 and 1969 (UP Forest Department 1989). Small-scale extractors were thus marginalized from forestry activities because they now lacked physical access to Revenue forests, and did not have access to credit for competing in commercial extraction of resources from Reserved Forests. Commercial extraction was dominated by merchants and traders from outside the region who had the ability to mobilize finances through extensive credit networks, and retain control over wholesale and retail markets for forest commodities. Attempts by small-scale extractors to continue in forestry activities through organization of forest-labor cooperatives failed as timber traders rationalized their labor costs by recruiting migrant workers from western Nepal and other impoverished regions (Rangan 2000).

This process of economic marginalization was compounded by a succession of natural disasters in the region. In 1971 and 1972, heavy monsoons caused floods, landslides, erosion, and extensively damaged terraced cultivation. Financial assistance from the UP State and Central governments barely trickled through the interstices of the various institutions and agencies. Village leaders assailed the state government for its negligence, demanded immediate compensation for flood victims, and development assistance for the Himalayan districts. But their demands went largely unmet by a financially constrained and sluggish state administration (Dogra 1980). Resentment against the state government escalated further in 1973, when the UP State Forest Department rejected a petition made by an artisanal cooperative in eastern Garhwal (*the Dasholi Gram Swarajya Mandal*, Chamoli District) to increase allotments of ash trees for promoting local manufacture of agricultural implements. Village leaders discovered that their request had been rejected in favor of honoring a contract with a sporting goods firm that had purchased rights from the Forest Department to extract 400 ash trees from nearby Reserved Forests (Bahuguna 1981; Dogra 1980).

The Forest Department's intransigence was regarded as a confirmation of governmental apathy towards the welfare and development of communities in the Garhwal Himalayas. Village leaders and small-scale extractors from nearby areas protested at divisional offices and auction sites of the Forest Department, threatening to obstruct all extractive operations if their demands were not addressed. During the felling seasons from 1973 through 1975, village leaders exhorted their communities to prevent forest contractors from extracting timber. A number of stand-offs proved successful. Men and women, regardless of age or caste, gathered in the Reserved forests adjoining their villages, physically obstructing the path of migrant laborers and urging them to return to their homes (Berreman 1989; Guha 1989; Jain 1984; Shiva and Bandyopadhyay 1987; Weber 1988). The Chipko movement was born.

THE EMERGENCE OF THE CHIPKO MOVEMENT

The protests in the eastern districts of the Garhwal Himalayas earned the name *chipko* (which means to adhere, or stick to) from the way in which people hugged trees to prevent them from being felled by migrant work-groups hired by forest contractors. Chipko's main focus centered on regaining access to small-scale forest extraction, and on pressuring the state government to provide developmental assistance to beleaguered communities in the region. The protestors were a heterogenous group (Upadhyay 1990), with multiple political affiliations and even conflicting goals. Some demanded abolition of large-scale extraction by forest contractors, others were for promotion of locally organized forest-labor cooperatives, expanding rights of access and giving more concessions to local communities; yet another group demanded a total ban on export of raw materials from the region (Aryal 1994a; Dogra 1980). Village leaders and student activists affiliated with the Communist Party of India, for example, demanded higher wages for forest laborers and a ban on exports of forest resources (Aryal 1994a; Rangan 2000); while those inspired by Gandhian ideals demanded timber subsidies and supply of other forest commodities at concessional rates for promoting local artisanal industries (Agarwal 1982: 42–3; Bhatt 1988; Government of India 1985). Conversely, small-scale forest contractors argued that the Forest Department needed to privilege local entrepreneurs and forest-labor cooperatives by regulating external competition (Bahuguna 1981, 1988).

Negotiations with the Forest Department continued without much success on any of these fronts. Forest officers defended their position by claiming they were bound by law to comply with the production targets set by national- and state-level plans. Demands for greater access to small-scale extraction were turned down by invoking policy documents which stated that forests were national resources and could not be left open to reckless exploitation by local communities (Government of India 1952, 1976). Local forest contractors were denied exclusive access to commercial extraction on grounds that such policies would result in production monopolies, contribute to inefficiency, and increase the price of raw materials for industry (Rangan 2000).

Faced with an impasse in negotiations, Chipko's leaders and local activist groups resolved to bypass the UP State administration and appeal directly to the Indian government. In 1975, Sunderlal Bahuguna, a local forest contractor, and one of several spokesmen of the Chipko movement, urged the central government to intervene before ecological problems threatened national security in the Himalayas. Employing the language of the state, interwoven with the vocabulary of popular protest (Mitra 1992), Bahuguna emphasized the importance of forests for strengthening national security and border defenses, and for solving ecological problems faced by communities in the Himalayan regions. Chipko, he asserted, was a groundswell of popular outrage against the relentless commercial forces deforesting the region (Bahuguna 1981, 1982, 1987). Himalayan peasants, he stated, depended on forests to meet their simple needs, but their subsistence was being undermined by forest contractors who denuded the slopes for private profit without concern for the nation's security or for the sacred mountains. Bahuguna's narrative cast Chipko's followers as victims of natural disasters, an uncaring state, and the market. Floods, poverty, out-migration, and women's sufferings, their daily struggle to collect fuel and fodder for their households were described as inevitable consequences of timber extraction by forest contractors.

Bahuguna's appeals proved effective. Politicians praised Chipko as the moral conscience of the nation, and promised immediate action to check the problems of ecological degradation and deforestation in the Himalayas (Government of India 1985). His ability to evoke a sympathetic response from the central government gained wide media coverage for the movement, which, in turn, brought support from scholars and activists in other parts of the country. Bahuguna was seen as the natural leader of the Chipko movement, the voice of a grassroots struggle that sought to protect the simple, peasant ways of life and restore the harmony between humans and nature in the Himalayas (Bahuguna 1987; Shiva and Bandyopadhyay 1986, 1987).

Chipko's ascent to fame, therefore, hinged on the central role played by rural elites like Bahuguna who could speak from interstitial spaces created by state institutions, markets and civil society, and seize the opportunities emerging from political and economic change. They were particularly successful when they articulated their demands in the state's vocabulary of national integrity, development, and democracy, and combining this language with symbolic acts of popular protest. Their protests gained wider audiences through simple, populist narratives that pitted peasants against the state and markets, but glossed over the heterogeneity of classes, interests, and constituencies within the movements. This skilled interweaving of state discourse and populist rhetoric made Chipko the unquestioned icon of grassroots environmentalism in India and international environmental circles. Environmental scholars retold the Chipko story as India's civilizational response to ecological crisis in the Himalayas (Shiva and Bandyopadhyay 1987). International NGOs and scholars praised Chipko as an inspiration for environmental activists around the world, claiming that the ideals it represented were far more important than the aims it initially had set out to achieve (Weber 1988: 128–9). Chipko assumed legendary status when, following appeals to the Indian government, and after drawing widespread attention from scholars and the media, several pieces of legislation and constitutional amendments aimed at forest protection were introduced between 1975 and 1980.

CHIPKO'S AFTERMATH

Chipko's growing reputation as an exemplar of grassroots environmentalism in the Third World diverted attention from the political and economic necessities undergirding the Indian government's willing capitulation. As the story of Chipko spread across the world and was retold time and again, it became detached from its specific demands regarding access to forest resources and local economic development. Details became difficult to remember. It seemed no longer important to recognize that Bahuguna's criticism of private forest contractors fortuitously coincided with a period in the 1970s when the Central government launched a program of public sector expansion and nationalization unprecedented in Indian history (Bardhan 1985, 1991). Nationalization began after an internal struggle within the Congress Party in 1969 which led to a split, and resulted in Indira Gandhi (Nehru's daughter) assuming the leadership of the newly formed Congress-I (Indira). Gandhi won the elections in 1971 by appealing to two dominant constituencies, the administrative classes and rural elites, for political support (Rudolph and Rudolph 1987). The Congress-I described nationalization in populist terms, allying itself with the masses by arguing that the government, as opposed to the narrow self-interests of private enterprise, would be more socially responsible and serve the public interest (Bardhan 1985: 58; 1991; Chakravarty 1987). Bahuguna's appeals on behalf of Chipko were received favorably by Indira Gandhi because the criticism of forest contractors added popular endorsement for her agenda of nationalization and public sector expansion. She urged state governments to listen to Chipko's criticism of forest contractors and respond to the recommendations of the National Commission on Agriculture. The Commission advocated extensive afforestation measures on public and private lands, and for the creation of public sector firms to replace private businesses engaged in forest extraction (Government of India 1976; Lal 1989).

The UP State legislature responded to Mrs. Gandhi's voluble support of the Chipko movement with Forest Corporation Act of 1975. The Act authorized the creation of the UP State Forest Corporation, which was to function independently from the UP State Forest Department. The Corporation was charged with the responsibility of improving production efficiency and stabilizing market prices for timber and other raw materials. The UP State Forest Corporation was to also provide opportunities for local employment by using a network of forest-labor cooperatives for carrying out its activities (Government of Uttar Pradesh 1975).

This piece of legislation was soon followed by the UP Tree Protection Act of 1976, which banned felling of all protected tree species (mainly valuable species identified by the Forest Department) on private lands. The Tree Protection Act overruled earlier regulations requiring landowners to pay a nominal tax to district-level revenue authorities for selling trees harvested on their property (Government of Uttar Pradesh 1976). Then the Indian government passed a constitutional amendment, also in 1976, which deemed that state governments would require its prior consent for embarking on any project that involved large-scale conversion of forests to other land uses (Upadhyay 1990). The legislation responded to the criticisms of environmental scholars and activists who argued that nearly 4.3 million hectares of forest areas in different parts of the country had been deforested by state governments within twenty-five years (1950–1975) under the pretext of promoting industrial development and hydro-electric projects (Agarwal 1982: 41). The indiscriminate destruction of forests by state governments, they claimed, needed to be regulated by the Government of India. Four years later in 1980, the Indian government created the Ministry of Environment for addressing problems of deforestation, environmental risk, and ecological degradation in the country. This action was undertaken alongside the passage of the Indian Forest Conservation Act of 1980. The Act stipulated that state governments required permission from the Ministry of Environment for converting designated forest areas of more than 1 hectare to non-forest land uses. It also imposed a fifteen-year ban on felling green timber in altitudes above 1000 meters in the Himalayan regions (Agarwal and Narain 1991; Lal 1989: 33–4; Upadhyay 1990).

Ironically, international repute and legislative successes bore down heavily on communities in the Garhwal Himalayas. Since the green-felling ban followed soon after the passage of the Forest Corporation Act, most communities living in altitudes above 1000 meters had little opportunity to find employment in forestry. Forest-labor cooperatives in higher elevations disbanded soon after the felling ban was imposed, while other cooperatives in the lower altitudes were riven by political rivalries and old disagreements within the Chipko movement that resurfaced among members. The UP Forest Corporation opted for labor contracts with Amates (migrant labor agents) who controlled bands of workers recruited from other regions. Mates were invariably from western Nepal, with a smaller proportion coming from the neighboring state of Himachal Pradesh. The Forest Corporation defended its choice of migrant workers over locally organized labor by arguing that Nepalese workers were preferred for their skills, reliability, and industriousness. Some Forest Corporation managers were more candid, freely admitting that contract-laborers from western Nepal were preferred because they were willing to work for lower wages and would submit to the control of their mates (Rangan 2000).

In addition to further reducing regional employment opportunities in forestry, the new legislation also weakened the UP State Forest Department. Its revenues steadily eroded along with its discretionary powers to settle local disputes over forest access, to allocate use of classified forest areas for small-scale infrastructure projects proposed by village- or block-level institutions. The Department's revenues declined in part due to the fixed royalties (adjusted periodically for inflation, but below market price) it received from the State Forest Corporation (UP Forest Department 1989). A large proportion of its forests in Garhwal came under the green-felling ban, and thus imposed constraints on carrying out the routine silvicultural activities necessary for maintaining valuable forest stock.

New forest legislation – which in some ways represented the heart of Chipko's reforms – effectively imposed a moratorium on most development activities in the region. Given that nearly two-thirds of the total land area in Garhwal was classified as forest (Government of Uttar Pradesh 1984), any development activity such as road-building, village electrification, or minor irrigation works requiring partial use or conversion of forest areas could not proceed without permission from the national Ministry of Environment (Government of India 1986). Numerous development projects were held up or shelved due to bureaucratic delays (Agarwal and Narain 1991; Mitra 1993a; Rangan 2000). All this provoked great anger towards the new forest laws and the Ministry of Environment. As one woman who had participated in Chipko protests angrily remarked, "Now they tell me that because of Chipko the road cannot be built [to her village], because everything has become *paryavaran* [environment]. We cannot even get wood to build a house ... our *haq-haqooq* [rights and concessions] have been snatched away. She added, "I plan to contest the *panchayat* [village administrative body] elections and become a *pradhan* next year.... My first fight will be for a road, let the environmentalists do what they will" (Mitra 1993b).

THE STRUGGLE FOR UTTARAKHAND

Throughout this period, Chipko's celebrity status abroad was paralleled by growing resentment over the lack of development in the region. By the late 1980s, regional political groups such as the *Uttarakhand Kránti Dal* (Uttarakhand Revolutionary Front) began publicly exhorting communities to start a *"jungle kato"* [cut down the forest] movement in defiance of all the forest protection laws. They declared their willingness to clear-cut areas on behalf of any community or village wishing to initiate development projects (Agarwal and Narain 1991; Rangan 2000), arguing that such acts were the only means by which the mountain districts of Uttar Pradesh would regain control over local resources and force the state and national governments towards granting statehood for the region.

The demand for statehood was not a new political cause. Even as far back as 1952, P.C. Joshi, a member of the Communist Party of India from Kumaon, had submitted an appeal to Prime Minister Nehru requesting the creation of a separate state of Uttarakhand. Although Nehru was himself against the division of Uttar Pradesh he forwarded the memorandum to the States Reorganisation Commission, which later rejected the appeal. The issue of granting administrative autonomy for the mountain districts was again discussed in the late 1960s with no success. The *Uttarakhand Kránti Dal* (UKD) was formed in 1979, and its representatives sought a meeting with Indira Gandhi to press their demands for a separate state of Uttarakhand. While they were not rejected outright, the discussions were indefinitely postponed by the UP government. The issue was carefully avoided by H.N. Bahuguna and N.D. Tiwari, both of whom were elected representatives of Mrs. Gandhi's party from Garhwal and Kumaon, and had served in her Cabinet (both had also been Chief Ministers of UP) in the 1970s and early 1980s (Dhoundiyal *et al.* 1993).

The sole concession made by the UP government during the mid-1980s was to create a new state agency called the Hill Development Agency which was to be responsible for planning and providing financial assistance for regional development. The agency's efforts were centered on the promotion of horticulture and tourism through price supports and investment subsidies, but without any investment in infrastructure expansion or institutional support for new marketing networks in the region. Consequently, much of the horticultural promotion in the higher altitude areas were rendered meaningless because most households could not afford the costs of transporting produce to regional markets. The combination of post-Chipko environmental legislation and lack of investment in roads and electricity in the mountain districts meant that investment subsidies for tourism concentrated in towns and cities in the Himalayan foothills (Rangan 2000).

Regional political groups pointed to these trends as typical examples of how development in Uttarakhand suffered from both bureaucratic apathy and *maidan* (plains) mentality towards the *pahar* (mountains), and repeatedly used these arguments to mobilize support for statehood. But even with growing support for their cause across the eight districts of Garhwal and Kumaon, there was little change in the stance of both the UP and central governments.

The simple reason behind the reluctance to grant statehood lay in the electoral arithmetic of Indian politics. For most national political parties, the attraction of ruling UP, one of the most populous states of the "cow-belt" electorate of northern India (also referred to as the Hindi heartland that contains close to half the country's population), was that it yielded the largest numbers of elected representatives to the national parliament. Control over UP effectively meant political control in New Delhi (a formula that has worked fairly well since Independence). Arguments for dividing UP into two or three states for purposes of administrative efficiency or effective development paled in comparison with the prospect of benefiting from the votes of a massive consolidated electorate. Thus most political parties at national and state levels routinely played on local sentiments, discussing the issue of statehood during parliamentary or state election campaigns, but avoiding the subject after assuming office. Between 1985 and 1997 almost every political party endorsed the creation of Uttarakhand, but did little else. The right-wing Hindu party, BJP, supported statehood for the region, during the election campaigns of 1991 and 1993, but did little more than change the name of the Hill Development Agency to the Uttaranchal

Development Agency after winning control of the UP legislature (*Sunday* 1994). The BJP's references to Uttaranchal (northern province) as opposed to Uttarakhand (northern region), was aimed at separating itself from the interests of regional political parties such as the UKD which had long been campaigning for the creation of Uttarakhand (Aryal 1994b; Mawdsley 1997).

FROM UTTARAKHAND TO UTTARANCHAL

The year 1993 marked a significant turning point in the electoral landscape of both Uttar Pradesh and national politics. No single political party succeeded in obtaining majority in the state. The old formula of control over UP equaling control over New Delhi foundered as the base of populist power shifted towards a coalition of the SP–BSP parties which represented the interests of kulak farmers and groups officially categorized as the "Other Backward Castes" (OBC), and "Scheduled Castes" (SC, meaning low- or "untouchable" castes) in northern India. The SP–BSP coalition supported the demands for Uttarakhand and went as far as commissioning reports investigating the practicalities of statehood for the region. But it also rapidly alienated itself from Uttarakhand supporters when it proposed a 27 percent increase in the number of public sector jobs that would be reserved for OBC and SC groups. Uttarakhand's activists argued that merely 2.5 percent of the region's population came under these categories, and charged the coalition government of using affirmative action as a ploy to retain jobs for plains-people and further reduce employment opportunities for hill-people. As the activists sought to protest their cause with the central government, the coalition government in UP ordered the police to physically prevent them from reaching New Delhi (see Mawdsley 1997 for more details). The SP–BSP coalition in UP justified its action by accusing Uttarakhand activists for being a front for upper-caste interests and against social upliftment for oppressed caste groups, and repeatedly characterized the Uttarakhand agitation of 1994 as anti-backward castes. Although the SP–BSP coalition did not remain in power for very long, the subsequent governments in UP continued to be formed through coalitions between BJP, BSP and other small political parties.

It was in the context of constantly shifting and unstable coalitions between the BJP and populist/caste-based or regional parties, both in UP and the center, during subsequent years that the statehood for Uttarakhand ultimately became possible. UP no longer offered national political parties the possibility of delivering a massive, consolidated vote; it had become a fragmented electorate. There was little advantage in holding the state together; national political parties stood a better chance of consolidating their presence in sub-regions of UP rather than across the state. It was this reasoning, then, that propelled the BJP-led coalition at both central and state level towards granting statehood for the region they now chose to call Uttaranchal. But the BJP (representing upper- and middle-caste/class interests) was also limited by its dependence on parties such as the BSP (representing lower- and religious minority class/caste interests) whose populist power base had substantially expanded in Uttar Pradesh. Thus the political negotiations over the boundaries of Uttaranchal resulted in the inclusion of two, heavily populated, sub-Himalayan districts of Haridwar and Udham Singh Nagar, ensuring that the electoral base of the BSP would not be undermined in the new state formation (see Figure 14.1).

UTTARANCHAL AND DEVELOPMENT

In November 2000, eight mountain and two sub-montane districts of Uttar Pradesh came into being as the new state of Uttaranchal. Much to the disapproval of Uttarakhand activists, the BJP-led coalition appointed their own party representative from the plains to head the interim government, and chose Dehra Dun, the largest city in the foothills of the Uttarakhand Himalayas, instead of a town in the heart of the mountain region, as the interim capital of the new state (Jayal 2000). In February 2002, much to the disappointment of both, the BJP-led coalition and the regional UKD party that had spearheaded the movement for statehood, the first state elections brought the Congress party into office. The new Chief Minister of Uttaranchal was N.D. Tiwari, an old Congress-I hand from Kumaon, who had previously been Chief Minister of Uttar Pradesh over four terms during Indira Gandhi's reign as India's Prime Minister. A survey of voters revealed that the new class/caste/territorial configuration of the Uttaranchal electorate resulted in the Congress party benefiting from a larger share of the diverse range of the lower caste/religious minority vote (which also went to the BSP). Nearly half the voters credited the people of the state, rather than either the UKD or the BJP, for the creation of Uttaranchal; close to one-third held the opinion that the UKD was a party of "opportunists" and "excited youth" (Kumar 2002).

There are many who fear that the *raison d'etre* of the state was lost at the very moment of its creation, and that the interests of the plains will still dominate development in the new state. Uttaranchal has inherited a debt of over Rs 300 million, but only generates an annual revenue of around Rs 42 million (Jayal 2000). It has requested the central government to be given special status for additional financial support, but this may no longer be forthcoming under the broadly neo-liberal economic climate and political dispensation in New Delhi. While the new Chief Minister proclaims that his government will pursue "sustainable development of the state through people's participation," he also admits that it would be necessary to create "the appropriate conditions" for attracting private investment for development of the state (Singh and Mittal 2002). This, for some, implies that the much-needed investment in infrastructure development is likely to concentrate in the foothills and sub-montane districts of the state, and that resource extraction in the mountains will proceed apace (Kumar 2001).

The journey from Chipko to Uttaranchal makes sense when seen through the lens of development, and the tensions of territory-function-populist power contained within it. To claim, as some have done, that Chipko was a new Third World social movement that opposed or sought alternatives to development, is to present a wildly romantic and improbable account that has little bearing on the geohistorical realities from which it emerged, and remains divorced from what has happened afterwards in the region. The creation of Uttaranchal does not herald the end of development and the beginning of a new anti, post- and alternatives to-development era. The struggle over development in this part of the Indian Himalaya has merely taken a different, yet familiar, turn.

REFERENCES

Agarwal, A. (1982) *The State of India's Environment*. New Delhi: Centre for Science and the Environment.

—— and S. Narain (1991) "Chipko People Driven to *Jungle Kato* [Cut the Forests] Stir," *Economic Times*. 31 March. New Delhi.

Amar Ujala (1991a) "Bahuguna's Statements Criticised," (trans. from Hindi). 8 February.

—— (1991b) "Inefficient Management of Forest Corporation causes wastage of timber in hill regions of UP," 18 February.

Ambedkar, B.R. (1945) *Annihilation of Caste: With a Reply to Mahatma Gandhi*. Bombay: Bharat Bhushan Press.

Aryal, M. (1994a) "Axing Chipko," *Himal*. 7 (1): 8–23.

—— (1994b) "Angry Hills: An Uttarakhand State of Mind," *Himal*. Nov–Dec: 10–21.

Atkinson, E.T. (1882) *The Himalayan Gazetteer*. 3 Volumes. Reprinted 1989. New Delhi: Cosmo Publications.

Bahuguna, S. (1981) *Chipko: A Novel Movement for Establishment of a Cordial relationship between Man and Nature*. 2 Volumes. Silyara, Tehri Garhwal: Chipko Information Centre.

—— (1982) "Let the Himalayan Forests Live," *Science Today*. March: 41–6.

—— (1987) "The Chipko: A People's Movement," in M.K. Raha (ed.) *The Himalayan Heritage*, New Delhi: Gian Publishing House. 238–48.

Bardhan, P. (1985) *The Political Economy of Development in India*. New Delhi: Oxford University Press.

—— (1991) "State and Dynamic Comparative Advantages," *Economic Times*. 19–20 March. New Delhi.

Bayly, C.A. (1976) *The Local Roots of Indian Politics: Allahabad 1880–1920*. Oxford: Clarendon Press.

Berreman, G. (1989) "Chipko: A Movement to Save the Himalayan Environment and People," in C.M. Borden (ed.) *Contemporary Indian Tradition: Voices on Culture, Nature, and the Challenge of Change*. Washington, D.C.: Smithsonian. 239–66.

Bhatt, C.P. (1988) "The Chipko Movement: Strategies, Achievements and Impacts," in M.K. Raha (ed.) *The Himalayan Heritage*. New Delhi: Gian Publishing House. 249–65.

—— (1991) "Chipko Movement: The Hug That Saves," *Survey of the Environment 1991*. Madras: The Hindu.

Bora, R.S. (1987) "Extent and Causes of Migration from the Hill Regions of Uttar Pradesh," in Vidyut Joshi (ed.) *Migrant Labour and Related Issues*. New Delhi: Oxford and IBH.

Brass, T. (1994) "The Politics of Gender, Nature, and Nation in the Discourse of New Farmer Movements," *Journal of Peasant Studies.* 21 (3): 27–71.

Chakravarty, S. (1987) *Development Planning: The Indian Experience.* New Delhi: Oxford University Press.

Corbridge, S. (2002) "The Continuing Struggle for India's Jharkhand: Democracy, Decentralisation and the Politics of Names and Numbers," *Commonwealth and Comparative Politics.* 40 (3): 55–71.

—— and J. Harriss (2000) *Reinventing India.* London: Polity Press.

Cowen, M. and R. Shenton (1995) "The Invention of Development," in J. Crush (ed.) *Power of Development.* London: Routledge.

Dhoundiyal, N.C., V.R. Dhoundiyal and S.K. Sharma (1993) *The Separate Hill State.* Almora, UP: Shree Almora Book Depot.

Dobhal, G.L. (1987) *Development of the Hill Areas: A Case Study of Pauri Garhwal District.* New Delhi: Concept Publishers.

Dogra, B. (1980) *Forests and People: The Efforts in Western Himalayas to Re-establish a Long-Lost Relationship.* Rishikesh: Himalaya Darshan Prakashan Samiti.

—— (1983) *Forests and People: A Report on the Himalayas.* New Delhi.

Dwivedi, R. (1998) "Resisting Dams and Development": Contemporary Significance of the Campaign against the Narmada Projects in India," *European Journal of Development Research.* 10 (2): 135–83.

Escobar, A. (1995a) "Imagining a Post-development Era," in J. Crush (ed.) *Power of Development.* London: Routledge.

—— (1995b) *Encountering Development: The Making and Unmaking of the Third World.* Princeton, NJ: Princeton University Press.

Esteva, G. (1987) "Regenerating People's Space," *Alternatives.* 10 (3): 125–52.

Friedmann, J. and C. Weaver (1979) *Territory and Function: the Evolution of Regional Planning.* London: Berkeley and Los Angeles: University of California Press.

Gadgil, M. and R. Guha (1992) *This Fissured Land: An Ecological History of India.* Berkeley: University of California Press.

Gore, C. (1984) *Regions in Question: Space, Development Theory, and Regional Policy.* London: Methuen & Co.

Government of India (1952) *National Forest Policy.* New Delhi: Ministry of Agriculture.

—— (1976) *Report of the National Commission on Agriculture, Volume IX: Forestry.* New Delhi: Ministry of Agriculture and Irrigation.

—— (1985) *National Forest Policy.* New Delhi: Ministry of Environment.

—— (1986) *The Environment (Protection) Act, 1986. (Act No. 29 of 1986).* New Delhi: Ministry of Law and Justice.

—— (1987) *Forest Rules Notification No. GSR14.* New Delhi: Ministry of Environment.

Government of Uttar Pradesh (1975) *The Uttar Pradesh Forest Corporation Act, 1974.* (U.P. Act No. 4 of 1975). Lucknow: UP State Legislature.

—— (1976) *The Uttar Pradesh Tree Protection Act, 1976.* Lucknow: UP State Legislature.

—— (1984) *Statistical Diary of UP State.* Lucknow: Economics and Statistical Department.

Guha, R. (1989) *The Unquiet Woods: Ecological Change and Peasant Resistance in the Himalaya.* New Delhi: Oxford University Press.

Hannam, K. (1998) "Ecological and Regional Protests in Uttarakhand, Northern India" *Tijdschrift voor Economische en Sociale Geografie.* 89 (1): 56–65.

Hart, G. (2001) "Development Critiques in the 1990s: *Culs de sac* and Promising Paths," *Progress in Human Geography.* 25 (4): 649–58.

Harvey, D. (1982) *The Limits to Capital*. Chicago: University of Chicago Press.

—— (2003) *The New Imperialism*. Oxford: Oxford University Press.

Hauser, W. (1993) "Violence, Agrarian Radicalism and Electoral Politics: Reflections on the Indian People's Front," *Journal of Peasant Studies*. 21 (1): 85–126.

Herring, R. (1989) "Dilemmas of Agrarian Communism: Peasant Differentiation, Sectoral and Village Politics in India," *Third World Quarterly*. 11 (1): 89–115.

Himachal Times (1990a) "Uttaranchal Demanded: Economic and Political Passions Aroused," 11 April.

—— (1990b) "Amendment of Forest Act for Developing Hill Areas," 2 April.

—— (1990c) "Consolidation of Hill Land Holdings Soon," 1 June.

Indian Express (1990) "UP to Amend Tree Protection Act," 28 March. New Delhi.

Jain, S. (1984) "Women and People's Ecological Movement: a Case Study of Women's Role in the Chipko Movement in Uttar Pradesh," *Economic and Political Weekly* 19(41): 1788–94.

Jayal, N. (2000) "Uttaranchal: Same Wine, Same Bottle, New Label?" *Economic and Political Weekly*. 35 (49). December 2–8. accessed through HTTP:<http://www.epw.org.in>.

Joshi, S.C. (1983) *Kumaon Himalaya: A Geographic Perspective on Resource Development*. Nainital: Gyanodaya Prakashan.

Khan, W. and R.N. Tripathy (1976) *Plan for Integrated Rural Development in Pauri Garhwal*. Hyderabad: National Institute of Community Development.

Khanka, S.S. (1988) *Labour Force, Employment and Unemployment in a Backward Economy: A Study of Kumaon Region in Uttar Pradesh*. Bombay: Himalayan Publishers.

Kohli, A. (ed.) (2001) *The Success of India's Democracy*. Cambridge: Cambridge University Press.

Kumar, P. (2000) *The Uttarakhand Movement: Construction of a Regional Identity*. New Delhi: Kanishka Publishers.

—— (2001) "Uttarakhand: One Year After," *Economic and Political Weekly*. 36 (51). 22 December. accessed through HTTP: http://www.epw.org.in>.

Kumar, S. (2002) "Uttaranchal Assembly Elections: Marginal Difference," *Economic and Political Weekly*. 37 (20). 18 May. accessed through HTTP: http://www.epw.org.in>.

Lal, J.B. (1989) *India's Forests: Myth and Reality*. Dehra Dun: Natraj Publishers.

Mawdsley, E. (1997) "The Uttarakhand Agitation and Other Backward Classes," *Economic and Political Weekly*. 31 (4): 205–10.

—— (1998) "After Chipko: From Environment to Region in Uttaranchal," *Journal of Peasant Studies*. 25 (4): 36–54.

Mies, M. and V. Shiva (1993) *Ecofeminism*. London: Zed Books.

Mitra, A. (1993a) "Chipko: An Unfinished Mission," *Down to Earth*. 30 April: 25–51.

—— (1993b) "There Can be No Development Without Women: Interview with Gayatri Devi," *Down to Earth*. 30 April: 50–1.

Mitra, S.K. (1992) *Power, Protest and Participation: Local Elites and the Politics of Development in India*. London: Routledge.

Naipaul, V.S. (1990) *India: A Million Mutinies Now*. London: Heinemann.

Pangtey, Y.P.S. and S.C. Joshi (eds) (1987) *Western Himalaya: Environment Problems and Development*. Vol. I and II. Nainital: Gyanodaya.

Pant, S.D. (1935) *The Social Economy of the Himalayans*. London: Allen and Unwin.

Polanyi, K. *The Great Transformation*. Boston: Beacon Press.

Raj, K.N. (1973) "Mechanization of Agriculture in India and Sri Lanka (Ceylon)," *Mechanization and Employment in Agriculture*. ILO: Geneva.

Rangan, H. (1997) "Property vs Control: the State and Forest Management in the Indian Himalaya," *Development and Change*. 28 (1): 71–94.

—— (2000) *Of Myths and Movements: Rewriting Chipko into Himalayan History*. London: Verso.

Rawat, A.S. (1983) *Garhwal Himalayas: A Historical Survey: The Political and Administrative History of Garhwal 1815–1947*. Delhi: Eastern Book Linkers.

—— (1989) *History of Garhwal, 1358–1947: An Erstwhile Kingdom in the Himalayas*. New Delhi: Indus Publishing.

Routledge, P. (1993) *Terrains of Resistance: Non-violent Social Movements and the Contestation of Place in India*. New York: Praeger.

Rudolph, L., and S.H. Rudolph (1987) *In Pursuit of Lakshmi: The Political Economy of the Indian State*. Chicago: Chicago University Press.

Said, E. (1994) *Culture and Imperialism*. New York: Vintage Books.

Sen, A. (1984) *Resources, Values and Development*. Cambridge, MA: Harvard University Press.

Shiva, V. (1989) *Staying Alive: Women Ecology and Development*. London: Zed Books.

—— (1991) *Violence of the Green Revolution: Third World Agriculture, Ecology and Politics*. London: Zed Books.

Shiva, V. and J. Bandyopadhyay (1986) *Chipko: India's Civilisational Response to the Forest Crisis*. New Delhi: Intach.

—— (1987) "Chipko: Rekindling India's Forest Culture," *The Ecologist*. 17 (1): 26–34.

—— (1989) "The Political Economy of Ecology Movements," *IFDA Dossier*. No. 71, May–June: 37–60.

Simon, D. (2003) "Dilemmas of Development and the Environment in a Globalizing World: Theory, Policy and Praxis," *Progress in Development Studies*. 3 (1): 5–41.

Singh, B. and N. Mittal (2002) "Uttaranchal opts for Eco-friendly Development," *Concept Chronicle*. 2 (8): 1.

Singh, S.P. and Anil Berry (1990) *Forestry Land Evaluation: An Indian Case Study*. Dehra Dun: Surya Publications.

Stebbing, E.P. (1932) *The Forests of India*. 3 Volumes. London: Bodley Head.

Sunday (1994) "The Sound and the Fury: Is the Uttarakhand agitation taking a violent turn?" November 6–12.

Swarup, R. (1991) *Agricultural Economy of Himalayan Region: With Special Reference to Kumaon*. Nainital: G.B. Pant Institute of Himalayan Environment and Development.

Thompson, E.P. (1993) *Customs in Common: Studies in Traditional Popular Culture*. New York: The New Press.

Trainer, F.E. (1989) *Developed to Death: Rethinking World Development*. London: Green Press.

U.P. Forest Department (1989) *Forest Statistics: Uttar Pradesh*. Lucknow: Forest Administration and Development Circle.

Upadhyay, C.B. (ed.) (1990) *Forest Laws, with Commentaries on Indian Forest Act and Rules, State Acts, Rules Regulations etc*. 7th edition. Allahabad: Hind Publishing House.

Upadhyay, H.C. (1990) *Harijans of the Himalaya: with special reference to the Harijans of Kumaun Hills*. Nainital: Gyanodaya Prakashan.

von Fürer-Haimendorf, C. (1981) *Asian Highland Societies in Anthropological Perspective*. New Delhi: Sterling Publishers.

Walton, H.G. (1910) *British Garhwal: A Gazetteer, being Volume XXXVI of the District Gazetteers of the United Provinces of Agra and Oudh*. Reprinted 1989. Dehra Dun: Natraj Publishers.

Watts, M.J. (1993) "Development I: Power, Knowledge, and Discursive Practice," *Progress in Human Geography*. 17 (2): 257–72.

Webber, T. (1902) *Forests of Upper India and Their Inhabitants*. London: Edward Arnold.

Weber, T. (1988) *Hugging the Trees: the Story of the Chipko Movement*. New Delhi: Viking.

Weiner, M. (1989) *The Indian Paradox: Essays in Indian Politics*. New Delhi: Sage Publications.

Wells, R. (1994) "E.P. Thompson, Customs in Common, and Moral Economy," *Journal of Peasant Studies*. 21 (2): 263–307.

Williams, G.R.C. (1874) *Historical and Statistical Memoir of Dehra Doon*. Reprinted 1985. Dehra Dun: Natraj Publishers.

Williams, R. (1983) *Keywords: A Vocabulary of Culture and Society*. Revised edition. New York: Oxford University Press.

Woodman, D. (1969) *Himalayan Frontiers: A Political Review of British, Chinese, Indian and Russian Rivalries*. London: Barrie and Rockliffe.

15

MOVEMENTS AND MODERNIZATIONS, MARKETS AND MUNICIPALITIES

Indigenous federations in rural Ecuador

Anthony Bebbington

This chapter attempts to make sense of the work of popular and non-governmental organizations in the Central Andes of Ecuador. Specifically, it considers how federations of indigenous (highland Quichua) communities have emerged in the period since land reform, the agricultural and rural development strategies they have pursued, and the ways in which they have engaged in formal political processes. Its rhetorical and analytical strategy is to sustain a conversation between the ways in which these federations seem to have pursued agricultural and political change with the ways in which indigenous agriculture and social movements have been conceptualized by certain authors. Exploring the dissonances and convergences between these conceptualizations and the strategies of these federations sheds light on the "liberatory" possibilities embodied by these federations as well as on the broader project of liberation ecology.

To open a conversation between the strategies of rural social movements and the analytical categories of those who write about them lies, I will conclude, at the core of liberation ecology (c.f. Bebbington and Bebbington 2001; Edwards 1994). Why? Because if the idea behind writing development stories is not simply to understand the world but to contribute to its change (Edwards 1994), then our understanding and our ability to contribute to such change are the greater if we build bridges across the gulf between the languages of social movements and popular organizations, and those of activists, intellectuals, and "policy makers." To say this is not to collapse into a "people know best" populism taken to task in the opening chapter. Indeed, to understand is not necessarily to agree with – *constructive* critique can be a form of support.

The case discussed in the chapter comes from two counties (*cantones*) in the province of Chimborazo. The phenomenon addressed – the emergence of indigenous federations and their engagement in rural development and rural politics – is, though, far broader in Ecuador (Bebbington *et al.* 1993; Ramón 2001). A national survey in 1998/1999 identified 155 federations concentrated in 43 cantones (Ramón 2001), and more recent data speak of 170 such federations (PRODEPINE, 2002). Not only is the existence of such organizations widespread, so has been their engagement in agricultural and rural development (Bebbington *et al.* 1992; Sylva, 1991) and more recently in formal local politics. While in the early 1980s there were no indigenous leaders who were municipal mayors in Ecuador, in the elections of May 2000 indigenous and peasant representatives were elected as mayors in 26 (12 percent) of Ecuador's 215 municipalities, and as prefects in five (24 percent) of Ecuador's 21 provinces (Cameron, 2001). The phenomenon is thus important and has radically changed the face of Ecuadorian society and politics over the last two decades.

The empirical analysis of these federations is thus a critical project. In this chapter I relate such analysis to the conceptual discussions of "alternative" development that draw upon concepts of "indigenous technical knowledge," "farmer-first" agricultural development, new social movements, and civil society. The assumption of much alternative development writing is that "alternative actors" such as indigenous peoples' organizations will carry forward these "alternative" agendas (Bebbington and Bebbington 2001). This is not always the case. In particular, a commitment to native, traditional, and agro-ecological techniques found in intellectual currents in social science and development activism is often missing among some of these organizations, who instead seem to focus on reforming dominant models of development while also keeping hold of principles of local control, democratization, and community-based sustainable development. Thus while intellectual concepts and popular practice may differ at the level of strategy, they converge at the level of wider political objectives. Looking more carefully at why this happens addresses questions raised by Peet and Watts in their introduction, about "the conditions under which knowledges and practices become part of alternative development strategies." It also problematizes the question by asking which knowledges and which practices?

In addition to analyzing the internal rationales of the strategies of social movements, questions of effectiveness must also be raised. To organize, be innovative, and create "decentered autonomous spaces" (Escobar 1992) is simply not enough – alternatives must also make a material difference in livelihoods. Peet and Watts (1996) are painfully correct in pointing out that the social movements literature says very little about how far, and under what conditions, these movements genuinely contribute to a more robust civil society and (perhaps more importantly) the basics of increased productivity, income, and employment opportunities in the popular sectors. If we are interested in "liberation ecologies," the alternative proposals of social movements and intellectuals alike fall well short of the practical challenge.

This chapter first looks at discussions of indigenous technical knowledge and farmer-first development, and points out – at the risk of caricature – the inherent weaknesses of such analyses. This leads to consideration of the political–economic context within which any alternative agricultural strategy must be pursued in Latin America. This in turn forces a reflection on the intersections between social movements, markets and the state. That reflection lays the basis for a short case study that allows us to sustain a conversation between the theoretical concepts of "outside" analysts, and the strategies of civil society organizations. The case study – which is based on fieldwork conducted discontinuously between 1988 and 1998 – discusses how a group of indigenous peasant federations and nongovernmental development organizations (NGOs) in the highland province of Chimborazo in the Central Andes of Ecuador emerged as actors in civil society and, from the early 1980s through to the early 2000s, composed local development strategies embracing technology, ethnic claims, and political engagement. These strategies have, over time, become increasingly eclectic, pragmatic, and modernizing – and yet the underlying vision of rural development still retains a very "alternative" agenda for an indigenous, grassroots-controlled modernization. This is particularly the case for Indian federations, whose agrarian programs have at different times incorporated modernizing approaches to agricultural development to promote a form of social change aimed at reinforcing indigenous culture and society. This suggests that what gives a strategy its alternative, indigenous orientation is not its content (i.e. that it uses indigenous technologies, etc.) but rather its goal (i.e. that it aims to increase local control of processes of social change). Indigenous people may well incorporate the techniques of those who have long been their dominators, and yet do so in a way that strengthens an indigenous agenda pitched in some sense against the interests of those dominating groups. The movement of these organizations and their leadership into elected positions in municipal, provincial and national government, only seems to confirm this interpretation.

The case also suggests that it is important to understand the agency of rural social movements as situated within political economic, socio-cultural and agro-ecological contexts that influence how, and why, they address the intersection between development and environment in the way that they do. This situatedness should be kept at the forefront of theoretical interpretations. It is one thing for theoretical analysis and development practice to recover the importance of these long-marginalized actors within civil society. But that does not mean we have to celebrate all that they do – for much of this may be ineffective, undemocratic, authoritarian, patriarchal and so on. However, if our analyses recover and understand the ways in which actors are situated, and how this affects their rationales, this might illuminate the limits on their capacities to compose viable and democratizing programs, and the reasons for their limited impacts. This in turn can be one step in the process of defining potentially more effective strategies that can incorporate "external innovations" but at the same time build from the rationales of the actors involved (rather than from imposed rationales). As will become clear, this means refining the conversation between agrarian populism and political ecology. It also means that any analytical and practical association of political ecology and civil society (which Peet and Watts suggest is central to the reconfiguration of liberation ecology) must be accompanied by a concern for the state and the conditions under which it might be reconfigured by rural social movements.

ALTERNATIVE PROPOSALS: INDIGENOUS TECHNOLOGY AND FARMER-FIRST AGRICULTURAL DEVELOPMENT

Agricultural revolutions, green and indigenous

The term "Green Revolution" is shorthand for an agricultural development based on new crop varieties, agrochemicals and machinery – a basic element of state policies in Latin America seeking to modernize the small and middle farm sector (Bebbington and Thiele 1993). Critiques of such approaches have served as a basis for much of the writing on "alternatives." These critiques argued that Green Revolution technologies aggravated rural poverty, undermined food security, damaged the biophysical environment, eroded local cultures and fostered further polarization in the countryside (Altieri 1987; Hewitt de Alcantará 1976; Lipton and Longhurst 1989), and indeed cultural and political-ecologists have often opposed the technological modernization of indigenous agriculture (Butzer 1990; Denevan 1989). Their writing is similar to those agro-ecologists who argue that agriculture should be grounded in ecological principles if it is to be sustainable (Altieri 1987; Altieri and Hecht 1990), and to Latin American writing around themes of eco-development, sustainable development, and ecological economics (e.g., see Esteva 1992; Leff 1994; Max-Neef 1991).

Out of these critiques have come persuasive and powerful proposals arguing that viable agricultural development strategies must be based on indigenous peoples' technical knowledge (ITK) of crops, animals, and the environment if they are to be viable (Denevan 1989) – on the grounds that indigenous knowledge is adapted to peasant production conditions, does not depend on external inputs, and is environmentally sound and culturally appropriate (e.g. Altieri 1987; Brokensha et al. 1980; Chambers et al. 1989; Richards 1985, 1986; Warren et al. 1995). This literature generated an alternative, so called "farmer-first," approach to orthodox approaches to agricultural development which: emphasized popular knowledge and participation (Chambers et al. 1989; Scoones and Thompson 1994); argued that agricultural revolutions are as indigenous as they are expert led (Richards 1985); and questioned the authority of professional knowledge in a way paralleling (perhaps ironically) certain post-structural approaches to expert knowledge (e.g. Escobar 1995).

The case for a "farmer-first" approach is motivated by concerns that are both political and theoretical. The political objectives are clearly to promote farmer participation in agricultural development, to encourage the democratization of agricultural organizations, to support ideas of social equity, and to challenge prevailing "taken-for-granted" power relationships in which the rural poor are always conceived of as "clients," recipients, and the objects of somebody else's development strategy. Theoretically, the concern is to relativize modernist rationality by suggesting that there are equally valid "native" points of view (Geertz 1983; Long and van der Ploeg 1994), to question grand evolutionary theories (Richards 1985), and to suggest that the political economy does not determine quite as many local outcomes as many radical approaches would suggest. The farmer-first project has achieved a great deal (not least through the work and life of Chambers). It has helped change attitudes to farmer expertise and indigenous peoples' knowledgeability; and it has undoubtedly helped put rural peoples' agency back into the picture, softening the pessimistic determinism of political economy. But it is also conceptually and practically problematic (Thompson and Scoones 1994).

First, the farmer-first approach is grounded in an exaggerated, over-generalized, and sometimes simplified critique of technological modernization. For there is other research, not written by apologists of the Green Revolution, which suggests that although agrarian modernization has had negative impacts in some cases this need not necessarily always be so. Grossman (1993) cautions against over-hasty generalization about export agriculture. He shows that while export agriculture may have undermined peasant food security in some cases (Grossman 1984), this is not a necessary consequence. In the Windward Islands the relationships between export production and food security are far more complex; in many cases export producers using a "modern" technological package have increased their food security. Similarly, there are cases in which small farmer adoption of Green Revolution crops and varieties has increased food security, offsetting crises that would have occurred without technological change (Goldman 1993; Rigg 1989; Turner et al. 1993). The implication is that generalized diagnoses of agrarian crisis, as well as generalized remedies, ought to be treated with caution (Richards 1990a). Alternatives in one context may not be appropriate for others.

Conceptual questions for the farmer-first alternative

The farmer-first approach often constructed an essentialized conception of indigenous agriculture that is homogenized, static, and easily taken out of socio economic, political, and cultural context (Fairhead 1992; Fairhead and Leach 1994; Scoones and Thompson 1994). By naming something called "indigenous technical knowledge," this literature created the sense that a body of knowledge exists in a coherent form. By discussing indigenous knowledge with a particular purpose in mind – namely to promote participatory agricultural development strategies building on farmer agronomic knowledge – the literature also emphasized the agricultural dimensions of rural life and the agricultural expertise of the rural poor. But it created the impression that all rural people are farmers, that agricultural technology is central to solving rural poverty, and that pre-modernized techniques are crucial to any solution. In addition, the emphasis on the "knower" (the farmer), and on the knower's capacity to invent and create, tended to remove agents from structures, and to replace determinism with voluntarism (Giddens 1979; Long 1990). Likewise, an emphasis on what knowers know about technology and ecology diverted attention from the myriad elements of political economy of which they knew less and over which they had far less control.

Recognizing this broader context of peasant livelihoods brings us back to a political-economic (or political–ecological) perspective on agrarian change. Some political-economic formulations may have had excessively deterministic overtones, but they at least kept the impact of wider social, political, and economic processes on farm resource management at the forefront of analysis (Blaikie and Brookfield 1987). They also countered both the populist argument and the dominant argument of the Green Revolution that technological fixes to social problems can be found. Against these arguments, political ecologists stress that the origins of the crisis of peasant agriculture lie in land tenure relations, market dependencies, the organization of the economy, the structure of the state, and the social relations of technological production (Bernstein 1982; Watts 1983, 1989; review in Bryant 1992). The implication is that if underlying causes of rural poverty are not addressed, promoting indigenous technology will achieve little – it may not even be an appropriate response.

Furthermore, if indigenous technical knowledge is as much indigenous, as it is technical, then it raises issues of ethnic identity and cultural politics. This is especially apparent when we consider social movements that incorporate ideas of indigenous knowledge and practice into their own alternatives. Some advocates of indigenous agricultural revolutions do deal with cultural politics (e.g. Richards 1990b), but by and large most writing focuses on the technological rationality of adapted peasant production practices (Brokensha *et al.* 1980; Knapp 1991; Warren *et al.* 1995). Agrarian technology is not merely an instrument for environmental manipulation, but is a symbol speaking to rural people of their social history and relationships, a sign by which they read their identities and their relationships with past, present, and future (Bebbington 1991). Similarly, when peasants incorporate new ideas and material technologies into their practices, this can become a sign that the group is now more distant from a past in which they were socially dominated, that their relationship with other social groups is changing, and that they now are claiming rights of access to resources and knowledges previously closed off precisely because of this domination. In short, the incorporation of modern technologies can be a sign of being liberated from a past of domination, at the same time as bringing new dependencies.

Indigenous revolutions under neo-liberalism: the problem of sustaining rural livelihoods

Discussions of small farm agricultural development strategies in contemporary Latin America make little sense unless they consider the economic transformations and livelihood crisis faced by poor people in large parts of the Andes and other fragile lands. Currency devaluations lead to rapid price increases in fossil fuel-based agrochemical inputs, making it essential that the use of Green Revolution alternatives be efficient and effective. Trade liberalization and the creation of regional trading blocs are opening agriculture to competitive pressures. These increase pressure on small farmer production to increase productivity, lower costs, increase competitiveness, and use inputs much more efficiently in technical and economic terms. In such a context, new proposals for the intensification of agriculture and livelihood possibilities are needed, especially in the Andes (Pichón et al. 1995; Reardon et al. 2001). Large parts of the rural economy remain stagnant and without significant improvement and diversification of livelihood opportunities in the region, a combined process of land subdivision, out-migration and resource degradation will leave large parts of the high Andes as little more than labor reserves (cf. de Janvry 1981).

What can an indigenous technology-based alternative contribute in such a context? At one level, rural people's knowledge of land and crops can make important contributions to technical responses to this challenge, particularly in the identification of lower external input options. Nevertheless, the economic and technical efficiencies demanded in this new context require capacities for numeracy, economic abstraction, market research (e.g. identifying niche markets), and identification of cost-controlling, productivity-enhancing genetic material that poor people rarely possess (Byerlee 1987). Research in Mexico, Brazil, Paraguay, and Peru in the 1980s found that formal and higher education had positive effects on productivity and income in rural areas precisely because it helped develop skills of abstraction and numeracy required to handle markets (Cotlear 1989; Figueroa and Bolliger 1986).

Livelihood intensification not only requires support for technical change, but also depends on the rural poor having improved access to product and input markets through relationships that allow wealth deriving from natural resource-based and agricultural activities to be captured and reinvested in the Andes (Bebbington, 1997). This improved and renegotiated market access is critical for the creation of incentives to the sustainable intensification of natural resource use and livelihood possibilities. Technological innovation alone, however environmentally sound and however grounded in traditional practices, will not achieve this.

Beyond questions of access are more basic questions of the existence of such markets. Indeed, with the progressive crisis of the Ecuadorian economy in the 1990s, in 2000 the country adopted the dollar as its national currency, in effect relinquishing control over significant parts of macroeconomic policy to the fiscal and monetary policies of the US Treasury and Federal Reserve. Costs of living and of production have increased dramatically, and many sectors of the economy have lost both export and domestic markets as cheaper imports have flooded in. In such a context, rural livelihoods depend increasingly on non-agricultural, often non-rural income sources (Barsky 1990; Escobal, 2001; Klein 1992; Lopez 1995;

Martinez, 1991; Reardon *et al.*, 2001). Indeed, partly as a result of these macroeconomic crises, international migration out of Ecuador to Spain and the US has increased to remarkable levels (Jokisch, 2001; Jokisch and Pribilsky, 2002). In many areas, and for many people, agriculture is therefore neither the only, nor the main, source of income (Martinez 2003). An adequate response to the Andean crisis must therefore go beyond the purely agricultural sphere. "Alternative" proposals must consider not only agricultural intensification, but also the expansion and diversification of off-farm rural income and employment opportunities. Indeed, there are many potentially synergistic links between agricultural intensification and expanding off-farm income opportunities (de Janvry and Sadoulet 2000; Reardon *et al.* 2001). Such observations imply that proposals for alternative agricultural development must go well beyond a focus on ITK in particular, and technology in general. Proposals must begin from the dynamics of the regional political economy.

ALTERNATIVE ACTORS:
SOCIAL MOVEMENTS AND LOCAL GOVERNMENTS

The challenge of a specifically indigenous, alternative agenda is to respond in a concrete, income-generating way to an Andean livelihood crisis, and in a way that simultaneously strengthens ethnic identity and politics. The challenge is thus material, cultural and political. How do movements carry forward these agendas so that they strengthen each other.

Resistance and identity in rural social movements

Some analyses see rural social movements in Latin America as forms of resistance to domination, exploitation, and subjection (e.g. Redclift 1988). Slater (1985) views them as a protest against traditional politics – in particular against excessive concentration of decision-making power and the incapacity of the state to deliver services. In this sense they are a consequence of the legitimacy problems of the state.

As Peet and Watts's introductory chapter points out, other authors focus more on ways in which movements are expressions of long-dominated and marginalized identities – identities which at the same time are reformulated through the activity of the movement (Evers 1985; Perreault 2003b). Such expression of identity is itself frequently a form of resistance. As Gledhill (1988) has argued for Mexico, the terminology of the "indigenous community" is often used as a way of resisting the all-pervasive intervention of the state in local processes of production and reproduction. Similarly, in Ecuador the recovery and projection of the idea of being Indian has been a form of resisting forms of white and mestizo domination, and of regaining a space for the values of being indigenous (Bebbington 1991; Ramón 1988).

These perspectives help us understand the nature, significance, and activities of popular organizations. The thornier question is what it means "to be an Indian" in the context of these new relationships with state and market. The integration of rural areas into the wider economy has brought many lifestyle changes to the Andean countryside. Modernity arrives variously in the form of fertilizers, radios, new textiles, bicycles, vans, school notebooks, school uniforms, and the clothes and vehicles of non-governmental and governmental extension agents. With these come new aspirations, access to which requires increased incomes. Furthermore, with the integration into a national political process and a new set of relationships with the state comes the idea that indigenous people are not only Indians, but are also national citizens with civil rights. Consciously or unconsciously, indigenous movements face two challenges: to reflect the multiple identities of those they represent, and to negotiate a relationship with the state in which they resist its predations and claim autonomous spaces, but at the same time make claims upon it as citizens.

The complexities of this balancing act were apparent in the early years of the umbrella organization for Ecuadorian indigenous people, the Confederation of Indigenous Nationalities of Ecuador (CONAIE) whose sociocultural and political strategy involved "the search for our own identity, or rather, the forging of an identity that continuously adjusts itself to this society and this supposed democracy which does not yet exist," according to one of its then leaders, Mario Fares. In June 1990, CONAIE called on Ecuador's Indians (a label the organization assumed explicitly) to support what was to become the first of a series of uprisings against government apathy toward indigenous peoples' needs and demanding government support for, and recognition of, Indian cultural difference (Macas 1991: 23). This act of protest asserted the values of "traditional" aspects of Indian identity. Yet, at the same time, CONAIE made demands for a full and fair incorporation of Indians into Ecuador's development process as their right as national citizens. These demands, each reasonable on their own, did not rest easily together. On the one hand, CONAIE wished to strengthen a conception of Indian identity largely grounded in past, more autarkic forms of production and social organization. Thus, it spoke of the recovery of indigenous crops, technologies, crop-environment theories, and cosmologies within larger strategies of ethnic self-determination and cultural revalorization (CONAIE 1989). On the other hand CONAIE demanded that Indians be allowed fairer access to markets, credit, research, and extension (Macas 1991: 26). It thus supported the perpetuation and recovery of cultural traditions, *and* demanded access to the means of rural modernization, the technologies and institutions of the cultural Other. These apparent contradictions point to the difficulty of defining an Indian identity in a modernizing economy.

In those early statements, a possible resolution to this tension existed in CONAIE's claim that because indigenous peoples are both Ecuadorian and Indian they are entitled to both community self-determination and rights of access to state resources (Macas 1991: 25–6). The implication was that communities themselves should decide the balance between traditional and modern markers of their ethnic identity. The case study argues that in some regions such a resolution has taken the form of a "bottom-up" self-management of the modernization process based on indigenous forms of organization, and also in a progressive engagement with the formal institutions of government (see below).

Identity to livelihood in rural social movements

Social movements may be expressions of cultural struggles over meaning (Escobar 1992; Peet and Watts, Chapter 1 in this volume), but the meanings over and for which they fight are not always clear. The struggle for meaning is all the more complicated when we consider the material struggles in which these cultural contestations are imbricated. If a material basis for the survival of Andean communities is not assured, a principal element of Indian identity will be eroded away, both metaphorically and actually. Should this happen, then the cultural struggles will have little significance in the longer term, as indigenous people will be unable to secure the material basis on which to sustain a cohesive cultural identity. In the current policy context, this material basis is genuinely threatened.

Can an ethnically distinct identity be sustained on the basis of transformed and modernized livelihood strategies? If so, how? The experience of the Quichua of Cayambe and Otavalo in northern Ecuador sheds light on this question (Bebbington 2000; Ramón 1988, 1995; Salomon 1981). In a context of severe land subdivision and erosion, local populations followed several strategies to intensify livelihoods. The most renowned is the development of a commercial textile sector, in which production and distribution is controlled by Otavaleño merchants and production is organized through a network of domestic units and small workshops (Korovkin 1997, 1998; Salomon 1981). A less remarkable, but therefore more relevant, experience occurred around Cayambe, where farmers developed commercial onion production (Ramón 1988). Another indigenous group, the Chiboleos, became known as producers, purchasers, and distributors of garlic. In all these cases, the intense commercialization of livelihoods and agricultural production has nevertheless been associated with the maintenance of strong markers of cultural identity in dress, language, kinship, networks, etc. (Bebbington 2000; Ramón 1988).

These groups' responses have involved more than simple adaptive, technological changes. Rather, indigenous people also changed the regional political economy so as to increase the accumulation of capital at the family and regional levels. Indigenous groups gained additional control over relationships of exchange by marketing their own products, enhancing the quality of those goods (e.g. the Chiboleos), and processing more of the materials leaving the region (e.g. the weavers of Otavalo). That said, such experiences remain more the exception than rule. As suggested by the title of a recent book "Rural Progress, Rural Decay" (North and Cameron, 2003), rural development patterns in

Ecuador are highly uneven, and many localities remain largely excluded from growth options and unable to exercise significant control over the markets in which they are most immediately embedded. If this was so when I began this research, it seems even more the case today as the combined effect of dollarization and trade liberalization leave very many local economies uncompetitive (Martinez, 2003). Poverty indicators are one clear measure of this. Between 1995 and 1999 poverty increased from 34 percent to 56 percent of the population – two million people became poor and inequality increased. Among indigenous and Afro-Ecuadorian groups fully 86 percent were deemed poor, and 92.7 percent lacked access to basic services (World Bank 2003).

From civil society to the state in indigenous organizations

Indicators such as these are indicative that both household and organized struggles to achieve more control over rural economy and markets face enormous obstacles. On the other hand, the influence of indigenous organizations on formal politics have been remarkable (Bebbington 2000; Cameron 2000, 2001). The case study discusses this in more detail for the case of Chimborazo, but similar processes have occurred at the national level. We noted the colonization of twelve percent of Ecuador's municipalities and 24 percent of its provincial governments by indigenous and peasant leaders in the 2000 elections. A similar process has occurred in national government.

Early involvement of indigenous organizations in national public institutions was concentrated in the education sector. With the election in 1988 of a center-left government, CONAIE negotiated with the government for the creation of a National Directorate for Bilingual Education (DINIEB) that would sit within the Ministry of Education and Culture but would be staffed and run by CONAIE itself (giving CONAIE and its affiliate organizations the considerable power of selecting teaching staff in bilingual schools: Selverston-Scher 2001). A similar phenomenon occurred in the late 1990s when the four year National Program for Indigenous and Black Peoples' Development, PRODEPINE went into operation following three years of preparation and piloting. PRODEPINE is a nationwide program (funded by the World Bank and IFAD, the International Fund for Agricultural Development) that offers support to indigenous and Afro-Ecuadorian organizations, and has the specific goal of fostering "ethnodevelopment" or "development with identity" (PRODEPINE 2002; Radcliffe 2001b; Uquillas and van Nieuwkoop 2000). Like DINIEB, PRODEPINE is directly managed by national indigenous organizations. With a $50 million budget, it gives indigenous organizations control over a remarkable volume of resources that are used for the following activities (most implemented by indigenous federations): capacity building in indigenous organizations; regularization of land and water tenure; rural investment in private and collective assets and rural enterprise; and institutional strengthening of the (governmental) Council for the Development of Nationalities and Peoples of Ecuador (CODENPE).

In some sense the culmination of this engagement with the central state came in 2002. Shortly after the formal dollarization of Ecuador's economy, national mobilizations uniting parts of the army, indigenous organizations and other popular sectors brought down the presidency of Jamil Mahuad. Following a caretaker government, in 2002 Ecuadorians elected the colonel who led those mobilizations as their new president, on a party ticket supported by the indigenous movement. Leaders from that movement – specifically, former officials of CONAIE – now occupy the positions of Minister of Agriculture (Luis Macas) and Minister of Foreign Relations (Nina Pacari).

Indigenous organization has thus affected national public institutions, policy and politics in Ecuador, opening up very significant spaces for indigenous participation and management. Whether such involvement is good, bad or just ugly is an open question. Indeed, since 1988, observes one pro-movement commentator, "DINEIB has been wracked by political and economic difficulties" (Selverston-Scher 2001: 89), clientelism has continued, and conflicts among currents within the indigenous movement have been played out through DINEIB. Similar tensions have been apparent within PRODEPINE. Whatever the case, these empirical phenomena demand analysis and explanation, and complicate greatly any notion of indigenous social movements as only acts of resistance, as only of civil society, or as only defense of culture. A whole range of complex and hybrid articulations and identities are clearly at stake, and evidently the state ought to occupy far more attention in liberation ecology, a point to which we return in the conclusions.

INDIGENOUS REVOLUTIONS IN CENTRAL ECUADOR

Economic and institutional change in Chimborazo

Located in the Central Andes of Ecuador, the *cantones* of Colta and Guamote in Chimborazo lie in a high altitude area of dominantly quichua people (also often referred to as Quichuas) that until the 1950s was largely controlled by rural estates, or haciendas. In many cases, Quichuas were linked to these estates through a land for labor arrangement, often overlaid with debt and other exploitative relationships. Land-use systems were broadly of two types: the hacienda-based production system, which, though low intensity, was the one arena in which aspects of technical modernization were introduced; and the intensively farmed small plots of the Quichuas. These Quichua systems were in many respects classic "indigenous" farming systems – diversified, intensive, based on food crops, and organic. Their sustainability, however, depended largely on manure from their animals grazed on the extensive pastures of the haciendas.

The social relationships underlying these labor and human–environment relationships were, however, contested, through more daily forms of resistance

(Scott 1985) with occasional land invasions and uprisings. Resistance became more organized and assertive in the 1950s and 1960s, as the national Indian movement became stronger and pressed for land reform. Such pressures, coupled with shifts in policy and political balance, led to land reforms in 1964 and 1973. Because Colta and especially Guamote were conflictive areas, they were defined as priority zones for the implementation of land reform legislation. This led to a series of agrarian, economic, and institutional changes (discussed in more detail in Bebbington 1990).

The textbook agro-ecologically-sound indigenous agriculture began to decline with land redistribution. Hacienda pastures were divided among Quichua farmers and turned to crops. Organic fertilization strategies became increasingly problematic. Also, as population increased, land was further subdivided and fallow periods reduced. No intensification, such as stall feeding of cattle or terracing, occurred, and soils degraded. The use of chemical fertilizers and pesticides increased during the 1980s and 1990s with their greater availability, bolstering yields from crops weakened by poorer soils.

Such agro-ecological changes were accompanied by socioeconomic transformations. Increased market orientation discouraged the cultivation of little-demanded, traditional crops and favored production of marketable food crops (such as potato and broad beans) and new horticultural crops (such as onions, garlic, carrot, and beetroot). Land subdivision in the context of local joblessness (itself an effect of regional underdevelopment and the failure of former estates to reinvest their surpluses productively) led to increased seasonal migration to urban and coastal areas. Quichuas associate periodic migration with mounting social problems and weakened cultural practices in their communities: participation in community activities declined; health problems, petty theft, and violence increased; and manners deteriorated. Post-reform changes suggest an increasing reproduction squeeze on the Quichua economy, with families ever less able to feed themselves and protect the ecological conditions allowing sustained production (Bernstein 1982). In this context, they have looked off-farm and beyond Chimborazo for livelihood options.

Post-reform actors in rural development: the state, NGOs, and the church

Institutionally, land reform marked the beginning of the increased prominence of a modernizing state in local development initiatives. This was reflected in the growth of agricultural extension and integrated rural development programs in the area, which still continue, although with declining resources. These operations, oriented to fostering a modernization of Indian production systems, worked with the basic toolkit of the Green Revolution, introducing new varieties (especially potato), chemical fertilizers, pesticides, etc.

Two other significant institutional changes occurred: an increasingly assertive Church – Catholic and Evangelical – that identifies with indigenous people, and engages in development work; and a growth of non-governmental organizations (NGOs) with a range of social origins, and all pursuing their conceptions of a development alternative. These NGOs are private development agencies, generally supported by European and North American funds, and staffed by people who for whatever reason prefer not to work with state agencies – or are unable to do so.

These different actors have played important roles in the emergence of Indian federations in the area, and in the strategies pursued by these federations. This is so in several regards:

1 Each actor encouraged the creation of federations, often for different reasons, but promoting basically the same organizational structure.

2 Although a number of NGOs and Catholic Church-based organizations were far more cautious than the state in fostering the use of agrochemicals, trying to promote the type of ITK-based alternatives discussed earlier, they have tended with time toward a similar technical packet, sometimes by choice, sometimes as a result of pressure from farmers – a packet based on the scaled-down use of modern varieties and agrochemicals.

3 All tended to use the same model of agricultural development, based on the delivery of inputs and services from an institution (state, NGO, or church), through extension agents working in communities. Similarly they mostly encouraged the idea that project management would later be taken over by the federations.

4 Each of them, even the state, fostered a discussion of the rights of Indian people: rights of equal access to the scriptures, equal citizenship rights as Ecuadorians, or rights to protect and project their culture.

5 They mostly promoted the idea of grassroots management of development processes, although, somewhat hypocritically they were not always willing to pass on project management to grassroots groups.

These common tendencies are important in several senses. They created the set of ideas and practices on which federations drew as they composed their own agrarian strategies. This alerts us to ways in which social movements reflect the institutional and intellectual environment in which they emerge. Neither movements, nor their strategies, are pristine or entirely self-generated. Not only are they actors "situated" in a political economic context that greatly influences the impact of what they do, they are also "situated" in a mesh of ideas and precedents influencing what they choose to do – the strategies they select to pursue their objectives. These experiences suggest also that organizations with a commitment to ITK-based strategies – such as the liberational Catholic Church and several indigenistic and agro-ecological NGOs – have found farmers largely uninterested in such proposals because of their limited livelihood impact.

The agrarian strategies of the Indian federations have a similar focus on technical modernization. But at the same time, this is linked to strategies to which the rural social movements literature draws attention: strategies of resistance and ethnic assertiveness, in which the role and social control of the state is challenged and reworked, albeit again in ways influenced by wider economic, political, and ideological relationships.

Indian federations: origins and strategies

Following the struggle for land, the emergence of stronger and more numerous communities as units of territorial administration has been one of the most significant sociopolitical changes in rural Chimborazo. At the same time, a novel form of indigenous organization – the federation of indigenous communities – emerged. Such federations engage in both political and developmental functions, negotiating with public agencies on the one hand, implementing development projects on the other. There are a number of federations in Chimborazo. The more radical of these trace their origins to disputes over land and other matters, often being linked to national indigenous and peasant movements and to the Catholic Church. The more developmentalist of the federations have origins in negotiations over access to resources from the state and funding agencies.

As just noted, the consolidation of the federations results in part from support from other organizations and donors. Neither the federations nor their programs are entirely "endogenous" innovations. However, if grassroots control, rather than technological content, gives strategy its "alternative" character, what matters is not that agricultural development strategies are endogenous, but that they are locally controlled. The emergence of these federations also reflects a further stage in the extended indigenous control of highland space. Going beyond the recovery of land as a means of production, the federations slowly are recovering the administrative control of rural space, taking back terrain once administered by the hacienda and questioning the very control of space by white–mestizo society. As white and mestizo presence declines, rural areas thus return to indigenous people as space in which to practice their culture and agriculture.

The federations' perspectives on relationships between technology and ethnic identity in an indigenous agricultural development can be understood within this increasing Indian control of rural areas. Rationales stemming from this conception do not alone determine the federations' strategies – socioeconomic and ecological processes are equally important factors. But this rationale gives meaning to such strategies. The result is a vision of Indian agricultural development embracing concerns for agrarian technology; a stronger Quichua cultural identity, and control of rural space. The form in which these concerns are combined varies over time and among federations. Nevertheless, the overall objectives are consistent.

In Chimborazo, an important point of debate among, and within, Indian federations is the extent to which they should work with modern agrochemical and crop technologies as opposed to traditional, low-input technologies. The more radical federations emphasized the recovery of traditional culture and technology in their earlier work. Their programs of agricultural development promoted the recovery of Andean crops, use of organic fertilizers, and replacement of pesticides with supposedly traditional methods of pest control. The rationale underlying this strategy was that it constituted a rejection of the agricultural technologies associated with white and capitalist culture, while at the same time affirming and validating indigenous identity (MICH 1989: 199). It also reduced market dependence, costs of production, and environmental pollution. Social and cultural empowerment, in these strategies, was to be based on an agrarian development based on traditional practices. However, it proved difficult to promote this alternative among Quichua farmers already producing for the market, who had little land from which to produce organic manures. Pressures from their members thus led the federations to work with chemicals, new varieties, and cash crops.

In making this shift, these radical federations approached the model of the more "developmentalist" federations which endorsed and promoted the use of modern technology through their own research, extension, and input supply programs. They argue that this technology can improve Quichua income. They also deem it a necessary technological response to the grazing crisis and soil degradation in Colta and Guamote. The cultural justification for such strategies is that modernization, far from being a cause of cultural erosion, was seen as a means of cultural survival. Periodic migration, and the problems associated with it, are seen as more of a threat to Quichua cultural coherence than the use of agrochemicals and new crop varieties. Technological modernization, along with the promotion of non-traditional cash crops, is therefore justified as a strategy that attempts to increase local income opportunities and reduce pressures to migrate. The principle is that indigenous cultural identity hinges on sustained and corporate rural residence, and not so much on retaining traditional technologies. The implication is that indigenous economy and culture must constantly adapt to survive and sustain group cohesion and forms of self-management.

At the same time, there has been a politically radical dimension to this bottom up modernization. Many Quichuas associate "traditional" technologies with the subjugation of the hacienda and wish to distance themselves from the practices associated with Indian life on the hacienda. In this sense, when Quichuas reject Quichua traditional technology, this is a metaphor for the rejection of the social relations through which they were dominated on the haciendas. By embracing modern technologies they are making a statement – as much to themselves as to others – that they now have the same rights to demand access to resources and benefits (including new technologies) that historically were the preserve of whites and mestizos. This use of modern technology is thus part of a wider discourse on citizenship rights.

In addition, the aspiration is that the hoped-for benefits of modernization – reduced migration, increased community cohesion – will strengthen indigenous organizations as sociopolitical vehicles for demanding change, access to resources, and a more prominent role for Indians in rural development and government. Such demands are given cogency by the fact that the Indian federations' management of rural modernization reflects an attempt to demonstrate Indian ability to use and manage modern administrative methods in a style similar to state programs. If Indians are able to administer rural modernization through their own organizations, the ethnic exclusivity of state rural development administration is no longer justifiable.

The decision to foster technological modernization thus has clear rationales. Modernization seems a necessary response to the realities of market production, soil degradation, and land subdivision, and is more in tune with farm family concerns than are strategies aimed at recovering traditional practices. It also has politically progressive resonances in that technological modernization need not be interpreted as cultural assimilation. At the same time, it is part of a strategy aimed at offsetting the underlying causes of sociocultural dislocation in communities and strengthening Quichua political organization.

By 1993 one federation – the Union of Indigenous Communities of Guamote (UCIG) – had sought to combine this technological agenda with an effort to renegotiate market relationships. With external financial and technical support it established its own marketing program, and opened a plant to process cereals and Andean grains in order to capture a higher price. The ownership of the plant was shared between UCIG and the communities providing it with cereals. This idea that small industries are important in rural development was not new, but once again the federation embraced it for more radical objectives: to exercise more ethnic control over food markets. In the process, an orthodox approach was being turned into an "alternative," locally controlled and indigenous strategy for rural development.

Around the same time, some Quichua leaders and federations opened a similar strategy vis-à-vis the local state, presenting themselves as candidates for election. Indeed, in 1992 and 1996 Guamote elected an indigenous mayor who had previously been the president of a federation. It also elected an increasing number of Quichua councilors – once dominated by mestizos, by 1997 six of its seven councilors were Quichua. And while the mestizo authorities had kept their distance from the federations, the new "indigenous municipality" began to collaborate with them on the implementation of various local development projects (Bebbington and Perreault 1999). By the mid-1990s then, after a series of experiments with modern technology, the federations and the forces on which they were based were also attempting to colonize the very modern institutions of the market and the state. The following section traces through some implications of these strategies, and also of the extent to which they have succeeded in the period since then.

IMPLICATIONS AND IMPACTS

This chapter began by suggesting that to open a conversation between analytical concepts and the rationales of social movements could improve those concepts, and point out weaknesses in the strategies of the movements. It aimed to do this for the case of Quichua federations in highland Ecuador. What has this discussion revealed about the rationales and adequacy of these strategies, the implications for theoretical concepts, and the questions raised in Peet and Watts' introductory chapter?

Movements, modernizations and indigenous agricultural revolutions

Indigenous people in Chimborazo perceive a close relationship between residing in rural areas and sustaining their identity as indigenous farmers. An "indigenous" agrarian development must therefore allow occupation of traditional Indian spaces. In the current context, productive strategies based on non-modernized technologies do not appear a viable means of ensuring this objective. The programs of Indian federations thus encouraged the incorporation of modern technology into local farming practice in an effort to offset migration and improve rural welfare. Contrary to the implications of some critical writings, this suggests that technological modernization can be a rational response to crisis in indigenous production and social systems, and yet at the same time have politically and culturally progressive overtones. This challenges thinking about indigenous agricultural revolutions and relationships among culture, technology, and politics. For an agrarian strategy to carry forward an "alternative, indigenous development" thus depends less on the technological content of that strategy than its social control and objectives. The objective in Chimborazo is to sustain livelihoods to allow the survival of other social practices that continue to mark these people as indigenous.

These experiences illustrate how analyses of indigenous agrarian strategies can benefit from a more critical look at the rationales behind indigenous strategies and factors underlying them. These strategies are not mere "adaptations to environment." They are influenced also by cultural and political logics and socioeconomic exigencies. Furthermore, they may take a form that on the surface seems counterintuitive – the incorporation of modern technology and administration as part of a strategy of cultural survival. Whether or not these responses are adequate is a secondary question. If research does not understand the reasoning underlying them, it will never make a useful contribution to an "alternative" rural development. Instead we run the risk of imposing our conceptions of what is "alternative" and of what it means to be "indigenous."

The implication is that indigenous technological strategies (of whatever hue) ought be understood as a dynamic response to changing contexts – a response constructed through farmers' practices as active "agents" "situated" within cultural, economic, agro-ecological, and sociopolitical contexts that are products of both local and non-local processes. Quichuas' understanding of their identity and agriculture in Chimborazo responds to wider socioeconomic processes. These processes challenge the viability of indigenous agriculture, but at the same time these wider processes provide resources and ideas that are taken in and reworked

411

by indigenous peoples. The federations' strategies, indeed the very existence of federations, are influenced by religious, nongovernmental and state organizations. Similarly, the ways in which federations and farmers interpret technologies are influenced by regional social history. Perhaps the most acute illustration of how the wider context can both constrain and enable the strategies of the rural poor is their insertion into the market. But within this context, indigenous people and their federations are capable of picking and choosing among the different resources, ideas, and technologies that these wider processes make available. They are selectively eclectic, composing strategies that don't necessarily fit the concepts analysts have of them, but which nonetheless are far more coherent and meaningful to Quichuas themselves than are those concepts.

Movements, modernizations and markets

Strategies, however, have to be effective in addressing the causes of the livelihood and socio-cultural crisis affecting large parts of the Andes, and here strategies based solely on technology (indigenous or modern) seem insufficient. A viable indigenous development requires a restructuring of marketing and other social relationships in order to place the production of higher-value and processed products under the control of rural people, thus increasing their incomes. Only then will some of the underlying forces undermining rural well-being begin to be genuinely addressed.

The challenge therefore is not to resist modernization, but to control it, take it further, and increase indigenous peoples' abilities to negotiate market relationships, administer rural enterprises and agro-industry, and compete in a hostile market. In Chimborazo, the case of UCIG reflected one attempt to begin treading this path. Experiences of other social movements and organizations in Ecuador and Bolivia likewise have suggested that a viable indigenous development, which at the same time respects and strengthens ethnic identity, can be based on such a modernization strategy (Bebbington 1996; North and Cameron 2003). But the path is not easy, and the successes of the federations' economic experiments have been modest. By 1998 UCIG's experiments in organized grain processing and marketing had each unraveled in the face of acutely limited capacity to manage economic enterprises and local power struggles within the federations and between them and the indigenous municipality (which argued that it ought to control the projects now it was under indigenous control). Indeed, these experiences suggest once again that for organizations to carry forward such strategies takes time and requires significant managerial capacity and formal organization in the federation (Carroll 2002). Such capacity building is a crucial technical issue with huge liberatory implications on which critical development studies and political ecology are largely silent.

Beyond capacity questions, the failure of these enterprises and the continuing stagnation of livelihoods is also related to the more familiar (for political ecologists) ground of political economy. Competitive product markets and the economic crisis clearly undercut many economic experiments in the region, and more generally Guamote and Colta continue to be horribly poor in material terms. Within North and Cameron's (2003) bi-partite mapping of Ecuador, they would find themselves firmly in an area of rural (economic) decay rather than progress. In the face of dollarization, rural decay, and remarkable outmigration in parts of Ecuador, it is therefore palpably evident that the political economy *must* occupy a central place in any liberation ecology, as must careful, detailed, pragmatic thoughts about the conditions under which this political economy might be reworked. This reworking implies more than the isolated initiatives of Chiboleos or Otavaleños invoked earlier on. At a practical level it means returning to debates on import substitution and targeted protection. At a conceptual level it means engaging with the literature in the sociology of development on the role of the developmental state, and the conditions under which political struggles might lead to forms of state embedding that foster alternative, more pro-inclusion forms of economic management (Evans 1995; 1997; Heller 1999).

Movements, municipalities and the state

While economic gains have been slight, the most interesting changes over the last decade have been political. In Guamote, the most dramatic change has been the shift in control of local government. Leaders linked to indigenous federations have come to occupy most of the elected slots in the municipal council (a phenomenon that has also occurred elsewhere in Ecuador). From this position in local government, they have sought ways of coordinating municipal relationships with the federations and with communities, and have experimented with a more direct form of democracy (known as an "indigenous parliament"). The case has generated far wider interest in the country and outside, and in a way that is now political and not merely demographic, Guamote is categorically an indigenous municipality.

This shift in the exercise of organized power and local control over development processes is perhaps a logical and successful outcome of the federations' struggles – but the shift has been even more dramatic than I would have anticipated when I began this research. The shift not only surprises, it also complicates the analysis because it "brings the state back in," in unexpected ways that challenge any easy counterposition of indigenous movement and the state, and demand theorization of this articulation. The political change in Guamote raises the question of whether the local state can be embedded in particular ways that will enhance its pro-poor, anti-exclusion developmental stance – and ultimately strengthen the role that it can play in facilitating more equitable forms of development. Indeed, the experience here (and elsewhere in Ecuador and Bolivia) suggests that a liberation ecology must think beyond some of the all too easy conceptual boundaries drawn between state and social movement in much political ecology – the practical and analytical challenge is not, then, one of associating *only* political ecology and civil society (Peet and Watts 1996) but also the state (Bebbington and Bebbington 2001).

At the national level, as we noted, Ecuador now has two Ministers drawn from the leadership of the indigenous movement. Politically the question obviously arises as to whether this was a wise decision – to become so intimately involved in a particular regime given the less than liberatory experience of indigenous leaders who have made similar decisions in other Andean countries. Whatever the case, the empirical fact of such Ministers also raises the question, again, of how to theorize a state capable of accommodating such changes (and to do so in a way that goes beyond simple narratives of cooptation), and how to analyze the potential ways in which such intra-state processes might contribute to a liberation ecology as outlined in the introduction to this book. Earlier on in this research program I was more concerned to argue that a liberation ecology had to look at the links between movements and markets – experience has shown that the need also to engage the articulations between movements and municipalities, movements and the central state.

In both the indigenous municipality and the indigenous ministry, the modernization dilemma traced in the chapter remains ever present. The new ethnic municipality of Guamote is very much a developmental one, continuing the federations' practice of fostering more or less modern, but municipally and communally controlled, development projects in the on-going search for economic alternatives (alternatives that in recent years have included projects for fish farming, organized tourism, irrigation, and reforestation to name but some). The strategy continues to be one of harnessing modern instruments (including World Bank loans via PRODEPINE) to pursue an alternative development.

The dilemmas inherent in this process are deepened at a national level, in spades. The agenda of the new Ministry of Agriculture, under one time President of CONAIE Luis Macas, bears little resemblance to the CONAIE statements of a decade ago (FAO 2003). The agenda focuses on: lobbying for increased budget to invest in agriculture; support to a range of local initiatives revolving around land titling, irrigation, targeted subsidies, technology development, agro-industry etc.; support to agricultural research by NGOs; strengthening inter-institutional coordination and support to peasant organizations; policies to improve market linkages, including financial services markets; and a program for watershed management (FAO 2003). The breadth of this agenda is clearly appropriate: Macas is responsible for the whole sector, not just indigenous farmers. But the implication is that the Ministry will aim to deliver a program fostering greater inclusion and aiming to build the capacity of local actors to compete within an existing market led model of rural development. An indigenous development strategy will have to be sought within this context.

At the same, the national indigenous organizations (led by CONAIE) themselves negotiated and ultimately decided to participate actively in PRODEPINE, the project with the World Bank, IFAD and the Government of Ecuador that is specifically intended to promote indigenous people's development. The resources that this program has made available to federations were without precedent, as was the notion that the federations would play a direct role in the management of a government program (which is what has occurred). But also without precedent were the contradictions to be negotiated: social movements that oppose neoliberalism managing a World Bank loan; market-oriented and infrastructure investments as a primary instrument for an ostensible "indigenous" development; the jockeying for power and resources among different indigenous movements, nationally and locally (just as federations within Guamote have jockeyed for power within the municipality); and so on. These changes – unprecedented, complex, contradictory – challenge our categories again and merit careful ethnographic study of the processes involved. The risk, of course, is that such analysis will fall into the old dichotomies of the like of the earlier debates on the Green Revolution and indigenous technical knowledge debates – indeed, to a degree this is what has happened in early efforts to make sense of these changes (Bretón 2001; Radcliffe 2001b). It is not immediately clear that such analytical tactics further the cause of a liberation ecology that is more than likely to be based on unprecedented hybrid practices and hybrid institutions – often contradictory hybrids that our concepts and instinctive political judgments are not necessarily well placed to address.

Alternative actors, alternative developments

What this chapter has not done is to take the discussion of alternatives a step further and ask whether we should be talking about an alternative to development rather than an alternative development (cf. Bebbington and Bebbington, 2001; Escobar 1995). The elaboration of such alternative utopias is a valid and important task, as part of a sustained questioning of dominant ideas and policies – and if the social and economic relationships in which the federations are enmeshed were different then it might be the case that other non-modernizing strategies would be appropriate and feasible. However, these utopias must also be constructed from practice, and grounded in the aspirations of popular sectors. The dilemma is that these aspirations and practices have now incorporated the experience of modernity and development. Those in the business of alternative utopias must be careful before rejecting popular aspirations for the benefits of modernity as some sort of false consciousness. Furthermore, there is an immediate problem of survival. Quichuas in Chimborazo do not have time to wait for the dawn of new utopias. They demand liberation from where they are now. The challenge then is to build short-term, pragmatic, and realistic responses that work from contemporary contexts, and do so in a way that is coherent with and builds towards longer-term utopias that are already immanent within the strategies and hopes of popular sectors.

Movements and the conduct of liberation ecology

Since my research began in Guamote there has been a mini-boom in publications dealing with indigenous movements, development and cultural politics in

415

Ecuador (Bretón 2001; Carroll 2002; Perreault, 2003 a, b, c; Radcliffe 2001a, b; Sawyer 1997; Selverston-Scherr 2001; Treakle 1998; Yashar 1998; to name just a few). This boom in intellectual production has various explanations: the rapidly increased visibility of indigenous movements in Ecuador; the growing concern for post-structural theory and the particularly interesting ways in which these Ecuadorian experiences could illuminate that theory; the interest in transnationalism; the normative hope that in the work of these movements might be found seeds for thinking through alternative forms of politics. These and more help account for this explosion. The literature is rich, theoretically and empirically and this is no place to review it. But there are two points about this boom worth making as a way of closing this chapter, and bringing it back to its beginning.

First, the boom came after the storm. Only a few analysts – mostly Ecuadorian – had any premonition of the sort of liberatory (if still very incomplete) political project that was set to unfold in Ecuador during the 1990s. Orin Starn (1991) has argued that anthropologists in Perú similarly failed to anticipate the cataclysmic rise of Sendero Luminoso (a very different sort of political project), in part because they were too busy conducting community studies of "traditional" practices, paying less attention to very modern forms of emergent politics. It is at least worth wondering whether social researchers missed the Ecuadorian revolution for similar reasons. In the final instance, the federations and confederations that led the Ecuadorian revolution are products of a quite modernizing development program led by churches, government agencies and NGOs. For too long the presumption has been that such a project is necessarily conservative, ethnically disarticulating and to be criticized – and perhaps for that reason researchers paid little attention to the new forms of organization and leadership to which these development interventions were giving rise. The parallels with the agricultural debates discussed earlier is clear.

Second though, and on a more positive note, much of this new research is being conducted in a way that helps sustain the conversation between movement and academic constructs that the introduction of this chapter called for. This must be positive, making available the products of intellectual production to the movements being studied, and also subjecting those products to the critique of those movements (or at least of their leadership). The point is important – for just as there has to be theory of liberation ecology, so there also must be method for its conduct. This new work speaks to the important role for research approaches that are more dialogical, more conversational, and more embedded in the processes of which they aim to make sense. Whichever way we cut it, liberation ecology has to be a committed project.

ACKNOWLEDGMENTS

My thanks once again to Dick Peet's editorial challenges and guidance, and to lessons over the years from Tom Perreault, Tom Carroll, Galo Ramón, Julio

Berdegué, Arturo Escobar and Jonathan Fox. The chapter invokes a notion of "we" and "our" at different times – the pronoun refers to what I perceive to be a collectivity of valued colleagues such as these friends and others within the political (liberation) ecology tradition who share a concern to write in ways that are critical, engaged and primarily motivated by concerns for social justice in development. Research has been funded at different times by the Inter-American Foundation, Fundagro, the University of Colorado at Boulder and the Pacific Basin Research Center at Harvard University.

REFERENCES

Altieri, M.A. (1987) *Agroecology. The Scientific Basis of Alternative Agriculture.* Boulder, CO: Westview Press.

Altieri, M.A. and S.B. Hecht (eds) (1990) *Agroecology and Small Farm Development.* Boston: CRC Press.

Barsky, A. (1990) *Políticas agrarias en América Latina.* Santiago: Imago Mundial.

Bebbington, A.J. (1990) "Indigenous agriculture in the central Ecuadorian Andes. The cultural ecology and institutional conditions of its construction and its change." Ph.D. dissertation, Graduate School of Geography, Clark University, Worcester, MA.

—— (1991) "Indigenous agricultural knowledge, human interests and critical analysis," *Agriculture and Human Values* 8, 1/2: 14–24.

—— (1996) "Organizations and intensifications: small farmer federations, rural livelihoods and agricultural technology in the Andes and Amazonia," *World Development* 24: 1161–78.

—— (1997) "Social capital and rural intensification: local organizations and islands of sustainability in the rural Andes," *Geographical Journal* 163 (2): 189–97.

—— (2000) "Re-encountering development. Livelihood transitions and place transformations in the Andes," *Annals of the Association of American Geographers* 90(3): 495–520.

Bebbington, A.J. and D. Bebbington (2001) "Development alternatives: practice, dilemmas and theory," *Area* 33(1): 7–17.

Bebbington, A.J. and T. Perreault (1999) "Social capital, development and access to resources in highland Ecuador," *Economic Geography* 75 (4): 395–418.

Bebbington, A.J. and G. Thiele (1993) *NGOs and the State in Latin America. Rethinking Roles in Sustainable Agricultural Development.* London: Routledge.

Bebbington, A.J., H. Carrasco, L. Peralvo, G. Ramón, V.H. Torres, and J. Trujillo (1992) *Los Actores de una Decada Ganada: tribus, comunidades y campesinos en la modernidad.* Quito: Abya Yala.

—— (1993) "Fragile lands, fragile organisations: Indian organisations and the politics of sustainability in Ecuador," *Transactions of the Institute of British Geographers* 18: 179–96.

Bernstein, H. (1982) "Notes on capital and peasantry," in J. Harriss (ed.) *Rural Development. Theories of Peasant Economy.* London: Hutchinson, pp. 160–77.

Blaikie, P. and H. Brookfield (1987) *Land Degradation and Society.* London: Methuen.

Booth, D. (ed.) (1994) *Rethinking Social Development: Theory, Research and Practice.* Harlow: Longman.

Bretón, V. (2001) *Cooperación al desarrollo y demandas étnicas en los Andes ecuatorianos.* Quito: FLACSO.

Brokensha, D., D.M. Warren, and O. Werner (eds) (1980) *Indigenous Knowledge Systems and Development*. Lanham, MD: University of America Press.

Bryant, R. (1992) "Political ecology," *Political Geography* 11: 12–36.

Butzer, K. (1990) "Cultural ecology," in C.J. Wilmott and G.L. Gaile (eds) *Geography in America*. Washington, DC: Association of American Geographers and National Geographic Society.

Byerlee, D. (1987) *Maintaining the Momentum in Post-Green Revolution Agriculture. A Micro-level Perspective from Asia*. Michigan State University International Development Paper 10. Department of Agricultural Economics, Michigan State University, East Lansing.

Cameron, J. (2000) "Municipal decentralization and peasant organization in Ecudor: a political opportunity for democracy and development?," paper prepared for the Latin American Studies Association meetings, Miami, 16–19 March.

—— (2001) "Local democracy in rural Latin America: lessons from Ecuador," paper prepared for the Latin American Studies Association Meetings, Washington DC, 6–8 September 2001.

Carroll, T. (ed.) (2002) *Construyendo capacidades colectivas. Fortalecimiento organizativo de las federaciones campesinas-indígenas en la Sierra Ecuatoriana*. Quito: Oxfam America.

Chambers, R. (1993) *Challenging the Professions*. London: Intermediate Technology Publications.

Chambers, R., A. Pacey, and L.A. Thrupp (eds) (1989) *Farmer First. Farmer Innovation and Agricultural Research*. London: Intermediate Technology Publications.

Confederación de Nacionalidades Indígenas del Ecuador (1989) *Nuestro proceso organizativo*. Quito: CONAIE.

Cotlear, D. (1989) *El desarrollo campesino social en las comunidades de la Sierra del Perú*. Lima. Instituto de Estudios Peruanos.

de Janvry, A. (1981) *The Agrarian Question and Reformism in Latin America*. Baltimore. Johns Hopkins University Press.

de Janvry, A. and E. Sadoulet (2000) "Rural poverty in Latin America: determinants and exit paths," *Food Policy* 25: 389–409.

Denevan, W.M. (1989) "The geography of fragile lands in Latin America," in J. Browder (ed.) *Fragile Lands of Latin America: Strategies for Sustainable Development*. Boulder. Westview.

Edwards, M. (1994) "Rethinking social development: the search for relevance," in D. Booth (ed.) *Rethinking Social Development: Theory. Research and* Practice. Harlow: Longman, pp. 279–297.

Escobal, J. (2001) "The determinants of nonfarm income diversification in rural Perù," *World Development*, 29, 3: 497–508.

Escobar, A. (1992) "Imagining a post-development era," *Social Text* 31.

—— (1995) *Encountering Development: The Making and Unmaking of the Third World*. Princeton, NJ: Princeton University Press.

Esteva, G. (1992) "Development," in W. Sachs (ed.) *The Development Dictionary: A Guide to Knowledge and Power*. London: Zed Books.

Evans, P. (1995) *Embedded Autonomy: States and Industrial Transformation*. Princeton: Princeton University Press.

—— (1997) "The Eclipse of the State? Reflections on Stateness in an Era of Globalization," *World Politics* 50: 62–87.

Evers, T. (1985) "Identity: the hidden side of new social movements in Latin America," in D. Slater (ed.) *New Social Movements and the State in Latin America*. Amsterdam: CEDLA, pp. 43–77.

Fairhead, J. (1992) "Representing knowledge: the 'new farmer' in research fashions," in J. Pottier (ed.) *Practicing Development: Social Science Perspectives*. London: Routledge, pp. 187–204.

Fairhead, J. and M. Leach (1994) "Declarations of difference," in I. Scoones and J. Thompson (eds) *Beyond Farmer First: Rural People's Knowledge, Agricultural Research and Extension Practice*. London: Intermediate Technology Publications, pp. 75–9.

FAO (2003) Seminario interno sobre la situación y perspectivas del desarrollo agrícola y rural en el Ecuador. 30–31 January 2003: accessed from HTTP: <http://www.rlc.fao.org/prior/desrural/ecuador.htm>.

Figueroa, A. and F. Bolliger (1986) *Productividad y aprendizaje en el medio ambiente rural. Informe comparativo*. Rio de Janeiro: ECIEL.

Geertz, C. (1983) *Local Knowledge: Further Essays in Interpretive Anthropology*. New York: Basic Books.

Giddens, A. (1979) *Central Problems in Social Theory*. London: Macmillan.

Gledhill, J. (1988) "Agrarian social movements and forms of consciousness," *Bulletin of Latin American Research* 7: 257–76.

Goldman, A. (1993) "Population growth and agricultural change in Imo State, Southeastern Nigeria," in B.L. Turner, R. Kates, and G. Hyden (eds) *Population Growth and Agriculture Intensification: Studies for Densely Settled Areas of Africa*. Gainesville: University of Florida Press.

Griffin, K. (1974) *The Political Economy of Agrarian Change*. London: Macmillan.

Grossman, L. (1984) *Peasants, Subsistence Ecology and Development in the Highlands of Papua New Guinea*. Princeton, NJ: Princeton University Press.

—— (1993) "The political ecology of banana exports and local food production in St. Vincent, Eastern Caribbean," *Annals of the Association of American Geographers* 83, 2: 347–67.

Heller, P. (1999) *The Labor of Development*. Ithaca. Cornell University Press.

Hewitt de Alcantará, C. (1976) *Modernizing Mexican Agriculture*. Geneva: United Nations Research for Social Development.

Jokisch, B. (2001) "From New York to Madrid: A Description of Recent trends in Ecuadorian Emigration," *Ecuador Debate*, no. 54, Quito, Ecuador.

Jokisch, B. and J. Pribilsky (2002) "The panic to leave: economic crisis and the new emigration from Ecuador," *International Migration* 40(4): 75–101.

Klein, E. (1992) "El empleo rural no agrícola en América Latina," in CEPLAES (ed.) *Latinoamérica agraria hacía el siglo XXI Centro de Planificación y Estudios Sociales*. Quito, Ecuador.

Knapp, G. (1991) *Andean Ecology: Adaptive Dynamics in Ecuador*. Boulder, CO: Westview Press.

Korovkin, T. (1997) "Taming capitalism: the evolution of the indigenous peasant economy in Northern Ecuador," *Latin American Research Review* 32: 89–110.

—— (1998) "Commodity Production and Ethnic Culture: Otavalo, Northern Ecuador," *Economic Development and Cultural Change* 47: 125–54.

Leff, E. (1994) *Ecología Política y Capital: hacia una perspectiva ambiental del desarrollo*. Mexico City: Siglo XX

Lehmann, A.D. (1990) *Development and Democracy in Latin America. Economics, Politics and Religion in the Postwar Period*. Cambridge: Polity Press.

Lipton, M. and R. Longhurst (1989) *New Seeds and Poor People*. London: Unwin Hyman.

Long, N. (1990) "From paradigm lost to paradigm regained? The case for an actor-oriented sociology of development," *European Review of Latin American and Caribbean Studies* 49: 3–24.

Long, N. and J.D. van der Ploeg (1994) "Heterogeneity, actor and structure: towards a reconstitution of the concept of structure," in D. Booth (ed.) *Rethinking Social Development: Theory, Research and Practice*. Harlow: Longman, pp. 62–89.

López, R. (1995) *Determinants of Rural Poverty in Chile. A Quantitative Analysis for Chile*. Washington: World Bank.

Macas, L. (1991) "El levantamiento indígena visto por sus protagonistas," in Instituto Latinoamericano de Investigaciones Sociales (ed.) *Indios*. Quito: ILDIS, El Duende, Abya-Yala, pp. 17–36.

Martinez, L. (1991) "Situación de los campesinos artesanos en la Sierra Central del Ecuador: Provincia de Tungurahua," manuscript. Quito.

—— (2003) "El campesino andino a fines de siglo (discusión sobre el caso ecuatoriano)," Quito. Mimeo.

Max-Neef, M. (1991) *Human-scale Development: Conception, Application and Further Reflections*. New York: Apex Press.

Movimiento Indígena de Chimborazo (MICH) (1989) "Movimiento Indígena de Chimborazo, MICH," in CONAIE (ed.) *Nuestro Proceso Organizativo*. Quito: CONAIE, pp. 195–202.

North, L. and J. Cameron (eds) (2003) *Rural Progress, Rural Decay: Neoliberal Adjustment Policies and Local Initiatives*. West Hartford. Kumarian Press.

Peet, R. and M. Watts (1993) "Introduction: development theory and environment in an age of market triumphalism," *Economic Geography* 69(3): 227–53.

—— (eds) (1996) *Liberation Ecologies*. London: Routledge.

Perreault, T. (2003a) "Making space: community organization, agrarian change, and the politics of scale in the Ecuadorian Amazon," *Latin American Perspectives* 30(1): 96–121.

—— (2003b) "A people with our own identity: toward a cultural politics of development in Ecuadorian Amazonia," *Environment and Planning D: Society and Space* (in press).

—— (2003c) "Changing places: transnational networks, ethnic politics, and community development in the Ecuadorian Amazon," *Political Geography* 22(1): 61–88.

Pichón, F., J. Uquillas and J. Frechione (eds) (1985) *Traditional and Modern Natural Resource Management in Latin America*. Pittsburgh. University of Pittsburgh Press.

PRODEPINE (2002) HTTP: <http://www.PRODEPINE.org> Ecuador. Project for the Development of Indigenous Peoples in Ecuador.

Radcliffe, S. (2001a) "Development, the State and transnational political connections," *Global Networks* 1: 19–36

—— (2001b) "Indigenous movement representations in transnational circuits. Tales of social capital and poverty," paper presented at the Annual Conference of the Association of American Geographers, 28 February–3 March 2003, New York.

Ramón, G. (1988) *Indios, crisis y proyecto alternativo*. Quito: Centro Andino de Acción Popular.

—— (2001) "El índice de Capacidad Institucional de las OSGs en el Ecuador," in A. Bebbington and V.H.Torres (eds) *El capital social en los Andes*. Quito. Abya Yala.

Reardon, I., J. Berdegué and G. Escobar (2001) "Rural nonfarm incomes and employment in Latin America: overview and policy implications," *World Development* 29, 3: 395–410.

Redclift, M. (1988) "Introduction: agrarian social movements in contemporary Mexico," *Bulletin of Latin American Research* 7: 249–55.

Richards, P. (1985) *Indigenous Agricultural Revolution. Ecology and Food Production in West Africa*. London: Hutchinson.

—— (1986) *Coping with Hunger: Hazard and Experiment in a West African Rice Farming System*. London: Allen & Unwin.

—— (1990a) "Local strategies for coping with hunger: Central Sierra Leone and northern Nigeria compared," *African Affairs* 89: 265–75.

—— (1990b) "Indigenous approaches to rural development: the agrarian populist tradition in West Africa," in M. Altieri and S. Hecht (eds) *Agroecology and Small Farm Development*. Boston: CRC Press, pp. 105–11.

Rigg, J. (1989) "The new rice technology and agrarian change: guilt by association?" *Progress in Human Geography* 13, 2: 374–99.

Salomon, F. (1981) "The weavers of Otavalo," in N. Whitten (ed.) *Ethnicity in Modern Ecuador*. Urbana. University of Illinois Press. pp. 420–49.

Sawyer, S. (1997) "The 1992 Indian Mobilization in Lowland Ecuador," *Latin American Perspectives*. 93: 67–84.

Scoones, I. and J. Thompson (eds) (1994) *Beyond Farmer First: Rural People's Knowledge, Agricultural Research and Extension Practice*. London: Intermediate Technology Publications.

Scott, J.C. (1985) *Weapons of the Weak: Everyday Forms of Peasant Resistance*. Yale: Yale University Press.

Selverston-Scher, M. (2001) *Ethnopolitics in Ecuador: Indigenous Rights and the Strengthening of Democracy*. Miami. North–South Center Press.

Slater, D. (ed.) (1985) *New Social Movements and the State in Latin America*. Amsterdam: CEDLA.

Starn, O. (1991) "Missing the revolution: anthropologists and the war in Perú," *Cultural Anthropology* 6: 63–91.

Sylva, P. (1991) *La organización rural en el Ecuador*. Quito. Cepp-Abya Yala.

Thompson, J. and I. Scoones (1994) "Challenging the populist perspective: rural peoples' knowledge, agricultural research and extension practice," *Agriculture and Human Values* 11: 58–76.

Treakle, K. (1998) "Ecuador: structural adjustment and indigenous and environmentalist resistance" in J. Fox and D. Brown (eds) *The Struggle for Accountability: The World Bank, NGOs, and Grassroots Movements*. Cambridge. MIT Press.

Turner, B.L., R.W. Kates, and G. Hyden (eds) (1993) *Population Growth and Agricultural Intensification: Studies from the Densely Settled Areas of Africa*. Gainesville, FL: University of Florida Press.

Uquillas, J.E. and M. van Nieuwkoop (2000) "Social capital as a factor of indigenous people' development in Ecuador," paper prepared for a workshop on social capital and the World Bank, Boulder, October 2000.

Warren, D.M., D. Brokensha and J. Slikkerveer (eds) (1995) *The Cultural Dimension of Development: Indigenous Knowledge Systems*. London: Intermediate Technology Publications.

Watts, M.J. (1983) "Populism and the politics of African land use," *African Studies Review* 26: 73–83.

—— (1989) "The agrarian crisis in Africa: debating the crisis," Progress in Human Geography 13(1): 1–41.

World Bank (2003) Second Indigenous and Afroecuadorian Peoples' Development Project. Project Concept Document. Latin America and Caribbean Region. Washington. World Bank.

Yashar, D. (1998) "Contesting citizenship: indigenous movements and democracy in Latin America," *Comparative Politics* 31: 23–42.

16

INDUSTRIAL POLLUTION AND SOCIAL MOVEMENTS IN THAILAND

Tim Forsyth

This chapter extends the discussion of liberation ecologies to Thailand's urban and industrial social movements – yet it also contains some words of caution. While acknowledging the crucial role played by social movements in livelihood struggles, the chapter argues that liberation ecologies may be too optimistic in assuming that social movements may successfully revise environmental discourses in favor of marginalized people. Instead of focusing only on the political agency of social movements, political ecologists must also assess how movements interact with, and even may become exclusive discursive structures. Movements may not be very "liberatory" if – rather than helping poor people – they replicate or impose new hegemonic discourses.

The chapter considers these arguments in relation to social movements concerning suspected lead and lignite poisoning in Thailand. By so doing, this chapter provides valuable attention to the "brown" environmental agenda of cities and factories versus the predominantly "green" agenda of forests and agriculture considered by other chapters in this volume. Yet, in addition, the chapter explores important epistemological challenges for liberation ecologies that may be more prevalent in the "brown" agenda than in "green" environmental topics. Many industrial and urban environmental risks are new to localities in developing countries, and consequently local experience and scientific certainty about their causes are low. Under such circumstances, local social movements may easily be dominated by outside expertise, or framed in ways that do not reflect local experience of risks.

The point of this chapter is not to dismiss liberation ecologies or the importance of brown environmental social movements. Instead, the aim is to understand each better. Much political ecology to date has focused, classically, on the marginalizing impacts of capitalist development, and on the struggles of poor people against alliances of industry and state. The case studies in this chapter certainly provide more evidence for this approach. But the case studies also show that such activism is linked to the production of discourses that do not always help the most vulnerable people. A science–policy approach – of understanding the coproduction of environmental activism and discourses – therefore needs to be integrated with the political economy approach of understanding social marginalization under rapid industrialization.

ENVIRONMENTAL DISCOURSES AND THE "BROWN" AGENDA IN THAILAND

"Brown" environmental problems occupy an uncertain place in Thai environmental politics. While environmental risks associated with urbanization and industrialization are undoubtedly growing in Thailand, some environmental activists and politicians still put more attention on the "green" agenda, and sometimes see environmental problems as essentially rural. "I come from a rural area," Thailand's prime minister, Thaksin Shinawatra, once said, "So it is only natural that one of my concerns has always been environmental issues."[1]

Such statements, of course, are not surprising. Thaksin is, after all, a factory owner and property developer, and may resist discussing topics that might attract environmental regulation. (As the billionaire founder of the Shinawatra Telecommunications group, he is the most dazzling representative of Thailand's new industrial elite.) Yet, Thaksin is also a politician and needs to win support from voters who share such views. Indeed, Thaksin's own constituents in the northern city of Chiang Mai once accused his wife of flouting planning regulations near a national park.

Clearly, Thailand has experienced immense – and at times devastating – environmental changes because of industrialization and rapid economic growth. But environmentalism in Thailand has also been shaped by social concerns at perceived corruption and state failures than by the nature of environmental hazards themselves. Indeed, environmentalism in Thailand has been closely linked to democratization, and the resentments against the state and new industrial elite, particularly under oppressive military regimes.

Hence, during the 1960s, early environmental groups – such as the Society for the Conservation of National Treasures and Environment – restricted their activities to fighting for the conservation of sites of architectural heritage or outstanding natural beauty. Increasingly, however, environmentalists began to oppose both state and industry. In the late 1960s and early 1970s activists help shelve plans to build a cable car to the Buddhist shrine at Doi Suthep in Chiang Mai. In the 1980s, an alliance of diverse social groups achieved environmentalism's greatest victories when the government canceled a proposed dam in the western rainforests of Nam Choan (1988) and then passed a national logging ban (1989). (These decisions coincided with Thailand's first formally democratically elected government in 1988.) In 1997, the relationship between environmentalism and democratization was further entrenched by the new Constitution, which affirmed the right of citizens to participate in decisions affecting natural resources and infrastructure development. Some specific environmental campaigns have also highlighted the impacts of development projects on marginalized poor – such as the construction of the Pak Mul dam in eastern Thailand, or the enforced resettlement of villages from reforested land in the northeast (Hirsch 1997).

Yet, despite these strong links of environmentalism with democracy and social welfare, environmentalism has not always been as socially inclusive as these other events suggest (Forsyth 2001). Some environmental campaigns have reflected divisions of ethnicity, social class, or the different perspectives of urban and rural people. For example, one NGO in northern Thailand has been called "racist" because it blames watershed degradation on mountain minorities alone, and seeks to exclude them from forest areas (Lohmann 1999).[2] Other NGOs adopt somewhat romanticized visions of nature. One other NGO director told the author in 1999 – with no sense of irony – that Thailand used to have 100 percent forest cover during the nineteenth century, and that she saw her job as the replacement of these trees. This statement stands starkly against evidence that Thailand never had this amount of forest, or that agriculture and human settlement may also have claims on land.[3]

Again, perhaps such trends are also not surprising. Social theorists such as Nash (1973) and Giddens (1994) have argued that the construction of wilderness as beautiful and threatened may emerge from a romanticized sense of lost heritage resulting from urbanization and modernization, and the rise of a middle class. But the underlying epistemology of such class-based environmentalism may be more fundamental than city-dwellers simply valuing wilderness more than rural-dwellers. Much environmentalism as a *new* social movement – following Marcuse and Habermas – has closely associated the destruction of environment with the simultaneous suppression of human nature under modernization. For example, Rosemary Ruether in *Liberation Theology* wrote: "Oppression of persons and oppression of environments go together as parts of the same mentality." (Ruether 1972: 18). Or, more recently, the German green activist, Wiesenthal commented: "The nub of the political objectives pursued in green politics … can be summarized in two postulates: preservation and emancipation." (Wiesenthal 1993: 56).

Such statements can have significant tensions when applied to rapidly industrializing societies. Clearly, many environmentalists in Thailand have assumed that people and environment have been oppressed by unfair development. But the words, "preservation" and "emancipation" – if applied as solutions – may have controversial implications. If preservation implies keeping landscapes stable, then this may imply restricting the actions of many small farmers (such as shifting cultivators) who regularly disrupt landscapes. (Indeed, this approach to "preservation" may also fall foul of new, non-equilibrium approaches to ecology that stress the role of disturbance and highlight how social norms have dictated what is seen to be "normal" landscape – see Zimmerer, this volume). And if emancipation is meant to imply the escape from the pressures of instrumental rationality under modernity (as discussed by critical theorists such as Marcuse and

Habermas), then how far does this denote anti-developmentalism, or the rejection of industrial growth, which then may also restrict development opportunities for the poor? Such concerns led Enzensberger in his classic critique of political ecology to warn: "in so far as it can be considered a source of ideology, ecology is a matter that concerns the middle class." (Enzensberger 1974: 10) Instead, critics have suggested that there should be a more inclusive, social-development orientation to environmentalism that may allow poor and marginalized people to participate in economic growth.

The liberation ecology approach has acknowledged these concerns. Peet and Watts wrote, "it is almost delightfully naïve to assume that the content of ... green movements is necessarily progressive," (Peet and Watts 1996: 268) and that, "critical social movements have at their core environmental imaginaries at odds with hegemonic conceptions." (Peet and Watts 1996: 263). But in placing faith with social movements, the approach may overlook how less powerful voices may be swept up into emerging hegemonic conceptions associated with middle-class environmentalism. The statements of Peet and Watts seem to imply that social movements make it possible for disaffected groups to communicate critical thoughts against powerful discourses. Yet, evidence suggests that social movements often do not exist independently of each other, but connect into overarching discourses, or that more powerful actors in alliances eclipse less powerful voices. For example, in the Philippines, Covey (1995) found that alliances between national environmental NGOs and grassroots organizations invariably became dominated by NGOs.

In response to these dilemmas, some critics have proposed that the focus of many political ecologists on the resistance of marginalized people to oppressive forces of development needs to be augmented with a more critical awareness of the discourses this approach evokes (e.g. Forsyth 2003). Instead of assuming that counter oppositions of society versus state and economy are necessarily progressive for social groups, there is a need to see how such activism may replicate existing environmental discourses, or replace one form of hegemonic discourse with another. For example, Hajer's (1995) discussion of "storylines" demonstrates how environmental discourses have emerged over time because of historic incidences of activism, and the framings associated with each. Frequently, storylines emerge because of "discourse coalitions" between different political actors, who have (often tacitly) agreed on a definition of risk because it allows each party to pursue other political objectives. A continual repetition of these definitions may therefore become a "songline of risk" (Jasanoff 1999), or a socially accepted construction of biophysical reality, even if there are uncertainties about this definition. In certain respects, these approaches to the historic coproduction of environmental activism and discourses reflect insights from Actor Network Theory, or the social influence on how reality is constructed.

Such social influences on the interpretation of risk may be particularly relevant to "brown" environmental conflicts in developing countries where material risks may have no appreciable history among affected populations, and where scientific uncertainty may be high. Under these circumstances, activists may seek to impose pre-existing forms of meaning onto environmental risks. In Thailand, for example, the first overtly industrial environmental dispute, concerning a proposed tantalum processing plant in the southern resort of Phuket in 1986, was framed in terms of impacts on tourism and livelihoods rather than health risks of pollution. Similarly, some of the earliest newspaper reporting of "brown" issues now seems quite naïve and sensationalist. One 1970 headline in *The Bangkok Post* announced: "Pollution may lead to famine,"[4] citing "serious vibration" as a national problem. A 1971 article about pollution was simply entitled "Ughhhhhhh!!!"[5] In 1970, Mrs. Eunice Martin, an American living in Bangkok, also suggested that citizens would have to wear anti-pollution masks by 1985. She added, helpfully: "perhaps in the future we can have red masks for firemen, white for doctors, masks with fringes for go-go girls and blue for workers."[6] (*The Bangkok Post* is Thailand's leading English-language daily, but is generally read and written by Thais).

Mrs. Martin's prediction almost came true: by the 1980s, facemasks were regular sights on Bangkok traffic police and factory workers. But have the understandings of these risks advanced beyond sensationalism? The point of this question is not to diminish the material and sometimes-deadly nature of "brown" agenda risks, but to indicate how social perceptions and activism have shaped policy responses to them. Social movements have been crucial to the identification of, and response to risks. But the influence of social movements on the epistemology of risks needs to be considered more critically in order to see how far common perceptions are biophysically accurate, and relevant to people affected. The following discussions now apply such consideration to the cases of lead and lignite poisoning in Thailand.

THE CASE OF LEAD

Lead poisoning next to lead mines has had a long history in Thailand. Industrialization and urbanization has increased these risks, with an increase in demand for lead, leaded gasoline, and factories (especially electronics) using lead for soldering. In 1989, ponds and streams in western Thailand were found to have up to 10,000 times the World Health Organization's suggested healthy limit of 10–15 micrograms per deciliter (mg dl^{-1}) of blood.[7] Too much lead in blood may lead to inhibited development in children, plus a range of chronic illnesses. Acute levels of lead may lead to death (Suwanna and Chaiphan 1994).

Partly in recognition of these trends, the Thai government established a new agency to deal specifically with the problems of industrial poisoning in 1990, the National Institute for Occupational and Environmental Medicine (NIOEM). A 36-year-old, US-trained female doctor, Dr. Orapan Methadilokul, headed the agency. The agency was the first step in the establishment of a national network of such occupational health offices.

Yet, this institute immediately fell into problems in 1991 when it investigated the unexplained deaths of four workers at a factory of Thailand's (then) largest overseas employer, Seagate, the American manufacturer of hard-disc drives in Samut Prakarn, an industrial province south of Bangkok. According to Dr. Orapan, she was suddenly called to the telephone while at the factory and was told: "What you are doing is hurting Thailand! How dare you investigate Seagate? I can have you fired!" Orapan alleged the caller was Staporn Kavitanon, the secretary to (then) Prime Minister Anand Panyarachun, and the director general of the Thai Board of Investment. When questioned, Staporn denied making the call. But Orapan later realized that her Institute had funding withdrawn, and a planned extension of the institute to regional offices cancelled (see Forsyth 1994).

The cause of deaths in the factory was never fully explained at the Seagate factory. On one hand, Orapan found that 36 percent of 1,175 workers tested at the plant to have blood-lead levels higher than 20 mg dl^{-1}. This compared with similar studies at the same time showing just 8 percent of Bangkok traffic police at or above this level, and just 2 percent for an average of all Bangkok residents. But on the other hand, such levels may be sufficient only to cause chronic health problems rather than sudden death (indeed, Seagate claimed that levels of 70–100 mg dl^{-1} would cause death). When the government replaced Orapan from the investigation, they sent in researchers from a separate agency under the Ministry of Industry,[8] which measured workplace environments – such as air and worktop levels – and found no evidence of unhealthy working conditions.[9] After the event, Orapan admitted that the initial findings of high blood-lead levels may have taken attention away from the potential impacts of solvents, or cleaning agents, used in the factory which can cause breathing difficulties and sudden death.

There were also more overtly political factors underlying the dispute about industrial poisoning. Orapan's inquiry came at a time when workers at the factory were striking for higher wages, and when Seagate – a company who had come to Southeast Asia to cut labor costs – was under increasing competition in the marketplace. Rumors that lead poisoning was also affecting workers inflamed the strike, and a number of other trades unions threatened sympathy action. Overnight, the Seagate strike became Thailand's most serious industrial dispute, with workers marching on parliament, and even petitioning President Bush (senior) to intervene on their behalf. Such factors may have encouraged the Thai government to cut NIOEM's funding when there was still uncertainty about alleged poisoning.

Two years later, however, this decision brought implications for a further case of suspected industrial poisoning. Between March 1993 and September 1994, an estimated 11 to 23 people working at the Northern Region Industrial Estate in Lamphun died from unexplained causes, but usually with symptoms of headaches, and sometimes inflamed stomachs.[10] Yet, following the closure of NIOEM, there was no agency to investigate these deaths.

Villagers were coughing; plants and livestock were dying. The leakage had occurred because sulfur filters at the power plant had failed, leading to the emission of sulfurous dust. Concern about lignite as a source of industrial pollution had started in Thailand.

The chief problem caused by lignite was excessive concentrations of sulfur dioxide produced during burning or mining, and the resulting condition of pneumoconiosis, a disease of the lungs caused by breathing in dust. The immediate impacts of the 1992 release were poorly recorded. But in following years, further incidents of toxic emissions were recorded, especially in 1997 when new filtering devices failed. Between 1996 and 2001, newspapers claimed that 48 people near Mae Moh had been killed because of lignite-related respiratory problems.[12] At times, concentrations of sulfur dioxide were measured at nearly 2,300 micrograms per cubic meter of air, compared with the government standard of 1,300 micrograms.[13] Villagers have been warned not to drink rainwater.

The impact of these problems was to create local campaigns for compensation and resettlement, and to inflame national consciousness of the ethics of industrialization. Families affected by the pollution filed a lawsuit seeking 3 million baht ($120,000 in 1996) in compensation from the government. Villagers also complained to the World Bank, as a main creditor of Thailand. Eventually, some 500 households were resettled in new villages away from the power plant. New flue-gas desulphurization units were installed (despite failing in 1997). The government also provided welfare support to sick villagers from an ongoing fund of 30 million baht-a-year ($750,000 in 2003).

Dr. Orapan was also appointed as an investigator of the causes of death. Some villagers still felt under-compensated, and still demanded resettlement.

Yet, many other themes underlay these immediate topics of protest. The main target of criticism was the Electricity Generating Board of Thailand (EGAT), the state-owned enterprise responsible for electricity supply, which also has a long history of proposing environmentally damaging infrastructure projects such as the aforementioned Nam Choan and Pak Mul dams, plus later projects such as gas pipelines from Myanmar and Malaysia. Activists and journalists therefore represented the Mae Moh dispute in terms of unaccountable and destructive development by this arm of the state. Some newspapers headlines read, "EGATs broken promises"[14] and "People's patience running out."[15] Yet, politicians too found this a convenient way to win support and perhaps avoid direct responsibility, by publicly distancing themselves from the actions of EGAT. For its part, EGAT also sued an Italian engineering company that had failed to deliver desulphurization equipment quickly enough.

There was also a sense that the dispute also represented the desire for some local people to participate in Thailand's sudden growth and wealth. One doctor working in the Mae Moh area later described how many claims for compensation were unrelated to the plant's pollution. He reported:

One old man came to see me in 1995 with a stomach illness but he asked me to diagnose a respiratory problem in order to gain compensation. When I refused, he angrily pointed out that he had already hired a truck to bring him to me!

(Prasert 1999: 2)

The dispute also tended to focus on lignite *per se* as the source of problems, rather than the sulfurous content, or the application of this fuel within specific contexts or technologies. Some scientists suggested that the unusual location of Mae Moh in the end of a valley sometimes ensured the occasional enclosed atmospheric conditions that resulted in emissions hanging above fields and villages. There was also confusion whether the most pollution was caused by the lignite mine or power plant. Similarly, some activists saw the introduction of flue-gas desulphurization as a climbdown from attempts to close the power plant or resettle the villages. Villagers said they did not want to "hold their breath" since no one can guarantee if technology will be effective all the time. Yet, Egat also implied that the villagers should not hold their breath waiting for the requested compensation. Rather than pay more compensation, the government has instead reduced chances of further catastrophic leakages, established a new emission level of 780 mg of sulfur dioxide per hour in 2000, and lowered the expected generating capacity of the plant. These emission levels are much lower than previous totals; yet still amount to some 11 tons of dust per hour.[16]

Similar to lead, therefore, the image of lignite in Thailand has evolved to symbolize dirty development and the failure of the state to listen to citizens' fears. This emotive heritage was to reemerge in a later dispute in the southern peninsular of Thailand in the coastal province of Prachuab Khiri Khan, where it was announced in 1997 that a consortium of foreign companies planned to build three new power plants, using coal imported from Australia. Immediately, local people feared that the plants would use lignite, and that the power stations would pollute fishing areas and tourist sites. One of the financial backers of the companies was also the ex-prime minister, Anand Panyarachun.

Legally, the proposed project was in a gray zone. The 1997 Constitution made it compulsory for local citizens to participate in such infrastructure projects; yet, the contracts with the companies had been signed before the Constitution was passed. Moreover, in keeping with many infrastructure projects in Thailand, the companies had secretly bought land in the selected areas before the project was announced as a way to get the land cheaply. The government used these arguments to justify persisting with the project and seeking to achieve public approval.

Such approval, however, was not forthcoming. In July 1997, villagers blocked main roads and marched on city hall. In December 1998, more than 60 people were injured as a further roadblock led to clashes between police and protestors. Some 5,000 people attended the rally, which also burnt effigies of the new prime minister. Meetings were called with journalists and academics, and protestors

Learning Resources
Centre